CONTEMPORARY IMRT
Developing Physics and Clinical Implementation

Series in Medical Physics and Biomedical Engineering

Series Editors:
C G Orton, Karmanos Cancer Institute and Wayne State University, Detroit, USA
J H Nagel, Institute for Biomedical Engineering, University Stuttgart, Germany
J G Webster, University of Wisconsin-Madison, USA

Other books in the series

The Physical Measurement of Bone
C M Langton and C F Njeh (eds)

Therapeutic Applications of Monte Carlo Calculations in Nuclear Medicine
H Zaidi and G Sgouros (eds)

Minimally Invasive Medical Technology
J G Webster (ed)

The Physics of Three-Dimensional Radiation Therapy:
Conformal Radiotherapy, Radiosurgery and Treatment Planning
S Webb

The Physics of Conformal Radiotherapy: Advances in Technology
S Webb

Intensity-Modulated Radiation Therapy
S Webb

Physics for Diagnostic Radiology
P Dendy and B Heaton

Achieving Quality in Brachytherapy
B R Thomadsen

Medical Physics and Biomedical Engineering
B H Brown, R H Smallwood, D C Barber, P V Lawford and D R Hose

Monte Carlo Calculations in Nuclear Medicine: Applications in Diagnostic Imaging
M Ljungberg, S-E Strand and M A King (eds)

Introductory Medical Statistics 3rd Edition
R F Mould

Ultrasound in Medicine
F A Duck, A C Barber and H C Starritt (eds)

Design of Pulse Oximeters
J G Webster (ed)

The Physics of Medical Imaging
S Webb (ed)

Of related interest

From the Watching of Shadows: The Origins of Radiological Tomography
S Webb

Series in Medical Physics and Biomedical Engineering

CONTEMPORARY IMRT
Developing Physics and Clinical Implementation

Steve Webb

Professor of Radiological Physics
Head, Joint Department of Physics
Institute of Cancer Research
University of London
UK

and

Royal Marsden NHS Foundation Trust
Sutton
Surrey
UK

CRC Press
Taylor & Francis Group
Boca Raton London New York

CRC Press is an imprint of the
Taylor & Francis Group, an **informa** business

CRC Press
Taylor & Francis Group
6000 Broken Sound Parkway NW, Suite 300
Boca Raton, FL 33487-2742

First issued in paperback 2020

ISBN-13: 978-0-367-45429-6 (pbk)
ISBN-13: 978-0-7503-1004-8 (hbk)

Visit the Taylor & Francis Web site at
http://www.taylorandfrancis.com

and the CRC Press Web site at
http://www.crcpress.com

British Library Cataloguing-in-Publication Data

A catalogue record for this book is available from the British Library.

Library of Congress Cataloging-in-Publication Data are available

Cover image: Multileaf Collimator, copyright Elekta, reproduced with permission (see also figure 3.2, page 42).

Series Editors:
C G Orton, Karmanos Cancer Institute and Wayne State University, Detroit, USA
J H Nagel, Institute for Biomedical Engineering, University Stuttgart, Germany
J G Webster, University of Wisconsin-Madison, USA

Cover Design: Victoria Le Billon

Typeset in LaTeX 2_ε by Text 2 Text Limited, Torquay, Devon

The Series in Medical Physics and Biomedical Engineering is the official book series of the International Federation for Medical and Biological Engineering (IFMBE) and the International Organization for Medical Physics (IOMP).

IFMBE

The International Federation for Medical and Biological Engineering (IFMBE) was established in 1959 to provide medical and biological engineering with a vehicle for international collaboration in research and practice of the profession. The Federation has a long history of encouraging and promoting international cooperation and collaboration in the use of science and engineering for improving health and quality of life.

The IFMBE is an organization with membership of national and transnational societies and an International Academy. At present there are 48 national members and two transnational members representing a total membership in excess of 30 000 world wide. An observer category is provided to give personal status to groups or organizations considering formal affiliation. The International Academy includes individuals who have been recognized by the IFMBE for their outstanding contributions to biomedical engineering.

Objectives

The objectives of the International Federation for Medical and Biological Engineering are scientific, technological, literary, and educational. Within the field of medical, clinical and biological engineering its aims are to encourage research and the application of knowledge, and to disseminate information and promote collaboration.

In pursuit of these aims the Federation engages in the following activities: sponsorship of national and international meetings, publication of official journals, cooperation with other societies and organizations, appointment of commissions on special problems, awarding of prizes and distinctions, establishment of professional standards and ethics within the field, as well as other activities which in the opinion of the General Assembly or the Administrative Council would further the cause of medical, clinical or biological engineering. It promotes the formation of regional, national, international or specialized societies, groups or boards, the coordination of bibliographic or informational services and the improvement of standards in terminology, equipment, methods and safety practices, and the delivery of health care.

The Federation works to promote improved communication and understanding in the world community of engineering, medicine and biology.

Activities

The IFMBE publishes the journal *Medical and Biological Engineering and Computing* which includes a special section on *Cellular Engineering*. The *IFMBE News*, published electronically, keeps the members informed of the developments in the Federation. In cooperation with its regional conferences, IFMBE publishes the series of IFMBE Proceedings. The Federation has two divisions: *Clinical Engineering* and *Technology Assessment in Health Care*.

Every three years the IFMBE holds a World Congress on Medical Physics and Biomedical Engineering, organized in cooperation with the IOMP and the IUPESM. In addition, annual, milestone and regional conferences are organized in different regions of the world, such as Asia Pacific, Baltic, Mediterranean, Africa and South American regions.

The administrative council of the IFMBE meets once a year and is the steering body for the IFMBE. The council is subject to the rulings of the General Assembly, which meets every three years.

Information on the activities of the IFMBE are found on its web site at http://www.ifmbe.org.

IOMP

The IOMP was founded in 1963. The membership includes 64 national societies, two international organizations and 12 000 individuals. Membership of IOMP consists of individual members of the Adhering National Organizations. Two other forms of membership are available, namely Affiliated Regional Organization and Corporate members. The IOMP is administered by a Council, which consists of delegates from each of the Adhering National Organizations; regular meetings of council are held every three years at the International Conference on Medical Physics (ICMP). The Officers of the Council are the President, the Vice-President and the Secretary-General. IOMP committees include: developing countries, education and training; nominating; and publications.

Objectives

- To organize international cooperation in medical physics in all its aspects, especially in developing countries.
- To encourage and advise on the formation of national organizations of medical physics in those countries which lack such organizations.

Activities

Official publications of the IOMP are *Physiological Measurement, Physics in Medicine and Biology* and the *Series in Medical Physics and Biomedical Engineering*, all published by the Institute of Physics Publishing. The IOMP publishes a bulletin *Medical Physics World* twice a year.

Two council meetings and one General Assembly are held every three years at the ICMP. These conferences are normally held in collaboration with the IFMBE to for the World Congress on Medical Physics and Biomedical Engineering. The IOMP also sponsors occasional international conferences, workshops and courses.

Information on the activities of the IOMP are found on its web site at http://www.iomp.org/.

Contents

Preface and acknowledgments

Intensity-modulated radiation therapy (IMRT), virtually non-existent 20 years ago and still embryonic 15 years ago is now an established clinical tool. Its development, once the preserve of physicists and engineers, whilst incomplete, is now at a stage when application specialists, oncologists, radiotherapists and radiographers can introduce IMRT to the clinic. Meanwhile physicists and engineers both continue research and development and support clinical implementation. Feedback from clinical use in turn influences R and D. *Clinical R and D* is the new IMRT specialty and randomized clinical trials of IMRT are still only a handful.

Previous book reviews of IMRT (from me and from others) have largely concentrated on its physics. This volume continues to review the developing physics but also covers the preliminary clinical implementation and the interplay between the two. The book can be read standalone and the chapters can be read in any order. This fourth book in a series is, however, better read, and should be more use, alongside the three complementary volumes. Only a limited repetition of detailed physics is included. Consequently this volume is less mathematical than its predecessors.

Multi-author books can draw on greater talent; the downside is that they are also often disjointed, overlapped and sometimes impossible to create because of so many competing priorities for authors' time. Single-author books ought to have a consistent style and cohesion but inevitably reflect the particular skills, special interests, but also limitations, of the author. In some areas of this book, I have reviewed clinical detail in which I relied on trusting my primary sources rather than my internal quality control as a physicist. There is therefore a greater weight of debateable opinion than objectively established fact than in the previous volumes; but I hope we would welcome that. If there were ever to be a fifth volume it should be a work jointly with a clinician. The references here are timestamped with an end date of June 2004.

I am deeply grateful to innumerable colleagues and friends both at the Royal Marsden NHS Foundation Trust and the Institute of Cancer Research and also in hospitals and research labs throughout the world. Their work essentially creates this review and they will recognize their contributions to the field herein.

I should like to thank John Navas, Simon Laurenson, and others at IOPP (and their reviewers) who have put their faith in my efforts. I am very grateful to Marion Barrell for much help with organizing figures, checking references and seeking permissions to publish. I also thank Marion and our other departmental secretaries (Rosemary Atkins, Sylvia Boucheré, Dana Roberts and Lesley Brotherston) for typing so many dictated tapes during the past four years of book writing. I thank all the authors and publishers who have allowed material to be used here for illustration[1]. Contact was made with all copyright holders and I apologise for those few cases where no response was received.

The material reviewed here represents the understanding and personal views of the author offered in good faith. Formulae or statements should not be used in any way concerned with the treatment of patients without checking on the part of the user.

This book is dedicated to Linda.

Steve Webb
June 2004

---

[1] Some figures only relate to their captions when viewed in colour. These are indicated in the figure captions and colour figures can be found at http:/bookmark.iop.org/bookpge.htm?&isbn=0750310049

Chapter 1

Intensity-modulated radiation therapy (IMRT): General statements and points of debate

1.1 Observations on IMRT at the current time

The year 2000, representing the entry into a new millennium, has been highlighted as a turning point in many contexts. Much of this is undoubtedly opportunistic capitalizing on a round-figure date but it conveniently was the year in which IMRT came of age and turned from a physicist's and engineer's interest into a rapidly expanding clinical technique whose prosecution in the clinical arena required the future collaboration of medical doctors, radiographers, engineers, physicists and others. This in turn set the scene for shifting the focus of attention to the expected clinical benefit. Meanwhile work continued to develop and refine the physics and technology of IMRT.

Morita (2000) delivered an address to introduce an International Takahashi conference that appropriately reminds us of this important scientist in the history of conformal radiotherapy (CFRT) and IMRT. Shinji Takahashi was one of the pioneers of radiological tomography (having invented various forms of axial-transverse tomography in the 1950s) and also of what he called conformational radiotherapy (a technique invented in 1960). Takahashi linked the definition of target volumes using axial-transverse tomography to his forms of conformational radiotherapy. In the 1960s, conformational radiotherapy was achieved in Japan at just a few centres using rotating multileaf collimators (MLCs) (with big leaf size) and rotating protectors whilst the patient simultaneously corotated (figure 1.1). These techniques were continually developed in Japan throughout the 1980s and largely in the 1990s have been overtaken by inverse treatment-planning techniques based on computed tomography and with the development of more sophisticated IMRT. Morita (2000) summarized the chain of tasks to be optimized to maximize the effectiveness of radiation therapy. Some specific cases were given to demonstrate the improved performance of CFRT. Whilst not providing

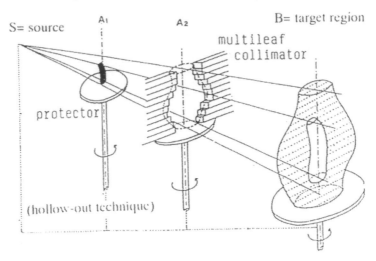

Figure 1.1. The principle of conformational radiotherapy as exemplified in the historical work by Takahashi. In this figure the patient section is shown rotating on the right. A MLC with large leaves, is shown at A2 defining a beam's-eye view of the target. The shape of the aperture is varied as the beam's-eye view of the target changes. At A1 on the left, a radiation-attenuating protector is also shown changing its disposition with respect to the collimator and the body as a function of angle. This is a form of binary ('on–off') modulation and, whilst it achieves some conformality, is not as successful as modern IMRT. (Reprinted from Morita (2000) with copyright permission from Elsevier.)

for high-spatial-resolution fluence variations, these Japanese developments are widely cited as early pioneering attempts to improve the physical basis of radiation therapy and should be regarded as part of the history of IMRT. Elsewhere in the world most radiotherapy used simple fields with or without shaping with blocks. Conformal therapy with automatic field shaping did not start until the mid 1980s and IMRT came a decade later (figure 1.2).

In the summer of 2000, I completed a summary of IMRT development to that time (Webb 2000d) and this book begins where that one left off, reviewing the work so far this century and looking to the future. That summary of IMRT (Webb 2000d) was completed three weeks after the 13th International Conference on the use of Computers in Radiotherapy (ICCR) in Heidelberg and, in the appendix, I wrote the last line to the book: 'Where will IMRT be at the time of the 14th ICCR (Seoul, South Korea)?'. This volume was completed a few weeks after this 14th ICCR and summarizes the answer to that question. There have been several overviews of the field in the last four years. I provided a response to an invitation at the turn of the millennia to crystal-ball gaze the future of the physics of radiotherapy (Webb 2002e). In Seoul, Ha (2004) also gave a set of predictions for what is important in the future of radiotherapy.

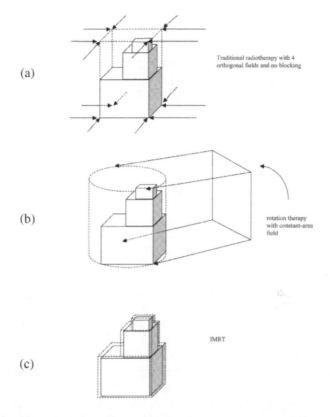

(a)

Traditional radiotherapy with 4
orthogonal fields and no blocking

(b)

rotation therapy
with constant-area
field

(c)

IMRT

Figure 1.2. This figure encapsulates the key differences between the major forms of photon radiotherapy. A representative PTV is shown as a solid comprising three cubic volumes. Imagine also that this PTV is surrounded by normal tissue which it is desired to spare from radiation. (a) shows how four uniform-intensity fields irradiating the volume normal to its planar faces would create a rhomboidal (box)-shaped region of high dose. Clearly volumes of normal tissue are irradiated. (It is accepted that the situation could be improved by geometric field shaping). (b) shows the principle of rotation therapy in which a field of uniform intensity and rectangular area is rotated through 2π around the PTV. A cylindrical high-dose volume would result, again irradiating normal tissue unnecessarily. (c) When fields are shaped geometrically using a MLC and also the intensity is varied across the plane of each field, the high-dose volume can be 'sculpted' to fit like a sheath around the PTV sparing normal tissue. This is the goal of IMRT. The figure is of course simplistically schematic. In clinical practice, the PTV may have a highly irregular shape. Whilst it can be argued that careful adjustment of field parameters might make conventional radiotherapy acceptable for *this* geometrical volume, for more contorted PTVs this would not be possible without CFRT or IMRT. Also for IMRT the degree of conformality will depend on the number and intensity structure of the fields as well as on the shape of the PTV and its proximity to normal tissue.

The IMRT Cooperative Working Group generated a lengthy review of IMRT in November 2001 (IMRTCWG 2001) which can be read as a tutorial on the subject and the IMRT Subcommittee of the American Association of Physicists in Medicine (AAPM) Radiation Therapy Committee has also produced a lengthy report offering guidance on IMRT implementation (Ezzell *et al* 2003). A joint document from the American Society for Therapeutic Radiology and Oncology (ASTRO) and AAPM has overviewed the whole process of implementing clinical IMRT (Galvin *et al* 2004). I have also provided a brief history of the developments in improving the physical basis of radiotherapy including comment on IMRT (Webb 2002d) and a longer history of IMRT (Webb 2003a). Table 1.1 provides a summary of key historical landmarks in the development of IMRT. In 2003 the AAPM held its annual Summer School entirely focusing on IMRT and resulting in a very comprehensive monograph (Palta and Mackie 2003). In the winter months of 2003/2004, the British Institute of Radiology published a seven-part set of review articles on IMRT in the *British Journal of Radiology* (Webb 2003b, Williams 2003b, McNair *et al* 2003b, James *et al* 2003, Guerrero Urbano and Nutting 2004a, b, Beavis 2004b).

At the risk of being quite controversial, I begin by standing back and taking a long view on IMRT and maybe on the whole subject of improving the physical basis of radiotherapy. I hope I am forgiven for sharing this somewhat personal view. It is offered as a reflection on the history of IMRT (Webb 2003a, b) and in the spirit of seeking a positive way forward. Perhaps it can be read as one might read a similar editorial to a journal packed with interesting articles on IMRT?

The principle of IMRT, that pencils of radiation emanating from different directions and with different intensities can be joined together to create high dose volumes with concavities in the outline and spatially-varying high dose, has been linked to an interesting historical analogy from Birkhoff (1940). He showed that, provided negative pencil blackness could be used, any picture could be represented by the linear sum of straight pencil lines of different blackness from different directions. Figure 1.3 shows his, now famous, 'cat' picture created this way. Of course, negative x-rays do not exist so the exact analogy breaks down.

Towards the end of the 1980s, very few physicists had begun to consider that the possibility to modulate the fluence of a beam would expand the options to create a high-dose volume around the tumour whilst sparing organs at risk (OARs). The notion of what today we call IMRT (which was incidentally not called that at the time) is generally credited to the Swedish medical physicist Anders Brahme. Some (maybe even he?) would argue that there were earlier attempts to modulate fluence via the tracking cobalt unit, the gravity-driven blocking mechanisms and other ways. However, these were very simple binary on–off modulations of the primary fluence (ignoring leakage) and it is to Brahme that the idea of inverse-planning for modulating fields is rightly credited. He solved one of the only problems amenable to analytic inversion in 1982 and his 1988 paper is considered the seminal first work on inverse planning. Thomas

Table 1.1. The development of IMRT—the one-page history. IMRT is poised to make a major clinical impact. How has the field reached the present position?

1895	The x ray was discovered on 8th November in Germany by Röntgen.
1896	Doctors understood the need to 'concentrate radiation' at the target but had means neither to do this nor even to know where the target was precisely.
1950s	Takahashi first discussed conformation therapy.
1959	Invention and patenting of the first multileaf collimator.
1960s	Proimos developed gravity-oriented blocking and conformal field shaping.
1970s	The Royal Free Hospital built the 'tracking cobalt unit' and MGH Boston did similarly.
1982	Brahme *et al* discussed inverse-planning for a fairly special case of rotational symmetry.
1984	First commercial MLCs appeared.
1988	Brahme published first paper on algebraic inverse planning.
1988	Källman postulated dynamic therapy with moving jaws.
1989	Webb developed simulated annealing for inverse planning. So did Mageras and Mohan.
1990	Bortfeld developed algebraic/iterative inverse-planning, the precursor of the KONRAD treatment-planning system.
1991	Principle of segmented-field therapy developed (Boyer/Webb).
1992	Convery showed the dMLC technique was possible.
1992	Mark Carol first showed the NOMOS MIMiC and associated PEACOCKPLAN planning system (now CORVUS).
1993	Tomotherapy (the Wisconsin machine) first described by Mackie.
1994	Stein, Svensson and Spirou independently discovered the optimal dMLC trajectory equations.
1994	Bortfeld and Boyer conducted the first multiple-static-field (MSF) experiments.
1999	First discussion of possible robotic IMRT.
2002	Commercial tomotherapy began.
2004	Large number of competing IMRT planning systems and systems for delivery. IMRT is well established in several geographically distributed centres.

Bortfeld, Radhe Mohan and I were independently working out ways to create modulated fluence in 1988 and by 1990 there were just a handful of papers on IMRT; the techniques were entirely concerned with planning not delivery. I do not think any of us imagined how fast and how greatly this research field would develop. Certainly I did not for I am no prophet.

The development of IMRT has been astonishingly fast. By the mid-1990s the essential features of all the main techniques for delivering IMRT had been established (see reviews in Webb 1993, 1997). Certainly there were details still to unravel, and important ones at that, but the key papers on the physics of IMRT delivery had been written. All the developments, with the exception of the NOMOS MIMiC, comprised one-off techniques developed in universities

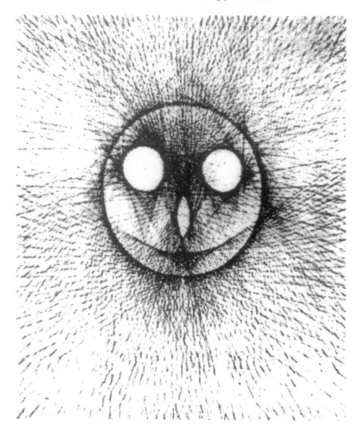

Figure 1.3. The diagram in the paper by Birkhoff (1940) showing how a black-and-white picture of a cat can be made up from a series of pencil lines of different thickness (density) provided these are both positive (normal pencil) and negative (unphysical erasure). This may be regarded as analogous to the principle of IMRT except for the fact that there are, unfortunately, no 'negative x-rays'. (Reprinted from *Journal des Mathématiques Pures et Appliqées* **19** 221–36.)

or hospitals. There was nothing available from manufacturers (again with the exception of the MIMiC).

By the turn of the millennia, all this had dramatically changed (see review in Webb 2000d). The major accelerator manufacturers were each offering IMRT delivery packages. Most of the treatment-planning system manufacturers were also offering IMRT planning facilities, usually by inverse planning. Focus began to shift from fundamental concerns about how to perform IMRT into the embryonic stages of clinical delivery. The fundamental developments of IMRT could be consigned to history, albeit very recent history (Webb 2003a).

All this is very satisfying because it shows that the transition from just a few workers doing fundamental physics through to widespread clinical implementation had been achieved in about 15 years which is a very short time against the background of the implementation of other technologies.

The scale of research progress has created a difficult position with respect to the literature. There are now literally thousands of papers and abstracts of work on IMRT and these can be hard to assimilate. Many papers report marginally different solutions to the same problems. One of the reasons for this and previous books in the series has been to try to overview these, filter them and present them to a new audience. This distillation is now, in my opinion, close to impossible.

Also there are hundreds of papers and abstracts showing how commercial inverse planning has been coupled to commercial IMRT delivery and used to initiate clinical IMRT. Again, this is very pleasing in that it indicates the widespread adoption of the concepts. But many of these reports are remarkably similar.

Is it too harsh to ask if there is a climate of over-reporting? The phrase 'write-only journals' has crept into the language! I am not old enough to know whether the same phenomena arose when, for example, the first linacs began to be used and others then copied the installations. But the slimness of back copies of the major journals together with the fact there were less meetings and conferences suggests that it did not.

There are also those who are orchestrating campaigns to challenge the clinical significance of implementing IMRT (not oppose it but question the possibly previously unquestioned speed of implementation and justification). There have been several important contributions from Schulz whose main thesis is that IMRT implementation is more due to manufacturer pressure, reimbursement and copycat fashion than from the basis of genuine proof of clinical benefit (see section 1.2). His perspective may be more American than European where phase-3 trials are being organized. This was also mentioned by Beavis (2003b) when describing implementation in an NHS clinic.

What is the way forward? Firstly I believe there are still many unsolved problems with the fundamental physics of IMRT and it is right to continue to investigate these. An example is the somewhat approximate nature of predicting the dose using analytic expressions and/or an incomplete knowledge of the behaviour of scatter dose with small-area segments. Another example is to challenge the assumption that one can ignore tissue inhomogeneities at planning and delivery (Papanikolaou *et al* 2003). Monte Carlo (MC) approaches will address that. Next, we still have less than adequate methods to check and measure absolute dose and to find monitor unit (MU) calibrations. There is the whole issue of planning for the moving patient rather than on some single imaging event (as if the patient remained frozen as imaged). One can identify that there could be a requirement to simplify the whole procedure of IMRT delivery by using purpose-built equipment rather than the MLC. Possibly robotic delivery could increase treatment flexibility. IMRT using cobalt units has hardly been addressed yet. This might open up IMRT for departments or countries without an adequate supply of linacs.

Regarding clinical studies, there are many who simply report the implementation of a technique on the basis that the predicted dose distributions will be better. This is admirable but somehow incomplete. What we really want to know is whether the implementation will make a difference and, even if it does, whether this difference is significant in terms of benefit to the individual and cost–benefit to society. According to Sharpe (2003) 'Confidence comes from evidence'. For this reason, I believe the clinical papers that merit most attention are those that are accompanied by documented evidence of either reduced side effects or improved probability of cure. Even better will be the reports of phase-3 clinical trials and, at the time of writing, there are not many of these (Jayaraman 2003). However, in chapter 5, an effort has been made to clearly identify the few papers reporting genuine clinical benefit and distinguish them from paper planning exercises or clinical implementations with no reported findings.

Perhaps the really long view comes from the unadorned observation that radiotherapy is 'only' a matter of placing some required dose at the 'right place' and avoiding doing so at the 'wrong place'. It is 108 years since people started to try to do this in 1896, the year after the discovery of the x-ray. No-one yet believes it is done 100% properly. Literally thousands of workers have contributed to aspects of the problem. So it would be unthinkable that any of us right now should regard the matter as not worthy of further study.

IMRT is just one way—and a very fine way—to relocate dose spatially. This hardly equates to having cracked all the problems of radiotherapy. It is important to focus on unanswered questions and novel opportunities.

The philosophy of IMRT and its ultimate dream scenario has been discussed (Webb 2001c), in particular noting that the goal has changed from seeking the ultimate conformality at the price of great complexity to achieving practical clinical gain. Current constraints are both technological, humanitarian and financial. A two-track approach is recommended in which there will be a continued development of fundamental concepts in tandem with clinically extending known techniques. Geographical widening of IMRT availability is an issue. Possibly the MIMiC-type IMRT may be overtaken by dynamic MLC techniques.

Bortfeld (2001) has reviewed physical and technological aspects of IMRT. Current issues at the cutting edge include: improvement of the accuracy of dose-calculation algorithms, Monte Carlo planning, reduction of the number of field segments and improved efficiency, use of MLCs with smaller-than-standard leaf widths, use of MLCs with small shifts of the isocentre or rotation and accounting for organ motion and inverse planning and delivery.

Suit (2002) in his L H Gray Lecture argued forcibly for the view that, in the next two to three decades, protons will entirely replace photons as the treatment modality of choice. This is predicated on the argument that dose distributions are incontrovertibly superior. There is no clinical debate. The only concern is financial. Proton facilities cost more. However, it may be argued they are in service longer. The Harvard Cyclotron operated continuously for 52 years. One accelerator can serve many gantries. All adjunct costs are the same as for photons. The improvement in disease definition through functional imaging is Suit's second prediction. A third major area of development will be that of techniques to compensate for internal organ movement. He highlights the history of 'doubting Thomases'. Almost all developments in improving the physical basis of irradiation therapy have had their critics. He urges us to follow the maxim of the pioneering physicist Lord Rutherford, known as 'the old Croc', because of his interest only in the future. Crocodiles cannot look backwards. He also writes: 'The history of medicine has repeatedly demonstrated that the perceived efficacy, and not the cost, primarily determine the fate of new technologies'.

Williams (2002a) has discussed how IMRT has moved from the sole preserve of the research community to a wider clinical application. His view is that the choice between dynamic and step-and-shoot delivery and the choice between forward and inverse planning has less to do with the overall outcomes as measured by the three-dimensional (3D) dose distribution than it has to do with local choices and local availability of equipment. Williams (2002b) has identified image-guided radiotherapy (IGRT) as perhaps the final missing link in our ability to deliver high-dose volumes to planning target volumes (PTVs) which are built with very small margins. One might comment that understanding the functional status of the target, sub-targets and OARs is also a key element. Accommodating the live moving patient is also still a problem to be solved.

Walker (2001) outlines the steps in performing and issues surrounding, the development of clinical CFRT and the role of IMRT. Balycyki *et al* (2001) discuss the issues underpinning the selection of new linear accelerators. One of the criteria highlighted is the need to support IMRT delivery. They note that this is difficult to arrange when the field is so rapidly moving. Cost should not be the main preoccupation. A study by Bate *et al* (2002) showed that when implementing IMRT of the prostate there was an increased workload at the level of simulation and planning but that, except in the first four days, the treatment time did not increase.

Most of us predicate the view that, as technology becomes more complex, yet more staff will be needed to run it, to check its accuracy, to ensure no errors and so on. A recent *Medical Physics* 'Point and Counterpoint' between Starkshall and Sherouse has debated this (Starkshall *et al* 2001). Sherouse believes in this. He believes the role of the medical physicist is to be much wider than some quality control engineer. He quotes Arthur C Clarke that sufficiently advanced technology is indistinguishable from magic and that this requires the qualified magician to be much more than a sorcerer's apprentice. The difference would be revealed in a radiation crisis of the kind of which there have been several noted examples. He feels the aspects of caring for machines have been overemphasized compared with caring for people.

Conversely, Starkshall believes new biological warfare on cancer will make radiation therapy redundant. He believes modern machinery can essentially check itself and requires little more than spot checks. He believes in 'treat-by-wire' like fly-by-wire. He does not believe medical physicists would avert disaster and should certainly not sit around waiting for the emergency which never arises. There is much to debate (see also Webb [2000d] on this same subject).

IMRT is coming of age. Around the world, courses are now regularly held. Among these are the (Siemens) MRC-Heidelberg IMRT Schools, the Royal Marsden School, the School at The Netherlands Cancer Institute, the School at the Mallinkrodt Institute, St Louis, the School at Stanford University and others. Other equipment manufacturers also run IMRT Schools. For example, a report on the first Elekta-sponsored course (March 2000) on IMRT appeared in *Wavelength* (2000d) in July 2000.

In summary, this introduction has highlighted:

(i)   the early beginnings of IMRT;
(ii)   the very rapid development in which history seems almost a misnomer;
(iii) the burgeoning literature and suspicion of overpublication;
(iv)   the focus on image-guided radiotherapy as the largest unsolved problem and imaging for plannning and imaging for accounting for the moving patient;
(iv)   the switch from technological development to the proliferation of clinical IMRT;
(v)   the burden of the cost of new technology, unlikely to diminish in the short term;
(vii) the growth of IMRT Schools;
(viii) the growth of an IMRT backlash.

1.2   Criticism of the philosophy of IMRT

In this section, we spend some time with the issues of IMRT philosophy and specially concentrate on those writings which have generated criticism of the clinical implementation of IMRT.

In November 2000, the journal *Medical Physics* pitted Sarah Donaldson against Art Boyer on the subject of the wisdom of introducing IMRT widely (Donaldson *et al* 2000). Donaldson argued:

(1) that there are little data to support general non-investigative use;
(2) that there are too many inaccuracies associated with patient positioning;
(3) that organ motion remains beyond physician and physicist control;
(4) that quality assurance procedures for IMRT are in their infancy;
(5) that IMRT requires unusually and unacceptably long delivery times that are disruptive in a busy clinical department;
(6) that IMRT involves excessive leakage radiation;
(7) that fusion of CT and magnetic resonance imaging is imprecise for IMRT;

(8) that IMRT start-up and maintenance costs are too expensive and

(9) that there is a costly learning curve for IMRT.

She argues that IMRT should remain an institutional investigative tool only.

Against the proposition, Boyer claims that the lack of biological data has no special role in IMRT any more than in any other form of radiation therapy. Today's therapy techniques have been developed against a background of lack of such knowledge and the same should happen for IMRT in the future. Secondly, he refers to the 'Catch-22' whereby, unless there are extensive IMRT facilities, the needed experience to enable clinicians to use 3D dose information and the additional control over the modulation will not be gained. He also points out that, at least in the United States, there is a wide variety of institutional techniques with a range from the very simple to the very sophisticated.

Donaldson then counters that IMRT adversely impacts the current working practices of the radiotherapy department. She concludes that the technique should only be applied in a research setting. In practice, many clinics have found that IMRT can be performed without increased burden once the initial stages of implementation have been worked through.

Conversely, Boyer points out an important issue, namely the preferred scope and speed of implementation of IMRT. If IMRT is introduced too early, this could precipitate experiences that will condemn the process. However, implementing IMRT too slowly will unduly postpone the realization of the advantages of the technique. Boyer argues that the wider the implementation of IMRT is, the more likely we are to answer the questions that concern Donaldson sooner rather than later. Similar arguments were made by Allen Lichter for the introduction of conformal 3D radiotherapy in the early 1990s.

This topical debate underlines concerns about the clinical implementation of IMRT that were heard hardly at all three to four years ago. Despite the fact that IMRT was being developed, little except positive comment had come from doctors and clinical physicists keen to take advantage of the approach. Most, fortunately in my opinion, still express this keenness and doubters are in the minority. This section highlights them to encourage the reader to enter the debate. The logic of the pro-IMRT-ers is not over-emphasized here because it is represented by the scientific contributions in the rest of the book. The remainder of this section may thus appear quite negative to IMRT.

Paliwal *et al* (2004) enter the debate on the wisdom of heavy use of IMRT. One protagonist in this *Medical Physics* Point and Counterpoint argues that its use (in the USA) has more to do with gaining profits from reimbursements and less to do with demonstrable improvement in tumour control. The counter protagonist argues that using IMRT for 20% of patients is justified and that IMRT properly continues the quest for improved precision.

Boyer (2001) has emphasized that implementing IMRT requires an assessment of primary and secondary barrier radiation safety as well as consideration of the risk of inducing a second carcinoma to the patient. Magnetic resonance images should be incorporated into the treatment-planning process and all staff should understand the, at least temporary, increase in work burden as a result of implementing IMRT. Boyer (2001) argues that the increased reimbursement for IMRT should be used to provide the necessary additional human resources. Good estimates of these costs have been provided by Gillin (2003).

Potters *et al* (2003b) question the definition of IMRT in the context of an editorial commenting on a specific paper. For example, they refuse to accept that breast irradiation using several static-MLC fields constitutes IMRT of the breast in the context of the definitions of process of care required for US Healthcare reimbursement. Vicini *et al* (2004) argue that concentrating attention on the terminology has detracted from the importance of improving breast IMRT in clinical practice. The argument is uniquely relevant to a healthcare system based on insurance reimbursement.

Purdy and Michalski (2001) have provided an editorial discussion on whether the clinical evidence for CFRT and IMRT supports the enthusiasm for the technology. They point to two important clinical trials. One, at the MD Anderson Hospital, compared delivery of 70 Gy or 78 Gy to the prostate with CFRT in a phase-3 randomized trial. The outcome showed increased freedom from disease failure and biochemical failure at the higher dose (Pollack *et al* 2000, 2002a, Pollack and Price 2003). A second study (Ryu *et al* 2001) showed favourable lower toxicity for increased dose. (Further hard clinical evidence may be found in chapter 5.)

Purdy and Michalski (2001) also emphasized that 3D radiotherapy does not mean using non-coplanar beams. It does mean taking more care over the 3D specification of the gross tumour volume (GTV), clinical tumour volume (CTV) and PTV. They feel that medical doctors should embrace this technology, become better educated and start to use it even before there is overwhelming evidence of clinical efficacy. Their view is that there is no other path to improving the physical basis of radiation therapy.

Levitt and Khan (2002) are, conversely, sceptical of the rush to judgement in favour of CFRT and IMRT. They point to difficulties in determining the volumes, problems with assessing the outcomes of radiotherapy and limitations with the few extant trials.

The several writings from Schulz address the same concerns about the speed of IMRT clinical adoption. A *Medical Physics* 'Point and Counterpoint' (Schulz *et al* 2001) has been given to the controversial arguments from Schulz that the pursuit of IMRT is falsely based and possibly a waste of time. Schulz argues that IMRT is highly profitable for manufacturers and good for employment of physicists but will not significantly reduce overall cancer mortality. He provides detailed statistics showing that IMRT at best can reduce the number of deaths by about 4900 in the United States, for which each centre would need an IMRT system, each of which would at best extend the lives of five patients per year. He concludes that this capital investment and operational expense cannot be justified.

Arguing against this proposition, Deye makes the point that, first of all, all new technologies have always had their critics and the arguments raised are not unique to IMRT. He also argues that outcome data cannot be used to predict which new technologies would be beneficial. He also argues that if IMRT is seen to be being introduced too quickly then this would imply that physicists are not spending enough time on the new developments rather than too much. In short, his argument is that it is the duty of scientists to get the science right and to let those who deal with the 'big picture' put things into prospective. Deye also remarked that there is clear evidence for reduced rectal toxicity in treating prostate cancer from the work at Memorial Sloan Kettering Cancer Center (see chapter 5). Schulz sticks to his guns and argues that it is competition between hospitals for patients that plays a far larger role in purchase decisions than do expectations of improved clinical outcome and that, in short, IMRT is market driven. This argument (Schulz *et al* 2001) is almost an exact repeat of one a few years earlier from the same author (see review in Webb 2000d).

Then, just a few weeks later in the literature (Schulz and Kagan 2002a), the same statements criticizing the overfast adoption of IMRT receive further printspace along with a general criticism of the outcomes of a workshop designed to find the most effective areas of medical physics research. The comment extends to criticize the programmes of work on image-guided radiotherapy and the Cyberknife. The rebuttal is less than firm (Cumberlin *et al* 2002) but cautious. Schulz and Kagan (2002b) have another attack at the enthusiasm for IMRT, this time with a reasoned discussion of the clinical issues in each tumour site. The final paragraphs intended as tutorial for radiation physicists are particularly barbed. However, the paper does provide a systematic comment on an organ-by-organ basis.

Schulz and Kagan (2003a), now becoming familiar critics of IMRT, comment on IMRT in cancer of the prostate. They state that, given the margins applied to prostate to create a PTV, the rectum must be significantly inside the PTV at least on its anterior side and therefore they find it hard to understand why sculpting the dose around the PTV will lead to an improvement in rectal complications. They argue that the rectum will wander in and out of the high-dose region due to normal physiological function. They even think that the reduced rectal toxicity might result as much from organ motion as from the precisely positioned dose distribution delivered by IMRT. In response to this, Zelefsky and Leibel (2003) re-present the evidence that IMRT of the prostate leads to decreased rates of Grade-2 rectal bleeding and they claim that this is explained by the reduced circumferential volume of normal rectum exposed to high radiation dose levels. They agree that prostates have always moved and will continue to do so and that this movement is not the cause of reduced rectal complications (see also chapter 5).

Kagan and Schulz (2003) and Pollack *et al* (2002b) further debate whether a change in biochemical failure is related to cause-specific death in prostate cancer. Pollack *et al* (2002a) found a change in post-treatment PSA levels when escalating CFRT from 70 to 78 Gy at Fox Chase Cancer Center (FCCC). Keall and Williamson (2003) have defended the development of IMRT technology in the light of further criticism from Schulz and Kagan (2003b) who claim that the ultimate radiation therapy (the 'infinitron') would still be inferior to surgery.

Fraass (2001) has commented that IMRT and inverse planning have been accepted somewhat quickly with few studies which quantitatively compare the benefits of this new technology with older CFRT techniques. Somewhat more positively he presented a number of types of comparison experiment which could be used to address most of the limitations that apply to making comparisons between unlike treatment modalities.

Glatstein (2003a) has written an interesting piece pointing out that the late 20th century has been a unique time for hyping technology (and drugs) directly to the public with almost no proof of their efficacy available. He cautions against this and is concerned that the 'dollar drive' may be behind it. Many patients are not able to assess the information they receive. The new techniques are billed as routine rather than research because only the former can be charged for in the USA. In response to Glatstein (2003a), Amols (2003) reminds him and us that whilst it may be convenient to 'blame' manufacturers, administrators, the internet, the mass media and patients themselves for the hype and the growth of IMRT, it is the clinician who 'holds the key to the IMRT suite' and thus controls the volume of use of this modality. Glatstein (2003b) reports that IMRT will not be subject to controlled clinical trials because to do so would challenge the notion that IMRT is clinically reimburseable and instead is 'research'. Also he points out we live in an era when direct appeal is made to the sick patient to make treatment choices they cannot comprehend. Some American insurance providers are refusing to pay for IMRT until the results of randomized controlled trials are known (Potters *et al* 2003b), a peculiarly perverse situation given both the paucity of such data at present and the view expressed previously that, in some circumstances, it may be unethical to conduct such trials. This is a peculiarly American perspective because clinical trials *are* planned in Europe. Nutting and Harrington (2004) and Dearnaley (2004) have joined this correspondence to point out that, in the UK, precisely the reverse economic conditions prevail. Unless a technology can be proven worthwhile, the National Health Service (NHS) will not fund it. They look forward therefore to the results of such trials and sharing this information with colleagues in the USA to address this issue. Amols (2004), however, emphasizes that even if (when?) proved valuable, IMRT issues of cost effectiveness will still remain. Also, not surprisingly, he points out it is convenient for the less well-funded nation not to have gone the way of the USA.

Halperin (2000) has also questioned the uncontrolled growth of radiation technology. His view is that healthcare costs will grow to 17% of GDP in the USA by 2010 just as the baby boomers retire. He is wary of unscrupulous salesmen who (sic) crank up sales or threaten to cut off the oxygen of patient custom. He questions whether overcomplex treatment plans are being prepared to recoup more costs when simpler ones would suffice. I quote 'The distinctions between evidence-based medicine and if-it's-reimbursable-we'll-do-it medicine have been lost'. 'Radiation oncology, as a specialty, is guilty of encouraging profligate spending, we will engender a backlash, and we'll deserve it.' He does comment that CFRT is a testable hypothesis. Indeed many groups are testing it. Hutchison and Halperin (2002) have demonstrated how decisions regarding radiation oncology purchases can be influenced by market-driven advertising.

Kavanagh (2003) wrote a provocative letter 'The emperor's new isodose curves' which, if I have understood correctly, is advancing the view that IMRT physicists are pursuing some clinically irrelevant goals. He likens the practice of IMRT to the selling of art and postulates that 'the art dealer fares better than the art critic'.

There has been considerable debate following the original paper by Levitt and Khan (2001) asking if the evidence supports the enthusiasm for CFRT of the prostate and dose escalation. The editorial accompanying the article felt their paper was too 'anti- technology' (Purdy and Michalski 2001). Levitt and Khan (2002) wrote a letter reiterating that their concerns lay with the inappropriate and untrained use of the technology rather than with the technology itself. Also they believed there is no clinical evidence for irradiating the prostate beyond 74 Gy and that indeed such irradiation may lead to long term radiation damage of normal tissues that will not be seen for many years. Purdy and Michalski (2002) re-emphasize their beliefs in the need for better training and randomized trials.

In Colorado Springs, in June 2003, the AAPM held its annual Summer School on the theme of IMRT. At that meeting there was a thorough exposition of the state of the art (Palta and Mackie 2003). Whilst the physics is globally universal, the emphasis was perhaps understandably from an American perspective so far as implementation is concerned. In the discussions some very quotable 'soundbites' were heard including the following:

(i)   'Is IMRT 'tailored treatment' or 'Emperor's New Clothes'?',
(ii) 'Is an IMRT plan really optimal or an optimal illusion?',
(iii) 'Remember there is a 'con' in concave' (Langer 2003).

These comments were not necessarily expressing the views of the speaker but said to provoke discussion. With regard to escalating costs in IMRT technology, Battista and Baumann (2003) defined some new principles in the 'ALARA mould'. Machines would be developed 'as long as it is technically achievable (ALATA)' but 'to operate the ALATA principle you need ALATA (a lot of) money'. Dose distributions do not necessarily have to be 'optimum' but should be 'as conformal as reasonably achievable (ACARA)'. Battista and Baumann (2003) reminded us of a quote from Harold Johns, one of the radiotherapy physics pioneers, that is as true today as when he said it: 'If you can't see it you can't hit it and if you can't hit it you can't cure it'. This was an exceptional School.

In concluding this chapter, which has been more about the politics of, and belief systems in, healthcare, we may note that the first few years of the new millennium have witnessed some published scepticism of the usefulness of IMRT.

Table 1.2. Arguments for and against the clinical implementation of IMRT.

FOR	AGAINST (As expressed but not necessarily true)
• IMRT creates concave high-dose distributions protecting organs-at-risk and enabling dose escalation to target.	• The value of clinical implementation is out of all proportion to the current clinical evidence for efficacy.
• IMRT builds on experience with 3D CFRT.	• It is implemented: (i) because it is there. (ii) because of manufacturer pressure. (iii) because of $ reimbursement.
• Early clinical evidence for IMRT supports the view that 'better dose distributions lead to better clinical outcome' (e.g. reduced xerostomia, reduced rectal toxicity, reduced dry-eye syndrome, reduced paediatric complications).	• There are too many 'difficulties': small physical segments, small MUs per segment, moving parts, hard-to-verify absolute dosimetry.
• Clinical implementation builds on years of physics development which has improved understanding of planning and delivery.	• The moving patient destroys the concept.
• It is thoroughly validated by professional groups and working parties.	• Workloads increase with IMRT.
• It is now relatively straightforward and widely commercially available.	• IMRT is hard to QA and verify.
• Widespread implementation will allow randomized clinical trials which will generate more evidence for benefit and eventually economies of scale.	• Delivery times are long; possible secondary cancers?
• Movement can be accommodated using IGRT.	• Mazes need redesigning.
• QA and verification tools exist.	• IMRT is cost ineffective.
	• The public are being asked to choose between treatments they cannot understand due to direct hype.
	• IMRT divides society. It is a rich man's technology.

This has ranged from what, at least to my eyes, looks like outright rejection of IMRT as a useful clinical tool through to more balanced questioning of its role. It is possible to be sympathetic to some of the underlying concerns about lack of evidence. However, a climate of reasonable expectation of benefit is often all one can 'put upfront' and ultimately the proof of the worth of IMRT will lie in the outcome of randomized phase-3 clinical trials. Without the resources to conduct such trials, the evidence can never come and, on this basis, the proliferation of technical installation and clinical implementation is justified. Many would say this same philosophy of implementation without hard evidence was applied to most of the significant developments in medical physics, e.g. the installation of linacs instead of kV radiation, the development and use of treatment planning, the coupling of 3D medical image data to treatment planning etc.

Table 1.2 summarizes some of the arguments for and against the clinical implementation of IMRT. The table is not meant to express incontrovertible truths but to offer arguments some have put. Readers can use this as the basis for their own thought and discussion.

The bottom line, however, is that IMRT has an unstoppable momentum.

Chapter 2

Developments in rotation IMRT and tomotherapy

The technologies of delivering IMRT in slices are known as rotation IMRT or tomotherapy. Purists may argue about the precise terminology but serial tomotherapy (MIMiC-based) and spiral tomotherapy (University of Wisconsin machine-based) are related and usually reviewed together. The history of what interaction there might be considered to be was best reviewed by Mackie *et al* (2003b). The concept of the binary modulator for slit-field radiation was first proposed by Swerdloff in Mackie's group in 1988. This led to the patent for the multivane intensity-modulating collimator (MIMiC) being held by the University of Wisconsin Alumni Research Foundation (WARF) and licenced to NOMOS who first made it a practical reality in 1992. The idea of spiral tomotherapy did not come until about 1992 when spiral CT became possible and Mackie saw the opportunity to reuse a gantry for this purpose. By then the NOMOS MIMiC was a reality (NOMOS is now part of North American Scientific). It was another 10 years before the spiral tomotherapy unit, a much grander engineering venture, became commercially available.

2.1 NOMOS MIMiC tomotherapy

2.1.1 Technology history

The NOMOS MIMiC technique for IMRT with its associated planning system, originally known as PEACOCKPLAN, now as CORVUS, was introduced 12 years ago in 1992. It has been stated that its inventor Carol had the concept as early as 1976 but did not take it forward at that time (Curran 2003). From the very outset, it was a commercially available device. When announced in the autumn of 1992 (ASTRO in Calgary and at a 3D radiotherapy meeting at the World Health Organization [WHO] in Geneva), it took the radiation physics community by some surprise because it was both unexpected and an unusual

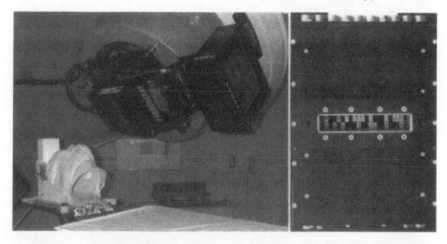

Figure 2.1. The serial tomotherapy approach to IMRT. This form of IMRT uses a temporally modulated mini MLC system such as the MIMiC (NOMOS Corporation) shown here mounted to a conventional low-energy megavoltage medical linear accelerator. Treatment to a narrow slice of the patient is delivered by arc rotation. The complete treatment is accomplished by serial delivery to adjoining axial slices. The right-hand part of the figure shows a patient's-eye view up into the MIMiC. (Reprinted from *IMRTCWG* (2001) with copyright permission from Elsevier.)

concept. The MIMiC (figure 2.1) was designed to be attached to any linac. It irradiated two narrow slices of the patient at any time by rotating a slit collimator through a series of gantry angles. In its simplest form, the modulation switched effectively every 5° (figure 2.2). From the patient's-eye view, the machine looked like a set of chattering teeth. The modulation was provided by varying the dwell time of the attenuating elements in the slit. The variation was controlled by an onboard rotating computer holding a floppy disk created from the planning system. The machine monitored the gantry angle independently and the sequence was recorded for verification. To irradiate a larger tumour, the patient was sequentially longitudinally traversed through the line of sight of the MIMiC. Until the first clinical IMRT made with an MLC in the late 1990s, use of the MIMiC was the only way to deliver IMRT clinically and the NOMOS Corporation could rightly style themselves as *the* Intensity Modulation Company. As MLC-based IMRT became more common, NOMOS diversified into planning for MLC-based IMRT. No-one really knows the exact number of IMRT treatments but, at the turn of the Millennium, it could be confidently stated that more IMRT treatments had been delivered by the MIMiC than by the MLC. One suspects that that situation may have now changed with the rapid proliferation of MLC-based IMRT. Detailed descriptions and photographs of the technology and developments from the NOMOS Corporation appear in earlier books (Webb 1997, 2000d) and

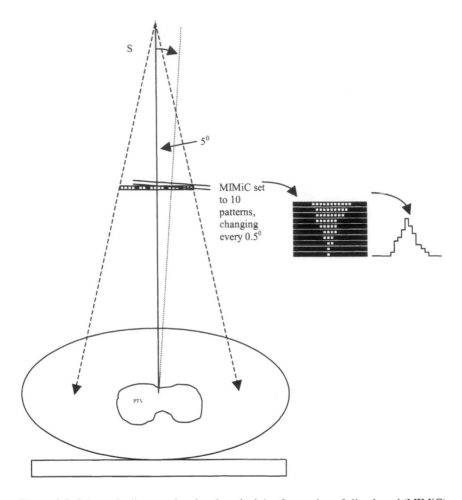

Figure 2.2. Schematic diagram showing the principle of operation of slice-based (MIMiC) tomotherapy. The irradiation of just one slice is shown and so also just one bank of the two-bank MIMiC is shown. The planning system works out the required 1D modulations at each 5° of rotation around the patient. A possible modulation is shown at the right of the figure. In order to realize this modulation in practice, the opening and shutting pattern of the vanes of the MIMiC is adjusted every 0.5° as shown in the ten schematic diagrams of the MIMiC (a dark bixel is meant to symbolize a closed bixel and a light bixel symbolizes an open bixel). In view of the 'long-lever effect' of radiotherapy, the PTV 'sees' the modulation as if it were at fixed 5° increments. After a rotation around the patient, each of the two slices irradiated simultaneously by the two banks of the MIMiC are appropriately irradiated. To irradiate the whole PTV, the patient is then shunted accurately to the next longitudinal position. The drawing is not to scale because the depth of the MIMiC vanes in the direction of the radiation is not properly represented here.

will not be repeated here. In the next subsections, we concentrate on issues discussed since 2000.

2.1.2 Matchline concerns and solutions

From the outset there was concern about the reliability of matching the slices and the Company provided the CRANE to ensure this. Even so, the technique may have been safest when applied to the head-and-neck where patient immobilization could be ensured. Dogan *et al* (2000a) addressed the match-line issue in serial-slice tomotherapy using the NOMOS MIMiC. They suggested that each treatment should be divided into two, one with the 'original' isocentre determined by CORVUS and the other with an isocentre shifted by 5 mm. Each treatment plan would be delivered on alternate days summing to the total treatment dose. They made measurements to show that the dose inhomogeneity in the PTV decreased from about 25% to about 12% on doing this. The trick in the technique lay in forcing CORVUS to offset the isocentre longitudinally by artificially adding a length to the PTV in the longitudinal direction. The new technique was shown to be more robust with respect to errors in longitudinal movement caused by the CRANE or the treatment couch being inaccurately controlled.

The original method proposed by Dogan *et al* (2000a, b) and Sethi *et al* (2000) to achieve this involved extending the target by a pseudo target region. In follow-up papers, Leybovich *et al* (2000a) have improved IMRT dose distributions by creating pseudo structures which are flagged as regions in which dose should be kept to a minimum. These have improved IMRT plans of the posterior neck and thoracic spine. Leybovich *et al* (2000b) have shown that there are alternative ways to create the second-phase treatment plan (figure 2.3). The following were considered:

(i) the addition of an adjacent small target extension instead of a full-width target extension;
(ii) the creation of a small 'pseudo target' some distance away from the original target;
(iii) the creation of a 'pseudo target' outside the irradiation volume.

They showed that each of these procedures creates a second-phase treatment plan whose set of abutment slices lie in different longitudinal planes to those without the target extensions. They then showed that the two-phase irradiation technique, even with precise positioning of the phantom with respect to the irradiation beams, leads to an improved dose homogeneity in the PTV using their technique. They then went on to show that, when matchline errors are deliberately introduced, either 0.5 mm or 1 mm, the two-phase treatment technique leads to an improvement in dose homogeneity to the target compared to the single-phase irradiation. It was found that the use of a pseudo target remote from the main radiation volume creates the least change in integral dose. Lehmann and Pawlicki

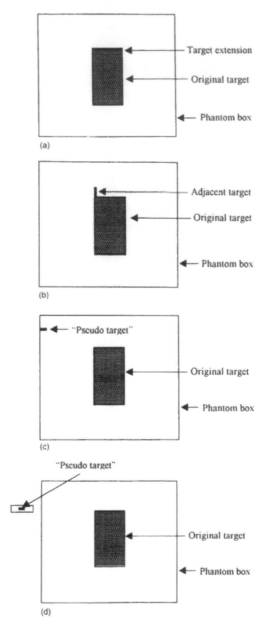

Figure 2.3. Methods of target modification to improve dose homogeneity and to minimize matchline errors: (a) a simple target extension, (b) addition of a small adjacent target, (c) addition of a distant pseudo target inside the phantom, (d) addition of a distant pseudo target outside the phantom. (From Leybovich *et al* 2000b.)

(2003) have also proposed adding 'tuning structures' to give extra flexibility on IMRT planning.

Dogan *et al* (2000c) have shown that the use of a pseudo structure positioned in the posterior neck region can improve the dose reduction to spinal cord when irradiating concave-shaped head-and-neck tumours. They also showed that IMRT improved dose distributions compared with 3D, conformal and two dimensional (2D) treatment plans.

Dogan *et al* (2002b, 2003a, b) have shown how the use of two pseudo targets in a CORVUS plan can lead to feathering the match line between two intensity-modulated fields that are required to build up a modulation pattern for a field area which is bigger than that coverable using a Varian accelerator.

Sethi *et al* (2001a) have further developed a solution to the problem of matching tomographic IMRT fields with static photon fields. They make the observation that in the longitudinal body direction, if static fields are matched to tomographic fields longitudinally and position errors of the order 3 mm occur, then hot and cold regions regions of the order 50% can result. This is the familiar field-matching problem that comes from trying to match two very sharply falling penumbra. Although the hot spots are significantly large, they do localize in a very small longitudinal space. The solution proposed was to create a buffer zone within the target volume, some 3 cm wide, and to arrange that both the IMRT planning and the static-field planning created a dose gradient in this region of about 3% per mm with the gradients in reverse directions with respect to each other (figure 2.4). This can be achieved by specifying a variable dose with distance prescription for IMRT and by using either a hard or a soft wedge for the static fields. Then, not surprisingly, when these two fields are just opposed with no error (exact match) this results in virtually no hot or cold spots. Moreover, if errors up to 3 or 5 mm occur, the hot and cold spots are still limited to a magnitude of less than 10% although the spatial area over which they arise can be up to ± 2 cm (i.e. across the whole width of the buffer zone). In summary, this paper simply exploits the fact that ramp distributions match more robustly than do sharp step function distributions. This is incidentally the fundamental reason why spiral tomotherapy has an advantage since it inherently creates triangle functions longitudinally.

2.1.3 Energy considerations

Dong *et al* (2000b) have studied the effect of beam energy in prostate serial tomotherapy. Plans were created using 6, 10 and 15 MV x-rays on the CORVUS inverse-planning system and it was shown that the treatment plans at the different energies exhibited very similar results. This had been shown three years earlier by Sternick (1997) (see also section 6.9).

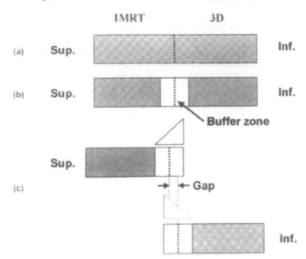

Figure 2.4. (a) Diagram of the target. The superior target is treated with IMRT and the inferior target with static fields. (b) A common 'buffer zone' around the matchline is used in treatment planning for both IMRT and static fields. (c) Beam-edge 'wedge modifiers' are used in IMRT and static-field plans to obtain the desired dose profile in the buffer zone. Combined dose profiles in the abutment region were measured for various gaps and overlaps. (From Sethi *et al* 2001a.)

2.1.4   Concerns about increased treatment time

The MIMiC IMRT technique generally requires the delivery of substantially more MU than a conventional therapy. This has implications. Dong *et al* (2000a) have studied the effects of induction of secondary cancers from IMRT delivered with the NOMOS MIMiC system. It was found that the risks were calculated to be 15%, 12% and 17% for 6, 10 and 15 MV for radiation-induced secondary cancers for a course of prostate treatment of 70 Gy. The more penetrating high-energy x-rays shorten the machine on-time so reduce the whole-body dose exposure but the high-energy x-rays produce significant neutron dose equivalent which greatly increases the estimated risk for radiation-induced secondary cancers. Tao *et al* (2002) have shown that IMRT delivers a greater integral dose to the body than conventional conformal plans. Conversely, Della Biancia *et al* (2002) have shown that IMRT can lead to a lower integral dose to the patient if a judicious selection of beam directions is made. The integral dose for different beam arrangements is a function of the average pathlength in the patient. Kry *et al* (2003) have calculated that the risk of secondary cancers roughly doubles with IMRT.

Mutic *et al* (2001) have shown that room shielding calculations, reassessed for IMRT, lead to the conclusion that conventional primary barriers are adequate for both dynamic MLC and serial-tomotherapy IMRT. However, the excessive head leakage produced by these modalities requires an increase in secondary barrier shielding. The calculations are based on a careful study of a series of efficiency factors for the delivery of IMRT. Shielding issues were discussed at length by Low (2003).

2.1.5 Machine features

The NOMOS MIMiC is known to have features associated with acceleration and deceleration phases of its vane transit. In particular, when vanes move from a completely open to completely closed position or *vice versa*, they bounce slightly against the end stops and this bouncing can be modelled as a sinusoidal behaviour with damping. Tsai *et al* (2000) have modelled this phenomenon with analytic equations and predicted the variation in dose profiles due to this damped oscillatory motion. The predictions were compared with experiment favourably. However, they concluded that the output dependence on the various parameters only borders on clinical significance and the paper is complex in relation to this.

Zheng *et al* (2002b) have repeatedly delivered the same PEACOCK plan to assess whether the leaves of the MIMiC are behaving accurately. It was found repeated dose point measurements agreed with each other to within 5%.

Hossain *et al* (2001) have studied the dose output as a detailed function of the switching frequency for the leaves in the NOMOS MIMiC collimator. It was found that the output is enhanced as the switch rate increases and therefore every patient plan needed to be renormalized based on the actual measurements taken during the delivery of a specified intensity pattern to a phantom.

Huang *et al* (2001a) have added a tertiary collimator called a 'Beak Slit' collimator to the MIMiC (Salter 2001) to improve the conformality of dose to a very small clinical treatment volume. This device truncates the bixel size at isocentre to 1.0 cm × 0.39 cm. Zinkin *et al* (2004) showed that the use of the 'Beak Slit' improved tomotherapy compared with the use of the MIMiC alone with 1-cm-wide slices.

Salter *et al* (2001) have used repeat CT scans to measure the repositioning accuracy of the NOMOS TALON removable head frame. Twenty-six repeat CT scans were used for nine patients. The mean magnitude of isocentre translation was 1.38 mm during a treatment course over six weeks. Average rotations were less than 0.5°.

Salter *et al* (2003) have made a comparison of the dose distribution delivered to a patient with and without correction for motion as deduced from BAT measurements. It was found that the mean dose to the tumour could be reduced from some 97% to 88.3% if BAT corrections were not made. Orton and Tomé (2003) have made a similar observation from planning studies with and without ultrasound-determined motion correction. For a detailed consideration of the use of ultrasound in assisting radiation therapy delivery, see section 6.10.5.

Huang *et al* (2002b) have reviewed the plans for 29 patients with prostate cancer treated with IMRT and planned using the NOMOS CORVUS system 3.0.

Some of these plans included three or four couch movements with gantry rotations from 240 to 120° clockwise. The plans were compared with conventional plans and the outcome showed adequate dose distributions to the prostate with sparser dose to the bladder, rectum and femoral heads.

Woo *et al* (2003) have described the evolution of quality assurance for IMRT delivered with the NOMOS MIMiC. This paper is a definitive description of the various tests both of the system and of individual patient dosimetry that are advisable when the MIMiC is in use.

Sanford *et al* (2000) have developed a technique to verify tomotherapy delivered serially using the NOMOS MIMiC. Dose distributions were evaluated with a spatial resolution better than 5 mm and errors in absolute dose were detected within a tolerance of about 6%.

Kapulsky *et al* (2001) have used the Peacock polystyrene film phantom and radiation therapy film dosimetry system for comparing planned and delivered dose distributions for the NOMOS MIMiC MLC system.

In summary, since the development of other IMRT delivery techniques the MIMiC continues to be widely used in the USA. However, the volume of technological development has, not surprisingly, slowed down since the major equipment developments were in the early to mid 1990s.

2.2 University of Wisconsin machine for tomotherapy

2.2.1 Development history

The tomotherapy machine of the University of Wisconsin is a purpose-designed technology which also uses a MIMiC-like collimator to deliver a single slit of variable modulation with the variation controlled by the dwell time of the vanes (figure 2.5). Instead of the double set of 20 vanes, this MIMiC has a single set of 64 vanes (figure 2.6). The gantry rotates more than once continually as the patient traverses longitudinally and slowly through the beam. From the patient's perspective, the radiation slit then appears to execute a spiral with fine pitch (figure 2.7). In spiral tomotherapy the fan-beam thickness (FBT) can be chosen by the operator to balance the fast treatment time and dose modulation in the superior–inferior direction. The pitch factor is the couch movement per rotation in units of the FBT. Typically used values between 0.25 and 0.5 mm result in advantageous overlap between adjacent helical rotations. The modulation factor (MF) is the ratio of maximum leaf opening time to the mean leaf opening time in any projection and can also be user-selected (Kron *et al* 2004a). IMRT delivery is complemented by the addition of onboard megavoltage computed tomography (MVCT). The fundamentals of the device were announced in 1992 and the first clinical deliveries were in August 2002. A commercial company, TomoTherapy Inc., has been formed to manufacture and market these machines (figures 2.8, 2.9 and 2.10). To put the developments in perspective, there are at present very few of these machines compared to the hundreds of MIMiCs in clinical service for

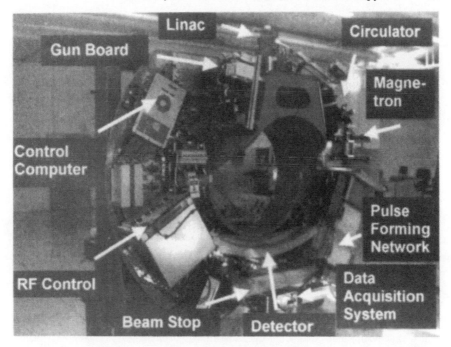

Figure 2.5. This photograph shows the helical tomotherapy unit with components labelled. This unit is the one installed at the M.D. Anderson-Orlando Clinic in Orlando, FL and represents an improvement in the manufacturability and serviceability as compared to the University of Wisconsin prototype; the linac jaw system MLC and detector system are nearly identical to the University of Wisconsin unit. (From Mackie *et al* 2003.)

sequential tomotherapy and to the large numbers of MLCs capable of performing IMRT (see section 2.5 for a discussion of the impact of this statement). This situation may change (Mackie *et al* 2003b, 2004). At the time of the 14th ICCR (Seoul, May 2004) four centres had treated patients with the commercial Tomotherapy machine (Kron *et al* 2004a). The perspective on the importance of this device is coloured by the fact that, for 10 years or so, it has been under development very much in the full gaze of the peer community (unlike the history of the NOMOS device). The design has also changed over the years; for example, initially an onboard x-ray kVCT system was planned. Even now when there are clinical commercial installations, its role is hard to assess because the numbers are small. This will probably change dramatically in the next few years as its clinical impact grows.

Figure 2.6. Photograph of the helical tomotherapy binary collimator as mounted on the helical tomotherapy unit. A bank of leaves is visible. Sixteen of the 64 valves are visible on both left and right. Local high-pressure air reservoirs are visible above and below. (From Mackie *et al* 2003b.)

2.2.2 Generation and use of megavoltage computed tomography (MVCT) images

It is important to distinguish that, in the last four years or so, the developments at Wisconsin have moved from simulations and laboratory measurements (the 'tomotherapy workbench') to the Wisconsin prototype and now to commercial models. As a result, when reviewing papers, the statements made must be interpreted in terms of which phase of development they refer to.

Ruchala *et al* (2000a) have described how MVCT images can be obtained on the University of Wisconsin tomotherapy workbench (a model of the tomotherapy system in which the phantom rotated on a spiral assembly in front of a stationary linac) using the radiation which is also used for treatment. Two major problems were addressed. The first is that only a limited number of MIMiC vanes are open, those directed towards the PTV, and so the datasets are necessarily incomplete. The second problem is that, in order to compute line integrals using the usual logarithmic attenuation method, it is necessary to know the input fluence to each line and this is also not a constant because the vanes are clearly moving to create the modulation. This latter problem was overcome by noting that there are periods at which the vanes are fully open and these can be used to 'normalize' the projection data for a limited subset of measurements thus

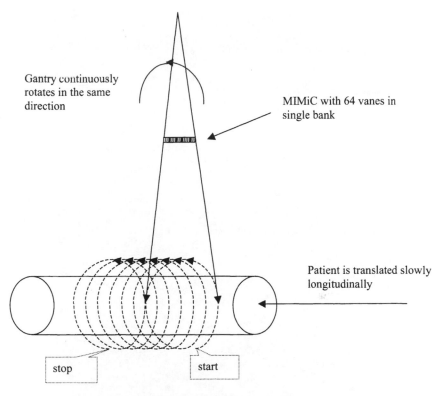

Gantry continuously rotates in the same direction

MIMiC with 64 vanes in single bank

Patient is translated slowly longitudinally

stop

start

Apparent spiral trajectory in the reference frame of the stationary patient

Figure 2.7. Schematic diagram showing the principle of operation of helical tomotherapy. The irradiation of a whole volume is shown. The MIMiC has only one bank of 64 vanes. The accelerator with this MIMiC rotates continuously around the patient without reversing direction whilst the patient slowly translates longitudinally. From the perspective of the stationary frame of reference of the patient, the beam appears to execute a helical trajectory.

permitting reconstruction. Because this is not a full dataset, the contrast resolution in the reconstructed results is necessarily not optimized. To overcome this, and also to address the problem of limited coverage of the data, Ruchala *et al* (2000a) proposed introducing a number of 'flashes' in which all vanes of the MIMiC are opened for a very short period of time at a regularly spaced number of angular intervals and these data used to assist with reconstruction. They showed the results of this technique on both phantoms and on a German shepherd dog cadaver. The name given by Ruchala *et al* (2000a) to the measurements made with the MIMiC vanes open in a stable position was 'spairscan'. They also comment on

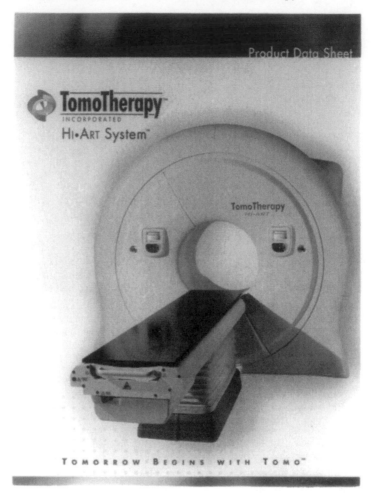

Figure 2.8. The cover of the first brochure for the Hi-ART TomoTherapy machine. Note that, as with the advertisement of CT scanners and MR scanners, the clinical equipment appears to be little more than a hole in a box, disguising the complexities beneath. Note the strapline 'Tomorrow begins with Tomo' (courtesy of Professor Rock Mackie).

the technique of reconstructing MVCT images using pure leakage (transmitted) radiation.

When a MIMiC MLC is used to deliver tomotherapy, the leaves close completely for a short time within each angular band defining a 1D intensity-modulated beam (IMB). So, for a nearly complete arc treatment, there are 69 such moments (of about 5 ms) in which the MIMiC is doing nothing other than leak radiation at about 0.3% of the open-field fluence. Ruchala *et al* (2000b) have shown that MVCT images can be computed using this entirely leakage

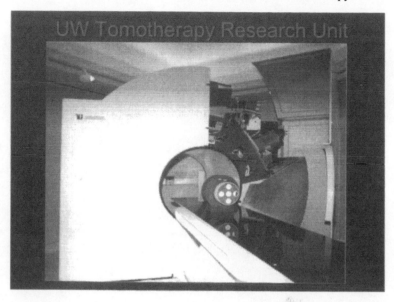

Figure 2.9. The University of Wisconsin Tomotherapy Unit with its covers off. (Courtesy of Professor Rock Mackie.) (See website for colour version.)

radiation (figure 2.11). Thus, no additional dose nor time is taken and the images are essentially 'free'. It was calculated that, on average, some 20% of the time the MIMiC leaves are all closed (this clearly depending on the inverse problem being solved). Thus, when a tomotherapy is set-up to deliver 175 cGy to the patient, '35 cGy leakage images' result. i.e. the images are as if the treatment would have given 35 cGy with all leaves open but in fact all-leaves-closed delivery had taken place. Density differences above 8% for 25.4 mm diameter objects and 3 mm air cavities could be observed. Ruchala *et al* (2000b) pointed out that these images are not really free because they result from unwanted leakage radiation which also provides patient dose. The laboratory benchtop test machine used a NOMOS MIMiC collimator so the '7 cGy image' in figure 2.11 is specific to this collimation. Typical MVCT images from the first-generation TomoTherapy machine were published by Ruchala *et al* (2002d) and, for the second-generation machines, images can be created with under 1 cGy due to the substantially lower leakage (0.3%) of the collimator on this machine (Ruchala, private communication). The reduced patient dose burden allows use of more dose with direct imaging techniques instead of the leakage method.

Ruchala *et al* (2002b) have created MVCT images and, with a 2 cGy scan, contrast of 3% can be clearly seen and air holes can be resolved at sizes of 1.2 mm for a 512 × 512 reconstruction matrix covering the 40 cm field of view. These images can be automatically registered to kVCT images to further assist with patient repositioning and verification. Ruchala *et al* (2003) have demonstrated

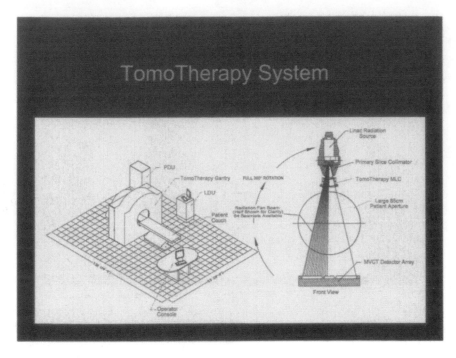

Figure 2.10. A schematic diagram of the University of Wisconsin Tomotherapy Unit showing the important portal imaging detector. (Courtesy of Professor Rock Mackie.)

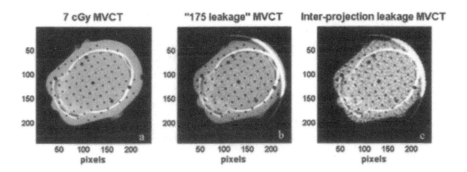

Figure 2.11. Three images of a Rando phantom created by megavoltage computed tomography using the University of Wisconsin prototype tomotherapy system: (a) image created with 7 cGy MVCT, (b) image formed with just leakage radiation and all beam elements closed and (c) an MVCT leakage image using the views collected solely during the 69 inter-projection regions when all of the leaves were closed. The air cavities in the phantom are 3 mm in size. These figures illustrate how MVCT images can effectively be obtained 'for free'. (From Ruchala *et al* 2000b.)

how the MVCT system can overcome image artefacts due to metal prostheses, patient size being overlarge and to aid planning and replanning.

Mackie (2001) and Olivera *et al* (2001) have described the developments in the helical tomotherapy prototype installed at the University of Wisconsin, Madison. The new Tomotherapy Unit in the University of Wisconsin radiotherapy clinic has a CT facility that, at low contrast, allows the resolution of objects of 2.5 cm in diameter that are 1% different in density from their background using a dose of just 3.5 cGy. Such CT images allow the patient positioning to be checked prior to tomotherapy. Mackie *et al* (2003a) has provided a detailed overview review of image guidance for precise CFRT.

Olivera *et al* (2000b) have addressed the issue of patient movement in helical tomotherapy. By measuring the movement parameters (shifts), one possibility to correct the treatment is to reposition the patient using the reverse direction shifts. However, this is difficult especially correcting for pitch, yaw and roll. Another possibility is to maintain the patient in the same position on the couch and modify the treatment delivery. They have created a delivery modification methodology and presented dose-volume histograms and dose distributions to analyse the capabilities and limitations of this method. Zhang *et al* (2004c) has described the most up-to-date method and this is reviewed in section 6.10.12.3.4. Zhang *et al* (2004c) point out that Tomotherapy is specifically disadvantaged in that it cannot use conventional gating methods nor methods which make the MLC 'breath' to mitigate motion effects. Fitchard *et al* (2000) have used canines to study the effects of mis-registration of the 'patient' in helical tomotherapy.

Kapatoes *et al* (2003) have used the onboard MVCT images to create a map of delivered dose using the input sinogram and showed how it compares with the planned dose.

2.2.3 Clinical application

Olivera *et al* (2000a) have presented four clinical conformal-to-target and conformal-avoidance optimization cases for tomotherapy. It was demonstrated that the complexity of delivery is almost independent of the complexity of the optimization.

The Wisconsin tomotherapy device is now a clinical reality (Mackie *et al* 2003b, Kapatoes *et al* 2001a, b, van Dyk *et al* 2003, Grigorov *et al* 2003). Mackie *et al* (2002) and Olivera *et al* (2002) have shown how the University of Wisconsin tomotherapy system can be used to deliver simple radiotherapy as well as complex IMRT. It can, in addition to being used for IMRT in a helical fashion, deliver conventional radiotherapy by topotherapy. Because there is no flattening filter, the open beam has an intensity that is highest at the centre of the beam and decreases nearly linearly towards the ends of the beam (Jeraj *et al* 2004). Hence, a modulation is required to produce a pseudo uniform beam. Mackie *et al* (2002) describe how straightforward this is to achieve. The machine is also being used to deliver complex whole body irradiation (Mackie *et al* 2003b).

Kron *et al* (2002) have compared plans created for tomotherapy with IMRT plans created using the Theraplan Plus treatment-planning system. The advantage of the former was shown to be improved sparing of normal structures and the ability to deliver high doses to more than one target simultaneously and/or to create in-target boost volumes. Grigorov *et al* (2003) in the same team have created prostate IMRT plans for the tomotherapy machine.

Welsh *et al* (2002a, b) have compared tomotherapy, conventional 3D CFRT and IMRT treatments for a specific case of a tumour close to the spinal cord and demonstrated that conformal avoidance using tomotherapy could leave the cord with a sufficiently small dose that future palliative treatment could be possible.

2.2.4   Commissioning issues

Kapatoes *et al* (2001c) have studied second-order effects in tomotherapy due to leaf bouncing and leaf latency. They show that although such features are complex they are dosimetrically insignificant.

Paliwal *et al* (2002a) have discussed the commissioning of the University of Wisconsin clinical tomotherapy unit. They pointed out the necessity to accurately align the MLC with the source and described a method of doing this by making images in which all the even leaves are open and comparing the images for two positions of the gantry 180° apart. It turns out that the megavoltage detector on the tomotherapy equipment has sufficiently good resolution to adequately reproduce the tongue-and-groove images of the accurately superimposed fields for alternately even and odd opened MLC leaves. Paliwal *et al* (2002b) have presented early experience on the University of Wisconsin tomotherapy unit. They described acceptance testing and commissioning and compared the system with another IMRT system. They compared measured and optimized planning dose distributions.

Balog *et al* (2003) have presented techniques for commissioning dosimetry for clinical helical tomotherapy. As helical tomotherapy machines do not have a flattening filter, this results in a treatment beam with a strong triangular shape across the transverse direction. This has to be modelled and measured as a precursor to dose calculation. The helical delivery process also superimposes inferior–superior dose distributions with slight translation offsets so any modelling error in the longitudinal axis dose profile would become significant over increasing translation offsets associated with more patient slices treated. As a result, this profile must also be measured accurately. The lack of the flattening filter is beneficial for many reasons. In particular, the beam output is more than twice what it would be along the central axis without the filter and, since all points at some stage in the tomotherapy delivery are receiving radiation along a central axis, this decreases the treatment time considerably. Finally, the effect of adjacent open beam elements has to be calibrated into the dose calculation. Jeraj *et al* (2004) have presented the detailed radiation characteristics of the machine.

2.2.5 Verification of MIMiC and University of Wisconsin tomotherapy

Paliwal *et al* (2000) have created a novel form of verification phantom for spiral IMRT. A solid-water cylindrical phantom 30 cm in diameter and 15 cm in height was machined to create an Archimedian spiral cavity in which radiographic or radiochromic film could be located. The arc length of the spiral cavity was 89 cm. The advantage claimed is that the film is then always at right angles to the direction of arrival of radiation and also can sample the 3D space with high resolution, something which is not possible with BANG-gel dosimetry. Paliwal *et al* (2001) have reviewed the use of the spiralogram for verifying the accuracy of tomotherapy. Richardson *et al* (2003) have provided further details of the spiral phantom which allows a 3D film-based verification of IMRT without recourse to the use of multiple orthogonal films. The ADAC PINNACLE TPS was adapted to predict the spiralogram for comparison. Additionally thermoluminescent dosimetry (TLD) validated the spiral film method.

The way the system works in practice is that whatever beams are planned onto a patient are then replanned onto the circular phantom and delivered. Then measured dose distributions are compared with those which are sampled along the Archimedian spiral from the planned dose distribution. The space involved is a so-called spiralogram. The film was calibrated in the usual way. An example was given in which it was claimed that the measured spiralogram closely followed the predicted spiralogram.

Lu and Mackie (2002) have developed a method to track the movement of a patient in sinogram space. They then proposed to use this tracking to reconstruct CT and single photon emission computed tomography (SPECT) images as if the patient movement had not taken place. They then extended the notion to explain that the concept could also be used to track intrafraction organ movement during tomotherapy.

Sen *et al* (2000) have devised a technique for recording the modulated intensity patterns of a MIMiC on a film as the gantry rotates which acts as a good quality-assurance tool for verifying the planned treatment.

2.3 Tomotherapy using a ^{60}Co source

Gallant and Schreiner (2000) are developing a tomotherapy apparatus using ^{60}Co sources. A large number of ^{60}Co sources are mounted in a ring gantry similar in design to a standard CT device and an optimization algorithm based on a modified genetic algorithm determines which beams to turn on, when and for how long, in order to deliver the desired 3D dose distribution.

Schreiner (2001) has discussed the practical potential for IMRT with ^{60}Co. A first-generation tomotherapy test jig with a rotate–translate stage was constructed in which a phantom moved through a 1 cm^2 ^{60}Co pencil beam. The beam intensity was then modulated either by turning the ^{60}Co unit on and off as the phantom was stepped through the beam or by varying the velocity of the phantom as it moved through the pencil beam. Early experiments showed that this could generate a fluence modulation for a ^{60}Co beam. A series of fluences with modulation have been applied to a phantom containing a polyacrylamide gel and a treatment plan for ^{60}Co IMRT generated with the MDS Nordion Theraplan Plus treatment-planning system was delivered to the gel successfully. Kerr *et al* (2001) showed that measurements and simulation agreed to 2.5%.

Barthold *et al* (2002) have described CoRA which is a design for a new ^{60}Co radiotherapy arrangement with multiple sources. This is a feasibility study. The apparatus designed comprises five ^{60}Co sources mounted on a U-arm which can pivot relative to a patient lying on a couch. The pivotal motion is such that the sources remain focused to an isocentre and describe arcs within the target volume. At the same time, the couch may take up different couch twists. The combination of the U-arm gantry movement and the couch twist can be modelled as an equivalent set of conventional gantry rotations and couch twists in order to study the treatment planning.

Barthold *et al* (2002) modelled the ^{60}Co source using the Bortfeld pencil-beam technique and used the VIRTUOS module inside the VOXELPLAN treatment-planning system to model four patient cases. The overall outcome of the study was to show that the ^{60}Co-based apparatus could generate conformal distributions almost as acceptable as those produced by a radiotherapy linac. The idea is that the ^{60}Co device could be of use to developing countries. They therefore laid to rest the potential criticism that the large source size of a ^{60}Co system could lead to deteriorated dose distributions.

Warrington and Adams (2001, 2002) have compared, using the HELAX treatment-planning system, IMRT plans for a linear accelerator and 'tele^{60}Co' beams. It was found that for thyroid, parotid, maxillary antrum, brain and breast tumours the ^{60}Co-IMRT modelling was satisfactory compared with the linear accelerator modelling. Less favourable sites were the deeply seated lesions in the chest and pelvis.

Singh *et al* (2002) and Iyadurai *et al* (2003) have constructed a manual prototype MLC with lead for the TH-780C tele^{60}Co unit. It consists of 15 pairs of lead leaves with projected leaf width of 8 mm at the isocentre producing a maximum field size of 12 cm × 12 cm. The leaf thickness was 8 cm and a tongue-and-groove (TG) design was used to reduce the interleaf transmission to 6% of the central-axis dose. The ^{60}Co beam characteristics such as the depth-dose, the beam profiles and the output factors were found to be very similar to those for conventional collimation and the MLC will allow beam shaping.

Kim and Palta (2003) have conceptualised/designed a system for using a plurality of ^{60}Co sources arranged in a grid matrix for delivering a 2D intensity-modulated beam (IMB) (figure 2.12). The sources may each be separately racked towards and away from the patient and the radiation intensity in each cell is thus determined by the inverse-square law change in fluence which will determine the dynamic range of intensities.

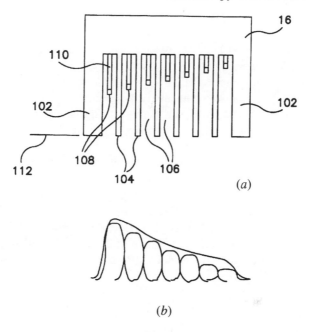

Figure 2.12. A figure from the patent for performing IMRT with ^{60}Co sources. The sources (e.g. 108) are housed in bixellated apertures defined by septa (e.g. 104). They can be moved to and away from a plane (112) so varying the intensity of radiation over that plane (shown in the lower part of the figure) by simple application of the inverse-square law. (From Kim *et al* 2002.)

It is important to note that these ^{60}Co-based studies and devices are either concepts or one-off developments and do not (as yet) rival other IMRT delivery techniques. They are included for completeness and for their potential.

2.4 Tomotherapy with an MLC

Achterberg and Mueller (2001) have developed a new multi-focal static tomotherapy system which incorporates an MLC and table movement. This conceptual study has been made using the Monte Carlo BEAM code and the ADAC PINNACLE^3 treatment-planning system and they presented an optimized design of the static tomotherapy device.

Deloar *et al* (2004) have produced a design study for kilovoltage tomotherapy in which a microMLC will be attached to a CT scanner allowing both imaging and treatment with the same machine. The dose rate is lower than for ^{60}Co tomotherapy and, as yet, the machine is only a concept.

2.5 Summary

The early years of clinical IMRT (1994–97) were dominated by the use of the NOMOS MIMiC. The technique is still widely used particularly in the USA, though not so much in Europe. MLC-based technology for delivering IMRT is widely considered to be a front runner rival and may have soon surpassed the use of the MIMiC at least on a pure number count of completed treatments. Spiral tomotherapy is in its infancy and like many other technologies that rival more conventional equipment for delivery, possibly faces an uphill struggle for adoption. In this sense, whilst it may present significant technological advantages, it shares this 'problem' with technologies such as proton therapy and the development of the Cyberknife, technologies believed to display advantages but clearly in the short-term expensive and numerically inferior to their main rivals. Against this background, a familiar 'Catch-22' arises of the difficulty of obtaining proof of the improvement for lack of the very equipment trying to be justified. We now turn to the MLC-based IMRT technique, the technique with the fastest growth in recent times.

Chapter 3

Developments in IMRT using a multileaf collimator (MLC) (physics)

The history of the MLC goes back at least to the patent by Gscheidlen in 1959 and the first commercial MLCs appeared in the mid to late 1980s. There are detailed reviews of the history, development, design and performance of MLCs by Webb (1993, 1997, 2000d). The major manufacturers and distributors of commercial MLCs, with a 10 mm leaf width at isocentre, are Elekta, Siemens and Varian. Many papers have detailed the dosimetric performances of each. Galvin (1999) and Bortfeld *et al* (1999) have provided useful reviews with tables of MLC properties. Huq *et al* (2002) have compared all three commercial MLCs for the first time using precisely the same criteria and experimental methods (figure 3.1). The different designs and positioning of the MLCs inevitably lead to performance differences. However, it was concluded that no single MLC was superior and that user choice is necessarily not just determined by radiation characteristics.

The Elekta MLC replaces the upper jaws (figure 3.2); the Siemens MLC replaces the lower jaws; and the Varian MLC is a tertiary add-on at the patient side (figure 3.3). The position of the MLC varies with respect to target and isocentre. The Elekta and Varian MLCs have rounded leaf ends but move on a plane. Conversely, the Siemens MLC has straight leaf ends and moves on the arc of a circle forcused at the source. The Elekta MLC has a single-stepped leaf side whereas the Siemens and Varian MLCs have double-stepped leaf sides. Key observations of Huq *et al* (2002) were:

(i)   the Elekta MLC generated the smallest collimator scatter,
(ii) the Varian MLC had the sharpest penumbra,
(iii) the Siemens collimator had the smallest radiation leakage,
(iv) the Elekta MLC had the deepest tongue-and-groove (TG) underdose,
(v)   the Varian MLC had the largest stepped-edge effect for a 45° edge.

However, these are statements somewhat out of context and a full study of the paper is needed to give the complete picture. Some properties of MLCs are given in table 3.1.

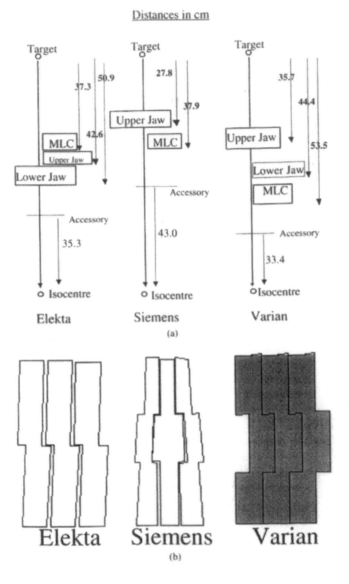

Figure 3.1. (a) A schematic diagram of the relative positions of the jaws, the 10-mm-wide multi-leaves and bottom surfaces of the jaws to isocentre for three different MLC systems (Elekta, Siemens, Varian). The distances are given in centimetres. (b) A schematic diagram of the end-on view of various leaves from different manufacturers showing the differences in leaf design that affect intra- and inter-leaf leakage. Other diagrams in this paper show the variation of head scatter with field size, the variation in penumbra for the three collimators and radiation leakage patterns for the three collimators as well as the variation in tongue-and-groove effect for the three collimators. (From Huq *et al* 2002.)

Table 3.1. Properties of commercial MLCs in Summer 2004.

Manufacturer	Name of MLC	Number of leaves	Maximum field size (cm²) at isocentre	Leaf width at isocentre	Leaf travel	Leaf height	Max leaf speed	Weight	Suits which accelerator	Focus
Elekta (1)	Integrated MLC	80	40 × 40	1 cm	32.5 cm (12.5 overtravel)		None specified		Elekta	Single focus
Elekta (2)	Beam Modulator	80	16 × 22	4 mm	11.0 cm overtravel		3 cm s^{-1}	50 kg	Elekta	Tilted
Varian	Millennium MLC-120	120	40 × 40	Central 20 cm of field: 0.5 cm Outer 20 cm of field: 1.0 cm	19.5 cm		2.5 cm s^{-1}		Varian	Single focus
	Millennium MLC-80	80	40 × 40	1 cm			2.5 cm s^{-1}		Varian	Single focus
	Millennium MLC-52	52	26 × 40	1 cm			2.5 cm s^{-1}		Varian	Single focus
BrainLAB	m3	52	10 × 10	3.0 mm, 4.5 mm and 5.5 mm	5 cm		1.5 cm s^{-1}	35 kg	Varian	Single focus
Radionics	MMLC	62	10 × 12	4.0 mm	5 cm	7.0 cm	2.5 cm s^{-1}	38 kg	All	Single focus
MRC Leibinger/Siemens	Mini MLC	80	7.3 × 6.4 on a Siemens	1.6 mm	1.4 cm	9.0 cm	1.2 cm s^{-1}	40 kg	All	Parallel
MRC/Siemens	Moduleaf	80	12 × 10	2.5 cm	5.5 cm	7.0 cm	2.0 cm s^{-1}	39.7 kg	All	Single focus
3D Line (Wellhöfer)	Mini MLC	48	11 × 10	4.5 mm	2.5 cm	8.0 cm	1 cm s^{-1}	35 kg		Double focus
Direx	Acculeaf	72	11 × 10				1.5 cm s^{-1}	27 kg		Two sets of leaf pairs at 90°

Figure 3.2. The Elekta MLC. (From Elekta website.)

The MLC can deliver IMRT in several modes. Historically, the first to be developed was the concept of linear addition of static fields of different shapes (Boyer 2003). This is referred to by many names including (i) step-and-shoot, (ii) stop and shoot and (iii) multiple static fields (MSFs). In principle, the subfields needed are simply a set of patterns summing to the full modulation and formed by methods which are well known. In practice, complications arise because of the need to factor in leakage radiation, head scatter and to take account of the MLC geometrical constraints, such as that, for some manufacturers' MLCs, interdigitation is forbidden (figure 3.4). These are known as 'hard constraints' because they cannot be violated. Historically, the determination of the leaf patterns was separated from the inverse planning (because the planning techniques were developed before methods for IMRT delivery were available). Hence, there was a period in which substantial 'fixes' were made to account for the physics of the delivery through collimation that was not perfect and with pieces of metal

Figure 3.3. A Varian accelerator and its MLC used for field shaping and IMRT (from Varian website).

that were forbidden to take up certain positions. Now there is a trend to take all this into account *at the time of planning* so that what is planned is genuinely what is delivered (see section 6.6). However, not all MLC-based planning and delivery has yet reached this stage. The determinators of leaf patterns are known as 'interpreters' or 'sequencers'.

Boyer *et al* (2000) have reviewed IMRT with dynamic multileaf collimation (figure 3.5). They described leaf-setting algorithms, leaf-transmission effects, leaf-position calibration, leaf-position offset corrections for the tongue-and-groove effect and dose-rate effects. They concluded that understanding the

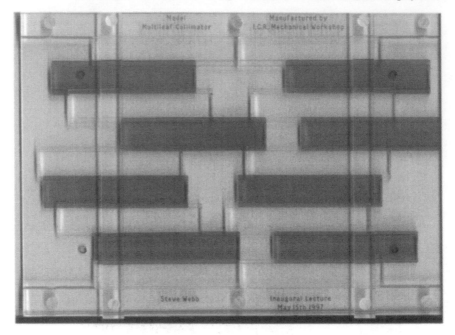

Figure 3.4. This illustrates interdigitation using a plastic model MLC. The MLC comprises nine leaf pairs with alternate pairs shown in blue and yellow. The tongue-and-groove regions show in green. Some leaf pairs are here arranged to show interdigitation. (See website for colour version.)

technology is essential for its successful application. Keall *et al* (2003) and Keall (2004) have provided the most thorough up-to-date review of dynamic MLC IMRT.

In this chapter, we follow a logical sequence of presenting the most recent developments in the use of an MLC for IMRT. We begin with the description of interpreters or sequencers which convert planned modulated beams into leaf patterns. Then we consider leakage which leads naturally to a discussion of how the delivery features of the machine affect the interpretation. This leads into a discussion of dose-calculating algorithms. Then there is a discussion of features of the MLC-based IMRT techniques before moving on to dynamic MLC delivery, combined therapies, IMAT and compensators for IMRT. The issue of the leaf width of MLCs appropriate for IMRT is discussed and grouped together with a description of new microMLCs. Finally, there are two sections on MLC-based IMRT verification. The first is concerned mainly with verifying individual treatment. The second is more concerned with quality assurance of the MLC equipment. However, many aspects of these latter two overlap and sometimes the placing of material could have been done differently.

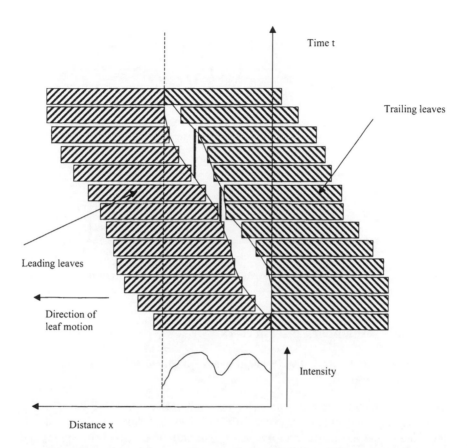

Figure 3.5. Schematic diagram to illustrate the principle of the dMLC technique. The 13 rows represent the positions of the leading and trailing leaves at 13 different times in the delivery of a 1D profile from just one leaf pair. The leaves move from right to left starting with their positions as shown in the 13th row and ending with their positions as shown in the first row (for convenience it is assumed here that the leaves can touch; this is not always possible in practice). The full pair of lines join the midpoints of the leaf ends and constitute the trajectories of the leading and trailing leaves $x(t)$ or $t(x)$ (these diagrams are shown in many orientations in practice and this does not matter). It is a simple matter to construct the intensity profile $I(x)$ from such a diagram since, considering only primary radiation, the primary intensity $I(x)$ is the difference in time of arrival of the trailing and leading leaves; i.e. $I(x) = t_{trailing}(x) - t_{leading}(x)$ where intensity and time are considered in the same units (they are proportional for constant accelerator fluence rate). Two sample intensities are shown by small vertical bars at two positions x and the whole profile is shown below the trajectory diagram.

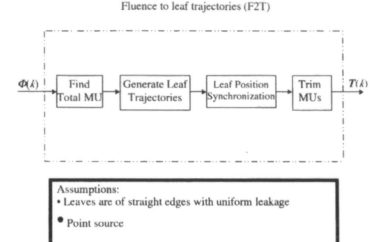

Figure 3.6. These are the procedures required to convert fluence to trajectories (F2T). (From Keall *et al* 2003.)

3.1 New sequencers/interpreters

3.1.1 General; dynamic IMRT sequencers with hard MLC constraints on leaves and jaws

There are numerous 'interpreters' which convert IMBs into MLC leaf profiles. That due to Stein *et al* (1994), Svensson *et al* (1994) and Spirou and Chui (1994) provided an idealized solution for one leaf pair with no machine constraints and for dynamic delivery. It simply stated that when the gradient of the fluence was positive, the leading leaf should move at the maximum velocity and when the gradient was negative, the trailing leaf should move at the maximum velocity. In both cases the complementary leaf provides the modulation. When there are multiple leaf pairs, TG underdose may be removed by leaf synchronization (Van Santvoort and Heijman 1996). The result does not, however, take any account of the minimum-gap constraints of the Elekta MLC, which has the strictest constraints of all the commercial MLCs. In this sense, these solutions are too limited. Convery and Webb (1998) provided a solution which does account for machine constraints by using both leaves and jaws to modulate the field for fluence patterns on coarse grids with uniform-across-bixel fluences. Interpreters are sometimes described as 'F2T' (figure 3.6) (fluence to trajectory) in contrast to 'T2F' (trajectory to fluence) (figure 3.7) describing the computation of *actual* fluence from the trajectories (Keall *et al* 2003).

Kuterdem and Cho (2001) have provided a new interpreter which works similarly but for a continuous fluence distribution in which there are gradual

$$\Phi_z = \ddot{O}_1 + \Phi_{LS}$$

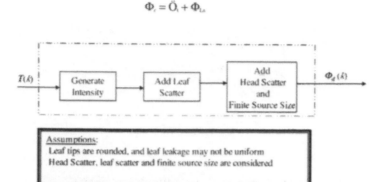

Figure 3.7. These are the procedures required to convert trajectories to fluence (T2F). (From Keall *et al* 2003.)

changes across the field. This also allows generation of zero primary fluence as well as using both X- and Y-jaws effectively. Kuterdem and Cho (2001) formulated the problem of optimizing the time delivery of dynamic IMRT including the machine gap constraints and firstly attempted a MATLAB solution using the Optimization Toolbox 2.0. With gap constraints temporarily switched off, the solution replicated the well-known 'three 1994-papers solutions' (see earlier). However, when gap constraints were switched on, the problem possessed local minima and was too sensitive to initial conditions. The solution also took approximately one day of computer time for a problem of leaf sequencing nine leaf pairs each with 41 control points.

They therefore started again as follows: Firstly the 'three 1994-papers solution' was adopted. TG effects were removed by synchronization. This creates gap violations. So an iterative process minimized reduction of leaf speeds as the trajectories were modified to remove gap violations. This (somewhat complex procedure) used 'look ahead' steps (as did Convery and Webb [1998]) to avoid overblocking by diaphragms.

A test pattern of one IMB from a nine-field prostate plan with 2 mm increments was used. It was shown that the use of the algorithm gave a rms agreement between delivery simulation and prescription better than 1 MU. When no secondary blocking was applied (or just use of Y diaphragms), the match was less good. The full algorithm operated in almost real time.

3.1.2 Sequencing multiple-static MLC fields—clusters

The alternative (to the dynamic) technique for using an MLC to deliver IMRT is to deliver the fields by step-and-shoot in which the beam is off between delivery of field components. The classic method (figure 3.8) to decompose modulations

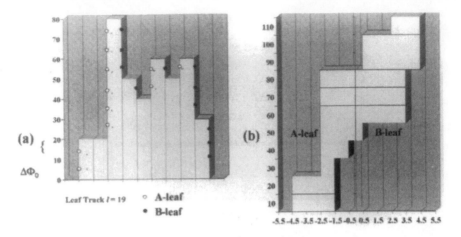

Figure 3.8. This figure shows the classic method of leaf sequencing for the 1D profile shown left in (a). The panel (a) indicates the means by which the sequence of leaf ends is determined and then panel b (right) indicates the resulting leaf sequence as a function of MUs using the sliding-window technique. (From Boyer 2003.)

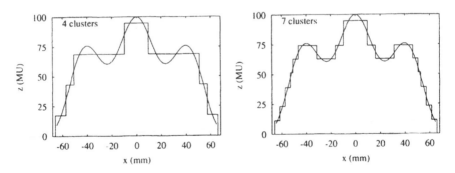

Figure 3.9. Optimized and clustered profile with four and seven clusters respectively. The cluster profile on the left has four clusters and no in-field minima, every cluster being a single segment. The cluster profile on the right has two in-field minima, seven clusters leading to nine segments. (From Bär *et al* 2001b.)

was worked out over 10 years ago (Boyer 2003; see also reviews by Webb 1997, 2000d).

Bär *et al* (2001a, b) have developed a new technique to sequence the fluence increments required to generate a beam profile in the MSF technique (figure 3.9). The sequencer combines features of clustering and smoothing to segment the IMBs. The outcome is that the fluence increments in the delivery are non-uniform. The sequencer is included in the iteration stage that actually generates the IMBs and is not applied *a posteriori*. It takes account of the (Elekta) MLC machine constraints. It was designed to be robust. The work was set against the background of the view from Que (1999) that no particular sequencer (up to that date) is universally preferable and also set against the views of Keller-Reichenbecher *et al* (1999) that combining sequencing and IMB generation as one process did not improve the treatment plans. They applied this to eight mathematical profiles and two prostate IMBs. They concluded that this leads to treatment plans with fewer field segments and a high MU efficiency. The conformity is a little better than the Bortfeld (equal-fluence-increment) algorithm, which as the authors state, is an inherently obvious outcome of using non-uniform increments.

Wu *et al* (2001f) have described a new algorithm for segmenting static step-and-shoot IMRT fields. The work describes a method for converting an idealized beam-intensity distribution to a deliverable sequence of static MLC segments. The first stage of the process is a multilevel approximation which groups the intensity values into minimum K-clusters with respect to a predefined approximation criterion. What this means is that the intensity levels are assigned, without change, to a set (cluster) up to a fixed total number of clusters; this is the value K. The second stage of the algorithm then predetermines the level that should be set uniformly within each of the cluster so as to minimize the discrepancy between the original intensity and the multilevel approximation. This technique is well known in the optimization literature and is known as the 'K-means clustering algorithm'. It has already been shown by Evans *et al* (1997) that this technique minimizes the total squared error in the problem.

The second part of the interpreter developed by Wu *et al* (2001f) is to start with the optimal multilevel approximations and to decompose the K multilevel approximations into L MLC deliverable shapes. This is done using a graphical user interface. The first segment to be delivered for each gantry angle is the one that covers the largest field to be irradiated allowing treatment verification. The authors show that the outcome of K-means clustering is to distribute the errors in fluence delivery over a wider geometrical area so that, whilst the total error is reduced, the effect of the error is spread more widely in space.

Much of this paper is taken up with discussing alternative techniques, in particular that of Evans *et al* (1997) who showed (with application to breast IMRT with high spatial resolution) that the use of a uniform incremental value was optimum. The authors here argue that they could not use this technique because the large bixel size, used for prostate radiotherapy would lead to a histogram of well under 100 values, not amenable to study by the technique of Evans *et al* (1997). They also give reasons why the technique of Galvin *et al* (1993) is less relevant. They comment on the proposal of Webb (1998a, b) to use a forced-baseline configuration with subsequent leaf sweep. Finally they discuss in great detail the physical limitations imposed by the Elekta MLC and, in particular, the way in which jaws and leaves must be combined to create a minimum number of segments. The features of the Elekta MLC are built into their decomposition scheme. They comment that MLC rotation can sometimes be advantageous in particular because it leads to a loss of superposition of cold slices due to the TG effect (see also section 3.4.2).

The number of clusters and therefore the number of segments depends critically on the error tolerance which is set during the initial clustering routine. They show, not surprisingly, that as this error tolerance is decreased (i.e. tighter tolerance) the number of segments increases. The technique is in use at the William Beaumont Hospital, Royal Oak, Michigan. Nioutsikou *et al* (2004) also showed the effect of trading number of segments against error tolerance.

Lehmann and Xing (2000) have used the CORVUS inverse-planning system to investigate the variability of conformality with use of a different number of intensity levels. The numbers chosen were 2, 3, 5, 10, 20, 50 and 100. Plans were sensitive to the number of levels when this was below 10 and less sensitive when the number exceeded 20. The number of segments of the treatment field goes up with increasing number of intensity levels and so the final decision on how many are needed has to balance the quality of the plan and the practicality of the delivery.

3.1.3   Non-uniform spatial and fluence steps

Beavis *et al* (2001) have studied the effect of optimizing both the intensity levels and the spatial increments of the step-and-shoot leaf sequencer for delivery of IMRT. They started by noting that, in general, the planning and delivery of IMRT are decoupled. Matrices of 2D modulations resulting from inverse planning were sequenced *a posteriori*. The first, and still one of the most commonly used techniques, is the Bortfeld technique of discretizing in space on a regular grid and based on discrete intensity using equal intensity increments. Provided, as for the Varian MLC, the pairs of leaves may move independently without hard leaf constraints, 2D profile sequencing reduces to a collection of decoupled sequencing tasks for 1D fluence profiles. Hence, Beavis *et al* (2001) concentrated on the solution of the 1D problem.

The rational for their work is to observe that the use of integral multiples of units step widths and integral fluence increments can compromise the accuracy of delivery. They postulated that using non-uniform intensity increments and non-uniform spatial increments will fully utilize the technical capability of the MLC delivery and result in more accurate treatment.

There are two features to their algorithm. The first is to create non-uniform step widths by finding a set of steps such that the difference between the largest and the smallest fluence within the step is less than some specified value. This specified value is increased until the number of steps collapses down to the user-desired number of steps. The second aspect of their algorithm is to select non-uniform intensity levels to minimize a cost function which is a sum-squared difference between the discretized and the continuous profile at all the points specified by the previous spatial discretization. They then investigated four solutions as follows:

(i) In the simplest optimization (so-called 'level zero'), the intensity levels and spatial increments were uniform, namely the solution proposed by Bortfeld *et al* (1994).

(ii) Optimization level one allows the intensity levels to be unconstrained but constrains spatially to a regular grid.

(iii) Optimization level two is the reverse. Intensity levels are constrained to be incremented equally but the step widths are optimally selected.

(iv) Optimization level three allows both intensity levels and step widths to be optimally selected.

They observe that only optimization levels zero and two in which the intensity levels are constrained to be incremented equally can be delivered by a classic step-and-shoot algorithm. This they did. They implemented optimization levels one and three using their own algorithm (Beavis *et al* 2000b). The study concentrated on just one profile and, not surprisingly, showed that optimization level three gave the least error between ideal and delivered fluence maps. No corrections were made for leakage or known dosimetric transmission effects. It was found that the most significant improvement in agreement was gained by allowing optimized step widths alone. They commented that this might lead to narrow segments but they did not think there would be too many, something clearly to be avoided because of the potential dosimetric uncertainty attached to introducing geometrically small segments. They also commented that, in the orthogonal direction, the MLC leaf width itself provides the fundamental limitation, something not studied in this paper. In conclusion, the new approach of non-equally spaced fluence levels improved the discretization as well as the use of non-uniform grid spacing.

3.1.4 Minimizing the number of segments

Dai and Zhu (2001a, b) have presented a new algorithm to minimize the number of segments in the delivery sequence for IMRT delivered with an MLC. It does this by selecting the candidate components that result in residual intensity matrices with the least complexity. The rational for minimizing the number of segments is that the delivery time for the MSF IMRT mode has three components; beam-on time, verification-and-recording-overhead time and leaf-moving time. In general, the verification-and-recording-overhead time is the largest such time, so minimizing the number of segments will minimize the dominant component of the delivery time and thus lead to a delivery time which is shortest. Chen *et al* (2004) make a similar study.

There have already been presented a large number of algorithms to sequence a 2D IMRT matrix into static field components. The best known ones are those of Galvin *et al* (1993), Bortfeld *et al* (1994), Xia and Verhey (1998) and also Siochi (1999) and these were all compared by Que (1999). Que (1999) found that the algorithm of Xia and Verhey (1998) most frequently resulted in the fewest segments but that no single algorithm was the most efficient for all clinical cases.

The algorithms all differ because they differ in the rule for selecting the intensity level of a segment. Therefore, the difference in performance of these algorithms is caused by these different rules.

The more complex an intensity matrix is, the greater the number of segments in the delivery sequence will be. Therefore, an algorithm that allows the complexity of an intensity matrix to decrease most quickly is the one most likely to produce a sequence with the least number of segments. This is the basis of the philosophy of Dai and Hu (1999) who proposed such an algorithm for determining jaw-setting sequences for IMRT with an independent collimator (i.e. no MLC). The work of Dai and Zhu (2001b) is an extension of that work for the MLC.

The bulk of the paper explains in detail the algorithm by which each segment is stripped away from a matrix until the residual intensity matrix becomes zero. The algorithm of Dai and Zhu (2001b) can operate in four different modes depending on which of the previously mentioned published stripping algorithms is used for determining the complexity of the residual intensity matrix. The heart of the MLC stripping algorithm is that they select at the kth segment the candidate that produces a $(k + 1)$th residual intensity matrix with the least complexity (calculated with one of the published algorithms). They do this under two major situation constraints, i.e. with or without an interleaf collision constraint.

To evaluate the efficiency of the algorithm and the influence of the different published algorithms used to calculate the complexity of an intensity matrix they tested 19 clinical IMBs generated using CORVUS version 3.0 revision 10. They used two indices to measure the efficiency of the algorithms. The first was the number of segments (the so-called major index) and the second was the total relative fluence in the sequence generated by the algorithm (the so-called minor index). When one algorithm generated a sequence with fewer segments than another algorithm, it was considered more efficient. If two algorithms generated two sequences with the same number of segments, the algorithm with less total relative fluence was considered more efficient.

The major conclusions were as follows:

(1)   For all beams the most efficient sequences were generated by one, or more than one, of the four variations of the new algorithm.
(2)   The improvement in the efficiency of their algorithm relative to that of the published algorithms varied from beam to beam.
(3)   No single variation of their algorithm was always the most efficient, although that in which the Bortfeld *et al* (1994) algorithm for complexity was used was more often than not the most efficient in terms of the number of segments.
(4)   For all beams, every variation of their algorithm was more efficient than the corresponding published algorithm itself.

These conclusions held both with and without interdigitation constraints.

The reason why these conclusions are reached is that the new algorithm searches a much larger solution space than the published algorithms. They refer to the work of Webb (1998a, b) who had previously published the algorithm for determining the possible number of combinations of shapes which can deliver a particular modulation and had previously demonstrated that it was not practical to find the global minimum. The overall conclusion was that multiple variations of the algorithm should be implemented in an IMRT treatment-planning system so that the most efficient sequence would not be missed.

Related to this, Gunawardena *et al* (2003) have designed an algorithm for minimizing the number of segments when segmenting a plan generated by the CORVUS planning system The concept is based on 'difference matrices' which guide the selection of flat regions from the modulated beam. Luan *et al* (2003, 2004) have also described a method to minimize the number of segments for a CORVUS plan, based on map theory. Advantages were demonstrated over the use of the Xia and Verhey (1998) algorithm and the Bortfeld *et al* (1994) algorithm.

3.1.5 IMFAST

Step-and-shoot IMRT requires specification of intensity-increment resolution and spatial resolution. The inherent accuracy of matching a segmented IMB to a planning-system-generated modulated fluence clearly improves as both resolutions are increased. However, this gives rise to an increased number of field segments with adjunct problems of increased leakage, radiation scatter and possibly fractional MUs. The Siemens product IMFAST provides a number of segmentation techniques each with adjustable constraints. Potter *et al* (2002) created continuous-level IMBs in the PLUNC planning system, then sequenced these into five or ten segments to create discrete-level IMBs. These were imported into IMFAST which uses five different segmentation techniques described in their paper. IMFAST also creates error maps, being the difference between the imported maps and the segmented maps. The segments were then reimported into PLUNC and doses recalculated. Quality factors were defined in terms of how closely the original planning goals were met by the segmented IMBs.

The authors found that the number of segments varied considerably between the several segmentation techniques. Furthermore, the quality of the plan deteriorated due to segmentation. The differences in quality between the several segmentation techniques became less as the number of fluence levels increased. No correlation was found between the high treatment efficiency (low number of segments) and high optimization quality. There was also no correlation between the average intensity-map error and the optimization quality. There was no consistent association between the total MUs and the total number of segments.

3.1.6 The best interpreter ever?

The subject of leaf-sequencing algorithms to create IMRT fields using the static step-and-shoot or MSF technique is ongoing and indeed there seems no end to the flow of algorithms being published. Langer *et al* (2000) have investigated three sequencing algorithms for MLC positions used to deliver IMRT (figure 3.10).

One of these is new; the other two are the Bortfeld technique and the areal step-and-shoot technique (Boyer 2003). The three techniques were applied to a 2D IMB for the prostate and also to an arbitrary map. It was found that the Bortfeld algorithm delivers the minimum number of time units (MUs) but does not minimize the number of segments. Conversely, the areal step-and-shoot technique generates a smaller number of segments but at the expense of increased number of time units. The new technique presented by Langer *et al* (2000) reduced the time units to those generated by the Bortfeld technique whilst simultaneously reducing the number of segments to below that generated by the areal step-and-shoot technique.

The approach of Langer *et al* (2001a, c) is called an integer search programme. It was shown, by investigating random maps to remove selection bias, that this technique yields better segmentations than the Bortfeld technique and better segmentation than the areal technique in terms of its ability to simultaneously reduce both the number of segments and the total MUs. No details of the new technique were given in these two abstracts but Langer *et al* (2001b) have described in detail the special integer algorithm devised to generate a sequence with the fewest possible segments when the minimum number of MUs are used and results were then compared to sequences given by the routine of Bortfeld *et al* (1994) that minimizes MUs by treating each row independently and the areal or reducing routines that use fewer segments at the price of more MUs.

It is important to reduce MUs because this affects the contribution from machine leakage and lengthening of each treatment session and also potentially leads to inaccuracies due to patient positioning errors. It is also important to reduce the number of segments because, for some machines, the time needed to switch the beam on and off between segments and to move the leaves can be lengthy.

The authors argue that an efficient sequence of leaf movements should take the fewest possible segments to generate a map with the number of MUs that is uses. It should also use the minimum number of MUs for the number of segments that it uses. Clearly, if either condition is violated, the sequence can be improved by either keeping the number of MUs constant and reducing the number of segments or by keeping the number of segments constant and reducing the number of MUs.

To date, most algorithms have concentrated on reducing one or other of these but not both quantities, i.e. it is well known that the algorithm due to Bortfeld *et al* (1994) uses the fewest possible MUs but may use more segments than the condition requires. Conversely, the areal algorithm is found to generate fewer segments than the Bortfeld technique but at the expense of increased MUs. These algorithms have usually been benchmarked against each other using random IMBs and clinical examples. However, their performance has not been gauged against an absolute standard. The absolute standard would be the minimum number of segments required to generate a sequence using the minimum number of MUs and, before the paper of Langer *et al* (2001b), this had not been found. Langer *et*

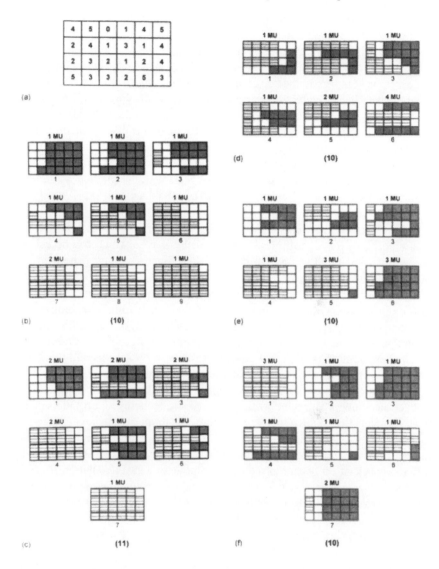

Figure 3.10. The figure shows five different ways of decomposing the simple IMB shown at the top left: (a) shows an arbitrary intensity map with integer entries. The leaves move along the rows; columns separate the positions at which the leaves can stop; (b)–(d) leaf sequences provided by three algorithms that generate the map; (b) the Bortfeld algorithm; (c) the reducing algorithm; (d) the integer (bidirectional mode) algorithm; (e) the integer (bidirectional) with leaf collision constraint algorithm; (f) integer with tongue-and-groove constraint. Individual segments are numbered sequentially at the bottom and the number of MUs for each segment is shown above its figure. The total number of MUs appears in parentheses below each identified sequence of segments. (From Langer *et al* 2001b.)

al (2001b)'s study aims to ask if it is possible to reduce the number of segments without increasing the number of MUs beyond the minimum required. They do this by designing an efficient integer minimization problem. The problem is set-up to take account of constraints, namely that no MLC leaf must have any holes, that leaf overtravel can be switched on or off, that leaf interdigitization can be switched on or off and that TG effects can be switched on or off. First of all, the integer programme is run in order to generate the minimum number of MUs required to generate the intensity maps when no restrictions are placed on the number of segments. Then the program is re-run with this number of minimum MUs to create the solution with the minimum number of segments. The sequencing problem thus yields the minimum number of MUs with the minimum number of segments. In other words, Langer *et al* (2001b) solved a problem previously said to have been unsolved.

They applied their new algorithm and benchmarked it to the Bortfeld and areal algorithms for, first of all, a simple algebraic case which can be checked by eye or by hand and secondly for a total of 19 intensity maps generated from planning IMRT of the prostate.

The conclusions were as follows. The Bortfeld algorithm used on average 58% more segments than provided by the integer algorithm with bidirectional motion and 32% more segments than an integer algorithm allowing only unidirectional sequences. In summary, the Bortfeld algorithm was generating the smallest number of MUs but generated too many segments and the integer solution could improve on this dramatically. Alongside this, the areal algorithm used 48% more MUs and the reducing algorithm used 23% more MUs than did the bidirectional integer algorithm, while the areal and reducing algorithm used 23% more segments than did the integer algorithm. In summary, the integer algorithm could deliver a solution with the smallest number of segments and the smallest number of MUs, thus improving the efficient delivery of static-field IMRT. The only drawback of this solution seems to be potentially large solution times but a further advantage is that it is possible to tailor the integer solution to work within constrained bounds of number of segments and number of MUs. In fact, this is a multidimensional optimization problem in which the user can have a large measure of control over the outcome. Li *et al* (2004a) have shown how to minimize the number of MUs and the number of field segments whilst also avoiding leaf-end abutments. With these observations it is tempting to say the MSF interpreter problem is solved for the optimum sequencing and that there is no need for further work in this area.

3.1.7 Sequencers exploiting MLC rotation

Beavis *et al* (2000a), noting that the resolution of IMRT beams is limited by the width of the elemental multileaf, have developed an algorithm to determine whether orienting the MLC in a direction other than parallel to one major axis of the 2D IMB would be preferentially beneficial. Intensity profiles were sampled at

1° intervals rotating the MLC about the centre of the field and for each collimator angle a set of profiles representing the required intensity surface under each leaf pair was computed with a width equal to that of a multileaf. In general, it was found that using non-zero collimator angles reduced the difference between the idealized 2D intensity surfaces and those produced by IMRT using the dynamic MLC technique.

The MLC can shape the region either with the edges or the sides of leaves resulting in a different area at the leaf base depending on single or double focusing of the MLC. Leal *et al* (2002) have shown significant differences in the treatment for isodoses enclosing the target depending on this phenomenon but that better dose distributions were achieved when the double-focused MLC from Siemens was simulated with a Monte Carlo technique.

Otto and Clark (2002) have designed an interpreter for delivering IMRT based on changing both the MLC leaf positions and the angle of collimator rotation. This is perhaps the first entirely novel delivery technique proposed for some time. The technique aims to overcome limitations in spatial resolution at right angles to the direction of leaf movement. It also aims to reduce interleaf leakage and TG artefacts since a variety of leaf orientation is called into play. Instead of the primary dose to a point being (largely) solely dependent on the behaviour of two leaves, it becomes dependent on many.

The algorithmic decomposition is an iterative calculation. Random orientations and random leaf patterns are offered sequentially and the dose contribution is found. If this makes the iterative-running calculation of dose closer to the prescription, then the contribution is accepted. Also contributions are accepted if the change falls the wrong way (i.e. worse) but within a small limit. This limit itself decreases as the iterations proceed. So in this way the technique has similarities with simulated annealing. To further tune the method, initially just a subset of the possible components are sampled. Then, as the calculation proceeds, more component possibilities are introduced being between the other set. Also, each possible component is checked that it can be physically delivered. So interdigitation and leaf collision can be avoided if required. The method involves sequential doubling of the number of segments and a gradual reduction of tolerance margins.

The technique was operated for both delivery of static segments (radiation off between rotations and leaf movement) and dynamic delivery (radiation on between movement of collimators and leaves). It was tested for 1-cm-wide leaves and for 5-mm-wide leaves (figure 3.11).

The method was applied to a five-IMRT-field thyroid case and also to a test 2D sinusoid. Dose distributions were also measured with film for comparison with calculations.

It was shown that the fit to the desired fluence prescription increased as the number of components increased; the error tolerance decreased as the iterations proceeded. Given the stochastic nature of calculations, some were repeated 100 times with different random number seeds and the results fluctuated with

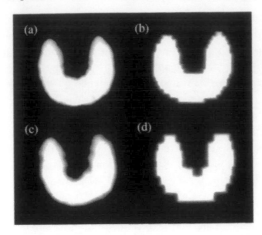

Figure 3.11. Dose conformity for this C-shaped fluence. Maps are shown for the 5-mm-leaf MLC using (a) rotational and (b) conventional techniques. One-cm-leaf MLC conformity results are displayed in (c) and (d). (From Otto and Clark 2002.)

a Gaussian distribution of fit error. However, the range of solutions was small. The fit to prescription increased as the range of possible orientations increased. With a range less than 90°, behaviour was poor. Conversely, behaviour hardly improved beyond 270° range.

Comparisons were made with 'conventional' IMRT both in static and dynamic mode and it was shown (for both leaf widths) that accuracy increased with the use of rotation. Measurements confirmed that both spatial resolution increased and interleaf leakage decreased with the introduction of rotation.

Hardemark *et al* (2003) have also exploited MLC rotation in sequencing IMBs to reduce the number of segments and to reduce errors due to leaf width.

Wang *et al* (2004) have developed a new sequencer in which the orientation of the MLC was an adjustable parameter at each gantry position. Then the collimator angles were chosen so that the maximum effective travel distance of any MLC leaf pair at each gantry position is a minimum. This was shown to reduce the number of segments and also the number of MUs leading to an efficient delivery without compromising conformality.

3.1.8   Comparison of dynamic MLC (dMLC) and multiple static field (MSF) techniques

The use of an MLC to perform IMRT by changing the leaf pattern with the radiation continually switched on is known as the dynamic MLC (or dMLC) technique (or 'sliding-window' technique) (Keall *et al* 2003). Some of the advantages of this technique are said to be (i) shorter treatment times and (ii) reduction of dosimetric errors due to fluence stratification in the MSF method. Disadvantages are said to be

(i)   less easy to understand,
(ii)   requires control of leaf speeds,
(iii) interruption is not easy to handle,
(iv) it is less easy to verify and
(v)   more MUs are needed (Keall *et al* 2003).

Not surprisingly, it is of interest to know how the dMLC technique compares with the MSF technique. Of course, if the fluence increments are small and numerous and the same is true of the spatial steps, then the two techniques are effectively the same as far as the radiation dose is concerned.

Chui *et al* (2000b, 2001a) have compared the delivery of IMRT using a conventional MLC by dynamic and segmental methods. The main advantage of the dMLC technique is that the continuous leaf motion enables the delivered intensity profile to closely match the desired one, so preserving both the spatial and intensity resolutions. Conversely, the MSF-MLC method (step and shoot) resembles conventional treatment with static fields and so is more easily verified. However, the MSF-MLC method approximates the fluence modulation by discrete levels of intensity at discrete spatial increments and thus it is expected that there will be some degradation in dose distribution. The authors set out to assess the degree of degradation depending on the spatial resolution and the number of intensity levels.

In their study, they varied the number of intensity levels from 5 to 20 and the spatial resolution in the leaf travel direction between 2 and 10 mm. Comparisons were made in terms of dose statistics from the resulting approximated plans. They also considered the total beam-on times in terms of MUs and the total delivery time in terms of minutes.

They reviewed the dMLC technique clarifying that the flatter the minimum slope is, the greater the maximum leaf speed and, thus, the total beam-on time is reduced. They also pointed out that an iterative procedure was required to convert the desired intensity (fluence) profiles into working fluence profiles such that when the working fluence profile was delivered and the effects of transmission and extrafocal radiation were taken into account, the overall fluence would approximate the desired fluence profile (see also section 3.2.2). These effects also took account of the effect of the rounded leaf end.

When implementing the MSF-MLC technique, equally-spaced fluence increments were studied. This distinguishes this study from those, for example, of Beavis *et al* (2001) who considered non-equal fluence increments. To study variable resolution, they merged multiple 2 mm-wide bins to create pseudo wider bins, for example, merging three bins of width 2 mm led to simulation of a bin width of 6 mm. The leaf-sequencing algorithm that they used minimized the total beam-on time and the particular MSF-MLC method could be viewed as a special case of the dMLC method with an infinite leaf speed. Although the derivation of the components was arrived at differently, the final arrangement was said to be equivalent to that due to Bortfeld *et al* (1994) with the same beam-on time. The total beam-on time was always less than that from any other MSF-MLC algorithms. There was no need to apply leaf-collision constraints for the Varian accelerator. There is no known method to construct a discrete level working profile to account for the contribution of extra focal radiation and so the authors used the working profile of the dMLC technique as a starting discrete profile for the MSF-MLC technique.

For the three prostate patients studied, the overall results of the dMLC plans were not very different from those of the MSF-MLC plans with 5, 10 and 20 intensity levels. It was concluded that a 10-level MSF-MLC plan with a spatial resolution of 2 mm in the leaf travel direction was comparable to a dMLC plan.

In contrast, three nasopharynx patients were studied and their conclusions were similar for each. The modulations contain sharp gradients for the protection of the cord and brain stem. It appeared that the number of levels mostly affected the target coverage for MSF-MLC plans whereas the spatial resolution had a more significant effect on critical organ sparing with a small spatial resolution yielding a better plan.

For the three breast patients studied, the results were also very similar and it appeared that a 10-level MSF-MLC plan for the breast was comparable to the dMLC plan and that further reduction in the number of intensity levels would not be acceptable. It should be noted, however, that because of the large non-modulated plateau in breast fields, a 10-level plan probably corresponds to only two or three fluence segments.

Regarding timing, the beam-on time required for the dMLC delivery was about 20% more than for the MSF-MLC delivery but the actual delivery time in minutes for MSF-MLC was 2 to 2.5 times that for dMLC due to the complex interaction of times taken for moving the leaves. The authors comment on the relation of their work to that of Budgell (1998) and that of Keller-Reichenbecher *et al* (1999) who also studied the fluence increment resolution effect on the precision of IMRT.

Ting *et al* (2000b) have compared step-and-shoot and sliding-windows leaf sequencing for IMRT. Over 130 patients have been treated using one or other of the two techniques. The conclusions apply to an accelerator in which there are no beam-on delays between beam segments. They made a number of conclusions including the following. Step and shoot uses 30–50% fewer MUs but takes between 50–100% longer to execute depending on the complexity of the intensity map. Step and shoot can deliver sharper dose gradients.

Linthout *et al* (2002) have compared the performance of dynamic *versus* step-and-shoot delivery for fields generated with the CORVUS inverse-planning system and delivered with an Elekta accelerator. They investigated the delivery of three fields varying from a simple open square to an inverted pyramid. They concluded that, in all cases, the dynamic delivery was preferred because it needed fewer field segments (control points), often (but not always) much fewer MUs, was less affected by the calibration accuracy of the MLC and resulted in dose distributions that more closely matched the calculations. Unfortunately, Elekta no longer support dynamic delivery.

Naqvi *et al* (2001a) have invented a technique to turn the step-and-shoot CORVUS fluence modulations into an equivalent dynamic modulation using ray tracing. They did this by generating a quasi-dynamic map by splitting the field segments into many sub-segments. They studied the errors incurred in the static-to-dynamic conversion.

Turian and Smith (2002) compared the MSF segment and sliding-window IMRT techniques for delivering IMRT treatments for head-and-neck cancer planned with the CADPLAN HELIOS system and delivered with a Varian accelerator. It was found that the sliding-window technique delivered a more uniform distribution than the MSF technique by approximately 5%.

Swinnen *et al* (2002) have compared static and dynamic delivery of IMRT. They showed good agreement between both techniques and calculation. Unlike Linthout *et al* (2002), it was found that the total number of MUs for the dMLC technique usually exceeded those for the MSF delivery but that this did not influence delivery time significantly because there were no interruptions of the beam required for the former method. King *et al* (2002) have also compared the beam-on time for sliding-window and step-and-shoot IMRT treatment plans.

3.1.9   Varian MLC and HELIOS planning system

André and Haas (2000) have implemented both the sliding-window algorithm and the step-and-shoot algorithm for the dMLC technique on a Varian Clinac 2300 CD accelerator. Plans were prepared using the HELIOS inverse-planning system.

Dirkx *et al* (2000) have compared two interpreters for the dMLC technique. The two interpreters are the LMC (leaf-motion calculator) in the inverse treatment-planning system known as HELIOS and the DYNTRAC algorithm developed in-house. The DYNTRAC algorithm includes leaf synchronization to remove the TG effect and also includes collimator scatter effects. LMC does not account for these and so is less accurate when compared with experiments. DYNTRAC requires more beam-on time by about 20% because of the inclusion of the limitations of the MLC.

Bohsung *et al* (2000) have commissioned the Varian IMRT system for delivering the dMLC technique at the Charité Hospital in Berlin.   They link the delivery system to the inverse-planning calculator HELIOS in the CADPLAN treatment-planning system, Before treating patients, a quality-assurance procedure is performed in which each intensity field is transferred to a rectangular phantom and the dose distribution in a reference plane and the absolute dose at a reference point is calculated and measured by film and an ionization chamber.   Agreement between the CADPLAN calculations and the measurements was good and, in addition, BANG-gel dosimetry was performed. In November 1999, they treated their first prostate patient using the dMLC system and claim this is the first patient so treated in Europe. The quality assurance procedures at this centre have been described by Boehmer *et al* (2004).

Bohsung *et al* (2000) have described BANG gel measurements to verify a simulated seven-field nasopharyngeal IMRT treatment plan. A number of performance tests were devised to check leaf-position accuracy and each IMRT plan has the interesting feature of including some orthogonal non-modulated fields with a few MUs. These fields are delivered every treatment fraction so that electronic portal images may be taken.

Myrianthopoulos *et al* (2001) have investigated the effect of delivery dose rate on the calculated fluence maps resulting from the HELIOS inverse treatment-planning system. They concluded that the error between planning and delivery increased with increasing dose rate and recommended the use of the lowest possible dose rate for IMRT.

Islam *et al* (2001) have used the CADPLAN/HELIOS inverse-planning system to plan IMRT for the 120-leaf Millennium dMLC system. Quality-assurance measurements were made with an ion chamber and with MOSFET detectors and films. It was found that pretreatment dosimetry showed agreement between measured and calculated doses to be within 2 mm in spatial dose distribution and 2% in absolute dose.

Chauvet *et al* (2001) have used the HELIOS treatment-planning system coupled to Varian 2300 CD and 23EX linacs equipped with dMLCs with 80 and 120 leaves, respectively. It was found that the treatment-planning software required considerable tuning before it could be accepted as having converged to an optimum solution.

Alaei *et al* (2002) showed that the static and dynamic IMRT techniques planned with the HELIOS planning system are comparable.

Oliver *et al* (2002) have compared conformal treatment plans from a Varian CADPLAN system and IMRT plans from a HELIOS system for complex tumour shapes located in the skull and close to the spinal cord and demonstrated the merit of IMRT.

3.1.10 Developments in Elekta IMRT

Elekta have developed IMRT through their International Consortium now comprising nine research centres (figure 3.12).

Elekta's integrated control system, RTDesktop, checks the linac MLC leaf positions every 40 ms and so effectively provides record-and-verify for every segment. The dose per MU is better than 1% after installation of the fast tuning magnetron. It was argued that this leads to a greater confidence in the use of the accelerator and thus a smaller workload and less resistance to the introduction of IMRT.

The FasTraQ MG6370 magnetron is the new tuning system for the Elekta accelerators (see also section 3.4.4). This has improved start-up performance and a shorter intersegment dead-time, projected to be 1 s. It has dose-per-MU

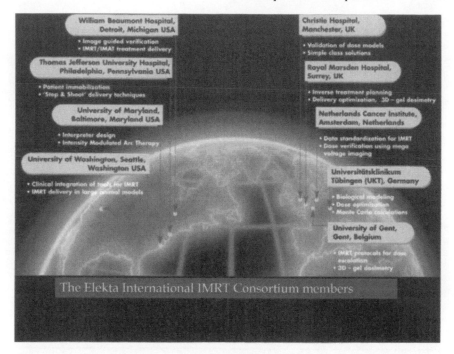

Figure 3.12. The constituent research centres that comprise the Elekta International IMRT Consortium. This consortium was formed by just 8 people in 1994 and has grown now to include many staff from nine centres.

accuracy within 1% for 2 MU or above. It can be retrospectively fitted to an accelerator head and the MUs per segment can lie between 1 and 1000 and the beam limit has been raised to 30 000.

3.1.11 Other interpreters

It should now be clear that there are a large number of options for choosing an interpreter and many considerations when implementing that chosen. These are summarized in table 3.2.

In this section are collected together statements about a variety of other interpreters that have been published, some with few details. Gustafsson (2000) have generated a general algorithm for sequential multileaf modulation which is able to take into account the restrictions on clinically available MLCs. This is the basis of the optimization in the HELAX inverse-planning system (now known as OnCentra).

Litzenberg *et al* (2002a, b) have studied the effects of incorporating the leaf-delivery constraints of a Varian 2100 EX accelerator equipped with a Millennium 120-leaf MLC into the delivery of IMRT. They find that, depending on the

Table 3.2. Options/considerations for interpreters/sequencers for MLC—based IMRT (when selecting or operating an interpreter these are some options/considerations).

Static or dynamic MLC delivery
Interpreting just primary fluence or including scatter, leaf transmission
Along-the-leaf spatial resolution
Across-the-leaf spatial resolution
Account of hard leaf constraints
Minimisation or not of tongue-and-groove underdose
Generation of minimum number of field segments and/or minimum MUs (MSF technique)
Inclusion of collimator rotation
Use of clustering algorithm (MSF technique). Specification of number of fluence levels
Uniform or non-uniform spatial and fluence steps (MSF technique)
Dose-rate effects
Use of leaves only or jaws-plus-leaves
Effect of leaf miscalibration
Idealised interpretation of just one 1D profile (one leaf pair) or all 2D IMB (all leaves)
A-posteriori interpretation or inclusion in inverse-planning stage
Performance on benchmark 2D IMBs compared with reference interpreter
Generation of fractional MUs
Ability to recomputed delivered dose through link to commercial TPS

parameters set for the control system, dynamic sliding-window leaf sequences can be produced which do not require beam interruptions nor doserate modulations.

Xia *et al* (2002c) have constructed a modified areal step-and-shoot algorithm that overcomes the limited overtravel distance of the Y jaws on a Siemens linear accelerator. Applying this to a variety of patterns, they showed that, for all cases, it was possible to extend the field length from 21 to 27 cm. The algorithm essentially uses the MLC to create corrections between parts of the field in such a way that the Y jaws are not required to violate their overtravel constraint.

Crooks *et al* (2000) have investigated several ways in which intensity maps (IMB matrices) produced in IMRT planning may be sequenced to optimize particular outcomes. For example, it was shown that two new algorithms could improve upon those known as the finite-differences method and Robinson's method. The first new algorithm produced, known as the absolute norm method, reduced the sum of the IMB matrix entries by the maximum amount at each stage. It was shown that this results in the smallest number of components. The second method, known as the two-norm or sum-of-squares method, reduces the sum of squares of the matrix entries by the maximum amount at each step and this results in the smallest beam-on time. The so called finite-differences method in which there is a reduction of a matrix element by the difference of the adjacent values and Robinson's method in which there is a reduction of the matrix by the removal of the largest rectangular block of numbers, lead to different results. It was shown, in particular, that beam-delivery time and the number of segments required were usually found to be complementary parameters, it not being possible to optimize both simultaneously. These investigations applied to step-and-shoot implementation of IMRT using static MLC segments.

Crooks *et al* (2002) have presented three new algorithms to decompose a 15^2 matrix of integer fluence values into static field components. These were two techniques which minimized the norm of the residual matrix after subtraction of a component compared with the norm before subtraction. The norm could be either an absolute-value norm or a quadratic norm. The third technique was a slab-dissection method. The implementations did not include MLC machine interdigitation constraints nor TG effects though it was argued that they could. The algorithms were tested on a large number of random matrices with different numbers of levels and different peak values. They were compared with the performance of the Xia and Verhey (1998) algorithm since this is known to perform well when compared with others (Que 1999). It was found that the minimum-absolute-norm method led to fewer segments statistically than the other algorithms. The same occurred when CORVUS fluence maps were decomposed. The slab-dissection method was effective in reducing the treatment time. No single method is consistently best and this is already well known from earlier studies. Certainly one cannot predict the best by inspecting the matrix. The authors stated that none of their three methods can simultaneously minimize both the number of fluence components and sum of the fluence values (treatment time). However, this is possible with at least one other algorithm (see section 3.1.6, Langer *et al* (2001b)).

Agazaryan *et al* (2001, 2002) have investigated the parametrization of 2D fluence profiles obtained from inverse planning using scoring indices. They characterized the profiles using a gradient index, the fraction of the field that is planned to receive less than the minimal level of transmission radiation (the so-called baseline index) and the efficiency which is the ratio of an average fluence level and the cumulative MUs. An in-house algorithm was used for sequencing the profiles. This could take account of the machine's hard constraints. They showed that there is a clear correlation between the deliverability indices and the accuracy of a delivered dose distribution.

Price *et al* (2002, 2003) have developed techniques to reduce the number of segments in the MSF-MLC IMRT technique and thus the overall subsequent treatment time. This was done by designing a series of concentric ellipsoids around the target and a gradient was then defined by assigning dose constraints to each concentric region. The technique was applied to 36 patients and led to an increase in the sparing of normal tissues. It also led to a reduction in the average number of beam directions and the average treatment time delivery.

Papiez *et al* (2001) have presented a new optimal step-and-shoot sequencer for unidirectional motion of leaves. They compared this technique with the Bortfeld algorithm and with the areal step-and-shoot algorithm (Boyer 2003).

Kamath *et al* (2003, 2004a) and Dempsey *et al* (2003) have made an in-depth mathematical analysis of segmented multileaf collimation. They showed that leaf sequencing based on unidirectional motion of the MLC leaves is as MU efficient as bidirectional leaf movement. They showed that if the optimal plan created by unidirectional motion violated the leaf separation constraints, then there is no plan (unidirectional or bidirectional) that does not violate the constraints. Kamath *et al* (2004b) showed that including the requirement to eliminate TG underdose increased the MUs a little but still led to the conclusion that the unidirectional movement gave the most MU-efficient leaf interpretation. These papers are very mathematical and rigorous.

Boman *et al* (2003) have presented a mathematical technique for optimizing leaf velocities for the dMLC technique.

3.1.12 The effect of removing the flattening filter

Most medical linacs use a flattening filter to create an almost flat uniform beam without modulation. Hence, all modulation studies are based on this flat beam. Fu *et al* (2004) have investigated the effect on IMRT of removing the flattening filter (flat beams if wanted could of course be made by appropriate modulation). The use of a Siemens, Varian and Elekta MLC for IMRT was simulated as well as a compensator with and without the use of the flattening filter. Ten nasopharynx cases and ten prostate cases were studied and the ratio of (i) beam on time, (ii) total segments, (iii) leaf travel time and (iv) total treatment time with and without the flattening filter was specifically computed. The total time is of course the sum of the beam-on time and the maximum of the record- and verify-time and the leaf-travel time. It was found that the beam on time typically reduced by about 43% but the number of segments and the leaf travel times were very similar (just a few % change). The absolute time reduction varied between about 3% and about 20% depending on (i) the type of clinical case, (ii) the accelerator manufacturer and (iii) most importantly the dose rate and maximum leaf speed and the record and verify time.

3.2 Radiation leakage and accounting for machine effects in IMRT delivery

This and subsequent sections describe variations on a theme of delivering IMRT with an MLC. Table 3.3 summarizes methods and factors to consider when implementing them.

3.2.1 General leakage issues

Lillicrap *et al* (2000) have pointed out that shielding of radiation from linear accelerators was historically never designed for IMRT and that the increased number of MUs needed for IMRT treatments leads to increased leakage doses.

Table 3.3.

Variations on the use of an MLC for IMRT

Classic multiple-static field (MSF) technique
Classic dynamic MLC (dMLC) technique
Penumbra sharpening ('rind therapy') (section 3.3.2)
Dynamic arc therapy (section 3.5)
Combined step and shoot and dynamic therapy (section 3.6)
Intensity-modulated arc therapy (IMAT) (section 3.7)
Modified intensity-modulated arc therapy (e.g. SIMAT) (section 3.7)
Aperture-modulated arc therapy (section 3.7.2)
Simulated tomotherapy (section 3.8)

Features to consider (applying to all or some of the above)

Dose calculation algorithm (section 3.3)
Leakage and scatter; measurement and use of interpreters (section 3.2)
Room shielding (section 3.2.1)
Effect of rounded leaf ends (section 3.2.3)
Dosimetry of small field components (section 3.3)
Monitor unit verification (section 3.3.3)
Field splitting and feathering (section 3.4.1)
Tongue-and-groove effect (section 3.4.2)
Leaf speed limitations (section 3.4.3)
Accelerator stability; delivery of small MUs per segment (section 3.4.4)

There are of course techniques that can reduce the MUs required (e.g. filtering the IMBs). Webb (2000b) has described how some aspects of the leakage in IMRT can automatically be included into the *wanted* fluence through an iterative feedback process. Essentially the leaf patterns are fitted to the transmission-corrected profiles. Williams and Hounsell (2001) argue that the IMRT inefficiencies are probably no more than those associated with wedge factors. Also the extra time taken to set-up a patient for IMRT and to treat, together with the effect of decreased MU efficiency, may lead to the total number of MUs delivered in the working day being much the same with, as without, IMRT. Hence, daily leakage radiation would be much the same for calculation of room shielding. (This statement was written at a time when prototype clinical IMRT generally took longer than conventional IMRT; this may now not be the case and this rider may no longer apply.)

Ipe *et al* (2000) have shown that the use of an MLC for the dMLC technique does not increase neutron production. Measurements of neutrons were performed at Stanford Hospital with a Varian Clinac accelerator delivering 15 MV photons.

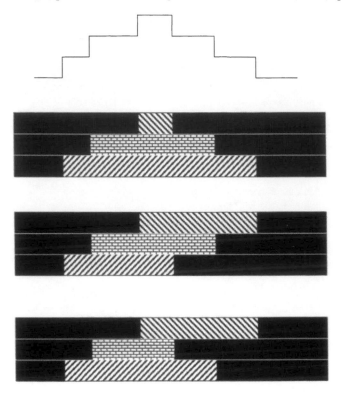

Figure 3.13. In this figure, a simple 1D IMB is shown comprising three fluence levels with equal fluence increments (upper panel). Below this are three possible practical realizations of the fluence profile. The MLC leaves are shown black and the three component irradiations are each shaded differently. It can be seen that for three levels there are 3!=6 possible configurations, three of which are shown. The fluence delivered in each case would be (slightly) different because of leakage and scatter variations (although the primary fluence would be the same in each case).

Höver *et al* (2003) have compared several MLCs and microMLCs and characterized their leakage radiation noticing differences between manufacturers and also differences between leaves of any particular MLC.

3.2.2   Factoring in delivery physics

3.2.2.1   Measurement and prediction of leakage and scatter

When an MLC is used for IMRT, it creates patterns with the leaves taking up different positions (figure 3.13). In order to compute the delivered dose distributions, a model is needed for x-ray transmission through and scatter effects in an MLC. These calculations are sometimes described as 'T2F' (trajectory to fluence' particularly when in conjunction with dynamic delivery) (Keall *et al* 2003). A numerical model has been presented by Chen *et al* (2000a). This analytic and numerical model was compared against a Monte Carlo calculation with good accuracy. In order to account for the finite size of the radiation source, the point-source transmission measurements were convolved with the distribution of source strength to generate transmission measurements that again were in agreement with Monte Carlo dose calculations. Arnfield *et al* (2000) have presented a method for determining multileaf collimated transmission and scatter for dynamic IMRT. The purpose of this work was to create the measurements which allow to take account of transmission when MLCs are used for creating IMRT using the dMLC technique. It is necessary to know the leaf transmission as a function of the field size. This is important because the field size varies during the delivery technique and the transmission can be a significant faction of the delivered dose. Measurements made by Arnfield *et al* (2000) were pertinent to two Varian MLCs, the Mark II, 80-leaf MLC and the so-called Millennium 120-leaf MLC.

The leakage radiation comprises two components: direct radiation and leaf scatter. The authors measured the direct radiation transmission by extrapolating the total transmitted radiation back to zero field size. So, for the 80-leaf MLC, they created a series of fields of length 5 cm with widths 0.5–10 cm for 6 and 18 MV. They then plotted the leakage transmission across the leaf direction and generated the usual sinusoidal patterns. They then line-integrated this transmission as a function of field size and, by extrapolating the graph back to zero field size, determined that the direct component of transmission was $1.48 \pm 0.1\%$. The transmission due to scatter from the leaves could then be worked out by taking this value away from the total transmission. For example, the leaf scatter of a 10 cm $\times$ 10 cm field was $0.20 \pm 0.01\%$.

Measurements for the 120-leaf MLC showed that the direct transmission was $1.34 \pm 0.03\%$ and the estimated leaf-scatter component was very similar to that for the 80-leaf MLC, not surprising given that the architecture of the central part of the MLC is very similar.

Naqvi *et al* (2000) have developed a dual-source model for the Elekta SL20 linear accelerator equipped with an MLC. This was used to predict output factors for complicated shaped fields so that MU calculations can be made. Prescription files from the CORVUS treatment-planning system were read into the programme and it was found that values for the target dose predicted by the model agreed with measurement to within 0.3% for any static field segment of arbitrary shaped size and off-axis position. For the entire delivery, the accuracy was within 0.6% of measurement using this calculation technique. However, CORVUS, not including the effects of scatter, overestimated the dose by between 1 and 4% when compared with measurement. Naqvi *et al* (2001b) have studied in detail the effects of head scatter in IMRT. They computed the form of the scatter source representation for an Elekta SL20 accelerator. They then used ray-tracing methods to compute the effect of scatter in a measurement plane. Calculations and measurements agreed to within 1%. They were particularly interested in the interaction of scatter with the complex collimation.

Baker *et al* (2002) have developed a head-scatter model for the Elekta Precise treatment unit which is capable of reproducing the measured head scatter to within 0.5% for fields ranging in size from 1 cm^2 to 40 cm × 40 cm. The model has been implemented for a PLATO treatment-planning system.

BäckSamuelsson and Johansson (2002) have shown that head scatter in IMRT treatments with the dMLC technique is not only dependent on the jaw field size but also on the size of the slit opening and the speed of the leaves.

Klein and Low (2000) have noted that, with IMRT treatments taking approximately five times as many MUs as conventional CFRT, leakage through leaf junctions becomes an issue. A comprehensive leakage study was performed for an MLC with both 1-cm- and 0.5-cm-wide leaf systems. They included in the calculation the effects of patient motion and concluded that interleaf leakage is a concern for IMRT therapy.

Klein and Low (2001) have observed that interleaf leakage as high as 4% is effectively magnified to 20% when account is taken of the five-fold increase in MUs required for many dMLC IMRT treatments. The concern is heightened for the 120-leaf system with more interleaf junctions. It is expected that patient motion will moderate the actual effects in practice. Studies were made for a 5-mm-leaf-width system (120 leaves on a Varian 23 EX) and for a 10-mm-leaf-width system (80 leaves on a Varian 2300 CD). Films were exposed perpendicular to the beams. Also a microionization chamber with an effective measuring width of only 1.5 mm was used to find the location of the maximum leakage intensity. Wedge-shaped IMBs were created and the intratreatment motion was simulated by adding translations and rotations and multiply irradiating the detector. Without movement, the peak leakage was some 4% of the maximum dose and decreased by about 1% with movement. Additionally, prostate IMBs computed by CORVUS were delivered under the same conditions including simulated patient movement. Specifically the parts of the fields where no primary dose was intended were analysed. Peak leakage dose was around 7–8% in these regions without movement and decreased by 1–2% with movement (details depending on amplitude of movement and the specific MLC). The effect of movement was to effectively spread out the leakage in space rather than reduce its overall contribution.

3.2.2.2 *Using leakage and scatter knowledge in the MSF-MLC technique*

The problem of incorporating the effects of leaf transmission and head-scatter corrections into MLC-delivered IMRT is different when the delivery is by a dynamic technique or by a step-and-shoot technique. For the dynamic technique, the effects can be incorporated by iteratively adjusting the MLC leaf positions. However, when IMRT is delivered with the step-and-shoot technique, the *a posteriori* segmentation has already determined the fixed leaf positions and these are no longer varied. This segmentation will have been done assuming that transmission is zero and that head scatter is zero, neither of which are correct assumptions. As a result, if the dose is delivered with these field patterns, without any further modification to the fluence values for each segment, the delivered dose will not be the planned dose because the delivered fluence will not have matched the planned fluence.

Ting *et al* (2000a) have developed a new stop-and-shoot leaf-sequencing algorithm for IMRT which accounts for leaf transmission, interleaf leakage, scattered radiation and leading-leaf penumbra. Outputs from the leaf-sequencing method for dMLC control have been compared with measurements made in water and solid phantoms. The sequencer has been used for more than 100 patients since July 1998.

Yang and Xing (2002) solved the problem of how to account for leaf transmission and head scatter in step-and-shoot (MSF) IMRT by the expediency of adjusting the MUs per segment until the delivered fluence matches the calculated fluence in a least-squares sense. The algorithm works by adjusting the fractional MU factors in the leaf-sequence file without considering the two effects until the delivery with fractional MUs matches the outcome of including the full physical effects into delivery of the leaf sequences with integer MUs . This is needed because, unlike in the dMLC technique, the fluence distribution cannot be continuously varied. The technique is iterative in the same way as that of Convery and Webb (1997) for the dMLC method.

Yang and Xing (2003) have presented further details. To modify the number of MUs applied to each segment such that, after modification, the fluence delivered including the effects of transmission and head scatter is most closely matched to the planned fluence, they used a downhill Simplex technique (figure 3.14). In order to implement this technique, they required to develop a three-source model for head scatter which they verified by predicting and measuring head-scatter factors for specific small fields and specific IMRT components, comparing the measured results with the predictions from Monte Carlo calculations. Leaf transmission was also estimated for a particular Varian accelerator as about 1.75%.

For both an intuitive test field (a simple 10 cm × 10 cm uniform open field) constructed from five consecutive 2 cm × 10 cm segments, and also for an IMB from CORVUS (part of a clinical prostate plan), they showed that the fluence maps obtained from their technique, when both transmission and leaf scatter were included, were in very close agreement with the calculated intensity map without these corrections. They also showed that measurements made with film agreed with Monte Carlo calculations very closely. In summary, these two complex physical effects are entirely taken care of by adjusting the number of MUs per segment.

Papatheodorou *et al* (2000a) developed a model to compute the forward dose for a set of interpreted dMLC field-shapes. This model was an extension of a primary-scatter separation used for conventional fields. It also took account of

Developments in IMRT using a multileaf collimator (MLC) (physics)

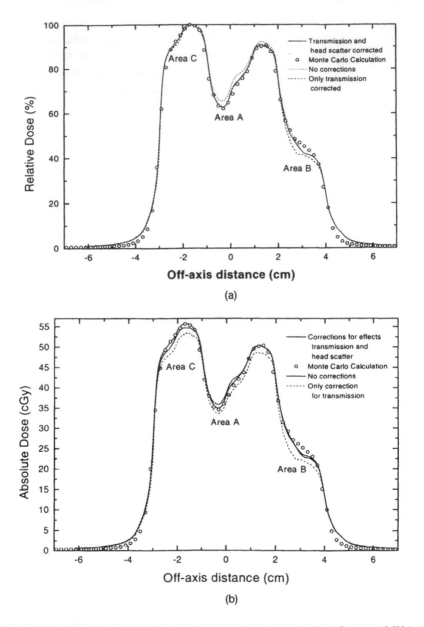

Figure 3.14. Illustrating the technique for correcting step-and-shoot fluences (MUs) to account for head scatter and leaf transmission. The profile is one line through a set of three measured dose distributions when the fluence component's MUs were either (i) uncorrected, (ii) corrected for transmission or (iii) corrected for both transmission and head scatter. Monte Carlo calculations are shown for comparison. These agree best with (iii). (Reprinted from Yang and Xing (2003) with copyright permission from Elsevier.)

the rounded leaf ends, the penetration through which becomes increasingly more significant as the size of the field segments becomes very small. The model used a single exponential for the head-scattered radiation with a pair of fitting parameters which could be different to account for the different leaf-end and side geometries. Calculations and measurements agreed to better than 3% in regions of uniform dose and to 3 mm in high-dose-gradient regions.

3.2.3 The effect of rounded leaf ends: light-field to radiation-field discrepancy in IMRT

There is a further phenomenon to be accounted for in the dMLC technique with the Varian accelerator. Because the leaf ends do not have a perfectly double-focused form but have rounded leaf tips, there will be an additional fluence through the tips depending on the position of the leaves in the field and this has often been described as equivalent to an effective increase in the window width between opposing leaves over and above the geometrical distance between leaf tips (figure 3.15). This equivalent shift has sometimes been called a leaf-gap offset and must be determined for each MLC. The reason is that the effective widening can be a significant fraction of the variable distance between the leaves. The 50% penumbra is pushed back in the direction of the tungsten from its position in the light field and thus it is necessary to reduce MLC leaf gaps in order to obtain agreement between calculated and measured profiles (Boyer 2003).

Arnfield *et al* (2000) measured this shift by delivering the radiation through a sweeping window of variable width. The leaves were set to sweep from one side to the other of a 10 cm by 10 cm field for a series of fixed separations between the leaf banks ranging from 0.5–10 cm. By making the appropriate calculations, they then determined that the equivalent shift was 0.114 ± 0.004 cm for the 80-leaf MLC and 0.088 ± 0.003 cm for the 120-leaf MLC. As a check on their measurements, it was found that when this was included in the calculation of dose, the predicted and measured values agreed to within 2%. Arnfield *et al* (2000) then backed up their experimental measurements with Monte Carlo calculations.

Kung and Chen (2000a) also pointed out that when using a (Varian) MLC to perform IMRT using the dMLC technique the radiation-field offset must be programmed into the machine instructions that drive the MLC. The radiation-field offset is the difference between the position of the 50% radiation boundary and the light-field boundary. Usually this is small, just a fraction of a mm. They investigated the dosimetric consequences of using the wrong value and experiments showed that the dose error sensitivity factor (percentage change in dose for each mm of radiation-field offset) was in the range 0–8% mm^{-1}. This demands that radiation-field offset be no more than 0.5 mm if the dose error is to be kept to better than 4%.

Cadman *et al* (2002) have made a study of this phenomenon and come to the conclusion that an MLC leaf-gap reduction of 1.4 mm is required to obtain agreement between calculated and measured profiles. This was found

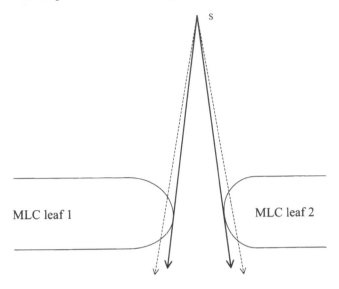

Figure 3.15. Illustrating the radiation field offset (RFO) phenomenon. The full lines show the edge of the lightfield defined by two MLC leaves. The dotted lines show connectivity to the 50% penumbra each side of the aperture. The distance between the two 50% penumbra points will not equal the lightfield width if the leaf ends are round.

by comparing profiles delivered with single open fields, profiles delivered using seven IMBs, each delivered separately, that correspond to an IMRT plan and they also showed that the full dose distribution in an IMRT plan agreed better with calculation from the ADAC PINNACLE system when this leaf offset was introduced. They comment that future releases of the ADAC PINNACLE system will have this discrepancy overcome.

Vial *et al* (2003) also measured the radiation-field offset and found that better agreement between calculation and experiment could be obtained by using an experimentally-determined value instead of the manufacturer-provided value.

3.3   Dose calculation for IMRT

Several studies have addressed issues arising from the types (size and irregularity) of field shapes generated in MLC-based IMRT.

Martineau and Manens (2001) have shown that there is good agreement when measuring the depth–dose distributions and profiles for very small fields such as those that will be delivered in IMRT when the following detectors were compared: diode, a pinpoint chamber, film, a Markus chamber and a diamond detector. They recommended the use of the diamond detector because of its good spatial resolution and near tissue equivalence for measuring depth-doses and profiles. For measuring output factors they recommended the pinpoint chamber. Sánchez Doblado *et al* (2003) have shown that small-field dosimetry is accurate based on reference 10 cm × 10 cm field dosimetry.

It is well known that in the step-and-shoot IMRT techniques some of the static field components are very small in area. Cheng *et al* (2003) have shown that, when lateral electronic equilibrium does not exist, the output factors for small fields significantly differ from those that would be predicted using Day's equivalent-square formulation. By making careful measurements, Cheng *et al* (2003) produced graphs of output factors as a function of field size under conditions of electronic disequilibrium.

Knoos *et al* (2001) have shown that the use of the collapsed-cone algorithm in the HELAX treatment-planning system led to only marginal changes in dose distribution from the use of a pencil-beam algorithm.

Cozzi and Fogliata (2001) have compared calculations from the HELAX treatment-planning system and measurements performed with ion chambers, portal imaging devices and films. Calculations and measurements agreed to 1% for output factors and 2% for depth doses and profiles outside high-dose-gradient regions.

Xia *et al* (2000b) have studied the influence of penumbral regions appearing in the middle of the target and dose inhomogeneity within the target introduced due to patient motion or inaccurate MLC positioning in the step-and-shoot IMRT delivery. Preliminary results showed dose errors between 2% and 5%.

Phillips *et al* (2001b) have shown that dynamic IMRT dose distributions should be modelled as a convolution of a scatter kernel with a continuous fluence assuming constant velocity leaf motion between control points. If, conversely, the beam modulation is modelled as a set of static fields, the results significantly differ from measurements.

3.3.1 Application of colour theory

It is well known in IMRT that the radiation fluence delivered through any one bixel depends on the state of opening of adjacent bixels. This was first shown for the NOMOS MIMiC by Webb and Oldham (1996) who provided an analytic technique for creating dose distributions in the so-called component-delivery (CD) mode. Essentially the same problem arises in the dMLC technique when bixels may be identified as being delivered by elemental movements of the multileaves. Concentrating on any one bixel, the state of the leaves covering or uncovering adjacent bixels affects the dose distribution through the bixel being inspected. For example, if one were to inspect a 3 × 3 panel of bixels and the central one is open, then there are 2^8 possible ways to select open or closed states of the eight neighbours of this central bixel. These 256 possible arrangements are (in group theory) known as colourings. The 256 arrangements lead to 256 different values for the dose from the inspected bixel.

Langer (2000) has analysed the colourings and shown that the number of independent patterns is less than 256. In fact, it was shown that this number is 51 distinct patterns, if no distinction is made between the method of collimation using the leaf faces and the leaf sides. If the symmetry between orthogonal coordinates is broken by collimator leaves whose ends and sides have different effects on bordering bixels, then the number of patterns increases to 84. However, it can be seen that both these numbers, being less than 256, lead to a simplification of the dose calculation. The (fewer number) elementary pattern dose calculations can be stored and then those which reduce to these through rotations and reflections, can re-use the same stored dose distributions. In this way, the calculation of dose distributions in IMRT can be speeded up by a factor of three or more. Langer (2000) gives examples of this technique. This is possibly now less necessary given that fast dose calculations are now fairly straightforward without this simplification.

3.3.2   Penumbra sharpening for IMRT

It has been known for some time that the use of margins, rinds of small spatial extent but higher fluence than the surrounding fluences, can sharpen the penumbra in radiation therapy. Sharpe *et al* (2000a) have performed a thorough investigation of the dependence of the effect of energy, the width of the rind and the added fluence. Both calculations and measurements were made at 6 and 18 MV and measurements were made using both rinds created using different apertures in cut lead sheets and using multiple multileaf collimated fields.

In summary, their work showed that the 50–95% penumbra could be significantly reduced by the use of high-fluence rinds. The precise fluence increase required depended on the radiation energy and on the width of the rind and on the nature of the surrounding tissue, whether water or lung. The effect was ubiquitous and only precise values depended on energy and material. Sharpe *et al* (2000a) went on to show that, for specific examples of a lung tumour treated with 6 MV x-rays, the margins could be considerably reduced and that penumbra compensation led to both an improvement in the dose-volume histogram of the PTV and also simultaneous increased sparing of the spinal cord.

3.3.3   MU verification

Dosimetrists are in the good habit of making routine MU checks of conventional therapy (non-IMRT). For the delivery of modulated fields (IMRT), this ability using conventional methods disappears. Other methods are needed and the following have been developed.

Kung *et al* (2000b) and Kung and Chen (2000b) have developed an MU verification calculation for IMRT as a quality-assurance tool for dosimetry. Huntzinger and Hunt (2000) have developed a manual technique for MU calculations in IMRT. Lujan *et al* (2001a) have compared the independent MU calculation from a modified Clarkson's method to the CORVUS dose calculation for a number of different IMRT fields and found that these agreed to within 2.5%.

3.4 Features of MLC delivery of IMRT

3.4.1 Large-field IMRT—splitting the delivery

The Varian MLCs are limited to creating fields of length 14.5 cm in the direction of the leaf movement when used for the dMLC IMRT technique. This is because no trailing leaf end may ever extend beyond the jaw position. Yet sometimes it is useful to be able to create larger IMRT fields, e.g. when delivering a simultaneous boost. Wu *et al* (2000c) have solved this problem by a simple technique (figure 3.16). A large IMRT field is 'split' into two contributions which overlap spatially. In the overlap region (generally a few cm in length), each IMB is feathered from half the value at the midpoint of the feathering length to zero such that throughout the feathering region the sum of the two split IMBs is equal to the unsplit beam. Wu *et al* (2000c) showed that there is no need to adjust output factors. They showed experimental agreement between computation and delivery using this method. They also suggested that one could split fields whose length is less than 14.5 cm as this would permit the use of back-up jaws (enveloping the split fields) to reduce radiation leakage.

Chui (2000) and Chui *et al* (2001b) have described clinical implementation of IMRT and specifically techniques for splitting a 2D intensity-modulated field into two, when the delivered field width exceeds the limit imposed by the MLC design. Chui *et al* (2001b) have shown, using an example from a nasopharynx field, that splitting should be done in the low-intensity region where cord protection is needed.

Hwang *et al* (2001) have also developed an algorithm to increase the maximum usable field size for the Varian dMLC system for IMRT. The IMRT field length is enlarged to the full length of 27 cm as appropriate to conventional multileaf collimated fields. The strategy employed was tested on 1000 random patterns.

3.4.2 Tongue-and-groove effect

The tongue-and-groove (TG) effect is a much studied unwanted underdose caused by the combined effects of irradiating through extended tongues and grooves along the sides of MLC leaves. In the delivery of IMRT by the dMLC technique, it can be entirely eradicated by synchronization. However, unwanted consequences of synchronization are: (i) increased treatment time and number of segments, hence increased leakage; and (ii) possible across-the-leaves dose error due to the requirement for very precise leaf positioning.

Deng *et al* (2000) have compared, using the EGS4/MCDOSE Monte Carlo code, the dose distributions using the fluence maps with and without TG effects.

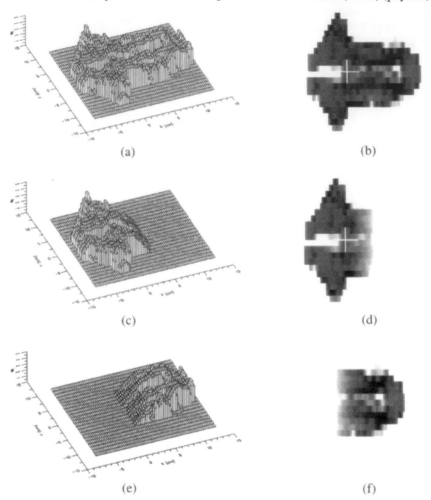

Figure 3.16. This diagram shows how to deliver a large-area intensity-modulated field either directly or by splitting it into two components in which the field boundary is feathered for each of the two components : (a) profile for the original large field; (b) simulated film for the large-field intensity distribution. The white cross denotes the isocentre. Darker areas represent higher intensity; (c) shows the profile for the left component field; (d) simulated film for the left component; (e) the profile for the right component; (f) simulated film for the right split field. (From Wu *et al* 2000c.)

They concluded that, when multiple fields were used and setup errors were included, there were no distinguishing discrepancies found, indicating that the TG effect can be neglected clinically (figure 3.17). Deng *et al* (2001a) conducted a computational study to show that the effect could, in general, be ignored for IMRT

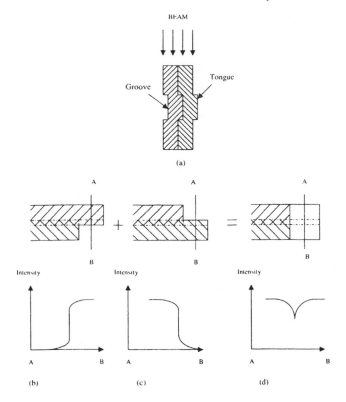

Figure 3.17. A schematic diagram of the tongue-and-groove effect in an MLC: (a) the design of the MLC tongue-and-groove is to reduce interleaf leakage; (b)–(d) schematic diagrams of two fields and their superposition defined by two adjacent leaves. The region centred between the two leaves in (d) is underdosed. This effect was shown to have no clinical significance when multiple individual fields are added together in an IMRT plan. (From Deng *et al* 2001a.)

treatments involving multiple ($\geq$ 5) fields and some modest patient movement. They came to this conclusion as follows. Firstly, IMRT maps were extracted from the CORVUS inverse-planning system for two cases. The first case was an eight-field coplanar treatment of the prostate using an 80-leaf Varian MLC on a Clinac 2300C/D accelerator delivering 15 MV radiation. The second case was a nine-field coplanar treatment of the vertebra using an 52-leaf Varian MLC on a Clinac 2100C accelerator delivering 4 MV radiation. The CORVUS maps were computed with the TG effect present. Then the MCDOSE Monte Carlo program was used to computed fluence and ultimately dose, taking into account leaf transmission and head scatter but ignoring the TG effect (so-called Plan 1). Alternatively, the files were taken through the MLC2MAP code to *include* the TG effect as well as the other effects mentioned (so-called Plan 2). Two more plans were then computed with these two plans blurred with typical patient movement. Results were shown as dose maps and DVHs for PTV and OARs.

They arranged for a statistical accuracy of some 1% with doses calculated on a 1 mm grid at isocentre. It was shown that the TG effect produced dose differences no more than 1.6% when the number of fields were eight or nine and patient movement was included. The doses were slightly larger in Plan 2 in low-dose regions and covered a smaller area in Plan 2 in high-dose regions. This small residual effect was considered clinically insignificant. It arises because the underdose regions do not coincide for each field contribution. This is especially true with the inclusion of patient movement. In the study, conversely, some single-field irradiation calculations were carried out showing much larger TG underdose greater than 10%. They concluded that if the number of fields is small (say $\leq$ 5), it may be necessary to conduct a study like this to check on the TG underdose. However, for IMRT with larger numbers of fields, they concluded that we can finally ignore the TG effect.

Whilst the technique of Van Santvoort and Heijmen (1996) and Webb *et al* (1997) can completely remove the TG undersdose for the DMLC technique, until recently no such equivalent solution had been proposed for the MSF technique. Indeed Webb (1998a, b) claimed that no such solution could ever be found via a direct search of the configuration options for the fixed number of components from a decomposition by the method of Bortfeld *et al* (1994). Que *et al* (2004) have presented a new method to decompose an IMB based on first applying the technique of Van Santvoort and Heijmen (1996) to create equivalent ('tentative') static-field components and then adjusting them (synchronizing right leaf positions) to remove the TG underdose in a step-and-shoot delivery. Applying this to a variety of clinical IMBs as well as random IMBs showed the problem was solved. However, the price paid is a very large increase in the number of field components as well as the solution having the smallest leaf openings (parametrized through their averaged leaf-pair opening [ALPO]) (see section 3.14.1). Of the more commonly applied decomposition methods, it was shown that the 'reducing level' method of Xia and Verhey (1998) had the smallest TG underdose with a realistic number of components.

3.4.3   Leaf-speed limitations

Sohn *et al* (2000a) have noted that the dMLC technique is most limited when requested to create shallow-gradient intensity profiles. This is because shallow-gradient profiles correspond to the leaves moving at high or maximum speed. The Varian Clinac 2300 CD accelerator automatically reduces the dose rate when the leaf velocity reaches its maximum value. The fidelity of generating different gradients as a function of dose rate and leaf speed were studied. The dose profiles were linear when the MLC was not moving at maximum speed but artefacts were generated when the leaf velocity limited the dMLC delivery and the profile became irregular.

3.4.4 Stability of accelerator and delivery of a small number of MUs and small fieldsizes

It is well known that a feature of IMRT, delivered with the step-and-shoot technique, is the creation of many small-area field segments and the requirement to deliver the segments sometimes with very few MUs per segment. This situation was extensively investigated by Hansen *et al* (1998). However, the conclusion of that paper was that the stability of an accelerator required to be measured for each individual accelerator and that measurements from one could not be translated directly to experience for another. Hence, Sharpe *et al* (2000b) have essentially repeated the work done by Hansen *et al* (1998) for two accelerators at the William Beaumont Hospital, Royal Oak, Michigan (figure 3.18). One of these accelerators was equipped with a slitless flight tube and a second machine, used for IMRT, initially had a slitted flight tube changed later to a slitless flight tube. Sharpe *et al* (2000b) showed that, provided the number of MUs was greater than one, the dose delivered per MU could be expected to be within ±2% of the expected value for the first (slitless flight tube) accelerator. This applied at both 6 and 18 MV. For the second accelerator, the corresponding figures were as follows. The dose per MU was within ±2% of the expected dose per MU when more than 3 MU were delivered at 18 MV and when more than 1 MU was delivered at 6 MV. It was found that changing from a slitted to slitless flight tube improved stability. The detailed plots shown are very similar to those in the paper by Hansen *et al* (1998) and Sharpe *et al* (2000b) presented curves for each energy and machine configuration measured at three different time periods of within a year.

Beam flatness and beam symmetry was also measured and found, for the slitless flight tube, to be within manufacturer's specification after 0.6 s corresponding to a delivery of 4 MU. For the slitted flight tube, the stability did not arise until 1 s after beam turn on or approximately 7 MU. Measurements were made in both the *GT* plane and in the perpendicular *AB* cross plane. Based on these experiments, the authors were confident that calculation algorithms for estimating relative output factors in a stable beam would apply to exposures as short as 4 MU. The rest of the paper presented detailed discussions of output factors as a function of field size and position and these were found to be very stable and predictable provided the field sizes were properly collimated to the expected field sizes.

Martens *et al* (2001a) have described the improvements consequent on adding a fast-tuning magnetron (FasTraQ) into three Elekta travelling-wave linear accelerators replacing the mechanical plunger with a solenoid-driven plunger in the magnetron. It was found that start-up and inter-beam segment times were typically reduced by between 2–4 s. Delivery efficiency for a variety of clinical and quality-assurance programmes was improved by an average of about 30%.

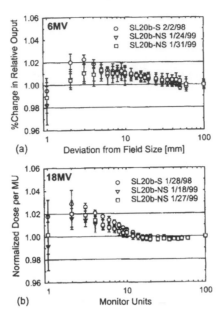

Figure 3.18. The dose delivered per MU by SL20b measured as a function of the total number of MUs delivered for (a) 6 MV and (b) 18 MV x-rays. This machine was initially equipped with a slitted flight tube (SL20b-S) but was later upgraded to a non-slitted design (SL20b-NS). The dose-per-monitor-unit acquired for the slitted flight tube (circles) is compared with data acquired at two points in time following the upgrade (inverted triangles and squares). (From Sharpe *et al* 2000b.)

The dose output was within 1% for a 2 MU beam segment. Dosimetric errors were small even down to a 1 MU beam segment. Beam symmetries and flatnesses were acceptable at all energies and dose rates showed no obvious degradation in low-dose beam segments.

Budgell *et al* (2001a) have also described FasTraQ. In Manchester, it has been fitted to a 7-year-old linear accelerator. It was found that the dose per MU *versus* MU improved to better than 1% at 2 MU compared with figures of 4% for the slitless flight tube and 5% for the slitted flight tube. For new accelerators, the performance was not changed by the introduction of the fast-tuning magnetron in terms of these parameters. The biggest effect was on treatment efficiency. The start-up time fell from 8 to 3.6 s and the intersegment time fell to 1.7 s at 6 MV and 2.2 s at 8 MV where the figures are both for a fluence rate of 400 MU min^{-1}. This was for stationary leaves so that leaf movement time was not part of the measurement. Budgell *et al* (2001a) reported that the total time per prescription fell by an average of 30.7%. The consequences are that one may now use more segments to improve the conformality or, keeping the number as before, decrease the overall treatment time.

Ezzell and Chungbin (2001) have studied, for a Varian linac, the effects of CORVUS-generated segments with less than 1 MU per segment. Komanduri (2002) has also commented that, as the number of segments for IMRT of the breast increases, the number of MUs per segment will decrease possibly leading to quality-assurance issues and concerns.

Rock *et al* (2001) and McClean and Rock (2001) have made measurements of small (1 cm × 1 cm to 3 cm × 3 cm) fields defined using both a Varian 2300 CD and an Elekta SL18 acccelerator and compared the data with that from the HELAX treatment-planning system (Version 5.1). Data were measured using an ion chamber, a diamond detector and film. The measured data included leakage through the leaves, linearity of dose with MUs, *x*- and *y*-profiles, depth–dose in water and, output factors. Good agreement was found between the treatment-planning system and measurements for depth–dose data for fields as small as 1 cm × 1 cm for the Varian linac and 2 cm × 2 cm for the Elekta linac. However, comparison of the planning-system-generated profiles with the measured profiles showed significant differences in both directions from both machines at 2 cm × 2 cm and 1 cm × 1 cm for the Elekta linac. Output factors also do not agree with measurements by as much as 20% for the Elekta linac at 1 cm × 1 cm and by 8% for the Varian linac at 1 cm × 1 cm. Preliminary investigations also showed that there may be problems with the use of both off-axis and rectangular fields. They concluded that caution should be used in the MSF technique when small-field segments are incorporated.

Jones *et al* (2003) have pointed out that, for the beam components in an IMRT plan which have a very small field area, the depth–dose curve will not necessarily follow traditional radiological scaling for tissue density inhomogeneities. Using the EGS4 Monte Carlo dose-calculation package, graphs were produced showing the depth–dose curves as beams of either 6 or 24 MV radiation passed through 3 cm of lung depending on the area of the beam element. The situation is very complex.

Mitra *et al* (2001a) and Cheng and Das (2002) have investigated the behaviour of two Siemens linear accelerators operating in IMRT mode at both 6 and 15 MV. It was found that the dose per MU was stable to 1% and linear with increasing MU beyond 5 MU at 6 MV and beyond 10 MU at 15 MV. Below these values, unacceptably large variations occurred. For example, at 15 MV and 3 MU, the dose per MU was 5% above that required. At 6 MV and 1 MU the dose per MU was 5% above that required. It was also found that profiles were varying more than 5% below 10 MU at 15 MV. Above this, they also showed that both accelerators displayed beam flatness and symmetry to better than 2%. This work was done by sequencing subfields using the Siemens LANTIS and PRIMEVIEW system. They then concluded these characteristics make the accelerators acceptable for performance of step-and-shoot delivery of IMRT. Saw *et al* (2003) found different experimental results for a Siemens accelerator in which the beam current was dephased from the microwave power whilst the field shapes were adjusted. It was found, conversely, that the dose per MU was stable to better than 2% above 2 MU.

Two consequences follow:

(i) IMRT delivery using this equipment should preferably have segments with more than 10 MUs. When planning with HELAX TMS, the number of segments should be kept low to ensure this condition.

(ii) Errors may occur in verification film dosimetry when MUs are divided down in order to avoid saturating some types of film. In a typical case studied, 7/22 segments studied had less than 5 MU in this divided-down mode of verification with film.

Bieda *et al* (2002) have shown for a Siemens PRIMUS accelerator that the beam fails to reach full intensity during the delivery of 1 MU and a little more than this is necessary for beam stability. This is relevant to IMRT with small MU segments. Aspradakis *et al* (2002) have shown that the Siemens Primus accelerator is ideally suited for step-and-shoot IMRT because its characteristics in beam uniformity, energy and dose linearity are virtually independent of the MU segment size. Li *et al* (2003) have made similar measurements to confirm the suitability of a Siemens accelerator for delivering small-MU fields for breast IMRT. Kulidzhanov *et al* (2003) also report on this problem.

Steinberg *et al* (2002) have improved the automatic frequency control circuitry of a Siemens linac to create a beam which is stable within 55 ms instead of the more usual 300 ms from onset of dose pulsing. This will create a situation in which the delivery of a fraction of a single MU becomes reliably possible.

Bayouth and Morrill (2003c) also characterized small fields for IMRT. Penumbra and leakage were remarkably consistent for the range of leaf positions studied.

Cheng *et al* (2002) have compared the beam characteristics of Varian, Elekta and Siemens accelerators, in particular noting differences in dark current irradiation (see also Bassalow and Sidhu 2003).

In concluding this review, we should note that the beam performance depends on the accelerator manufacturer, the specific accelerator, the segment size and the version of the hardware. There are some inconsistencies between reports from different groups on similar equipment.

3.5 Dynamic arc therapy

Dynamic arc therapy is the technique whereby, as the gantry rotates, through 2π, the field shape and intensity across it are changed by either using MLC movement (figure 3.19) or some gravity-oriented device (GOD).

Pignoli *et al* (2000) have made measurements to compare with calculations performed by an in-house-developed treatment planning system (TPS) for

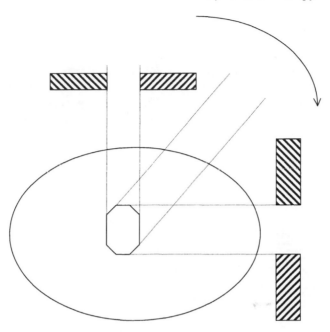

Figure 3.19. Illustrating the principle of the dynamic arc technique. As the gantry rotates, the open MLC leaf aperture specified by a leaf pair defines the beam's-eye view of the target from that direction. Other leaf pairs operate similarly for the other transaxial planes. This is little more than the principle of conformational radiotherapy of Takahashi shown in figure 1.1.

the dynamic arc therapy technique delivered with the 3D-Line International microMLC (see section 3.11.7). This technique constructs either three or five arcs with the field shape of the microMLC tailored to the beam's-eye view of the target at each gantry location. Absolute isocentre dose was accurate to about 2% and the calculated and measured isodoses matched to within 2 mm in transaxial and sagittal planes.

Nakagawa *et al* (2000b) have developed a new accelerator capable of dynamic conical conformal therapy—they called this dynamic therapy. The idea is that, instead of using couch rotation with respect to gantry rotation in a fixed plane, the C-arm of the accelerator can itself rotate by up to 60° relative to its default position, thus allowing non-coplanar conical conformal therapy. It was shown that this improves treatment with respect to coplanar CFRT for large tumours.

Boyer *et al* (2002a) have developed a sweeping window arc therapy (SWAT) technique. In this, the leaves move along horizontal leaf tracks parallel to the axis of a single rotation of the gantry.

Rivard *et al* (2002) have developed a doubly dMLC-IMRT technique in which the gantry moves at the same time as the dynamic leaves.

Keane *et al* (2000) delivered total body irradiation using an arc and a gravity-oriented compensator to achieve an intensity-modulation pattern. They also changed the technique to use an MLC to obtain similar uniform dose distributions through intensity-modulated-arc therapy (IMAT). By dynamically changing the field sizes using the MLC as the gantry rotates, the equivalent to the custom compensator can be produced.

Frencl *et al* (2001) have discussed how IMRT can be performed without an MLC. The idea is to use a special treatment technique for head-and-neck lesions. This comprises many static fields modified with a gravity-oriented block for shielding the OAR, the spinal cord. Extra compensating filters are added to improve the dose homogeneity in the PTV. Recently Danciu and Proimos (2003) have compared various materials for making gravity-oriented absorbers.

Oh *et al* (2003) and Kim *et al* (2004c) proposed a new IMRT method based on a scaled model of the PTV. The model hangs under gravity in a mercury box as the gantry rotates. During this rotation, the beam's-eye view of the PTV changes conforming to the target and the fluence is modulated automatically. This is a kind of variable compensator. Webb (2004a) has suggested an alternative to the use of mercury and an inverse-planning method to calculate the optimum gravity-oriented attenuator (figure 3.20).

3.6   Combining step-and-shoot and dynamic delivery for dMLC

In general, IMRT is delivered with an accelerator equipped with an MLC in either dynamic or step-and-shoot mode. Both of these could be considered embodiments of the dMLC technique (the latter provided it comprised a large number of segments). As pointed out by Martens *et al* (2001b), most departments choose to work with one or other of these techniques to keep the investment costs sensible. Alternatively, the equipment provided by a manufacturer may dictate the choice. Dynamic delivery is more time efficient than step and shoot but the step-and-shoot technique is less complex and has a lower cost of quality assurance. Martens *et al* (2001b) have combined the best aspects of the two techniques in a hybrid method. They call this 'interrupted dynamic sequences'. It is not unlike the work proposed by Beavis *et al* (1999, 2000b).

The way this works is as follows (figure 3.21). Consider, for example, prostate IMRT delivered from three fixed gantry directions. A simultaneous boost dose is delivered to the PTV minus the volume of overlap of the PTV in the rectum using smaller segments and the whole prostate is also irradiated. This leads to the combination of what the authors call a convex and a concave dose distribution and is essentially an anatomy-based segmentation. When the segments are stacked

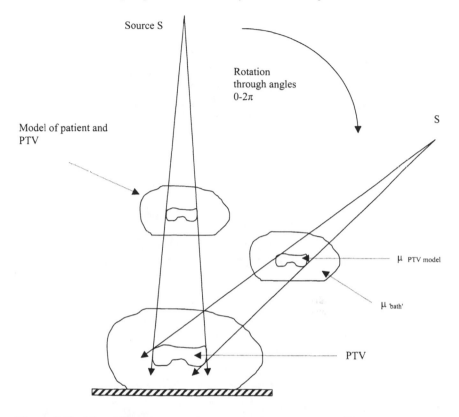

Figure 3.20. The delivery concept for a gravity-oriented device (GOD). The source is represented at S. A section of the patient is shown in the lower part of the figure and a demagnified version of this as a GOD between patient and source. The beam is collimated to the beam's-eye view of the PTV and modulated by the variable pathlength through the absorber. Two attenuation coefficients $\mu_{\text{'bath'}}$ and $\mu_{\text{PTV-model}}$ are determined by inverse planning. This cannot effectively rival other methods of IMRT delivery.

together, this leads to a large dose gradient at the anterior rectal wall and typically each of the beams may have a number of segments.

Considering just the segments required to realize just the intensity modulation from one beam direction, the segments are classified into those that belong to the part of the treatment delivering the convex dose distribution to the whole prostate and the other part of the treatment which relates to the delivery of the concave dose distribution. These two groups are called classes of segments. The major step forward made by these authors is then that, for segments which belong to the same class, dynamic transitions are allowed. The reason for this is that for concave dose distributions corresponding to the segments that deliver these, the intensity needed must continuously increase with decreasing distance

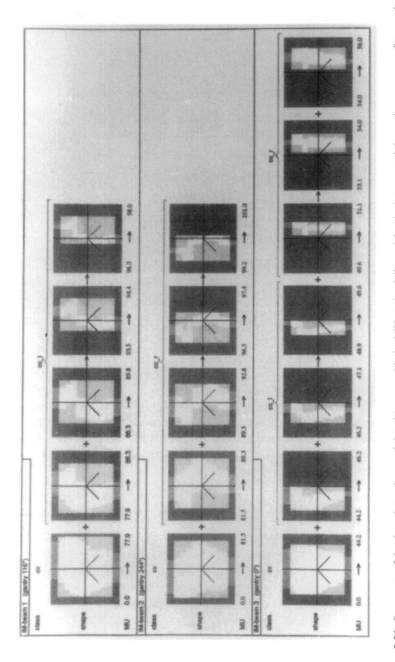

Figure 3.21. Segments of the three intensity-modulated beams with the MUs to be delivered for the interrupted dynamic sequences. Step-and-shoot and dynamic transitions are represented by a plus sign and by an arrow respectively. (Reprinted from Martens *et al* 2001b with copyright permission from Elsevier.)

from the rectal wall. Intuitively, the authors then argue that dynamic transitions might actually be desirable because they blur the stepwise effect of segment edges on intensity. However, a segment edge which is required to create a sharp in-field intensity gradient should not be used as a dynamic sequence because blurring of such an edge is undesirable and therefore step-and-shoot transitions are performed between the segments for one class (those delivering the concave volume of high dose) and the segment delivering the convex dose distribution. This is the basis of the technique.

For two successive segments suitable for dynamic transition, an equal number of MUs is transferred to the transition. The maximum number of such MUs allowed for dynamic transition was 10. The accelerator was set to operate at 100 MUs per minute and a complex detail in the algorithm relates to ensuring that the dynamic transitions are selected in such a way as to minimize the overall transition time. This actually means that some segments inside the concave class cannot be delivered with dynamic transition but must be delivered with step-and-shoot transition. Martens *et al* (2001b) then go on to make measurements using a PTW-Freiburg LA48 linear ion chamber array in which the centre-to-centre distance in the chamber is 8 mm and the whole array is shifted longitudinally by 1 mm seven times to result in a measurement of a profile with a measurement every 1 mm. They then make measurements for the same delivery using either the step-and-shoot-only technique or the so called interrupted dynamic sequences technique and they show that the two techniques are dosimetrically equivalent. The new technique, however, is some 10% more time efficient. As part of this work, they also make measurements of the relative dose output per MU as a function of the MU delivered per segment and show that, provided that 3 MUs or more are delivered, the deviations in dose output are less than 1% irrespective of the dose rate. They also make measurements of field flatness under different dose-rate conditions. In summary, they find the interrupted dynamic sequences technique maintains the intended sharp gradients near the rectum but leads to no major effects on the dose distribution in the regions of dynamic transition.

3.7 IMAT—technical issues

Yu and Li (2001) have developed a delivery optimization scheme with combined rotational and fixed-gantry IMRT. This is because they do not believe that either of these separately are optimal. An optimization algorithm was used to generate beam intensities or beam angles, then one by one each beam was deleted from the initial set and the intensities for the remaining beams were reoptimized. The difference between the new and the baseline cost represented the importance of the deleted beam. In this way, importance factors were evaluated for all beam angles. The approach was then to deduce the optimal delivery technique such that, where the importance factors peak at certain angles, fixed fields are used at these angles, whereas when the importance factor as a function of angle is

Original IMBs planned at fixed gantry angles, in this example separated by 3^0

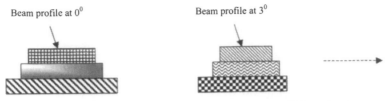

Beam profile at 0^0 Beam profile at 3^0

Angle for beam profile delivered via IMAT labelled to left of each subfield; multiple rotational arcs

$0\text{-}3^0$ rotation 1 $3\text{-}6^0$ rotation 1

$0\text{-}3^0$ rotation 2 $3\text{-}6^0$ rotation 2

$0\text{-}3^0$ rotation 3 $3\cdot6^0$ rotation 3

Figure 3.22. This illustrates the principle of IMAT. Imagine the IMRT plan has been made with IMBs at a series of fixed gantry angles (for illustration here they are at 3° separation in angle. Imagine that each beam has three intensity levels. The logic of IMAT is that there are three arcs and each of the individual fluence components is delivered spread out over 3° of arc. The MLC leaves move as minimally as possible between arc segments. In practice, the individual IMBs would be planned at a wider spacing $\Delta S°$ with more (say N) fluence levels and there would be N arcs.

flat, rotational delivery will be chosen. The authors claimed this has a significant clinical impact.

Yu *et al* (2004) have further developed the scheme for optimizing hybrid IMRT. The first step in the scheme is to compute the angular cost function (ACF) being a measure of how useful a particular beam orientation is with respect to treating the target and sparing normal structures. The basic idea then is to treat this function as the required delivery quanta at each of the possible angles and to approximate it with delivery sectors. A delivery sector is a rectangle that can be fit under the angular cost function curve. Long sectors are delivered by arcs and short sectors are delivered with fixed fields. It was shown in a planning example that the hybrid IMRT technique gives a better dose distribution than either fixed field IMRT or IMAT.

IMAT, the multiple-arc IMRT technique invented by Yu (1995), usually operates by first computing field modulations using an inverse-planning system and subsequently and separately chopping up these modulations into arc patterns for an MLC (figure 3.22). There are as many arcs as there are number of intensity levels in the plan. A lay-scientist account of the work at the University of Maryland on IMAT appeared in *Wavelength* (2000c) emphasizing the simplicity of the technique, the requirement for quality assurance (QA) and that the method is useful when it is desirable to spread out the unwanted normal-tissue integrated dose. Wong *et al* (2002a) have presented an intriguing extension of the concept.

Their idea was to create field patterns for multiple-arc therapy in a single planning step.

Their concept was to create a sequence of arcs with each arc having a different planning goal. They also specified three levels of complexity for the new technique. They rooted their notions in the concepts of arc therapy and (the italian) bar-blocking therapy. All three strategies presented have one common property, namely that the first arc creates MLC-shaped field outlines which entirely span the PTV. The differences between the three strategies then centre on how they cope with the organ-at-risk (OAR) sparing and in the balance of dose to PTV and OAR.

Strategy 1 achieves the following:

(i)   One arc uses the MLC to conform to the beam's-eye view of the PTV in each gantry orientation,
(ii) One additional arc for each OAR then uses the MLC to conform to the beam's-eye view of the PTV but shielding the OAR with the bank of leaves that covers the least PTV view.
The total number of arcs is one for the PTV and one for each OAR.

Strategy 2 instead achieves the following:

(i)   One arc uses the MLC to conform to the beam's-eye view of the PTV in each gantry orientation,
(ii) Two additional arcs for each OAR use the MLC to create an 'island shield'. The arcs use the MLC to conform to the PTV in each beam's-eye view of the arcs and shield the OARs in the beam's-eye views with opposing banks of leaves for the two arcs.
The total number of arcs is one for the PTV and two for each OAR.

Strategy 3 instead achieves the following:

(i)   one PTV arc as before,
(ii) two additional arcs for each OAR as in Strategy 2 and
(iii) Two additional arcs per OAR that use the MLC to compensate dose non-uniformity in the PTV whilst shielding OARs. The extra arcs shield the critical organ as before but irradiate the portion of the PTV only that wraps around the critical organ.
The total number of arcs is one for the PTV and four for each OAR.

The techniques are operated in practice by weighting the contributions differentially for each arc and selecting the weights that lead to an acceptable plan. Wong *et al* (2002a) show examples of the MLC configurations and the resulting plans. The indications are that the modified IMAT technique has great flexibility and can provide a better therapeutic ratio compared with conventional IMAT or conformal therapy.

Chen and Wong (2004) have developed an MLC leaf optimization scheme for IMAT. This work started from the premise of two arcs, one conforming to a PTV and the other to the PTV avoiding the OAR (called simplified IMAT or SIMAT). Then, by dividing arcs up into 10° gantry angle increments, the optimization variables become the MLC leaf positions with user-specified leaf-movement constraints. Examples are shown of the improvement consequence on this planning technique.

Shepard *et al* (2002a, 2003a) have developed a new inverse-planning technique for IMAT that begins with a specification of the number of arcs and incorporates the MLC and gantry constraints into the optimization. For this reason, there is no loss consequent on the more traditional method of resequencing inverse-planned modulated fields for IMAT. All constraints are observed on the fly as the calculation proceeds.

Shepard *et al* (2003b) have put on the web a 'toolbox' comprising data for pencil-beam irradiation of a uniform cylinder together with tools for constructing structures and beam's-eye views of structures. The elemental beams are 1.0 cm × 0.5 cm in size in a 20 cm × 20 cm area with associated depth–dose data stored in a 125 × 125 × 31 cube. Data can be rotated and combined to field shapes. It is argued that this permits the investigation of different optimization algorithms.

3.7.1 IMAT in clinical use

Yu *et al* (2000, 2002) have described their clinical implementation of IMAT. This has been implemented at the University of Maryland School of Medicine in Baltimore and by May 2001, 50 patients had completed their treatments with the IMAT technique. The distribution of cases was 22 head-and-neck cases, 19 central nervous system (CNS) cases and nine prostate cases. As far as planning for IMAT was concerned, forward planning with a commercial system RENDERPLAN 3D was compared to inverse planning using CORVUS. Arcs were approximated as multiple shaped fields spaced every 5–10° around the patient. The number and ranges of the arcs were then chosen manually. It was found that between two to five arcs were needed to achieve highly conformal distributions and that the dose homogeneity to the PTV was high with IMAT. It was noted that the CORVUS system has an inherent tendency to generate highly modulated fields even when there are a very large number of fields, since there is nothing built in to preclude this. This does not matter for the delivery of CORVUS plans using the NOMOS MIMiC but inherently seems counterintuitive. In theory, the more beams that are in use, the less the modulation needs to be required for each beam and Yu *et al* argue that, for this reason, forward planning for IMAT might be preferable to inverse planning for IMAT. They also compared the two treatment-planning techniques described with a commercial forward planning of 3D conformal treatment plans not using intensity modulation. Not surprisingly, they showed that the IMAT technique generates more conformal plans than the conformal technique with just geometrical field shaping. They argued that, provided the modulation does not change greatly from angle to angle, the field shapes of the MLC also do not change much from angle to angle and so the limiting feature is not the maximum leaf speed but the maximum gantry rotation speed and the output of the accelerator. For 32 patients, comparisons were made between measurements and predicted doses, using phantom measurements and these were all less than 1% to the dose at isocentre. Care had to be taken to use a dual-source model with consideration of leaf thickness and shape to calculate the head-scatter factors for irregularly shaped fields. One feature that was emphasized was that it was not a good idea to perform a full inverse plan and then to chop it up for IMAT since the degree of modulation would generally lead to a large number of field components and a large change of fields shape between gantry angles. In summary, it was better to tailor the inverse planning to the IMAT technique.

Yu *et al* (2002) repeated the many advantages of IMAT. It can be implemented on existing linacs equipped with an MLC. IMAT also takes only 10–15 min whereas five to seven gantry-fixed-angle IMRT takes 20–30 min. Because it does not collimate the beam into a slit, most of the target is in the field all of the time, thus maintaining a high efficiency. No additional patient-transport mechanisms are required to index the patient from slice to slice and therefore there are no beam-abutment problems. They restated that the different MLC constraints have to be built into the inverse-planning process. In summary, IMAT has been successfully implemented using Elekta accelerators.

Li *et al* (2000b) have compared IMAT with fixed-gantry IMRT. It was found that the use of seven fixed-gantry fields with ten intensity levels generated worse results than IMAT for most of the cases studied even when only three intensity levels were used in IMAT. The IMAT inverse planning was performed by approximating the motion to fixed-gantry angles with increments of 5°. The results should be evaluated in the context of studies by Rowbottom *et al* (2001).

Ma *et al* (2000d) have shown that IMRT of the prostate is possible using IMAT with just two conformal arcs of 120° and 140°. The IMAT plan was computed using the RENDERPLAN treatment-planning system and the treatment was compared with a seven-field IMRT treatment generated using the CORVUS commercial inverse-planning system. Not surprisingly, IMAT and IMRT were shown to be superior to conventional treatments but the IMAT treatment was shown to give less rectal dose than the full IMRT treatment. IMAT led to 30% of the rectal volume receiving over 50 Gy and 16% receiving over 58 Gy whereas IMRT led to 33% receiving over 50 Gy and 14% receiving over 58 Gy. IMAT was also simpler to deliver.

Cotrutz *et al* (2000) reported on a related form of IMAT. Wu *et al* (2001c) have developed a new leaf-sequencing algorithm for IMAT based on the use of graph theory. Iori *et al* (2003) have introduced IMAT in Reggio Emilia when some dMLC IMRT techniques would be too complicated or lead to too great a workload.

Original IMBs planned at fixed gantry angles, in this example separated by 3^0

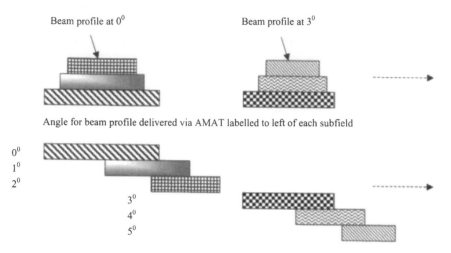

Beam profile at 0^0 Beam profile at 3^0

Angle for beam profile delivered via AMAT labelled to left of each subfield

0^0
1^0
2^0

3^0
4^0
5^0

Figure 3.23. This illustrates the principle of AMAT. Imagine the IMRT plan has been made with IMBs at a series of fixed gantry angles (for illustration here they are at 3° separation in angle. Imagine that each beam has three intensity levels. The logic of AMAT is that the three component intensities and field shapes are delivered at three separate angles spaced by 1° with virtually no change in the resulting dose distribution. In practice, the individual IMBs would be planned at a wider spacing $\Delta S°$ with more (say N) fluence levels and the individual fluence components would be delivered at intervals of $\Delta S/N^0$.

3.7.2   IMAT modified to aperture-modulated-arc therapy (AMAT)

Crooks *et al* (2003) have introduced a radically new method of IMRT delivery known as aperture-modulated-arc therapy (AMAT). The acronym is deliberate to draw an analogy with intensity-modulated arc therapy (IMAT). The logic of the proposal is as follows (figure 3.23). Typically, IMRT using an MLC is delivered at a set of discrete angular orientations with the MLC leaves taking up a series of positions at each orientation. The new proposal is based on the logic that says that the dose distribution from a radiation beam will change very little if that beam is moved by a small angular displacement. The authors then propose that one might imagine that, for an IMB with say ten component apertures defined by an MLC, this could be represented as ten separate beams coming from angles very slightly separated in asimuth with just one aperture at each of those angular orientations. By extending this logic, a number of beams, each with a number of apertures, is entirely reduced to a single aperture at a much larger set of angular orientations. Unlike IMAT which uses several arcs, AMAT needs only one. The logic continues that, by delivering the radiation in arc therapy mode with a constant output rate from the accelerator, all that then has to be done is to modify the time for which a particular aperture is exposed (and, in practice, this means modifying the angular orientation over which a particular aperture is exposed) and then, by delivering the set of apertures, so modified, using a constant output from the accelerator, the overall delivered dose will be the same as that which would have been delivered had a series of static components been delivered each from a number of fixed beam orientations.

Crooks *et al* (2003) have written code to convert NOMOS CORVUS distributions into such AMAT sequences. This was done for a number of planning situations and shown, by delivering both conventional and AMAT deliveries to film and mosfets, that the dose distributions are not vastly different. However, they consider a 'close agreement' to be of the order 7–10% and it is debatable whether others would agree. Also there is very little change to the overall number of MUs used nor to the time of delivery. However, the technique has a certain elegance. Also it must be commented that the technique is limited by the ability of the MLC leaves to change their shape rapidly during small arc rotations. All these aspects are being investigated by the authors in further work.

3.8   New ideas related to the dMLC technique

MacKenzie and Robinson (2002) have presented a technique to deliver 24 fields each modulated using the sliding-window dMLC technique and spaced at 15° intervals to simulate the effects of tomotherapy. This is essentially an extension of the small-number-of-fields dMLC techniques to a larger number of fields. They claim that this will overcome some of the drawbacks of tomotherapy, for example the requirement to have specialized delivery equipment. An iterative filtered-backprojection algorithm was used to create IMRT fields which were interpreted for the dMLC delivery technique. These were then delivered to a homogeneous cylindrical phantom, initially constraining the problem to deliver a single axial slice using a large number of projections. The delivery was performed with a Varian Clinac 2300 CD equipped with a 52-leaf MLC and verification film was placed along the transverse-axial plane coincident with the rotation plane of the gantry inside a series of slices of the cylindrical phantom. By analysing the experimental measurements in Fourier space and correlating with the predictions, it was found that, for a variety of distributions, the measurements agreed with the predictions to within about 3% which they argue is roughly the uncertainty attached to the measurement technique. They then repeated the measurements delivering the same modulation to a variety of slices by synchronizing all the leaves.

Figure 3.24. This shows the basic building blocks of an Ellis compensator with some arranged to make a 1D modulated profile by the compensation technique. The area of the base of the blocks determines the spatial resolution and the height of the blocks determines the intensity resolution.

3.9   Compensators and comparisons of compensator and MLC-based IMRT

3.9.1   Do we need the MLC for IMRT?

A traditional way to create a modulated fluence over a shaped area has been the use of a compensator. The first compensator was the so-called 'Ellis compensator', a modern example of which is shown in figure 3.24. This comprised Lego^{TM}-like blocks of varying sizes and attenuating thicknesses. Another way to make a compensator is to mill a metal block to the required shape. Another way is to cut sheets of lead of different shapes and glue them together to make a pattern (figure 3.25). A fourth way is to use a hot-wire cutting device to cut a mould of the required shape out of tough Styrofoam and then to fill this with either lead or tungsten balls or a hot liquid melt which cools to the required shape.

Sherouse (2002) has written a perceptive letter extolling the virtues of the metal physical compensator compared with the use of an MLC for IMRT. A few sentences are worth quoting: 'There is a widely held misconception that IMRT and MLCs are the same thing.' 'IMRT without an MLC is like a fish without a bicycle' (think about it!). 'MLCs are not inherently an enabling technology for IMRT but some form of computer optimization surely is.' 'Inexplicably the

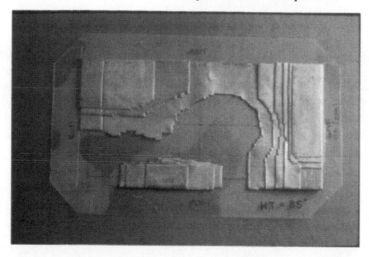

Figure 3.25. This shows how a compensator was constructed from thin lead sheets glued together in order to modulate the intensity of a tangential beam for improving the homogeneity of dose in breast radiation therapy. (Courtesy of the Breast Technology Group, Royal Marsden NHS Foundation Trust.)

MLC enthusiasts seem to always conclude that the moving patient is the entire problem, rather than question the wisdom of their choice to unnecessarily use time-varying fluence patterns.' He argues that the use of MLCs generate massive re-education and training requirements and that the literature is filling rapidly with reports on subtle and gross problems with the use of MLCs for IMRT. Finally, in the US, IMRT with a compensator is billed less than with an MLC, thus driving centres towards the latter—this is surely wrong. In short, Sherouse was, and still is, a fan of the compensator. Chang *et al* (2004a) emphasized the advantages of collimators in terms of static delivery with fine-scale spatial resolution and the disadvantage as a limited dynamic range.

Welch and Harlow (2001) controversially have stated that a department with just one MLC is likely to have excessive stress due to the uncertainties consequent on downtime for this MLC. They argue that departments should have at least two MLCs and, if the funding for only one is forthcoming, it might be better to improve the block-cutting and compensating facilities instead of purchasing a single MLC. This paper somewhat kicks against the trend for the increased use of technology in radiation therapy.

3.9.2 Use of compensators for IMRT

It is well known that the use of IMRT improves the homogeneity of dose to the breast particularly with large-breasted women (see section 5.7). Compensators provide a particular delivery technique. The use of CT is largely avoided in the

UK because of the difficulty of access to a CT scanner for this large patient group. Also, following work at Institute of Cancer Research/Royal Marsden Hospital, compensators can be designed from portal imaging data. Compensators can also be designed from measurements of the optical contour of the breast with inferred lung positions. Wilks and Bliss (2002) have taken this a stage further and proposed that a library of compensators can be reused for subsequent patients. They planned 94 patients between June 1999 and May 2001. The first 28 led to the production of 28 compensators used for treating those patients. The subsequent 66 patients were treated with the most suitable compensator selected from the library.

Wilks and Bliss (2002) and Bliss and Wilks (2002) have produced compensators by taking seven outlines using OSIRIS (QUADOS—Sandhurst). A check film determined the lung thickness. Each compensator was built from sheets of lead of thickness 0.5 mm. The technique for compensator selection is as follows. A compensator could be individually fabricated. Alternatively, a compensator could be manually selected from the library. More adventurously, a compensator could be selected from the library by calculating the dose distributions for using all field-size-appropriate compensators and selecting that which produces the best dose-volume histogram (DVH) for the breast. The compensator selected is then used to fully plan and evaluate the dose distribution to the individual patient with inspection of isodose plots. The goal is not to produce the best plan but an 'acceptable' plan. On average, it was shown that the use of the library compensator did not statistically degrade dose compared with the use of an individually fabricated compensator.

Xu (2002) and Xu *et al* (2002) have invented a mechanism to generate an automatic compensator for IMRT that is reusable and have characterized its performance (figure 3.26). This is something genuinely new. A deformable attenuating material comprising 50% tungsten powder, 35% silicone rubber and 15% paraffin was constructed. The linear attenuation coefficient of the material was measured to be 0.41 cm^{-1} and the maximum thickness of the compensator was 10.2 cm which allows a transmission of 1.6% for 6 MV x-ray beams. A set of 16×16 pistons were made to stamp down into this material to form a compensator of previously calculated variable thickness. Each piston has a cross section of 6.37 mm $\times 6.37$ mm which projects to 1 cm $\times$ 1 cm at the isocentre. The size of the pistons and location of the compensator were such as to create a 16×16 matrix of 256 cm^2 bixels at the isocentre. The procedure was to stamp the material out of the beam (but in the accelerator head) and then actuate the compensator into the beam. Following use, the compensator was restamped to a uniform thickness prior to the whole process repeating for another stage.

A limitation of the prototype was the non-divergent nature of the beam leading to a penumbra increasing in width with distance from the central axis and ranging between 5.5–10 mm. The thickness was determined for each bixel using a simple logarithm of the ratio of input-to-output intensity and a division by the linear attenuation coefficient of 0.41 cm^{-1} at 6 MV. The percentage depth–dose

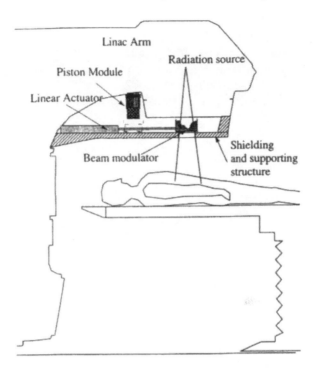

Figure 3.26. The automatic physical compensator for IMRT mounted in the head of a linear accelerator. The piston module stamps on the deformable attenuation material and generates a physical compensator. This is then pushed into the line-of-sight of the beam by a linear actuator. (From Xu 2002.)

curves determined using slabs of the material of thickness up to 4.4 cm showed beam hardening at depth up to 4%.

Compensators were created and IMRT delivered to film. Calculations of the delivered dose were made assuming a Gaussian spot and by single ray tracing through the metal, scatter ignored. Comparisons of profiles between measurements and calculations agreed to about 3%. Extracting data from each bixel and plotting against thickness of compensator showed an independent measure of μ agreed to $\simeq$ 3% with that above.

This was preliminary work with a prototype and it may be imagined how more detailed modelling of beam hardening and scatter could improve the fit. The next stage is automation. The authors claim IMRT can be performed as easily with the automatically reusable compensator as by the better-known MLC-based methods. Xu *et al* (2003) report on a 32 by 32 model with improved spatial resolution.

Yoda and Aoki (2003) have also developed a very neat technique for producing a multiportal compensator system for IMRT delivery (figures 3.27–

Figure 3.27. Developed compensator mount having a diameter of 55 cm and a height of 10.87 cm, which is placed on top of the linac head. The mount has six circular holes having a diameter of 10 cm, so that six cylindrical compensator enclosures can be positioned. The mount is rotated by a built-in stepping motor (which is seen behind the mount) and the motor is controlled by a PC installed in an operating room next to the treatment room. (From Yoda and Aoki 2003.)

3.29). Their system comprises a carousel which fits below the head of the accelerator and contains six circular holes each of a diameter 10 cm, into which preformed compensators can be inserted. The compensator carousel is rotated by a motor and this is done automatically from outside the treatment room. The estimated time for treated using five fields is just a couple of minutes for an isocentre dose of 2 Gy with a 2 Gy min^{-1} beam intensity. The rationale for this development was that it is reported that currently available MLC-based IMRT techniques have several clinical disadvantages and the multiportal compensator system overcomes these. It has a shorter treatment time; it has less radiation leakage; and there are no difficulties with MU calculations. Aoki and Yoda (2003) have described apparatus using a 10×10 piston arrangement to stamp out the compensator shape prior to pouring heavy alloy granules and vacuum forming the compensator. Note this is distinct from the development of Xu (2002) which uses piston stamping in the treatment room.

The individual compensators are made as follows. Firstly a set of 10×10 rods is forced into an appropriate shape to make the curved 2D surface that will eventually become the final surface for the compensator. The rod heights are determined from a 2D intensity map from a treatment-planning system. The resolution is 0.5 mm. Each rod element has a square section of

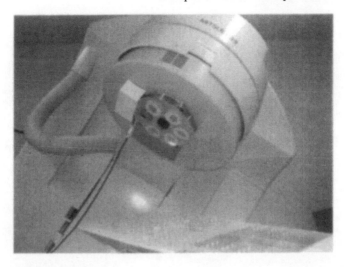

Figure 3.28. Compensator system with Mitsubishi C-arm linac. Only five compensators are installed so that the empty hole can be used for rotation angle calibration. (From Yoda and Aoki 2003.)

5.97 mm × 5.97 mm so that the spatial resolution of the intensity modulation is 10 mm × 10 mm on the isocentre plane. After the rod heights are set-up, a pre-heated thermoplastic sheet is placed over the rod elements and vacuum forming is performed. Then heavy alloy granules with a density of 10.1 g cm^{-3} are poured into the vacuum-formed shape. These are tightly packed in by screwing down a pressure plate and then plastic tape is applied to hold all the granules in place. Once prepared, all five compensators are placed in the carousel and the whole process takes about 20 min. The authors have delivered an IMB using 6 MV x-rays from a Mitsubishi C-arm linac to a film and shown that the required dose distribution roughly matches that measured. The linear attenuation coefficient of the material was 0.0515 mm^{-1}. This represents a very nice technique for compensator-based IMRT.

McCurdy *et al* (2002) have developed an automated compensator-exchanging device which enables compensators to be used for IMRT, the new compensator being selected at the same time as the gantry angle changes.

Warrington *et al* (2002) have developed a carousel suitable for housing up to four conformal lead-alloy blocks of maximum field dimension 10 cm for selected conformal or IMRT prescriptions delivered with an MDS Nordion Elite 80 tele^{60}Co machine (figure 3.30). This has a 2-cm-diameter source and an 80 cm source-to-isocentre distance. Studies showed that the system could deliver beams with a satisfactory accuracy of between 1 and 2 mm at isocentre.

Peltola (2003) has designed compensators using tin-granulate and low-melting-point heavy metal alloy, using the CADPLAN HELIOS system and

Figure 3.29.   View of the 10×10 rod elements after adjusting to rod height according to the desired intensity profile. Each rod element has a square cross section of 5.97 mm $\times$ 5.97 mm so that the spatial resolution of the intensity modulation is 10 mm $\times$ 10 mm on the isocentre plane. The rod has slightly cutout corners so that vacuum forming can be performed by drawing air downward through the cutouts. (From Yoda and Aoki 2003.)

(a)

(b)

Figure 3.30. (a) The carousel equipment for delivery of radiation of different geometrical shape and fluence using a ^{60}Co machine. In the form shown here, only the geometrical shape can be varied but the inserts could instead be compensators. (b) The components of the apparatus shown dismantled. The two plates on the right are a jig for pouring the cerrobend block with the result shown to the left. (Courtesy of J Warrington, Royal Marsden NHS Foundation Trust.)

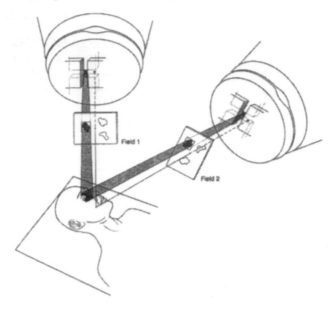

Figure 3.31. Delivery of multiple IMRT treatment fields with one multi-field modulator. The gantry isocentre (small filled circle) is placed outside the target volume (paranasal sinus cancer). The gantry and collimator rotate to their first positions as the jaws collimate the first field. Additional collimation can be provided by the MLC leaves, blocks or full thickness of the modulator. The beam is switched on and the first field is treated. Then the gantry and collimator rotate to their second positions and so on to complete the treatment. (Reprinted from O'Daniel *et al* 2004 with copyright permission from Elsevier.)

showed that, with suitable correction factors, beam-hardening effects can be overcome.

Gurgoze and Rogers (2002) have described what they call MLC replicators (MLCRs). By this they mean compensators. They show that the multileaf replicator plan and the MLC plan agree to within 3% and that the multileaf replicator can deliver IMRT in 7 min compared to 17 min for the MLC.

Papatheodorou *et al* (2000b) have developed a virtual compensation technique using the dMLC technique. The primary-scatter separation dose model of the treatment-planning system was extended to account for the intensity modulation generated by the dMLC.

O'Daniel *et al* (2004) have also constructed a rotating multifield compensator. For each field, the compensator rotates to have exposed a different pattern of metal thickness. It was shown that use of this device gave conformal distributions for treating paranasal sinus cancer that were as good as the use of an MLC (figure 3.31).

3.9.3   Comparison of compensator and MLC-based IMRT

Chang *et al* (2000b, 2002) have compared the dosimetric outcome of performing IMRT with either a compensator or the MSF-MLC technique. They considered two tumour situations and used the inverse-planning system PLUNC to create modulation maps, having first specified the number of fields. They then, for the MSF-MLC technique, stratified these maps into a discrete number of fluence intervals. They varied this number and studied the effects. The compensators were constructed from tin-granulate poured into moulds whilst the stratified 'skyscraper-like' maps were fed to the Siemens IMFAST sequencer to determine the field components. Dose distributions were computed and compared for the two techniques.

They introduced an interesting index, the Uniform Dose Volume (UDV), being the volume of the PTV raised to 95% or more dose. They showed that this UDV index was largest (of course) for the unstratified and spatially continuous compensator technique. The UDV for the MSF-MLC method increased as the number of modulation intervals increased. From the behaviour of this index, they determined that five levels were sufficient for the cases studied. They also showed that the number of field segments generated by IMFAST increased in proportion to the number of segmentation levels and was typically between one and two segments per level per field. Interestingly, it was found that the biggest differences between the two classes of IMRT technique were due to the finite spatial resolution of the MSF-MLC technique and not to the number of fluence levels selected. This was particularly so for the protection of the OARs which they quantified by studying the integral DVHs. As well as using IMFAST, they computed segments independently using PLUNC and obtained the same number of segments but with slightly modified MUs due to the two-Gaussian model used by PLUNC for head-scatter calculations compared with a one-component model in IMFAST.

In conclusion, Chang *et al* (2000b) found that the compensator could be regarded as a gold-standard IMRT treatment against which the 'degradation' introduced by the MSF-MLC techniques could be measured. Chang *et al* (2003) continued this theme pointing out the superior fluence and spatial resolutions of compensators for 340 patients. However, they also point out the limitations in the dynamic range of fluence using some tin-based compensators.

West and Jones (2000) have also compared the dose distributions delivered from physical compensators and multileaf compensation with the MSF technique. Not surprisingly, it was found that the MSF technique degraded as the number of fields applied was reduced.

Grigereit *et al* (2000) have developed compensators for IMRT and showed that delivered dose distributions were within 1% of measured dose distributions.

Jursinic *et al* (2002) have delivered IMRT using five fields with compensators milled from brass with a spatial resolution of 2 mm and shown that the delivered IMRT plans are better than those using MLC modulation.

Kermode and Lawrence (2001) have developed a practical system to use the HELAX treatment-planning system to design simple 2D compensators to account for changes in patient contour or specific OAR shielding. The method was verified by comparing the predicted and measured beam profiles. Kermode *et al* (2001a) gave data on the improved dose distributions in the head-and-neck regions due to the use of compensators.

Schefter *et al* (2002) have shown that dose distributions created with a compensator and with MSFs were very similar when treating cervix cancer using five fields. Increasing the number of fields did not improve the treatment. They argued that the increased simplicity of using compensators gives this technique an advantage.

3.10   Optimum width of leaves for an MLC

Oelfke (2001) has described the features of MLCs that are required for IMRT. Bortfeld *et al* (2000) have applied sampling theory to the problem of determining the optimum leaf width of an MLC for radiation therapy. The work of Bortfeld *et al* (2000) is predicated on ensuring that the sampling is optimized in relation to the physical width of the point-spread function of the radiation. They showed that this can be derived from measurements of the 20–80% penumbra at the edge of a step-function fluence and it turns out that the standard deviation is equal to the width between the 20% and 80% dose contours divided by 1.7. From this, it could be concluded that, for a 6 MV beam in soft tissue, the optimum sampling distance is 1.5–2 mm. The most obvious next step is to make the physical size of the MLC leaves equal to this sampling width and examples shown later in the paper for both IMRT and for the delivery of fixed static fields show that, when this is done, the effects of the collimator size are not observed in the dose distributions. They pointed out that it is actually an adverse step to try to reduce the width of the collimator below this as this would, due to sampling considerations, lead to aliasing in the distribution.

They drew a distinction between the sampling distance and the leaf width and showed that, provided the sampling distance is maintained as before, the leaf width can double and the same results can be achieved by a double delivery of radiation with the collimator moved by half a leaf width between the two components of the irradiation. This is the so-called shift technique. Using an IMRT example, they showed that this leads to a dose distribution which is much the same as that delivered by a single delivery with leaves of the original small 2 mm leaf width. However, examples with the leaves considerably increased in size demonstrate that the dose distribution deteriorates. Finally, they commented that sampling theory can be used to find adequate beam element sizes and voxel sizes for dose calculation and inverse-planning algorithms.

Nill *et al* (2002) have compared IMRT dose distributions delivered with MLCs of different leaf widths. They studied two particular so-called internal

MLCs and two particular external MLCs and the overall conclusion was that the smaller the leaf size the better the homogeniety of dose to the target. However, reducing the leaf width to 2.75 mm, whilst it resulted in further enhancement of the target coverage, did not change the dose to OARs. They are still questioning whether this is of any clinical significance.

Wang *et al* (2003b) have compared the use of two MLCs of leaf widths 4 and 10 mm to sequence a CORVUS plan and showed improved OAR protection consequent on the use of the former.

Drzymala *et al* (2001) have studied the geometric aspects of IMRT planning. They particularly looked at the variation of dose conformality with changing the MLC leaf width, collimator angle and couch rotation. They concluded that it is always best to use narrower leaf widths where possible and that when treating with multiple coplanar beams it is desirable to avoid parallel-opposed beams.

3.11   MicroMLCs for IMRT

There are techniques for using a conventional MLC with leaf widths projecting to 1 cm at isocentre with shifts and multiple deliveries. These simulate the use of a microMLC and are here referred to as virtual microMLCs. Planning the required modulations has been described by Hou *et al* (2004) who also showed improved smoothing properties of the developed IMBs (see section 6.5.9.) Reports of using three manufacturers' MLCs follow and after that specific microMLCs will be reviewed. There is no clear distinction between a microMLC and a miniMLC so the former name will be used here although some companies prefer the latter.

3.11.1   Siemens (virtual) microMLC

Siochi (2000a) has developed an algorithm for minimizing the number of segments and junctions in virtual microMLC IMRT. This is the technique in which a 10 mm × 5 mm bixel intensity map is combined with that of a 5 mm × 10 mm bixel intensity map delivered at orthogonal collimator settings. The aim was to minimize the number of segments and junctions for dosimetric purposes.

Siochi (2000c) has presented in great detail the way in which a conventional MLC projecting to a leaf size of 10 mm at isocentre can be used to deliver an intensity modulation over a grid of 5 mm × 5 mm. The technique relies on delivering two intensity maps at collimator settings that are 90° apart. The superposition of these two component maps in which the leaves can be discretized to 5 mm along their length in each direction will yield the required outcome. The solution is non-unique and most of the study concerned the specific choices that need to be made to do an efficient decomposition.

Mitra *et al* (2001b) have studied the Siemens HD270 MLC which is a virtual microMLC that smooths the defined field edges by a series of coordinated incremental couch movements. The movements can be in increments of 5, 3 or 2 mm over a 27 × 27 cm treatment field size corresponding to 1, 2 or 4 shifts.

The study showed that, from looking at dose at depth, negligible improvement in target coverage resulted from increasing the leaf resolution to better than 3 mm.

Cheng *et al* (2000) have studied the dosimetry of a virtual mini multileaf and its clinical application. The virtual mini multileaf in question was the Siemens HD270 which combines the use of a large-leaf MLC with table translation perpendicular to the leaf plane of either 2, 3 or 5 mm. Film dosimetry showed that the 80–20% penumbra width for a 45° block and a circular block were reduced using the HD270 compared with the use of the wide-leaf MLC. At 2 mm resolution, the scalloped isodoses virtually disappeared, although the penumbra was still somewhat larger than for a cerrobend block. The HD270 was also shown to improve the dose-volume histogram of the rectum compared with the use of a broad-leafed MLC.

Twyman and Thomas (2002) have also described the Siemens HD 270 which uses three couch shifts in conjunction with the use of a 1-cm-wide-leaf MLC to simulate the action of an MLC with 2-mm-wide leaves.

3.11.2   Varian (virtual) MLC

Greer *et al* (2003) demonstrated experimentally that an improvement in fluence resolution perpendicular to the direction of leaf movement could be obtained by combining the fluence patterns from two Varian MLC deliveries shifted by half a leaf width with respect to each other. This reduces the leaf sampling distance and improves dose resolution. They made film measurements for several test patterns: a bar pattern, a clinical field from a parotid plan and a doughnut of fluence. The results were in line with the conclusions of Bortfeld *et al* (2000). Two different sampling schemes were investigated.

3.11.3   Elekta (virtual) microMLC and microMLC

A simplified account of the work at the Royal Free Hospital using an Elekta MLC appeared in *Wavelength* (2000b) emphasizing that pseudo high-resoution MLCs can be fabricated by the isocentre shift method. This has been in use since 1994 for nasopharynx cases and for all routine prostate cases. A side effect of the phased-field technique (PFT) is the reduction of interleaf leakage by about 50%.

The new Elekta MLC is a high-resolution MLC that will be IMRT compatible (figure 3.32). It will be optically verifiable. It has a 4 mm leaf pitch and is fully integrated into the treatment of the Elekta accelerator head. It can generate a 16 cm × 22 cm field. It has no back-up jaws because the leaves are set in a solid millstone. It will allow interdigitation. Mosleh-Shirazi (2002) has characterized the beam properties of this prototype Elekta high-resolution MLC. The along-the-leaf and perpendicular-to-the-leaf 20–80% penumbral widths were 4 and 3 mm respectively. The MLC replaces the lower jaws and the whole field size is modulatable for IMRT.

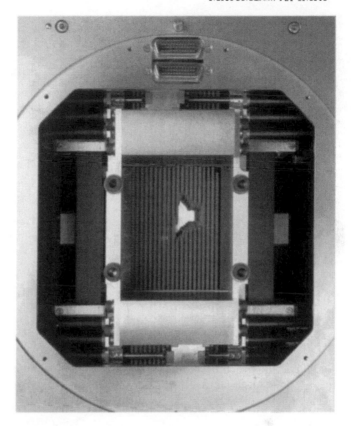

Figure 3.32. The Elekta Beam Modulator MLC. (Courtesy of Neil Harvey, Elekta Oncology Systems, UK.)

3.11.4 Radionics microMLC

Hoban *et al* (2001) have used a Radionics microMLC for delivery of IMRT (figure 3.33). By March 2001, five patients had been treated. The microMLC was attached to a Siemens Primus linear accelerator. The Radionics X-plan treatment-planning system with the integrated planning module from KONRAD was used for inverse planning. Field sequences and intensity maps were then delivered to a phantom for comparison of planned and delivered dose distributions. TLD measurements were also made. Hoban (2000) has developed a model for dosimetry for the micro-dMLC technique. This uses a Gaussian source, head-scatter function and a pencil-beam kernel convolved with the opening density map with the addition of radiation leakage. This calculation matches measurements quite well. Hoban (2002) has used a microMLC to deliver a simultaneous IMRT boost in the prostate.

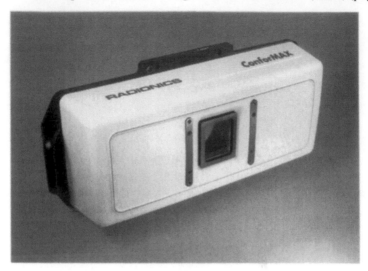

Figure 3.33.   The Radionics microMLC. (Courtesy of Vicky Aerts at the Radionics Company.)

Parker *et al* (2000) compared dose distributions from geometric conformation with IMRT for intracranial lesions. They also made use of the Radionics microMLC and found that IMRT was required when sparing nearby normal tissue is a high priority. Otherwise geometric conformal radiosurgery was sufficient.

Solberg *et al* (2003) have used a microMLC to treat prostate cancer by IMRT. Wang *et al* (2003b, c) showed that the rectal-and-bladder sparing was improved by using the Radionics microMLC compared with the use of a conventional Siemens MLC. Wang *et al* (2003b–d) have also demonstrated the dosimetric advantage of the microMLC in treating prostate cancer with IMRT.

3.11.5   BrainLAB microMLC

Barker *et al* (2001) have compared the dosimetric characteristics of a standard Varian 52-leaf MLC and a BrainLAB M3 microMLC for square, rectangular and irregular fields at 6 MV using a Varian Clinac 2300 CD linear accelerator (figure 3.34). Conventional percentage-depth–dose data were unchanged for most situations. The microMLC provided a sharper penumbra. The relative dose factors for a given MLC or microMLC field size depend on the jaw settings as well as on the tertiary field size.

Benedict *et al* (2001) compared IMRT delivered with the BrainLAB M3 microMLC with stereotatic radiosurgery performed using 15 fixed non-coplanar fields and arc-based stereotatic radiosurgery with 4–5 arcs per isocentre. The

Figure 3.34. The BrainLAB microMLC. (Courtesy of the BrainLAB Company.)

IMRT technique for three of the four patients studied significantly reduced the ratio of treatment volume to target volume.

Yenice *et al* (2001) have made plans for static and IMB delivery to be delivered with a BrainLAB M3 microMLC. It was found that the use of the microMLC consistently improved upon static CFRT.

Agazaryan and Solberg (2003) have written a sequencer for both static MLC and dMLC versions of IMRT for the BrainLAB MLC on a Novalis accelerator. They incorporated leaf transmission and a parameter to vary the number of leaf patterns.

3.11.6 DKFZ-originating microMLCs

3.11.6.1 The MRC systems GMBH/Siemens 'Moduleaf'

The Deutches Krebsforschungszentrum (DKFZ), Heidelberg, have designed and constructed a new MLC for both stereotactic radiotherapy and IMRT. It has leaves which project to 2.5 mm at the isocentre and so is intermediate between the large-scale MLCs and the microMLCs. It is marketed by MRC Systems GMBH/Siemens under the name 'Moduleaf' (figure 3.35). There are 40 leaf pairs. Each leaf is 7 cm high and the maximum field size at isocentre is 12 cm by 10 cm.

The leaves are driven by motors which are arranged in banks to drive alternate leaves from the top or the bottom of the carriage. The leaves carry position-sensitive potentiometers. They can overtravel by 5.5 cm (see table 3.1). Hartmann and Föhlisch (2002) reported measurements to characterize the performance.

The leaves are focused in the y-direction by the convergent shape of the leaf cross section although they are not precisely focused to the source in order to remove the need for TG arrangements. The 'focusing' along the leaf (x) is quasifocusing through a three-sided leaf-end pattern. This leads to a mean penumbra of 3.2 mm for the x-profiles and 2.9 mm in the y-profiles. The effective penumbra for a 45° edge is 3.5 mm. The actual leaf positioning is accurate to 0.1 mm. The mean transmission is 1.3%. The small penumbra and accurate positioning implied suitability for stereotactic radiotherapy and for IMRT.

3.11.6.2   New DKFZ microMLC

Pastyr *et al* (2001) have designed a new prototype MLC at DKFZ, Heidelberg. This has physical leaf thickness below 3 mm allowing a leaf resolution of below 5 mm at isocentre. It also has linearly-guided double-focusing leaves to ensure that the edges are always directed towards the focus and yielding a very small position-independent penumbra at all positions of the leaves (Forced Edge Control). There is a simple high-precision linear leaf-guidance system. The MLC was designed for IMRT. It does this by attaching a rotating cog to two racks alongside each other but with slightly different teeth numbers per unit distance. One rack drives the leaf. The other drives a pivoting leaf edge (figure 3.36). Hartmann *et al* (2002b) further described the mid-sized MLC developed at Heidelberg which has a mechanism for automatically focusing the leaf in the direction of leaf motion irrespective of leaf position. The MLC also has extremely flat potentiometers which are used as pairs embedded in each leaf to control and additionally check the leaf position with high accuracy. Tücking *et al* (2003) at Heidelberg have shown that the use of the microMLC impoves dose conformality.

3.11.7   3DLine microMLC

Mapelli *et al* (2001) have described a new microMLC from the 3DLine company (figure 3.37). This has a redesigned leaf shape to reduce the TG effect. Leaf transmission is less than 0.4%; leakage between leaves is less than 0.5%; penumbra is less than 4 mm and the accuracy and dose delivery at the prescription point in dynamic therapy is better than 2%.

3.11.8   Comparison and use of microMLCs

Höver *et al* (2000) have compared three different microMLCs for delivering IMRT. Their results show that not all the collimators could be used for IMRT.

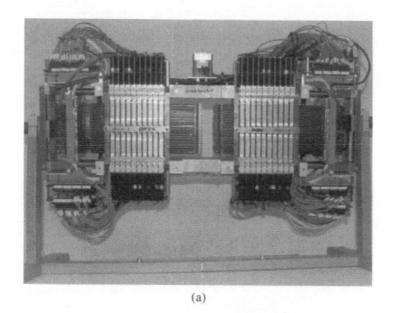

(a)

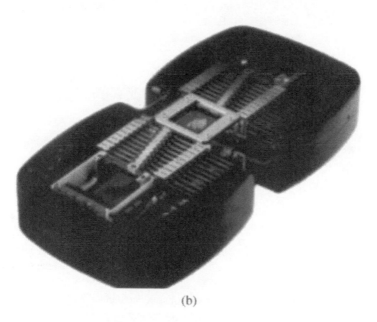

(b)

Figure 3.35. The DKFZ/ MRC-Systems/Siemens 'Moduleaf' MLC (a) with its covers off, (b) with its covers on. (Courtesy of Professor Wolfgang Schlegel; Heidelberg.)

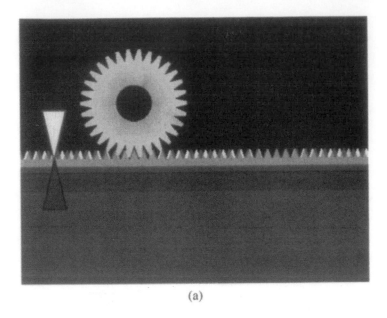

(a)

(b)

Figure 3.36. (a) Showing how one cog can operate simultaneously on two linear gear trains to drive one train a little faster than the other and so make the two move relative to each other. This has the effect of (b) turning the end of the leaf so its face is always tangential to the divergent radiation from the source. (Courtesy of Professor Wolfgang Schlegel; Heidelberg.)

Figure 3.37. 3D Line microMLC. (Courtesy of the 3DLine Company.)

Leakage measurements showed a wide variation up to a factor of four between the MLCs. Hoẅer *et al* (2001) have further reported on the dosimetric characteristics of five add-on MLC systems with small leaf widths. These were from Radionics, BrainLAB, MRC, MRC/DKFZ and DKFZ. Measurements showed that IMRT can best be done with double-focused systems whereas stereotactic treatment may be done with cheaper single-focused or even parallel-sided MLC's.

Gibon *et al* (2000) have presented a method to optimize microMLC irradiation for IMRT. The positions of the fields were optimized using a genetic algorithm and the leaf position and the weighting factors of fields were optimized by a simulated annealing method. The results showed a conformal improvement by using IMRT and/or by increasing the number of fields. However, in many cases, an optimized three-field microMLC technique with IMRT obtained results which were almost equivalent to those obtained with non-optimized five-field techniques. The results depended on the shape of the target.

Yang *et al* (2002) have shown that the use of a microMLC implemented as a tertiary device can improve the high-precision requirements of stereotactic radiotherapy.

3.11.9 Multi-level MLC

Topolnjak *et al* (2002) have described a new collimation system being developed in Utrecht. One aim is to produce a 40 cm × 40 cm field with the same resolution as could be obtained with a microMLC. This will have six banks of a multileaf system in three levels with each paired bank at 60° to the paired bank above and below it. The banks of leaves are of conventional width but this allows the simulation to mimic the microMLC. Because the tongues and grooves of one particular bank are always overshadowed by leaves from a bank above or below, it might be possible to abolish the TG mechanism altogether, thus simplifying the design. Interleaf transmission is always blocked by at least one leaf from another bank. Also there would be no need for collimator rotation. The group have performed a computer design study. Topolnjak *et al* (2003a) has reported on Monte Carlo modelling of the six-bank multileaf system for IMRT and Topolnjak *et al* (2003b, 2004a, b) have described a recursive sequencer for this three-level MLC which is still at the design stage.

Potter *et al* (2003a) have alternatively described a two-level microMLC (Alanya Enterprises Corp, Paris) to improve spatial resolution.

3.12   Increasing the spatial resolution of a conventional MLC

Williams and Cooper (2000) have presented a study of the effect of the use of a tertiary grid slit collimator applied to improving the spatial resolution of a conventional MLC. The collimator comprises a plate with slits of width Δ spaced at intervals of W where W is the width of a conventional MLC leaf. W/Δ sequential irradiations take place with the collimator successively indexed by Δ for each irradiation. The prototype tertiary collimator was made of brass of thickness 8 cm with parallel-walled slits focused to the radiation source. It was mounted in the blocking tray of an Elekta SL25 accelerator.

The advantages of this device over a micro or miniMLC were said to be:

(i)   high spatial resolution can be obtained over a field as wide as can be collimated by the conventional MLC;
(ii)   no seriously increased penumbra;
(iii) no requirement for the pseudo microMLC methods based on the use of a conventional-width MLC.

Limitations included:

(i)   the requirement for extremely precise incremental relocation of the tertiary collimator;
(ii)   a small increased irradiation time;
(iii) degradations due to transmission through the shielded parts of the collimator; and
(iv)   a potential reduction in uniformity due to multiple-subfield superposition.

Test irradiations at 6 MV were made to characterize the elemental radiation profile of a single irradiation. Measurements were made at a depth of d_{max} and at a depth of 10 cm for slits of with 2.5 mm and 5 mm. (For a conventional MLC with $W = 10$ mm use of these tertiary collimators to provide a shaped field would therefore require four and two sequential irradiations respectively.) The following were observed:

(i)   The variability in peak height of irradiation within the slits was greater for the 2.5 mm collimator possibly due to difficulty of accurately machining the slits;

(ii) the lowest intensity reached in the shielded regions was higher for the 5 mm collimator.

The effect of combining multiple irradiations with sequential advances of the collimator was determined by two methods. Firstly, the data for a single irradiation were reused computationally to calculate the effect. Secondly, measurements were made. Both showed that the variation in intensity was within $\pm 3\%$ for the 2.5 mm collimator and somewhat worse for the 5 mm collimator.

When shaping a sinewave-shaped field of varying amplitude and periodicity, the expected field-shaping benefits of the tertiary collimator were observed with little degradation in penumbra (figure 3.38).

The effect of *misregistering* subfields was shown to be about 18% mm^{-1}. Cooper *et al* (2001a) have investigated whether radiation dosimetry parameters change when the grid slit collimator is in use, concluding that no significant differences were identified. The grid factor was found to be 2.95.

Cooper *et al* (2001b, 2002) have further discussed the grid slit collimator. The prototype collimator had a leaf pitch of 10 mm equal to the width of a conventional multileaf. It was found that artefacts could arise due to mispositioning of the slit collimator with respect to the edge of a leaf when this was defining part of a field. These artefacts were of two types. The first arose because of the precision with which the grid could be manufactured, the slit width used and the accuracy of the indexing. A second more important artefact was due to the partial overlap of a slit and the junction between two leaves (which set the field width for that slit). In this situation, the two leaves 'seen' essentially define two different widths leading to narrow fingers of overdose just outside the field or underdose just within the field.

These artefacts were overcome by constructing a second prototype with 5-mm-wide slits on a pitch of 15 mm in such a way that, if one slit were located at the centre of one leaf, the next would straddle the junction of the next two. With both adjacent MLC leaves being used and set to the same position to define the slit length the edges of any of the slits were never required to be adjacent to a leaf edge and this removed the artefacts.

Cooper *et al* (2001c) have studied the issue of the best way to 'index' this tertiary collimator to ensure that abutting irradiations correctly junction with each other (figure 3.39). They considered two methods. In the first, the tertiary collimator remained stationary and the patient couch was incremented by the pitch of the open slits of the collimator (consider a tertiary collimator with mark-to-space ratio of unity, made of brass 8 cm thick and with a slit width projecting to 5 mm at isocentre). The second indexing technique was to arrange for the tertiary collimator to move on the arc of a circle focused back to the source location. Again, to achieve an increment of 5 mm at isocentre the tertiary collimator

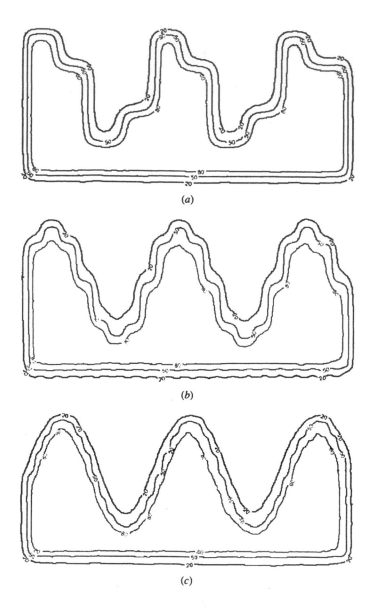

(a)

(b)

(c)

Figure 3.38. The three images show the isodose patterns (20%, 50% and 80%) for a 4-cm-wavelength high-dose area created under three separate delivery conditions: (a) created with the conventional MLC, (b) created with two grid irradiations of size 5 mm and (c) created with four grid irradiations of size 2.5 mm. Notice that the isodose lines are more faithfully reproducing the desired pattern of high dose using the tertiary collimator. (From Williams and Cooper 2000.)

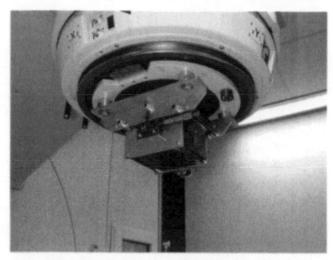

Figure 3.39. Photograph displaying the current form of the tertiary collimator from the Christie Hospital, Manchester and its method of attachment to an Elekta SL accelerator. (From Cooper *et al* 2002.)

needs to increment by a rotation of 0.005 rad or 0.29°. Clearly the first method can only lead to a good match at the isocentre because the open slits diverge. Hence, a variation in dose at some other depth would be expected. Cooper *et al* (2001c) made measurements of the radiation profile along the direction of potential collimator movement, at both isocentre and at several depths above and below. They then digitized the film measurements and computationally added the appropriately shifted profiles. Whilst, for pure translation, the dose inhomogeneity was only about 8% at isocentre, it rose to some 20% at planes 3 mm above and below. However, when the profiles were computer summed using the rotation method of indexing the dose inhomogeneity remained at about 9% for all planes including the isocentre. This method of computational addition was used to simulate perfectly accurate additions. Then the whole experiment was performed by actually double-irradiating film with the tertiary collimator rotating in its cradle. The dose inhomogeneity rose to about 15% but was constant for all planes.

The originators of this concept had hoped to be able to produce a device which generalized the use of an MLC with conventional-width leaves and interest the appropriate manufacturer. The use of mini- and microMLCs somewhat overtook the concept but at a considerable price (Williams, private communication).

3.13   Verification of MLC-delivered IMRT

With the rapid proliferation of systems for delivering IMRT in a clinical setting, using an MLC, the topic of verification has become very important. There has been a general trend to widen the scope of techniques from one-off to those which can be usefully applied in many clinics. Low (2001) and Budgell (2002) have summarized verification methods for IMRT. In this section, verification of specific clinical IMRT deliveries is described. The subject of routine verification of the performance of an MLC is described in section 3.14. This is often referred to as QA rather than verification. Boyer *et al* (2002b) also identified aspects of QA of IMRT that are related to the delivery equipment and those aspects which are patient-specific.

Ramsey *et al* (2003) have debated (in a *Medical Physics* Point and Counterpoint) the wisdom or otherwise of individually verifying each IMRT treatment plan. An argument on one side is that measurement only validates a specific treatment plan and does not address issues of planning assumptions, movement etc. An argument on the other side is that time freed from such individual verification could be better spent on system QA to the benefit of all IMRT patients.

3.13.1   Electronic-portal-imager-based IMRT verification

Partridge *et al* (2000b) have used electronic portal imaging (EPI) to verify both the leaf tracking and also integrated intensity measurements. Modifications have been made to a commercially available system to give accurate knowledge of the cumulative delivered dose in every frame of a sequence for leaf-tracking measurements and secondly to reduce optical crosstalk in the camera-based systems. Portal imaging was demonstrated to be a useful tool for the verification of the dMLC delivery technique.

Partridge *et al* (2000a) have made a comparison of three particular implementations of the use of electronic portal imaging for leaf-position verification during the dMLC IMRT delivery technique. The three systems investigated were in three different centres namely those at the Royal Marsden NHS Foundation Trust, the Christie Hospital NHS Trust and The Netherlands Cancer Institute, Amsterdam. The goal of the work was to characterize the performance of the leaf-and-jaw motion and to compare measurements with prescription predictions. The first is an inhouse development; systems at the second and third centres are commercially available. This investigation is different from the investigation of total integrated fluence during the dMLC technique since this latter does not provide information about what caused any potential error. The measurements described by Partridge *et al* (2000a) give information about the actual trajectories of each leaf and jaw. All measurements were made using the Elekta dMLC delivery technique as existed at that time. Ten different prescriptions were used. In any one prescription, either a jaw or a set of leaves moved. In the first eight prescriptions, the jaws and leaves moved at constant speed and in the ninth and tenth prescriptions the jaws and leaves moved in a sequence: slow, fast, slow. This was to test whether the measured positions of the leaves and jaws stayed within tolerance, noting that the tolerances for the Elekta dMLC technique depended on the leaf speed to ensure that the dose error did not exceed 2%. For example, if a leaf moves 50 mm in 20% of the total dose prescribed, a position tolerance of 5 mm gives a maximum dose error of 2% and so on.

Partridge *et al* (2000a) described all three systems and referred to the primary papers in which they had been previously described. The integration time varies between the three systems: for the Royal Marsden NHS Foundation Trust system, it is 0.2 s; for the Amsterdam system 1.3 s and for the Manchester system 0.14 s. The Royal Marsden solution was to record ten camera frames with an integration time of 200 ms. Since each frame was locked to a particular accelerator pulse, the precise number of delivered MUs corresponding to the measured frames was precisely known. For the Amsterdam solution, in which a complete frame takes about 1.3 s and each of the 256 lines are read out sequentially, the time is stamped into the 257th pixel of a 257 × 256 array. The Manchester solution is to trigger the image acquisition at known percentages of the delivered dose. Partridge *et al* (2000a) demonstrated how important it is to quantify and eliminate all systematic measurement errors and they described the technique for doing this in some detail. They also described performance measures in terms of absolute errors in leaf position per leaf per frame, means over this quantity and rms errors and total rms deviations. The rest of the study summarized specific examples taken on the three systems which demonstrated that the dMLC technique for all prescriptions and for all fluence output rates and for all leaf speeds was working within the manufacturer's tolerance. Note that this is not a study of dosimetric verification.

Partridge *et al* (2000c) have evaluated the capabilities of another different commercially available camera-based electronic portal-imaging system for IMRT verification. The system used was the TheraView camera-based electronic portal imaging detector (EPID) system (Cablon, NL). Because it was not possible to interact with the hardware and software they attached a special set of light-emitting diodes which coded for the number of MUs delivered. The monitor chamber circuitry of the Elekta linacs produces a digital pulse every 1/64th of an MU and, in this way, each image was tagged with the cumulative dose by a series of dots running down one edge. Software was written to decode these dot patterns to yield the precise number of MUs corresponding to each image. It was shown that the system could be used to monitor the dynamic leaf tracking during a breast radiation therapy. By adding frames together, the integrated intensity was also shown to be a good approximation of the incident fluence.

Partridge *et al* (2000c) also attached an anti-scatter grid to the scintillator to ensure that calibrations made with a large field were also appropriate to small fields and they showed that this was so by making measurements of the radiological thickness of a target using both small and large fields. They also showed that the stability of the imaging system was good over both short and long periods of time. The overall geometrical accuracy of the system was determined to be ± 2 mm with a dosimetric accuracy of ± 1.2 MU. It was shown that the major geometric distortion in the system was a rigid-body rotation of $1.2 \pm 0.4°$ about the optical axis of the system but that this could be corrected. The system was also shown to develop afterglow but, given that this was present at both calibration and measurement stages, the degradations cancelled out. In conclusion the system was suitable for IMRT leaf-tracking verification.

Partridge *et al* (2001b) have performed megavoltage computed tomography (MVCT) in cone-beam geometry using an amorphous silicon flat-panel portal imaging detector prior to the delivery of IMRT. The same detector was then used to measure transmitted primary fluence. This fluence can then be backprojected through the CT model of the patient to form an effective input fluence map for each beam which can be compared with the actual fluence map from the planning system for verification.

Partridge *et al* (2002) extended this and have conducted a proof-of-principle study for performing 3D *in-vivo* dosimetric verification of IMRT (figure 3.40). An Alderson RANDO phantom was CT scanned, a horshoe-shaped target was defined with an embedded OAR and a five-field, five-levels-per-field IMRT plan was made at DKFZ using KONRAD.

A flat-panel imager was used at treatment time to make an MVCT scan. Each projection dataset required correction for detector sag and an iterative deconvolution of the significantly large radiation scatter. The same flat-panel detector was used to create both individual images of the step-and-shoot segments and summed exit images of beam fluence. After again correcting for scatter, these images were used to reconstruct the *plane* of input primary fluence (see also Partridge *et al* 2001b). Good agreement was found with the predicted input fluence from the planning system.

The predicted fluence was then projected into the MVCT scan to calculate, using Monte Carlo-generated dose kernels, the predicted delivered dose map. When compared with the dose map from the planning system, the agreement was $\simeq 3\%$ in low-dose-gradient regions, 3 mm for high isodose lines. The small size of the detector limited potential clinical implementation. Also the time taken to gather the projection data from MVCT (17 min) and dose (from 120 MU) was too great. The technique makes assumptions about the conversion to Hounsfield Units, the exponential attenuation and the authors comment at length on the potential systematic and random errors.

Woo *et al* (2003) have built an automatic verification system for monitoring step-and-shoot IMRT field segments using portal imaging. The signal from a BEAMVIEW portal-imaging system were extracted to a frame-grabber and PC. Overlaid on this is the corresponding field shape from the CORVUS planning system. Edge-detection software could then pick up errors. Warkentin *et al* (2003) have presented a method for using a flat-panel portal-imaging detector to verify the dose in IMRT. The essence of the technique is that the fluence measured by

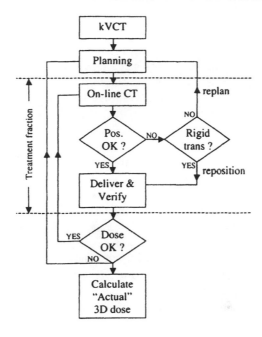

Figure 3.40. A schematic overview of the adaptive radiotherapy process. If the patient geometry measured by an on-line CT system is out of tolerance, a decision must be made as to whether it is possible to simply re-position the patient using a rigid-body transformation and continue with treatment or re-plan. Note that the on-line CT facility permits a calculation of the delivered dose distribution. (From Partridge *et al* 2002.)

the flat-panel detector has to be converted into a measure of detected dose by deconvolving two physical phenomena. The first is the dose kernel due to build up in the material of the detector and the second is the glare kernel due to optical scatter of photons. The first of these was created using Monte Carlo techniques and the glare kernel was created using an empirical fitting of a double Gaussian to experimental data. Also the flood-field correction has to be built in to the portal-imaging measurement. They showed that this technique allowed a prediction of dose that agreed to better than 2% with film measurements and the differences between the measurements created by portal imaging and those predicted from a treatment-planning system were much higher than this showing the errors in the planning system.

Yin *et al* (2000) have made portal measurements of IMRT beams and backprojected these through reconstructed CT images to create a map of the relative delivered dose distribution. These were compared with TLD measurements and verified the accuracy of IMRT delivery.

3.13.2   Other EPID designs

Mornex *et al* (2000) have developed IRIS, a portal-imaging system which is based on large-area-pixel-matrix technology using amorphous silicon detectors. The IRIS system is capable of producing up to ten images a second and therefore monitoring the beam steering in almost real time for dynamic IMRT.

Keller *et al* (2000) have studied the properties of a SLIC EPID for verification of IMRT delivered by the dMLC technique and have determined that the applicability depends on the nonlinear response of the system. For a maximum leaf speed of 2 cm s^{-1}, they find that a sampling rate of 2 Hz is required, somewhat faster than is presently available.

Ploeger *et al* (2002) and Sonke *et al* (2003) have developed a method for geometrical verification of dynamic IMRT that makes use of a scanning EPID. They show that motion distortion effects of the leaves are corrected based on the treatment progress. The method was tested on three different dynamic prescriptions. Specifically Sonke *et al* (2004) showed that, for the amorphous silicon imager, the potential causes of motion distortion and motion blurring were ghosting and the fact that the flat panel integrates over the frame time. Neither effects were shown to be clinically significant and the position of a leaf moving at 0.8 cm s^{-1} was verifiable to an accuracy of 0.25 mm.

Chang *et al* (2001) have used a Siemens BEAMVIEW, camera-type, EPID to rapidly make measurements for step-and-shoot IMRT. The recorded intensity maps were then compared with computed maps visually in real time. In order to do this, they had to develop a simple solution for image distortion, a method to handle histogram differences and image-analysis tools.

3.13.3   Technical aspects of EPID imaging for IMRT

Evans *et al* (2001) have studied the effects of spatial and temporal sampling on the recorded dose distribution of an EPID monitoring an IMRT delivery using the dMLC technique. The following physical phenomena were modelled:

(i)   the delta function pulsed nature of the linac radiation,
(ii)   the time interval between radiation pulses,
(iii) the finite acquisition time for each image frame,
(iv)   the time interval between image frames (thus defining the imaging duty cycle),
(v)   the spatial sampling size of the EPID,
(vi)   the imaging point-spread function of the EPID.

They studied three IMBs under a variety of imaging conditions. These were a spatially continuous hemispherical IMB, a spatially sampled IMB from a CORVUS inverse-planning system and a spatially sampled IMB from the technique to improve breast dosimetry using IMRT.

The magnitude of the error in recording distributions was a function of the sampling properties. Evans *et al* (2001) showed that provided images could be recorded in 0.33 s every 1–2 s interval the errors were below 2%. Detailed explanations were provided to substantiate the nature of the errors. Sometimes these were oscillatory with the periodicity relating to the sampling interval and the phase relating to the offset between the start of irradiation and the start of sampling.

3.13.4   Extraction of anatomical images from portal images generated during IMRT

Fielding and Evans (2001) have described the use of a camera-based EPID to make measurements of the moving leaves in a dMLC treatment. Image frames were acquired at a typical rate of one every 2 s with an integration time of 1.5 s. These frames were then summed to produce an integrated fluence profile to compare with the prescribed fluence profile. The portal images were calibrated pixel-by-pixel using the previously published quadratic method and this enabled maps of anatomical thickness to be produced which are useful for both dosimetry and patient set-up.

Fielding *et al* (2002a) have shown that anatomical information can be extracted from electronic portal images created using IMBs (figure 3.41). They described two techniques. In the first, a series of frames was summed to produce an integrated portal image of the delivered beam with each image corresponding to the multileaves in different positions. Following this, a second set of images was created in exactly the same way and summed but with a PMMA block similar to the thickness of the patient in place instead of the patient. The first summed image was then divided by the second image and this led to a visualization of patient anatomy. A modification of this, involving a second calibration known as the quadratic method and involving four calibration measurements each delivering the IMB to a different thickness of PMMA blocks, led to somewhat improved results. This technique was applied to imaging prostate and pelvic nodes using the Elekta/Theraview system.

Fielding *et al* (2002b, c) have also applied the method that they developed for imaging patient anatomy using a portal-imaging detector and an IMRT beam to the Varian amorphous silicon flat-panel imager. The technique relies on adding together images of the patient with the multileaves in different positions and then calibrating for the variable input intensities. Calibration successfully removes the intensity modulations of the beam so that the skeletal anatomy of the upper thorax and neck region can be seen in the calibrated image. Fielding *et al* (2003a, b) have shown that, with this technique, it has been possible to detect patient (phantom) movement as small as 2 mm, leaf mispositioning of the order 2 mm and a 1% error in MUs.

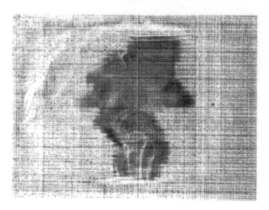

Figure 3.41. A clinical image of the pelvis extracted from an EPID measurement created with modulated fields. The quadratic calibration method was used with manual registration. (Reprinted from Fielding *et al* (2002c) with copyright permission from Elsevier.)

3.13.5 Blocking-tray-level measurement

Xu (2001) has developed a device which goes in the blocking tray comprising a film and a diode to make measurements of fluence distributions and a single-point dose measurement for IMRT verification. For 107 IMRT fields, 95% of the point doses were within 5% of the CADPLAN calculated doses and all fluence map films visually agree with plots from CADPLAN.

Glendinning *et al* (2001) have developed a novel strip-ionization chamber comprising ten opposing pairs of plates which are placed in the blocking tray of the accelerator. Each subchamber is 80 mm × 3 mm with 3 mm separation. This chamber was used to monitor the positions of leaves during a dMLC treatment when set-up to deliver dynamic wedge and longitudinal and lateral sinusoidal intensity modulations. The same patterns were monitored using an EPID and it was found that this demonstrated that the Elekta SLi dMLC system in its research release had a systematic lag with respect to that prescribed of 0.67 s. Thomas and Symonds-Tayler (2003) have developed a block-tray mounted ionization chamber also for IMRT verification.

Eberle *et al* (2003) have built a test liquid ionization chamber with 20 × 20 pixels. This resides in the output port of the linac head and can measure MLC leaf positions accurately and fast for static and dynamic therapy. The performance of different liquids was evaluated, tetramethylpentane giving the highest charge yield.

Ma *et al* (2003a) have used a film in the accelerator-blocking tray to measure the fluence of a modulated and open field. Given that the detector response is the same for both fields, it cancels out in the determination of the actual delivered fluence through a Fourier division method.

Renner *et al* (2003) have described a new way to verify radiotherapy including IMRT. Film was used to measure the entrance fluence from beams. These films were so calibrated as to give the bixel-by-bixel MUs for a modulated beam. Then a pencil-beam dose calculation was made using the planning CT data to create an independent measurement of the dose in a patient that could be compared with the planned distribution. It was recommended that the film (with build-up) be in the blocking tray of the accelerator so measurements could be made at any gantry angle in case the accelerator performance changed with orientation. Alternatively, EPIDs could be used.

3.13.6   The two-level MLC

Greer and van Doorn (2000) have proposed a novel design for a dual-assembly MLC. The collimator is, in all respects, the same as a conventional MLC except that every other leaf pair is in one of two alternate banks which are arranged one above the other. So, the upper bank, for example, comprises all the even-numbered pairs of leaves with spaces in-between and the lower bank comprises all the odd-numbered pairs of leaves also with spaces in-between. The leaves can be moved by motors in the same way as the conventional MLC. In the configuration described, movement of the leaves in both the upper and lower bank (see figure 3.42) creates the usual geometrical field shaping and, for example, if all leaf pairs in both banks are closed, the field is completely shielded.

Conversely, if and when one bank is moved laterally perpendicular to the direction of leaf movement by just one leaf width, the attenuating elements in the two banks line up and the spaces line up. In this way, the collimator can be used to create a slatted view of the geometrically collimated radiation field whatever the position of the leaf pairs along their direction of travel. Now imagine the reverse phasing in which, once again, the attenuating elements in the two banks line up but they now take the positions that were previously taken by the spaces whilst the aligning spaces take the position previously taken by the aligned leaves. This enables the taking of a second slatted portal image which, when combined with the first, yields a full image of the patient in the treatment position.

Greer and van Doorn (2000) observed that if the slit width is small, for example 2 mm (and the corresponding leaf widths are also 2 mm wide), then just the single-exposure slatted image shows most of the detail and the true portal image can be recovered via an image-processing technique. Conversely, if the leaves and spaces are wide, for example 10 mm, then a single exposure is not good enough to show patient detail and two single exposures, one phased with respect to the other as described earlier, must be combined to form the image.

Performance of this collimator was quantitatively evaluated by creating experimental single slits of radiation with the attenuating elements having different spacings, different attenuating thicknesses and being at different source-to-collimator and source-to-detector positions. Greer and van Doorn (2000) showed a large number of very detailed graphs showing how the profile of

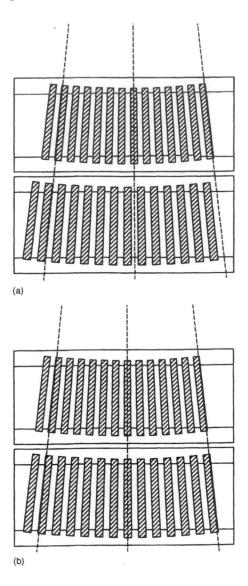

(a)

(b)

Figure 3.42. (a) A schematic diagram illustrating the multileaf collimator design in the shielding position. Each bank of MLC leaves (viewed end-on) is split into two vertically displaced levels, with each level consisting of every second leaf. The leaves in the upper level shield the leaf width spaces or slits in the lower level. (b) A schematic diagram illustrating the multileaf collimator design in its imaging position. One of the levels is moved laterally by a leaf width so that the leaves are aligned with the other level. Radiation is then transmitted through the collimator as multiple-slit fields. (From Greer and Van Doorn 2000.)

radiation delivered through a slit depends on these properties. They also modelled these profiles by the well-known technique of determining the profile of radiation as the sample of a dual Gaussian source 'seen' by sets of points in the detector space. In general, the analytically-modelled profiles closely matched the experimentally-measured ones.

They ascribed the origins of their concept to the well-known patent by Swerdloff *et al* (1994a,b) for a tomotherapy fan-beam intensity-modulating collimator with alternate vanes in two vertically displaced banks. However, the concept proposed is quite different being appropriate to the geometry of a conventional MLC and the new proposal is not for tomotherapy.

A key observation was that, as the slit width decreased, the profiles become broader than the geometric slit projection resulting in increased overlap of adjacent profiles. They also studied a configuration in which the leaves in the two banks were slightly larger than the spaces in the adjacent bank and could thus remove the need for any machined 'tongue and groove'.

They also proposed that lateral movement of one bank with respect to another by a partial leaf width could be used to improve the spatial resolution of multileaf collimation. As far as can be told, no device of this nature has been constructed. The experiments reported in the paper were made with single slits although a single bank collimator comprising poured lead slats had been constructed for experimental purposes.

The periodicity inherent in dual exposure images even with a 10 mm slit implies that adjacent slit profiles do not superpose according to the laws of perfect superposition. This may cast some doubt on whether the Manchester grid collimator will be able to collimate fields that are equivalent to those collimated by a microMLC.

Chang *et al* (2003) have studied a four-bank MLC in two levels. The jaws of the accelerator are each replaced by a microMLC and they show that this leads to a better ability to segment modulated fields when compared with the use of the conventional two-bank MLC.

3.13.7   Water-beam-imaging system (WBIS)

Li *et al* (2001a) have designed, built and implemented a verification system for measuring delivered IMRT dose distributions using a water-beam-imaging system (WBIS) (figure 3.43). This system comprises a 30-cm-diameter plexiglas cylinder containing a 5-mm-wide single slice of plastic scintillator with 20 cm of clear water on either side of the scintillating screen. The scintillator is backed using reflective paint. The scintillator is viewed end-on through an optically transparent window by a charged-coupled device (CCD) video camera. This camera can capture 512×512 10-bit pixel images of the screen with an integration time adjustable from 1 to 120 ms. The images are processed by a frame-grabber that can sum between 1 and 2000 images. It was determined that the output response of the WBIS was linear with delivered MUs. Background images are subtracted

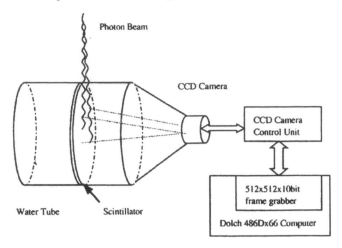

Figure 3.43. A schematic diagram of the Wellhöfer water beam imaging system (WBIS). The data line from the control unit interfaces via a frame grabber board with a portable personal computer. This system was developed at Stanford University School of Medicine. Dose distributions recorded with it were compared with those created by Monte Carlo planning techniques. (From Li *et al* 2001a.)

from each field measurement. The measured dead times were found to be less than 1 ms at all delivery dose rates and therefore negligible.

The image captured by the CCD camera is blurred by the scattered light represented by a point-spread function which was fitted to a multiple Gaussian. Through a calibration procedure, the parameters of the multiple Gaussian were determined. Then all images including those with background subtraction were deconvolved using this multiple Gaussian function and the dose distributions so captured by the device were then shown to agree much more closely with Monte Carlo-predicted dose distributions than those without the deconvolution process.

The equipment was used to make measurements of the integrated dose from a series of beam-intensity-modulated irradiations from a variety of gantry angles by capturing the effect of each delivery and adding frames together to create a picture of dose in a transverse-axial plane that was then correlated to the predicted dose distributions from a planning system including a Monte Carlo dose-calculating engine. Agreements between the deconvolved WBIS images and the prediction of the Monte Carlo method were very good. The key feature of the device is that data processing and comparison between processed measurement and the reference prediction could all be performed within 5–10 min, much faster than the use of films, BANG-gel or TLD measurements.

3.13.8   Integrated portal fluence and portal dosimetry

Chang *et al* (2000a) have used the Varian PortalVision Mark-1 System to record frames of portal fluence as the leaves move during the dMLC irradiation. They used a high (900)-MU measurement so that the leaves could be slowed down so as not to move too far during the 9 s it takes to record and store a high-resolution image. The portal image pixel values can be related to a measure of fluence at each pixel. Then, by summing the images weighted by the time interval for each image, a measure of the total integrated fluence at each pixel was obtained (figure 3.44). This image comprised both primary and scattered fluence and Chang *et al* (2000a) assumed a linear combination of these, weighted by two 'fitting constants'. Then they defined a space-varying error term as the mean-squared deviation between the expected measurement and the actual measurement. For a so-called perfect IMRT delivery, the fitting constants were determined to minimize this mean-square deviation.

This formed the basis for estimating the importance of genuine delivery errors, for example those due to a stuck leaf, a wrong modulation or erroneous jaw positions. They showed that the mean-square difference between expected and measured integrated portal fluence was a good parameter to establish that an error may have occurred and the action level for it. They also made absolute dose predictions to act as a verification tool for the dMLC IMRT technique.

Chang (2001) has used the Varian aS500 amorphous silicon EPID for verifying IMB delivery. Twenty-five prostate IMRT subfields were recorded independently as images and converted to dose rates using a calibration curve. These were then integrated over time to obtain a measured dose profile which was then compared with the intended profile for treatment planning. Using a 15 MeV beam, the correlation between measured and intended dose profiles was acceptable within 1% but strong energy dependence was observed for the 6 MeV beam making the calibration curve deviate from a straight line for that energy.

Pasma *et al* (2001) have converted portal images into measures of transmitted fluence and compared these with predicted fluence for the dMLC technique. The technique developed has been implemented inside the Varian CADPLAN treatment-planning system and the method was tested on a Varian 2300 CD treatment unit, showing that, outside narrow peaks and steep gradients, measurements agreed within 2%.

Van Vliet-Vroegindeweij *et al* (2001) have used an EPID to measure beam profiles and compared with those produced by a 3D treatment-planning system.

Kirby *et al* (2002) have made measurements of integrated exit fluence in IMRT when delivered with the Elekta Precise linac and when measured with the Elekta iView (fluoscopic) and iViewGT (amorphous silicon) portal imager. It was found that summed fluence patterns agreed well with the fluence data extracted from the inverse-treatment-planning system PLATO.

Van Esch *et al* (2001a, b) have shown that predicted portal dose maps agree with the measurements from an EP10 (Varian PortalVision) to better than 3%.

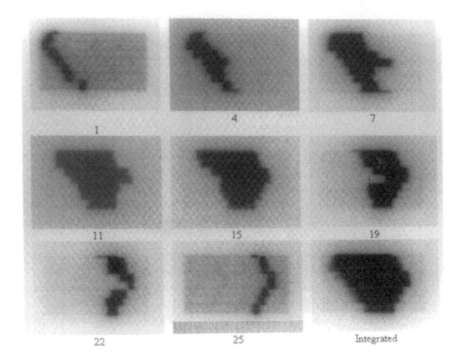

Figure 3.44. A moving picture of an intensity-modulated beam for prostate treatment using the sliding-window technique in use at Memorial Sloan Kettering Cancer Center. Twenty-five images were taken during the course of treatment and eight of them are shown here labelled 1, 4, 7, 11, 15, 19, 22, 25. Because the Varian Portal Vision is a dose-rate imager, each image is converted to a dose-rate map, then multiplied by a weighting factor corresponding to the time interval for that image. The weighted dose rate maps are then summed to obtain the integrated dose map shown in the lower right-hand part of the figure. (Reprinted from Chang *et al* (2000a) with copyright permission from Elsevier.)

Kim *et al* (2002) have predicted, using Monte Carlo techniques, the portal image that would be created using an IMRT delivery for a clinical head-and-neck case using dMLC delivery. They showed that the Monte Carlo simulations successfully reproduced the measured pre-treatment and the transmission dose images.

3.13.9   IMRT verification phantom measurements

A very large number of reports have been made of techniques to verify IMRT delivery using phantoms.

Ma *et al* (2001a) have described techniques for verification of IMRT at Stanford where IMRT has been implemented clinically since the end of 1997.

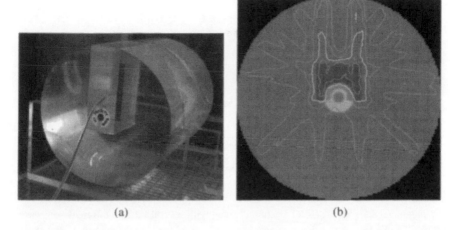

Figure 3.45. IMRT plan verification using a PMMA phantom with a bone insert (left) and isodose distributions for the plan generated by CORVUS using 15 MV photon beams on a Varian Clinac 2300 CD accelerator (right). (From Ma *et al* 2003a.)

Leaf sequences were verified using film or the WBIS. Monte Carlo methods were also used to verify the dose distributions in the patient. Two phantoms were constructed (figure 3.45), both cylinders of 30 cm diameter, one of PMMA, the other of water. Monte Carlo and CORVUS dose calculations agreed with measurements to within 2% for both but differed by 4% when bone and lung heterogeneity inserts were placed in the phantoms (Ma *et al* 2000a, 2003a). The water phantom must, of course, be irradiated in the upright position requiring a complicated conversion of gantry and collimator angles to match the planning of a horizontal cylinder. The advantage of the PMMA phantom is that it can be irradiated horizontally on the couch.

Radford *et al* (2000, 2001) have designed an anthropomorphic IMRT verification phantom. This phantom was designed to accept one insert which would be in place when acquiring CT data for treatment planning and a second insert, which would provide precisely placed TLD and radiochromic film for measurement of dose from intensity-modulated fields. The phantom was designed to be mailable and so is lightweight and waterfillable. It is part of the multi-institution NCI cooperative clinical trial effort.

Oelfke *et al* (2001) have reported on the clinical IMRT technique at DKFZ Heidelberg. Planning is performed with KONRAD. The plan is transferred to a phantom and special detectors, buried in the phantom, are able to make point-dose and volumetric-dose measurements to compare with predictions (figure 3.46). Verification of 67 patient treatments showed an agreement between calculation and experiment to $0.6\% \pm 2.2\%$. By November 2000 143 patients had been treated and the number in 2004 greatly exceeds this. Rhein and Häring (2001)

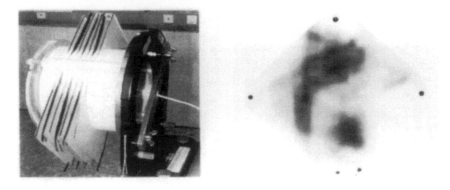

Figure 3.46. This shows two steps of a dosimetric verification process for IMRT. The left-hand image shows the IMRT phantom loaded with several films in different transverse planes. On the right-hand side is shown one of the developed films whose optical density is calibrated to absolute dose values. These are then compared to the predicted dose values provided by the treatment-planning programme. (From Oelfke *et al* 2001.)

described how 170 patients had been treated with IMRT at DKFZ Heidelberg using a Siemens PRIMUS linear accelerator by this later date.

Kermode *et al* (2001b) have verified HELAX-planned multisegment fields using the LA48 linear detector array and compared measurements with predicted linear dose distributions. In the majority of cases, the HELAX system modelled segmented-field radiation treatment to an adequate degree of accuracy.

Agazaryan *et al* (2000) and Agazaryan and Solberg (2003) have developed a dMLC technique and built a phantom for 3D dose verification.   The sequencer includes transmission, interleaf and intraleaf leakage and includes leaf synchronization to reduce the TG effect. It was found that output-factor correction for small fields was the most challenging phenomenon for dosimetry. Measured dose at the position of each diode in the phantom was compared with calculation of dose from the sequenced leaf positions.

Chuang *et al* (2002) have investigated the use of metal oxide semiconductor field effect transistors (mosfets) for clinical IMRT dose verification. The mosfet was stable over a period of two weeks to within 1.5%. The mosfet presented a linear response to dose between 0.3 and 4.2 Gy. The percentage-depth–dose measurements measured with mosfets agreed to within 3% of those measured with an ionization chamber. A phantom was constructed and IMRT plans were delivered to the phantom comprising 81 possible mosfet-placement holes. It was found that the measurements from the mosfet and from the CORVUS planning system agreed within 5%, whereas the agreement between ionization chamber measurements and the calculation was 3%. It was concluded that the mosfet detectors were suitable for routine IMRT dose verification. Their main advantage is their extremely small detector size (0.4 mm^2) and they can immediately be re-used. They can make multiple point-dose measurements and they can also be used for *in-vivo* dose measurements. Some concern over the angular dependence of early mosfets has been disposed of now that the design and characteristics of new mosfets have significantly improved.

Cosgrove *et al* (2001) reported on the verification measurements made in a phase-1 prostate and pelvic node IMRT trial at the Royal Marsden NHS Foundation Trust (figures 3.47–3.53). Treatment planning was (at that time) carried out with a NOMOS CORVUS Version 3 inverse-planning system and the leaf sequences were interpreted using interpreter code developed by Convery and Webb (1998). To verify a plan prior to the start of a treatment course, a volumetric approach was adopted using the pelvic region of an Alderson Rando phantom. The IMRT beams, calculated for the patient, were recalculated onto the Rando phantom geometry and then delivered to the phantom which contained sheets of XV film loaded within it. Once developed, the films were scanned using a Vidar scanner and 2D isodose distributions were generated and compared with the plans. The quality-assurance goal was to ensure that high isodose lines agreed to within 3 mm and/or doses in low-gradient regions agreed to within 4%. Additionally, 80 TLD chips were used to verify a treatment plan. These were positioned in appropriate places within the phantom and agreement in both low- and high-dose regions was typically within 2% of that calculated. Additionally, films strapped to the gantry of the linear accelerator showed that the fluence distributions were reproducibly delivered on subsequent fractionation occasions (figure 3.54). Further details of this trial have been presented by Adams *et al* (2004).

Lee *et al* (2001b) have designed a home-made cylindrical phantom which can contain a microionization chamber and a thimble chamber as well as films placed transversely to make verification measurements for IMRT. A slab accommodating TLD chips can also be incorporated. This system is very similar to that described by Cosgrove *et al* (2001).

Saw *et al* (2001) replanned CORVUS-generated IMRT beams designed for a MIMiC-based IMRT delivery on to two phantoms. One phantom was 7 cm in diameter and 12 cm long and the other was a 12 cm × 30 cm × 30 cm phantom and the third was a 22 cm × 30 cm × 30 cm phantom. It was found that the difference between the replanned dose calculations and the measured dose distributions were smaller when the phantom had a comparable size to that of the patient.

Short *et al* (2001) delivered IMRT computed with the Radionics X-plan planning system with plans exported to a Siemens Primus linear accelerator. In an autosequence 30–50 subfields were delivered. These were checked using film. Additionally, the fields were exported to be replanned on to a phantom containing TLDs and measurements were compared with predictions. The agreement was to better than 5%. Rose *et al* (2001) showed that, at the same centre, the use of a water phantom could demonstrate good agreement between planning and measurements.

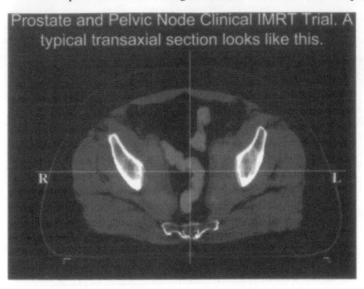

Figure 3.47. A typical transaxial section of the prostate and pelvic nodes illustrating how the small bowel loops into the concavity creating a case for clinical IMRT. (Courtesy of Drs Viv Cosgrove and David Convery.) (See website for colour version.)

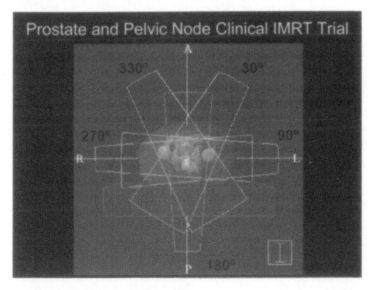

Figure 3.48. The five gantry angles used for pelvic IMRT at the Royal Marsden NHS Foundation Trust. (Courtesy of Drs Viv Cosgrove and David Convery.) (See website for colour version.)

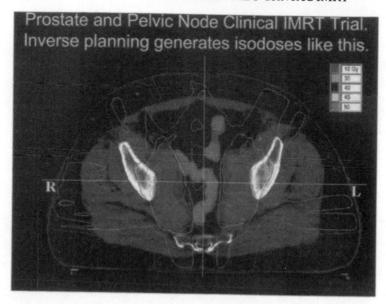

Figure 3.49. IMRT planning spares the small bowel looped into the concavity shown in figure 3.47. (Courtesy of Drs Viv Cosgrove and David Convery.) (See website for colour version.)

Mazurier *et al* (2001b) have used the inverse treatment-planning system HELAX Version 5.1 to generate intensity-modulated field segments for the PRIMUS accelerator from Siemens. Film measurements showed that the leaf-position accuracy was better than 1 mm. The monitor chamber linearity was adjusted so that the difference between the dose-per-MU obtained for 1 MU and for 200 MUs was just 3%. The difference between the calculated and measured output factors was less than 3%. Full IMRT was verified by comparing calculated and measured dose distributions in an axial plane of an IMRT phantom with an agreement better than 5%.

Gluckman and Reinstein (2000) and Gluckman *et al* (2001b) have developed a computer tool to correlate predicted and measured dose distributions and provide systematic measurements of equivalence or lack of this. Gluckman *et al* (2001a) have adapted an IMRT prostate phantom for full 3D dose verification using polymer gel and also multiple-plane radiochromic film inserts.

Depuydt *et al* (2002) have extended the 'gamma index' tool of Low (2001) to compare calculated and measured IMRT dose distributions in a phantom.

Sorvari *et al* (2000) have commissioned a linear accelerator delivering the dMLC technique. Calculations were compared with measurements in anthropomorphic phantoms containing TLDs and it was found that results compared well except in low-dose regions where the delivered dose was about 5% lower than that expected by calculations.

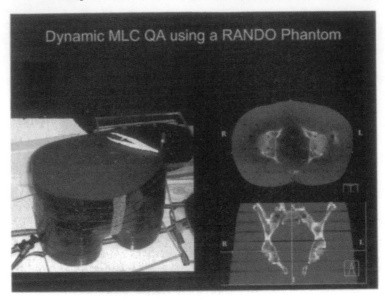

Figure 3.50. A section of the RANDO phantom arranged for TLD loading together with x-ray images. (Courtesy of Drs Viv Cosgrove and David Convery.) (See website for colour version.)

Cardarelli *et al* (2001) have evaluated the MED-TEC quality-assurance phantom for IMRT.

Bakai *et al* (2002) have described QA for IMRT in which point measurements were made with the PTW LA48 linear diode array or a very small ion chamber. Létourneau *et al* (2004) have used the MapCheck 2D diode array for IMRT verification and quantified its performance.

Leybovich *et al* (2002) have shown that the use of small ionization chambers does not improve the accuracy of IMRT dose verification and that larger ion chambers (0.6 cm^3) are recommended for absolute dose verification.

Boehmer *et al* (2004) described the IMRT quality-assurance tests performed at the Charité Hospital in Berlin.

Reports summarized here are, by necessity, something of a mixed bag of experimental arrangements. Further reports are expected because IMRT verification programmes have to be established in all centres beginning clinical IMRT.

3.13.10 Verification by software techniques

Lee *et al* (2001c) have developed a virtual pre-treatment verification for IMRT. Essentially a computer program models the leaf motion of the moving MLC and factors in all the needed physical processes. The fluence thus generated is then

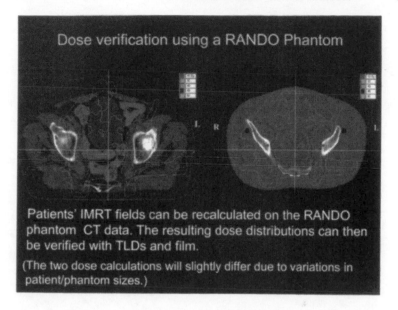

Figure 3.51. Recalculation of intensity-modulated fields onto a RANDO phantom for IMRT verification. (Courtesy of Drs Viv Cosgrove and David Convery.) (See website for colour version.)

compared with that from the inverse-planning system. Agreement to within 2% was obtained. This is surprising unless the physical processes are also included into the planning system.

Moran *et al* (2001a) have developed a semi-automated verification system for the dMLC technique. The log sequence files from a Varian 2100 EX system with a 120-leaf MLC were used to create the expected and the actual positions for each leaf. From these data, the leaf trajectories were calculated and the expected and actual dose images were created. The images were then compared with digitized films or with digital fluence images from a portal imager. Moran *et al* (2001b) have compared the dosimetry of static conformal therapy, static segmental MLC therapy and the dMLC technique.

Teslow (2002) has developed practical extraction and report language (PERL) programmes within the CORVUS treatment-planning system to summarize the outcomes of inverse treatment planning and to provide tools for supporting confident patient treatment decisions.

Chen *et al* (2000b) have developed a verification process for MLC-based IMRT which makes use of CT virtual simulation. Treatment planning was performed using CORVUS and the planned fields were then translated onto a virtual simulation workstation which is able to compute a digital reconstructed radiograph in the beam's-eye view of each field. A fixed MLC field with the same boundary as this IMRT field was then also created for portal imaging so

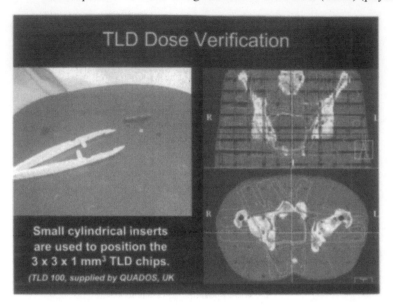

Figure 3.52. TLD dose verification for IMRT. (Courtesy of Drs Viv Cosgrove and David Convery.) (See website for colour version.)

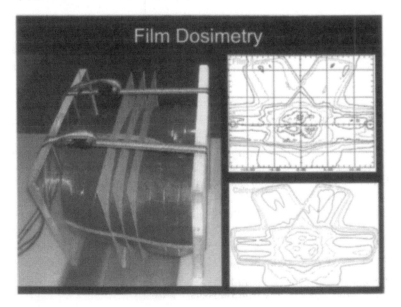

Figure 3.53. Film dosimetry for IMRT verification using a RANDO phantom. (Courtesy of Drs Viv Cosgrove and David Convery.) (See website for colour version.)

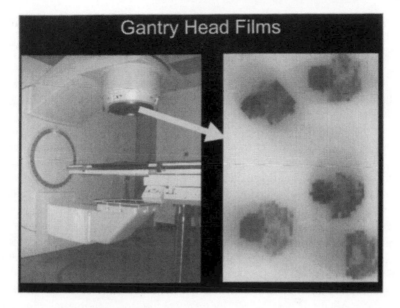

Figure 3.54. Film strapped to the gantry exposed to show the delivered fluence. (Courtesy of Drs Viv Cosgrove and David Convery.)

that the field boundary can be verified for each field. A dose calculation was performed using the intensity pattern and compared with that produced by the inverse treatment-planning system. Agreement was to within 4%. The authors claim that this makes redundant the need for in-phantom measurements to verify IMRT.

3.13.11   Comparison of delivered modulated fluence profile with plan-predicted modulated profile

The IMRT-planning process usually comprises three stages. The first and second are the computation of fluence maps and the coupled calculation of dose using these fluence maps. The third is the use of a leaf sequencer to give the pattern of MLC leaf movements which will deliver the required fluence. The whole process is far more complex than the delivery of traditional or even geometrically-shaped CFRT and so its QA is a major issue. Most workers to date have concentrated on making measurements of the delivered dose and compared these with the calculated dose plans. Techniques have involved using an EPID, film, multiple-point dosimeters and the WBIS system. Xing and Li (2000) pointed out that such quality-assurance techniques are dependent on the combined performances of both the leaf translator (or interpreter) and the functionality of the MLC equipment. They have decided to view this quality-assurance problem differently and to separate the two elements of the process. So, in their study, they did not consider the QA of the leaf movements directly but, instead, concentrate on whether the leaf interpreter will produce the required fluences to match those produced by the planning system.

The interpreter translates a 2D map of bixel fluences into a set of leaf positions. In doing so, it needs to take into account the specific equipment limitations imposed by each manufacturer's MLC. It also needs to take into account the transmission through the leaves and head scatter. In essence, the interpreter must generate a set of leaf patterns which will produce a 2D intensity modulation in which the combined effects of both direct radiation and leakage radiation yield the desired fluence pattern that was generated by the treatment-planning computer. This is an iterative process. It relies on the use of a specified dose model (see section 3.2.2). The crux of the work presented by Xing and Li (2000) was to show that, when fluence modulations from the CORVUS treatment-planning system were passed through an interpreter with a suitable dose model, the computed fluence modulations differed by no more than 0.5% anywhere in the field from the required prescription of fluences. They showed that this close agreement was found for 20 specified patients and for prostate plans comprising eight 2D modulations per plan. They then went on to show that there was also an agreement within 2% between the fluence modulations so computed and the measurements which were made with the WBIS system thus showing that the agreement did not critically depend on the model for dose deposition.

These conclusions are the same as those from Convery *et al* (2000) and Cosgrove *et al* (2000). It could be argued that in separating the QA into these two components and only looking carefully at one of them, the process is less rigorous than those based on the use of portal-imaging systems.

Xing (2001) further reported on the separation of the problems of quality assuring the leaf movement and the QA of the dose distribution. In this work, the leaf sequences were fed to software which simulated the motion of the MLC and generated a new fluence map which was then compared quantitatively with the reference map of the treatment-planning system. This approach was used to validate the CORVUS and HELIOS treatment-planning systems and separated out potential errors due to leaf motion from those due to other delivery effects.

Azcona *et al* (2002) compared IMRT plans made with a HELAX TMS system with measurements for several test modulated fields and several clinical cases. They concluded that 'fudge factors' in the planning system need to reflect the size of the field components. Errors can still arise due to inadequate modelling of transmission, head scatter and output factors. Some of these modelling errors are due to different assumptions made about the jaws between planning and delivery. Planning assumed the jaws were set at the envelope of the summed field components whereas Siemens Mevatron Primus IMRT delivery set the jaws to the envelope of each field component in step-and-shoot mode.

3.13.12   Verification of canine and human IMRT using *in-vivo* dosimetry

Phillips *et al* (2000, 2001a) have emphasized the requirement for QA of all aspects of the IMRT chain. Paraspinal and nasal tumours in dogs were treated using the Elekta IMRT techniques. Biological endpoints were used in live dogs to assess the dosimetry. The clinical models were purpose-bred dogs treated using an Elekta SL20 accelerator. Of 15 dogs, nine had experimental paraspinal tumours and six were controls. Twenty-one fractions were delivered with 4 Gy per fraction.

The validation of dynamic IMRT for head-and-neck tumours was as follows (Parsai *et al* 2001a, b). CT contours were first defined using drawing tools in PRISM. Planning was performed using projection onto convex sets (POCS) and the macro-pencil-beam algorithm and delivery was, in the early years, by dynamic delivery and later by step-and-shoot (Elekta changed their supported technique). The data from POCS fed into a translator which was verified using a forward dose calculation and provided the segments needed for linac delivery. IMRT was delivered using two 6 MV Elekta SL linear accelerators fitted with MLCs. It was found that, on average, there were about 100 segments per field, about seven fields per plan, about 400–500 MUs per beam and an average delivery time of about 20 min. A linear diode array was used to check beam profiles. Parsai *et al* (2001a, b) showed profiles of dose delivered to a phantom which showed good agreement between measurements made with a linear array of diodes and calculations of the dose profile. Measurements were also made with films. Delivery of individual beams to phantoms determined the MU calibration. Dog cadavers were also irradiated with TLDs in the spine and the readings from TLDs and the calculations matched quite well. It was determined that there was no need for Monte Carlo calculations. Dose measurements were made using *in-vivo* dosimeters. The precision and degree of accuracy with which IMRT can be delivered was assessed by creating highly localized dose distributions to tumours near critical structures. *In-vivo* and *in-vitro* measurements were made. Physical measurements of the dose distribution were made in live dogs with targets adjacent to the spinal cord in the head-and-neck region and also in deceased cadaver dogs.

Parsai *et al* (2001a, b) found that, as well as dose profiles measured in cadaver dogs closely following the pattern of the predicted dose profiles, measurements using TLDs contained in the critical normal tissue structures of live canines also matched calculations within 3%. Overall the survival rate for both control dogs and dogs receiving IMRT were found to be markedly different favouring those irradiated with IMRT. After a year's follow-up of IMRT *versus* non-IMRT treated dogs, it was found that the control animals showed severe normal-tissue damage, severe tetraparesis, compared to the IMRT group. This veterinary work has led to a successful implementation of an IMRT programme.

This group also developed step-and-shoot compensators using a special tool within PRISM. This tool optimizes beam weights for field segments and can eliminate segments with small MUs. Overall it was concluded that they had validated the dose-calculation model and also the means of delivery using the dMLC technique.

Engstrom *et al* (2002) present what they believe to be the first TLD *in-vivo* (human) dosimetry of IMRT. A naso-oesophageal tube filled with TLD rods was inserted into the patient before treatment and data from eight patients and almost 299 measurements agreed well with the planned dose from the treatment-planning system.

Higgins *et al* (2003) have investigated *in-vivo* diode dosimetry for the routine QA of clinical IMRT. For the type of diode selected, it was found, for delivering a wide variety of intensity-modulated fields to a water phantom, that diode and ion chamber agreed within 2.5%. The variability between diode and expected (i.e. planned) delivery on average was less than 5% with some outliers rising higher. They then conducted *in-vivo* measurements using diodes and found again that the ratio of diode to expected dose was generally within 5% but could not be guaranteed to better than 10%. The rationale behind this development is the intention to use standard patient diode systems employed for conventional radiotherapy translated across to IMRT.

3.13.13   Polyacrylamide gel (PAG) dosimetry for IMRT verification

Polymer gels (or polyacrylamide gels [PAGs]) are gels which, when irradiated, change several physical properties that are amenable to readout using medical imaging techniques. They change their transverse magnetic resonance (T_2 MR) relaxation time, their x-ray linear attenuation coefficient and their ultrasound acoustic properties. Lepage *et al* (2001) have proposed a method of using Raman spectroscopy to measure the concentration of proton pools and so to understand the processes which take place when monomer gels are polymerized by irradiation.

3.13.13.1   Review

MacDougall *et al* (2002) have conducted a review of the precision and accuracy of dose measurement in photon radiotherapy using polymer gel dosimetry. This was not a simple literature review but instead aimed to put together the data that would form a meta-analysis for certain questions concerning the accuracy of gel dosimetry with respect to other techniques. The authors surprised themselves by finding that there were very little data of a firm nature, most authors having compared polymer gel dosimetry with film or TLD measurements or specific planning calculations. The systematic review by MacDougall *et al* (2002) on the precision and accuracy of dose measurements in photon radiotherapy using polymer dosimetry was criticized by De Deene *et al* (2003). They made the points that the authors of the report were not experts in the field and they felt that the authors had made incorrect and misleading impressions about the accuracy of gel dosimetry. MacDougall *et al* (2003) then replied that much of the criticism was based on a misconception that exists around systematic reviews and evidence-based medicine. The review was not meant to be a topical review in the traditional sense but one in which the accuracy of a particular technique was evaluated purely in terms of the evidence produced by the peer-review papers.

Polyacrylamide gel (PAG) dosimetry has much promise and is being actively explored by many groups. It is promising for IMRT verification. It is probably fair to say it has not become a serious rival yet for other more traditional dosimetry methods. Firstly we review new methods for reading out PAGs.

3.13.13.2   PAG readout techniques

Hills *et al* (2000a) have shown that PAGs change their x-ray attenuation and so may be imaged with CT. This is simpler and cheaper than using magnetic resonance imaging (MRI) readout. The dose response was linear and reproducible and insensitive to temperature but has poor dose resolution. Hills *et al* (2000b) applied x-ray CT readout to gels from stereotactic radiosurgery. The high-dose region was CT scanned in 3 mm intervals with 3 mm slice thickness. To improve image quality, 16 images were averaged per slice and a background subtraction was performed. Then the dose images were imported into the treatment-planning software and registered with planning CT images using image fusion. The two distributions agreed to within 1.5 mm. A second application was made to proton therapy to characterize depth-dose. Trapp (2003) has made an in-depth study of gel dosimetry readout using x-ray CT techniques.

Mather *et al* (2002) investigated the changes in ultrasound acoustic parameters of irradiated gels. It was found that using a fundamental frequency of 4 MHz, a pulse repetition rate of 1 kHz and using the far field, the following were observed. The inverse acoustic speed varied quasilinearly up to about 20 Gy with a speed-dose sensitivity of $1.8 \times 10^{-4} \pm 1 \times 10^{-5}$ s m^{-1} Gy^{-1} before reaching a plateau with further increase in dose (figure 3.55). The variation of attenuation with absorbed dose was quasilinear up to about 15 Gy with acoustic-attenuation-dose sensitivity of 3.9 ± 0.7 dB m^{-1} Gy^{-1} (figure 3.56). The variation in transmitted signal intensity with absorbed dose was also quasilinear up to about 15 Gy with inverse acoustic transmitted signal-dose sensitivity of 3.2 ± 0.5 V^{-1} Gy^{-1}. There was considerable scatter on the curves and further work was planned. However, ultrasound readout may provide an alternative to the use of MRI, optical tomography or x-ray CT.

Oldham *et al* (1998) presented a new technique to improve the calibration accuracy in BANG-gel dosimetry. They made use of a 16-echo Carr–Purcell–Meiboom–Gill multispin echo sequence and used echoes 3–14 to determine the sample value of T_2. By irradiating a large number of individual long vials irradiated end-on to take advantage of the depth-dose distributions to obtain a range of doses and increase the calibration accuracy, they came to the conclusion that the error on dose was due to the error in the intercept of $R_2 (= 1/T_2)$ at low doses and was dominated by the uncertainty in the gradient for larger

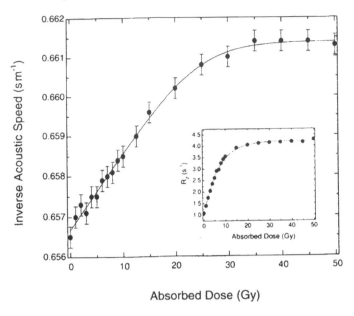

Figure 3.55. The inverse of acoustic speed of propagation in an irradiated gel as a function of absorbed dose shown in a comparison with the MRI result of the dependence of the R_2 relaxation parameter as a function of absorbed dose. Notice that the former is linear over a much wider range. (From Mather *et al* 2002.)

doses. The following year, Baldock *et al* (1999) showed that using a two-echo technique, the error was dominated by the error in R_2. Most recently Low *et al* (2000a) have repeated the measurements made by Oldham *et al* (1998) but using the two-echo sequence. In a detailed analysis of the errors in dose, it was concluded that their results supported the conclusion of Baldock *et al* (1999) that the limiting calibration precision is the voxel-to-voxel standard deviation. It was concluded that systematic errors due to MRI scanning artefacts limited the use of the technique to an overall error of roughly 0.2 Gy. Trapp *et al* (2004a) have also studied error propagation in gel readout.

Oldham *et al* (2001) have shown how optical CT techniques can be used to readout irradiated gels. It was found that high precision (better than 1.3% noise at high dose) and a spatial resolution of 1 mm^3 were achievable (figure 3.57). Oldham *et al* (2003) have shown further developments to automate scanning (figure 3.58) and Oldham and Kim (2004) have developed techniques to correct for distortion (based on scanning a matrix of needles) and for scatter.

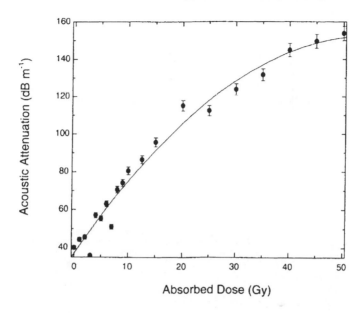

Figure 3.56. The acoustic attenuation of irradiated gel as a function of absorbed dose. (From Mather *et al* 2002.)

3.13.13.3   *Use of PAGs for IMRT verification*

Groll *et al* (2000) have implemented IMRT using the Varian dMLC system and delivered treatment plans to an Alderson phantom and BANG-gel phantom, the plans being calculated by CADPLAN. Calculations and measurements were in sufficient agreement that the technique proceeded to clinical implementation.

De Deene *et al* (2000) have investigated the chemical stability of monomer and polymer gels and two different types of long-term instabilities have been detected. Gel dosimetry has been validated as a 3D dose-verification method for CFRT treatments in an anthropomorphic phantom. Future developments include the development of lung-equivalent gels.

Gustavsson *et al* (2000) have used polymer-gel dosimetry to verify the delivery of a C-shaped target dose distribution by IMRT. The treatment was planned using the HELIOS software from Varian and delivered using the sliding-window technique on a Varian 2300 CD accelerator with a 120-leaf dMLC. Good agreement was found between the dose distribution recorded by the gel and the corresponding data from the treatment-planning system.

De Wagter *et al* (2001) have used polymer-gel dosimetry to verify the dosimetric accuracy of IMRT to typically 2% with a spatial resolution of 1 mm in three dimensions.

Love *et al* (2003) have verified that IMRT of the breast delivers a smaller target dose inhomogeneity than conventional tangential fields.

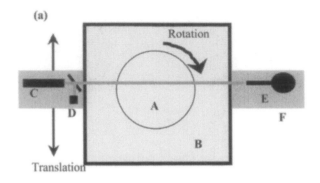

Figure 3.57. Optical CT scanning apparatus for irradiated gels: (a) schematic view from the top, (b) left-hand side of scanner, (c) right-hand side of scanner. Labels are: A = gel flask, B = optically matched water bath, C = laser, D = reference photodiode, E = lab-sphere-field photodiode assembly, F = baseplate. (From Oldham *et al* 2001.)

Figure 3.58. Modified optical CT scanner for irradiated gels incorporating automated multislice acquisition. Components are: A = laser, B = mirrors, C = beamsplitter, D = reference diode, E = linear stepping translation stage, F = optical bath containing immersed gel dosimeter, G = vertical translation stage, mounting rotation stage and suspending gel dosimeter in the optical bath, H = field photodiode, and I = X95 (structural mounts) support. The two mirrors B on opposite sides of the water bath translate the laser beam across the optical bath for acquisition of each projection. (From Oldham *et al* 2003.)

Novotny (2002) has used polymer-gel dosimetry for verifying dose distributions delivered using a Leksell gamma knife.

Vergote *et al* (2004) used polymer-gel dosimetry to verify a clinical IMAT treatment for whole pelvic radiotherapy. They obtained agreement to 3% in regions outside high dose gradients.

Trapp *et al* (2004b) have made the first PAG verification measurements that have influenced the choice of clinical delivery.

3.13.13.4   New PAGs

Gustavsson *et al* (2001a) have optimized the composition of a polymer gel for 3D verification in radiotherapy.   Acrylamide was exchanged for 2-hydroxyethylacrylate.   Two series of gels were prepared in which the concentration of monomer and the bisacrylamide (BIS) per total amount of monomer was varied percentage by weight. It was found that the maximum concentration of the monomers was determined by the solubility of BIS in water. Increasing the fraction of BIS lowers the absorbed dose sensitivity but extends the dynamic absorbed dose range.

Gustavsson *et al* (2001b) showed the comparison between calculated and measured IMRT dose distributions where the latter was a volumetric measurement with a new polymer gel using copper and ascorbic acid called MAGIC (Fong *et al* 2001). Calculated and measured dose-volume histograms agreed. The MAGIC gel can be used under normal levels of oxygen which is a major advantage since excluding oxygen in the manufacture of other gels is problematic.

Gustavsson *et al* (2002) has evaluated the MAGIC polymer gel for IMRT verification. IMRT was delivered using 6 MV photons from a Varian 2300 CD linac equipped with a 120-leaf dMLC and a kidney-shaped target was defined with OARs. It was found using the (Low) γ criterion that 95% of the points in the volume of interest fulfilled the criterion at the 3 mm or 3% deviation level. The gel was read out both at 2 and 14 days after irradiation using an MR sequence with 32 multiple spin echoes.

Fricke gels have not entirely been abandoned. Gum *et al* (2001) have developed a special phantom comprising Fricke gel to measure an IMRT treatment plan generated by KONRAD with seven fields. The gel phantom was then immediately read out using an MRI system and agreed with the plan to better than 5%. Parker and Brodeur (2002) have used ferrous-sulphate-gel-based dosimetry to verify IMRT. The gel was placed in a head-shaped phantom and preliminary results indicated agreement with calculation to better than 10%.

3.13.14   Film as an IMRT dosimeter

Sohn *et al* (2000b, 2003) have studied the dosimetry of small fields, created using the NOMOS MIMiC, of size 0.5 cm × 0.5 cm, 1 cm × 1 cm, 2 cm × 2 cm … 6 cm × 6 cm. They measured the central-axis depth–dose curves, the profiles at depth and the output factors and compared these with the predictions of a CORVUS planning system calibrated using a 4 cm × 4 cm field. Measurement were made with radiochromic film. It was found that the planning system accurately predicted the dose per MU at depth for all field sizes greater than 1 cm × 1 cm. However, disagreements were found for small depths (build-up regions) and in the penumbra regions. The output factor reduced with field-size reduction but was accurately predicted down to 1 cm × 1 cm but wrongly predicted for 0.5 cm × 0.5 cm field size. The implication for IMRT is not clear but it is apparent that, when using small fields, the need to make measurements of dose and to at least question if not disbelieve the predictions of planning systems is advised.

Minken and Mijnheer (2001) have investigated the potential of radiographic film for measuring output factors of small fields for IMRT. It was found that radiographic film behaved more like a solid state detector for small fields than a water-like detector and led to an over-response due to beam hardening. Mijnheer (2001a) showed that the photon energy spectrum changes with depth and the film is spectrum dependent. It is already known that film irradiated side-on has a different response to film irradiated face-on and relative dose distributions are to be preferred over absolute dose measurements with film. IMRT verification can also be performed with TLDs, ionization chambers and *in-vivo* dosimetry. Portal imaging is also being developed to verify IMRT with the latest development being the use of flat-panel imagers. This group concluded that radiographic film in dose verification of IMRT plans should be handled cautiously.

Martens *et al* (2002) have studied whether film is an acceptably accurate dosimeter for IMRT. There are concerns that radiographic film has an energy-dependent response, that sensitivity increases with field size and depth due to an increased contribution from low-energy photons and that the photon energy spectrum also changes with position from the central ray. It was observed that the response varied by no more than 3% as a function of field size with respect to a standard 10 cm × 10 cm field. The film response varied with the dose rate (and so fractionation) due to the well known failure of reciprocity and an intermittency effect. These variations can be as much as 9% of the OAR dose in an IMRT plan. However, if a maximum of five beams contribute to a plan, then the overall effect is good to 2%. It was concluded that film is an acceptable IMRT detector.

Ju *et al* (2002) have investigated the low-energy response of film as an IMRT dosimeter. Film over-responds to low-energy photons that are more present in penumbra regions and regions of low photon attenuation. They demonstrated this phenomenon by making film measurements of pyramid and inverted-pyramid fluence distributions showing differences between one-off film measurements and the sum of film measurements for the constituent field components. They made investigations with film both perpendicular and parallel to the direction of incidence of the radiation. They recommended the use of thin lead filters that improved the match between film measurements and ion chamber measurements. Bucciolini *et al* (2004) implemented this suggestion and showed that with the lead filters the dose-*versus*-density curve was independent of field size for both normal and tangential irradiation. They showed good agreements between IMRT plans and film dosimetry as a result.

Fayos *et al* (2001a) have delivered IMRT with the Varian dMLC system with plans created using CADPLAN. Verification measurements were made by irradiating films in a phantom using patient-specific IMBs. When the dose on a particular plan was analysed, the low-dose gradient and high-dose gradient areas were studied separately. Fayos *et al* (2001b) use Kodak XOMAT-V film and ionization chambers to verify the movement of leaves on a Varian 2100 CD Clinac accelerator for IMRT.

Nguyen *et al* (2002) have presented the characteristics of Kodak EDR2 film which allows it to be used for IMRT verification due to its extended dose range and Huang *et al* (2000) have also used Gafchromic film in the verification procedure of IMRT.

Low *et al* (2002a) have described a system in which they use a relatively inexpensive document and transparency scanner as a densitometer for routine verification of IMRT. They compared the performance of this instrument, which was never designed for this purpose, with that of a confocal scanning laser densitometer and found that the maximum-mean and standard deviation of the measured dose differences between the document scanner and the confocal scanner were 1.48% and 1.06% respectively. They argued that this system provides accurate and precise measurements up to an optical density of two, sufficient for routine IMRT verification and thus widening the scope of verification of IMRT to smaller community clinics.

Perrin *et al* (2003) have developed a software tool to compare the measurements of film from phantoms irradiated with modulated beams with the planning predictions. A variety of visual analysis methods were available including isodose overlays, dose-difference maps, profile overlays, difference isodoses and dose-difference histograms as well as the γ image. Schreibmann *et al* (2004a) have also presented a software package for dose visualization in IMRT.

3.14   Quality assurance (QA) of MLC delivery

In this section, attention is focused on the performance of the MLC itself rather than on the verification of clinical IMRT delivered with the MLC (see section 3.13).

3.14.1   Average leaf-pair opening (ALPO)

Zygmanski *et al* (2000) and Zygmanski and Kung (2001) have presented a method to identify dMLC irradiation sequences that are highly sensitive to a systematic MLC calibration error. The concept of an average leaf-pair opening (ALPO) is defined. This is, as its name suggests, the mean value of the distance between leading and trailing leaves weighted by the MUs for each segment where the average is taken over all segments and all leaf pairs. The denominator calibrates for the total number of MUs summed over all leaf pairs. The paper gives some simple geometrical examples (see also Kung and Chen 2000a).

There are two sources of a systematic MLC gap error. The first is centreline mechanical offset (CMO) and the second is the radiation field offset (RFO) in the case of a rounded-end MLC (see section 3.2.3). It was shown that a systematic gap error of $\Delta x = 2\Delta$RFO $+$ CMO leads to a fractional average fluence error of $\Delta x/(\text{ALPO} + 2\text{RFO} + \epsilon)$ where ϵ is generally of the order of 1 mm. Using this result, any set of 2D dMLC sequences, either step-and-shoot or dynamic, can be used to compute, firstly the ALPO and secondly, for given RFO and CMO, the corresponding fractional average fluence error. Close agreement was shown between analytically calculated values of this error and numerical values of this error. All that remains is to specify the threshold beyond which this error must not grow in order to obtain a binary 'yes, no' distinguisher between sequences which are acceptable and sequences which must be rejected. This paper includes the sentence that the term dMLC technique means both step-and-shoot and sliding-window delivery technique. Zygmanski *et al* (2003a) capitalize on this concept by designing dMLC patterns that are least sensitive to patient motion.

Zygmanski *et al* (2001b) suggested a tolerance of 2 mm for the Varian 52-leaf dMLC treatments. There is a trade-off between having a leaf tolerance which is too tight (which leads to very accurate dosimetry) and the consequent increase in delivery time due to tripping the accelerator (see also Zygmanski *et al* 2003b).

3.14.2   Routine QA of MLC leaf movement

LoSasso *et al* (2001) have described in detail their comprehensive QA for the delivery of IMRT with an MLC used in the dynamic mode in the Memorial Sloan Kettering Cancer Institute, New York (see also LoSasso 2003). They described specific QA procedures that are followed to assure the quality of operations. These resulted from an incremental improvement to the overall QA programme started when IMRT treatments commenced in 1995. It was identified that the calibration of the gap between opposing leaves was a critical component of QA. Very small gap errors could lead to large dose inaccuracies. Two methods were used to identify encoders that exhibited excessive count loss. Firstly, specific training modulations were delivered to film from which it could be determined visually which leaves had motors that were going out of control. A second technique used a software utility to exercise leaves through a calibration cycle, then an exercise cycle and then a second calibration cycle and the change in counts between the two calibration cycles assessed the ability of individual leaves to maintain their calibration during dMLC operation. Diagrams were provided showing how many motors had been replaced over a period of time and geographically which motors of the specific sets had been replaced (figure 3.59). It was noted that the main replacements occurred in the central sets of motors, those that were used most regularly for IMRT purposes and therefore over-exercised. Some 77 motors had been replaced over a five-year period to September 2000. The changes in the calibration of the leaf coders gave a very visual indication of a failing motor and the technique at Memorial Sloan Kettering is to replace motors long before they cause specific dosimetric problems. This study requires detailed reading to understand the specific QA checks.

In addition to these machine checks, specific dose-based verification was made for specific patient IMRT treatments (for a general discussion of this topic,

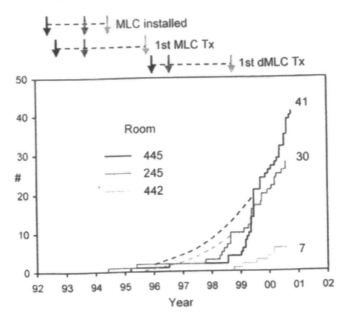

Figure 3.59. The chronology of leaf motor replacement in three treatment rooms at Memorial Sloan Kettering Hospital, New York. The full lines indicate the actual cumulative motor replacements for each MLC and the broken lines show the shapes of the curves that would arise from applying the current criteria for replacing motors prior to 1998. Time lines indicate the approximate dates of MLC installation, the first static MLC treatment and the first dMLC treatment for each MLC. The motors that were most likely to be replaced were those that were used in the dynamic mode for prostate treatments. These were the motors close to the central axis because these fields rarely exceeded 10 cm in length. This is a snap-shot picture as the situation was in November 2001. (From LoSasso *et al* 2001.)

see section 3.13). One pre-treatment check measured point doses to 2D dose distributions in a phantom for comparison with calculated results. The ratio of measured and prescribed doses for approximately 400 patients was within 1%. It was noted that individual points that were beyond this mean ratio were usually lying in the TG regions or were due to small set-up uncertainties. In summary, LoSasso *et al* (2001) pointed out the three-stage process of IMRT QA. The first is routine acceptance testing and commissioning. The second is the use of a routine QA process and the third is a patient-specific QA process to minimize the possibility of human error.

Dirkx and Heijmen (2000b) have implemented the dMLC technique on the MM50 Racetrack Microtron in Rotterdam. Inverse-planning techniques were developed to create intensity modulations and interpretation techniques to develop leaf trajectories that take into account scatter, penumbra, transmission and other dosimetric effects. Before introducing the technique clinically, they went through a careful process of QA following guidelines previously given by Chui *et al* (1996) and LoSasso *et al* (1998a, b, 2001).

The way in which the control of the MM50 is operated was described in detail. The MLC is calibrated by successively positioning the leaves at five positions, 0, ± 7, −14 and +15 cm, storing the corresponding readings of the potentiometers in a file. Then, to position a leaf at an intermediate position, linear interpretation between the stored calibration values was used. It had previously been demonstrated that the MLC control is precise in terms of static-field definition to within 0.05 cm.

The first test to be carried out swept a 0.1 cm by 40 cm slit beam across the field stopping the leaf pairs at 7 cm intervals, a technique known to generate a bright line of radiation at right angles to the direction of the leaf travel for perfect leaf travel (sometimes referred to as a garden fence test). It was concluded over an 80-day period that the leaf positioning was stable to within 0.05 cm. A second test swept a 0.4 cm by 10 cm slit across the beam and measured the dose deposited at the centre of a uniform 10 cm by 10 cm square thus built up and it was found that the variation in dose was only 0.4% pointing to the stability of the gap width in time being better than 0.01 cm. When the same test was carried out with the gantry at different orientations, the same observation was made indicating that the movement of the leaves was not affected by gravity.

Specific intensity-modulated fields created for a patient with prostate cancer were then measured using a portal-imaging device and again the variation in the dose profiles over a long period of time was found to be less than 1% and, on average, the leaves were positioned correctly within 0.01 cm. There was a small increase in inaccuracy to 0.03 cm when the leaves were entirely travelling against gravity at a gantry angle of 270°.

Dirkx and Heijmen (2000b) then introduced deliberate interrupts to the treatment and showed that these introduced changes in dose distribution, which were less than 1%. They thus concluded that the acceleration and deceleration of leaves could be effectively ignored. Overall the results of the QA investigation showed that the dMLC technique executed with the MM50 Racetrack Microtron was acceptable for clinical introduction.

Essers *et al* (2001) have commissioned the Varian CADPLAN (HELIOS) IMRT planning system and linked it to delivery of the dMLC technique using a Varian 2300 CD machine. Quality-assurance tests were performed, some similar to those recommended by Chui *et al* (1996) and LoSasso *et al* (1998a, b, 2001). By sweeping small slits to define fields and deliberately accelerating and decelerating leaves they determined that the accuracy of leaf positioning was within 0.5 mm and the effects of acceleration and deceleration were negligible with dose errors less than 2%. IMRT profiles were delivered on separate time occasions with a stability better than 2% even when leaves travelled against gravity.

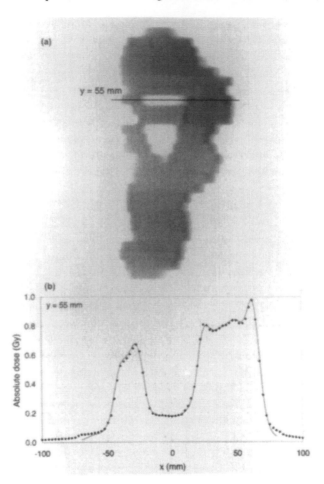

Figure 3.60. The figure shows (a) an IMB delivered to film measured at 5 cm depth in a 6 MV beam and (b) a measurement using a diamond detector and calculated (line) dose profile in the direction of the leaf motion at $y = 55$ mm. (Reprinted from Essers *et al* (2001) with copyright permission from Elsevier.)

A number of different 1D and 2D profiles were delivered and dose calculation compared with measurement (figure 3.60). The discrepancy between the two depended on the values assumed for certain parameters. It was found that the best agreement was obtained for a leaf transmission of 1.8% and a 'leaf separation' of 2 mm. Agreements were within 2% or 2 mm in regions of high dose gradients. Conversely, the agreement worsened when 'manufacturer recommended' values were alternatively set.

Some deliberate attempts to create somewhat clinically unrealistic narrow peaks led to larger 6% dose discrepancies. TG underdose was not specifically excluded but tends to be minimized because all leaf travel times are constrained to be the same.

Xia *et al* (2002a) have studied the precision of the correlation between the MLC positions and the cumulative MUs for Varian-equipment-delivered IMRT. It was shown that, when a profile was delivered with small MUs and/or at a high dose rate, significant artefacts could be generated for test fields. This was explained in terms of the sampling rate (20 Hz) and the communication time lag between the control systems. Bayouth *et al* (2003b) have commented on the relation between leaf-position accuracy and dosimetric accuracy and presented film tests to determine accuracy.

Duan *et al* (2003b) determined that, for a Varian accelerator fitted with a 120-leaf MLC, the delay time due to communication with the dMLC was about 90 ms. This leads to discrepancies in dose delivered with IMRT, which increase with dose rate and with the frequency of beam interruptions. Low dose rates, slow leaf speeds and low frequencies of beam interruptions reduce the effect of the delay and catch-up cycle.

Klein *et al* (2000) have presented techniques for assuring the quality of IMRT delivered using multiple-segmented fields. The origin of this requirement was their goal to deliver segmented MLC IMRT, so that the prostate could be given a dose that was different from that to the prostate and seminal vesicles combined. The comparison of calculated with experimentally delivered irradiation showed the dose distributions differed by no more than 4%.

Ramsey *et al* (2001a) have made a study of the leaf-position error during conformal-dynamic and intensity-modulated arc treatments. The characteristic of their technique is that they make measurements of leaf-position accuracy taken with leaf velocities arranging from 0.3–3 cm s^{-1}. It was found that the average position errors ranged from 0.03–0.21 cm with the largest deviations occurring at the maximum achievable leaf velocity.

Low *et al* (2001a) have studied the performance of a Varian MLC for delivering IMRT in both static and dynamic mode. In static mode, it was found that the error introduced by field abutment mismatches was $16.7 \pm 0.7\%$ mm^{-1} at 6 MV and $12.8 \pm 0.7\%$ mm^{-1} at 18 MV. They also found that the leaf-positioning accuracy was 0.08 mm and so concluded the static MSF IMRT technique was accurate enough. In dMLC mode, they studied the degradation in smooth fluence delivery at high leaf speeds.

Woo (2000) has used the Siemens BEAMVIEW portal imager to verify IMRT segments delivered by the dMLC technique for QA purposes.

Vieira *et al* (2001) have developed specific modulation patterns which, when measured using an electronic-portal imaging device, can indicate whether sliding leaves have been mispositioned. Measurements were performed both with a Varian 2300 CD Unit and a MM50 Racetrack Microtron.

Bruinvis and Damen (2001) have described the various QA tests that must be made for a treatment-planning system for IMRT applications.

Olch (2001) has described the QA tests that have to be employed in a small paediatric American clinic in implementing IMRT using the Varian accelerator and the Nucletron PLATO treatment-planning system.

Chen *et al* (2002a) have developed a set of calibration tests for MLC leaf positioning in IMRT. They find that when the gantry is at 90° or 270° with a collimator angle of zero larger discrepancies than expected were observed due to the effect of gravity.

Mavroidis *et al* (2000) have presented the results of the framework BIOMED1 EC project 'Dynarad' showing the variability of radiotherapy techniques in use throughout Europe and illustrating the benefits and quality requirements of the more conformal techniques such as IMRT.

Bate *et al* (2000) described how the post of Quality Control Officer was established in the University of Ghent IMRT treatment programme. During the start-up phase of routine IMRT treatments, it was considered necessary to have such a rigorous quality-control programme to monitor errors in medical treatment prescription, simulation, treatment planning, the treatment data transfer and daily set-up.

De Brabandere *et al* (2002) have studied 12 patients whose IMRT treatments were planned with CADPLAN/HELIOS. They then characterized the beams in terms of their skewness and asymmetry. They attempted to see whether these quantities were generally the same for all 12 treatments. The means and standard deviations were found and then a QA tool was proposed whereby if a plan deviated from these, it was suspected of error. They admit their proposal is somewhat counterintuitive given that IMRT treatments are designed to be customised to each patient.

Beavis *et al* (2002) have implemented clinical IMRT using the CMS FOCUS treatment-planning system and delivering the treatments with a Varian 600 CD linac using a step-and-shoot method with the Millennium-120 MLC. They describe the QA that is required and show that measured MU comparisons agree within 1.5% with predictions when 'dosimetry unfriendly' segments are removed.

Valinta *et al* (2001a) implemented IMRT using a Siemens PRIMUS accelerator linked to a CORVUS inverse-planning system and they describe the quality control procedures developed.

3.14.3   Modelling the effects of MLC error

Cho *et al* (2001b) have studied the influence of random field perturbations on dynamic IMRT. They did this by assuming a Gaussian distribution of the position of leaves relative to the expected true position and quantitated the outcome in terms of changes to the dose distribution and dose-volume histogram.

Liu and Xing (2000) have used CORVUS to evaluate the effects of both random and systematic errors of MLC leaf positions in IMRT delivery. They studied three sites—prostate, nasopharynx and brain—with five patients for each site. They found that the systematic errors were a larger influence on the dose distribution than the small random errors which tended to smear out. They determined that a leaf-positioning accuracy of 0.8 mm was necessary if the target dose was to remain within 3% of its specified value.

Parsai *et al* (2003) have investigated the effects of random and systematic beam-modulator errors in dynamic IMRT via a modelling study (figure 3.61). The leaf positions of MLC leaves during dynamic delivery and also of jaws during dynamic delivery were randomized using a Gaussian distribution of specified width. During this process, the mechanical limitations of the particular MLC (the Elekta MLC) were respected in the leaf-sequencing algorithm so that if leaves were randomized to closer than 5 mm they were pushed apart by half the required difference to establish half a centimetre gap. At the same time, the maximum velocity allowed by the hardware of 2 cm s^{-1} also provided a constraint that was not violated during the randomization. Hence, the statistical randomization led to an error distribution that was a clipped Gaussian rather than a true Gaussian distribution. Beams were selected from a clival meningioma case because of the close tolerances imposed by the sensitive normal structures. It was shown that the target-dose uniformity decreased with increase in random errors. When both MLCs and jaws were perturbed, the minimum target dose began to deviate by 5% of the prescription when a standard deviation of 1 mm was used in the randomization. In comparison, if MLCs or jaws alone were perturbed, 5% deviation did not occur until the standard deviation was increased to 1.5 mm. In addition to these random variations, systematic perturbations were modelled and even errors of the order of half a millimetre were shown to result in significant dosimetric deviations.

Somewhat counterintuitively, it might be expected that, if the results from a large number of such randomizations were averaged, this statistical uncertainty would all wash out but in fact this was shown not to be the case and target doses were always less than those predicted without randomization. The authors offer an explanation of this in terms of the fact that on average randomization of both jaw and leaf positions will decrease the radiation aperture in 75% of the cases.

3.15 Summary

The MLC has an established place in the delivery of IMRT. Considerable ingenuity has been applied to sequencers and to factoring in the effects of machine constraints on delivery. Attention has been given to considering the measurement of and role of leakage through the collimation and through the gaps in the collimation. Both MSF and dMLC techniques are now widely established and in clinical practice and we may expect these to seriously rival the use of the NOMOS MIMiC, particularly in Europe. IMAT is now a commercially available method and variations are appearing. Hybrid IMRT delivery methods have been

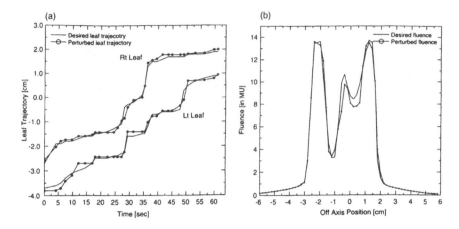

Figure 3.61. (a) Leaf trajectory for two opposing leaves in a dynamic delivery; (b) corresponding fluence profiles. The dotted lines in (a) and (b) represent the state of the MLC / diaphragms when perturbed randomly with a standard deviation of 1 mm. The bold lines are desired deliveries. (From Parsai *et al* 2003.)

developed to include catering for the mix of MSF and dMLC techniques in one delivery. Hybridicity also led to the use of rotating collimators to increase the flexibility of MLC-based methods. There have been more microMLCs designed and built as well as attention given to the theory of leaf width. A huge effort has been given to QA of the fundamental equipment and also to techniques which verify specific clinical deliveries.

Chapter 4

Developments in IMRT not using an MLC

4.1 The Cyberknife

The Cyberknife is a robotic device holding an X-band linac, which is capable of irradiating tumours stereotactically with feedback of organ motion to the robot (see also chapter 6). The robot allows six-axis control. The use of the robotic arm allows the linac to be moved with a high degree of freedom. Levin (2001) has described in detail the scientific and business history of Accuray, the company which markets the Cyberknife (figures 4.1 and 4.2).

The initial vision of the person, John Adler, who started the development was that this would extend the concept of frame-based head-and-neck stereotactic radiotherapy and radiosurgery to frameless whole-body radiotherapy and radiosurgery. Adler had worked with Lars Leksell in Sweden, the developer of the Gamma knife (now manufactured by Sweden's Elekta AB Company). Adler first worked at the Harvard Medical School in the Massachusetts General Hospital in Boston. He then left Harvard in 1990 to go to the Stanford University School of Medicine where he founded Sunnyvale, California-based Accuray with the goal of carrying Lars Leksell's vision from the brain to the rest of the body. The development is described as the third major step in radiotherapy technology last century (the first being the use of computed tomography data post 1972 to better define radiation shapes), the second being the development of IMRT in the 1990s and the third being the future of radiotherapy centred around the Cyberknife. The main thrust of the approach is to take account of patient movement, breathing and other involuntary movements so that the exact location of the tumour is at any one moment tracked by the robotic linac. Adler's philosophy was that, because radiosurgery works so well in the brain, there was no biological reason why it should not have the same clinical application throughout the rest of the body. The development of the Cyberknife became possible when one of Adler's patients introduced him to a small California-based company that made portable linear accelerators originally designed for industrial inspections, which were relatively lightweight and therefore easily attachable to a robot. The financial history of

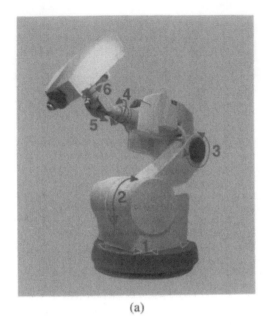

(a)

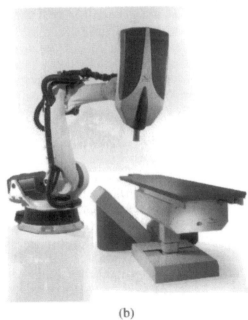

(b)

Figure 4.1. (a) The most recent version of the Cyberknife and its six motions (from Accuray website); (b) also showing the x-ray image acquisition detectors. (Courtesy of the Accuray Corporation.)

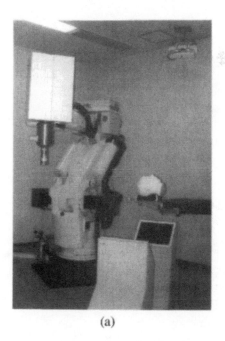

(a)

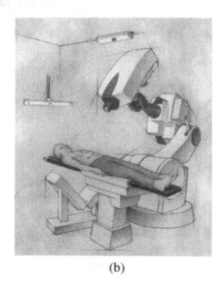

(b)

Figure 4.2. (a) shows the Accuray Cyberknife at Osaka, Japan. The 6 MV X-band linac is mounted on the arm of a robotic manipulator and the treatment couch is positioned between the two x-ray cameras and their respective diagnostic x-ray tubes. In (b) a schematic diagram of the real-time tracking system is shown. By using the data from the infrared tracking system and stereo x-ray imaging system, the robot arm moves the linac to the ideal position in real time. This is an earlier version of the equipment shown in figure 4.1. (From Shiomi *et al* 2001b.)

the Company is somewhat tortuous. Originally Adler looked for funding sources through angel investors and by 1994 Accuray had five sales on the books and had completed the installation of the first system. There was a period just after this of (eventually abortive) negotiations with Picker International, a unit of General Electric Co plc, and the recent success of Accuray has largely been due to Asian-based financial investment mainly from Japan and Taiwan. Accuray's present Asian distributor is the Meditec Corporation, a wholly owned healthcare subsidiary of the Marubeni Corporation.

The key feature, described as the 'crown jewels' of the Cyberknife system, is its use of minimally invasive on-the-fly imaging to continually feedback the position of the tumour to the robot, now known as Accutrak or Synchrony (figures 4.3 and 4.4). Essentially the procedure ensures that the tumour is in the same place with respect to the robot at the beginning, during and at the end of the treatment. Effectively the robot 'chases the tumour'. The procedure is painless but can last anywhere from 30–90 min. Accuray received 510(k) approval for the

Cyberknife for treatments of head-and-neck tumours in July 1999 and in 2001 lodged an application with the Food and Drug Administration (FDA) for approval to treat extracranial tumours such as those of the prostate, breast and liver. One of the main difficulties of marketing this equipment is the difficulty of selling a $3 million piece of capital equipment in today's cost-constrained environment, particularly against larger competitors that are committed to the technology status quo. (This is a general issue pertaining to the introduction of any new technology which essentially challenges the role of existing technology. The argument is that there can be no evidence for the benefit of the new until there are many clinical installations and there are unlikely to be these until there is clinical evidence, a 'Catch-22 situation'—see also section 2.5). However, by the end of 2000, it was reported that Accuray had sold and shipped a total of 18 Cyberknife systems, five in the United States and 13 in Japan. This number is growing (Thompson 2003, private communication). Currently, European regulatory approval is still to be sought. It is claimed that the main advantage of the Cyberknife is that it can be used for both radiotherapy and radiosurgery with the equivalent of intensity modulation whereas most photon IMRT systems can only be used for radiotherapy. In particular, it is expected that with patients having wide access to the internet, they will look for centres with this kind of equipment and the marketing side of Accuray certainly believes that one cannot afford not to have a Cyberknife.

Schweikard and Adler (2000) have described how patient motion can be tracked and fed back to compensation using the robotic arm which controls the Cyberknife system. The Cyberknife is equipped with two fluoroscopic cameras, which constitutes an image-processing system and, based on the information about the patient's location, the robot 'chases the moving patient'. The essence of the technique is that stereo x-ray imaging determines the position of precise internal markers which are correlated with the spatial location of external markers placed on the patient's skin which are viewed by infrared detecting systems (figure 4.5). Then, during the treatment, the external markers are continually monitored. The internal position of the target is computed from this monitored position data and, if the target is not in the correct position with respect to the beam, the patient or the beam may be realigned as treatment continues. Yan (2003), using different equipment, have shown a similar reproducible correlation between the motion of internal markers detected by x-rays and the motion of external markers detected by infrared. Ramaseshan *et al* (2003) have also correlated the movement of internal and external markers for lung.

Webb (1999, 2000a,e) has shown how robotic IMRT might become the ultimate IMRT delivery technique in that it can sculpt contorted dose distributions with high precision. He compared the technique with the delivery of IMRT using more conventional methods and showed that, whilst these should remain the clinical norm in the current climate of research into alternatives, it is useful to know the degree by which these methods fall short of the ultimate.

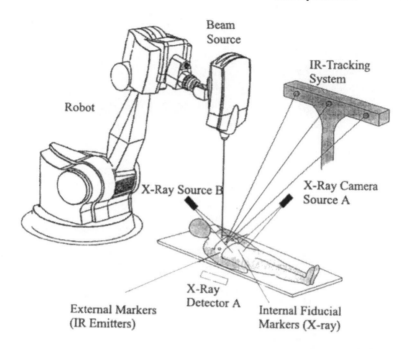

Figure 4.3. A system overview of the Accuray Cyberknife and also the motion detecting system. Infrared tracking is used to record the motion of external markers attached to the patient's abdominal and chest surfaces. Stereo x-ray imaging is used to record the 3D position of internal markers (gold fiducials) at fixed time intervals (e.g. 10 s) during treatment. The relationship between the motion of the external markers and the internal markers is determined and, from this, the continuous measurement of the motion of external markers is used to translate to a pseudo continuous measurement of the motion of internal markers. (From Schweikard *et al* 2000.)

Rodebaugh *et al* (2001a) have reported on the use of the Cyberknife for treating lung lesions, some of which are partially mobile during the irradiation time. This makes use of the Accutrak/Synchrony infrared light-emitting-diode tracking system that is used to indirectly track target movement during respiration. Rodebaugh *et al* (2001b) have examined the accuracy of the Accutrak system in conjunction with the Cyberknife. It was found that the robot has an inherent delay in responding to instructions from the measurement of position using skin-located light-emitting diodes and the Accutrak uses an adaptive filtering algorithm to predict the tumour location 0.8 s into the future. It was determined that the means of the errors ranged from 0.6 to 3.4 mm. Murphy *et al* (2001) have shown that, in feeding back motion to the Cyberknife, it is necessary to take account of non-stationary motions and time-changing phase relationships.

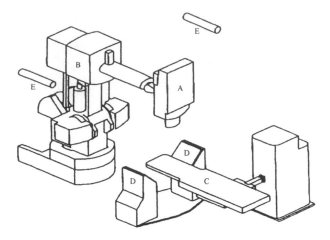

Figure 4.4. A schematic diagram of the Cyberknife, illustrating (a) the X-band linac, (b) the robotic manipulator, (c) the treatment couch, (d) the diagnostic imaging cameras, (e) the x-ray sources. (Reprinted from Murphy *et al* (2003) with copyright permission from Elsevier.)

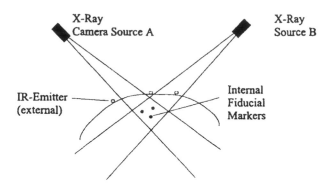

Figure 4.5. This shows the way in which the Accutrak imaging system records the internal movement of the tumour and feeds it back to the robot to correct beam direction. X-ray sources A and B are viewed by x-ray detectors A and B (not shown, see figures 4.1–4.4). These record the position of the fiducial markers every so often (e.g. every 10 s). The infrared tracking system operates continuously recording the positions of the external infrared markers. These positions of the internal and external markers are correlated and used to predict the motion of the internal markers when only that of the external markers is recorded. (From Schweikard *et al* 2000.)

The Cyberknife can be fitted with circular collimators of different radii. Deng *et al* (2001b) have used the EGS4/Beam system for computing the elemental dose distributions for various cone sizes. Ma *et al* (2001b) have reviewed the role of the Cyberknife in radiation therapy emphasizing image guidance. Specifically, detailed beam phase-space data are now available for accurate patient dose calculation using the Monte Carlo method and treatment planning and beam delivery have been validated using various phantoms and the dose distributions have been verified using Monte Carlo simulations.

Guerrero *et al* (2001) have used the Monte Carlo package EGS4/MCDOSE to make a prediction of dose distribution using the Cyberknife with and without feedback of motion. When motion was not fed back to the robot, dose coverage of thoracic tumours significantly decreased. With feedback of the motion, the coverage was restored to close to the planned value.

Shiomi *et al* (2001a) have developed a method of QA for an image-guided frameless radiosurgery system which uses radiochromic film. The QA method was applied to the Cyberknife. In order for the Cyberknife to maintain high accuracy, precise radiation is only possible if the coordinate systems, which the Cyberknife keeps, i.e. the patient's coordinates, the robotic coordinates and the image-processing coordinates are all matched accurately. Before performing irradiation in individual cases, Shiomi *et al* (2001a) calculate the overall irradiation error using gafchromic film. The errors are then fed back into the Cyberknife planning to improve accuracy. The technique comprises replanning the patient treatment onto a phantom comprising a box phantom of a film set between 3 mm-wide spacers. The treatment is then delivered to the phantom and the films are read out to provide a measure of the dose distribution. The centre of gravity of the dose distribution and that of the planned distribution is measured for each slice and also in three dimensions and mislocation between the two is assumed to be an error. A Cyberknife was installed at Osaka University Hospital in April 1998 and has passed through three stages of commissioning. In its current stage, the overall median error is known to be about 0.7 mm. Since the planning is performed with voxels of size 3 mm^3 and the range of brain movement within the skull under normal conditions is estimated to be about 1 mm, the Cyberknife was assumed to be adequately accurate. The particular Cyberknife at the Osaka University Hospital was installed in such a way that the angle between the couch and the robot was 45°. These measurements were made each time a patient was to be treated. In August 2001 (Hemmi, private communication), there were just nine Cyberknife systems in the world. The five in America are at Stanford, Cleveland Clinic, the University of Texas, the University of Pittsburgh and Newport Beach, CA and the four in Japan are at Osaka, Okayama, Yamaguchi and Kumamoti. Shiomi *et al* (2001b) have also provided a description of the Cyberknife (in Japanese).

Murphy (2002, 2004) has discussed three facets of frameless radiosurgery using the Cyberknife. The first is the accuracy with which it can direct the dose to the treatment volume. The second are the methods that can be used to deliver radiosurgery doses to extracranial sites that cannot be easily immobilized by a frame and the third is the pattern of movement that has been observed in patients during frameless treatment.

Gibbs *et al* (2002) have presented the first clinical results of using the Cyberknife system for treating spinal lesions. Gall and Chang (2003) have shown that the dose calculation of the planning system coupled to the Cyberknife is accurate to 2% provided tumours are not close to large tissue inhomogeneities. If they are, the dosimetry can be compromised, an area for future investigation.

Related to the development of robotics and other special purpose accelerators, a recent *Medical Physics* 'Point and Counterpoint' argues the case for and against a purpose-built linac for IMRT. The case 'for' includes the view that, for IMRT, large fields with flatness are not required since in general small-field components make up the IMBs. Scanning low-energy linacs could suffice. The argument 'against' is based on the objection to a purpose-built machine, given that conventional linacs with MLCs can do the job perfectly well.

4.2   The design of the shuttling MLC (SMLC)

Webb (2000c) proposed a radically new concept for an MLC for a photon linear accelerator for delivering IMRT with high MU efficiency. The concept is to consider each M (rows) $\times N$ (columns) 2D IMB as a $N/2$ set of M (rows) $\times 2$ (columns) areas of modulation. Each area is then delivered by a set of M shuttling attenuating elements (called here the shuttling MLC) with a very high MU efficiency. The elements shuttle between each of the two rows comprising the $M \times 2$ area and the modulation is provided by the dwell-time variation of the elements (figures 4.6 and 4.7). The SMLC was extensively modelled and the principles of this SMLC were discussed and examples illustrating its intended operation were given. The main achievement reported was the development and robust testing of an interpreter which describes the position–time course of movement of the elements as a function of MUs. This interpreter fully accounts for leakage transmission through the elements. It completely avoids the across-the-rows TG underdose. A large number of 1D and 2D IMBs have been subjected to this interpreter and it was shown that, for random patterns of fluence, the SMLC is more MU efficient than the Bortfeld–Boyer technique (the most efficient with a conventional MLC) when the modulation is highly structured. SMLCs can increase the MU efficiency of IMRT compared with the multiple-static-field (MSF-MLC) technique or dMLC technique with conventional MLCs.

In a companion study (Webb 2001a), two new arrangements similar to that described in the earlier study, but with less mechanical complexity, were shown to be constructionally simpler but less MU efficient (figures 4.8 and 4.9). Additionally another new concept of SMLC was shown which also increases the MU efficiency compared to the MSF-MLC technique and often improves the MU efficiency compared with the previously reported SMLC for highly modulated intensity distributions (figure 4.10). It also leads to zero TG underdose in the direction orthogonal to that of the shuttling elements (so-called across-the-rows). None of these designs have been turned into practical devices yet.

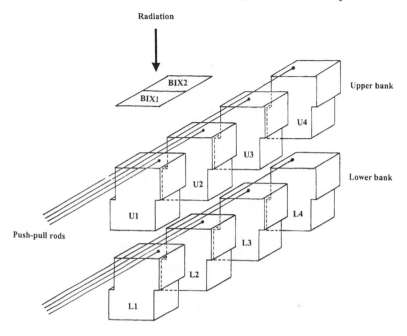

Figure 4.6. Showing the principle of the shuttling MLC (SMLC) or 'push-pull' or 'flip-flop' intensity-modulating device. The device comprises multiple rows of attenuating elements (such as U1) arranged in two banks, one immediately above the other. Just one row is shown in the diagram and just four elements are shown for each of the two banks. In practice, there would be many more elements per bank per row and many more rows. The elements U1 and L1 can move into the spaces shown between them and U2, L2 respectively. By varying the dwell time of the elements in these spaces and in their positions as shown the intensity in bixels BIX1 and BIX2 can be modulated, e.g. U1 and L1 can either be above each other as shown or both above each other in the adjacent space or one can be as shown and the other in the adjacent space (essentially blocking both BIX1 and BIX2). The other pairs of elements behave independently the same way (e.g. U2 and L2, U3 and L3 etc). Tongue-and-groove arrangements couple the rows (shown). Different tongue-and-groove arrangements are needed at the faces of the elements along the rows (not shown).

4.3 IMRT with the 'jaws-plus-mask' technique

IMRT generally requires complex equipment for delivery. Just one study has investigated the use of 'jaws-only' IMRT with not discouraging conclusions. However, the MU efficiency is still considered to be too low compared with the use of an MLC. Webb (2002a) proposed a new IMRT delivery technique which does not require the MLC and is only moderately more complex than the use of jaws alone. In this method, a secondary collimator (mask) is employed together

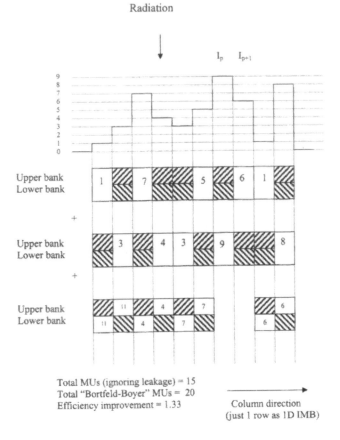

Figure 4.7. This shows a 2D IMB and the technique to deliver it with the shuttling MLC (SMLC). The 1D IMB has five pairs of non-zero bixels. Each pair is delivered independently using an elemental shuttling arrangement of just two attenuators, one in the upper bank and the other in the lower bank. These are shown shaded (and imply zero fluence). Open elements with numerals indicate the open bixels and the MUs delivered with the SMLC in this state. I_p and I_{p+1} are the two adjacent bixels whose sum is the maximum of the sum of all such pairs in the sct. Here, e.g., bixels 7 and 8 require just two states of the SMLC (component deliveries). The other bixels require three states. Note the SMLC attenuators do not necessarily change at each monitor unit interval. They have one of three states (the patterns shown) with a different number of monitor units assigned to each state. The time taken to change states is ignored.

with the linac jaws (figure 4.11). This mask may translate parallel to the jaw axes. Two types of mask have been investigated. One is a regular binary-attenuation pattern and the other is a random binary-attenuation pattern (figure 4.12). Studies showed that the MU efficiency of this 'jaws-plus-mask' technique, with a random

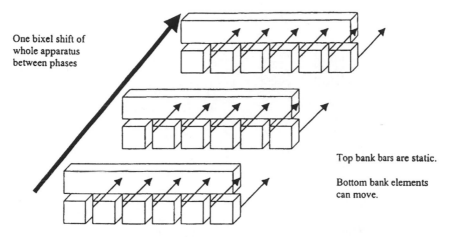

One bixel shift of
whole apparatus
between phases

Top bank bars are static.

Bottom bank elements
can move.

Figure 4.8. 3D view of a HC(i)2-W-SMLC (HC = half complexity) with one static bank and one movable bank plus an overall 1-bixel shift to move from irradiation phase 1 to phase 2. The arrangement to deliver a two-dimensional IMB (6×6 elements are shown) is shown.

binary mask, is more than double that of the jaws-only technique for typical 2D IMBs of size 10 bixels $\times$ 10 bixels and with a peak value of 10 MU (or quantised into 10 fluence increments). For 2D IMBs of size 15 bixels $\times$ 15 bixels with a peak value of 10 MU (or quantised into 10 fluence increments), the MU efficiency of the 'jaws-plus-mask' technique with a random binary mask is almost triple that of the jaws-only technique. Some further extensions to this concept were presented showing that some more practical mask arrangements are possible but with somewhat compromised MU efficiency. Some comments were provided on practicalities and on delivery times.

Webb (2002b) showed that a much simpler relocatable mask than previously conceptualized can lead to very similar improvements in MU efficiency and a decrease in the number of field components compared with the use of jaws only (JO). The new concept comprises a set of relocatable single-bixel attenuators (SBAs) which can be moved into the field components otherwise collimated by jaws only (figure 4.13). Typically for a 15 bixels $\times$ 15 bixels 2D matrix of fluence with a peak value of $I_p = 10$ MUs (or, equivalently, 10 stratified fluence levels) and using just four SBAs the MU efficiency is nearly three times that of the JO technique and the number of field components is reduced to about 0.6 the number required by the JO technique. These gains become greater by using more SBAs or for larger I_p values. Component reordering was achieved to minimize the total delivery time including intersegment deadtimes. Practicalities were discussed.

For IMRT, a simple technique was described by Webb (2002c) which split any $N \times N$ matrix I of fluences into two mathematically equivalent matrices I_A and I_B summing to $2I$, each of which requires considerably fewer segments to

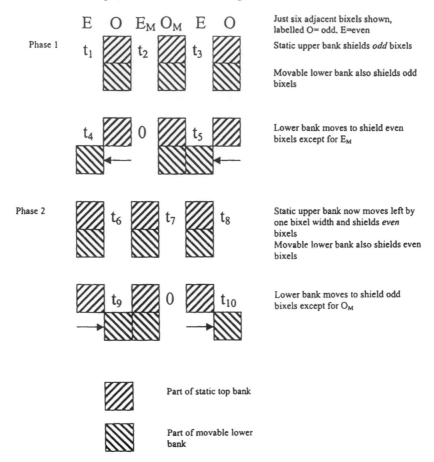

Figure 4.9. Principle of operation of a HC(i)2-W-SMLC (HC = half complexity) with one static bank and one movable bank plus an overall 1-bixel shift. Shown is the arrangement to deliver a one-dimensional IMB (six elements are shown corresponding to one row in figure 4.8). The elements are labelled odd (O) and even (E). O_M and E_M are the odd and even elements which have the greatest adjacent (odd + even) sum. The irradiation is in two consecutive phases. For phase 1 (upper two parts of diagram), the static bank of attenuators (part of a static bar when there is a 2D IMB) shields the odd bixels. The movable elements shield either the odd or even elements in the configurations shown for the times t_i shown. The time t_2 is the overall time for the first phase. For phase 2 (lower two parts of diagram), the static bank of attenuators (part of a static bar when there is a 2D IMB) shields the even bixels. The movable elements shield either the even or odd elements in the configurations shown for the times t_i shown. The time t_7 is the overall time for the second phase. The total irradiation time is $T = t_2 + t_7$. All the times can be determined from the bixel values.

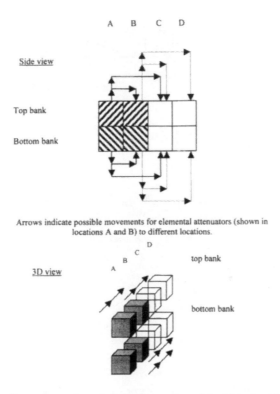

Arrows indicate possible movements for elemental attenuators (shown in locations A and B) to different locations.

Figure 4.10. The figure shows the principle of operation of the 4-bixel-wide shuttling MLC (4-W-SMLC). Just one component of the 4-W-SMLC is shown indicating the mechanism which can generate four bixels of a one-dimensional IMB. This mechanism is replicated in the direction of shuttling to create a row of such mechanisms which would deliver a 1D IMB. By stacking multiple such rows in columns, a 2D IMB can be delivered. In the elemental mechanism, there are two elemental attenuators in the top bank and two in the bottom bank. They are not shown to scale because they would be some 10 cm deep and only 4–5 mm wide. The arrows indicate the possible directions of movement of the attenuators. The elemental attenuators shown in position A can move discretely to positions B or C. Those shown in position B can move to positions C or D. The movement within the two banks is entirely independent. This generates $3^4 = 81$ possible formations but there is much redundancy (degeneracy). There are only six ways to create an aperture with two open bixels, four ways to create a 1-wide aperture and 1 way to create a 0-wide aperture (complete closure), i.e. 11 non-redundant formations. The degeneracy only affects the total attenuation in the blocked bixels. In the 3D view, the attenuating elements are shown shaded and the spaces into which they can move are adjacent. The gaps are for clarity and not present in practice.

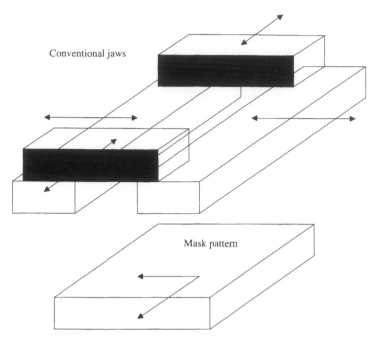

Conventional jaws

Mask pattern

Figure 4.11. Schematic diagram showing the jaws and the mask and the directions of translation. The diagram is not to scale and, in practice, the mask would be a section of a spherical annulus.

deliver IMRT using the jaws-plus-mask technique (Webb 2002a, b). When I_A and I_B are delivered alternately at fractions throughout the treatment the total number of field segments required is between one-half and two-thirds that of delivering matrix I when unadjusted (see also section 6.3.19).

Earl *et al* (2003b) have provided a direct aperture optimization (DAO) technique that can create JO IMRT with as few as 50 segments per field. This number will depend on the matrix size, the number of fluence levels required and the complexity of the modulation required.

4.4　The variable aperture collimator (VAC)

Webb *et al* (2003) extended these concepts of using a tertiary mask plus jaws for delivering IMRT without an MLC. The new concept was to sweep a variable aperture collimator (VAC) across the space of the IMB to be delivered and to strip this IMB down into MSF components, each deliverable with the VAC (figures 4.14 and 4.15). The stripping algorithm was described and it was shown, for several designs of VAC, that the mean number of field components and mean number of MUs was less using the VAC than would have been required for a

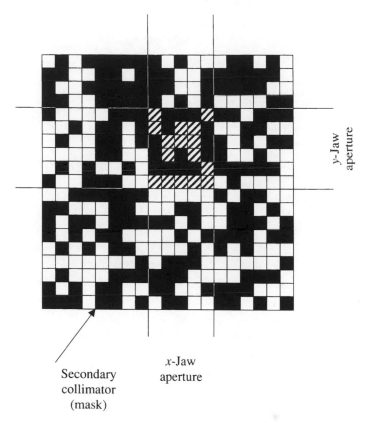

y-Jaw
aperture

x-Jaw
aperture

Secondary
collimator
(mask)

Figure 4.12. A schematic diagram showing how field components are collimated by jaws plus mask. The secondary collimator is shown as a matrix of black (attenuating) and white (transmitting) squares. Each square corresponds to each delivered bixel. The four locations of the jaws are shown as full black lines (made white in places for clarity). In the arrangement shown, those bixels shown shaded are open in the secondary collimator and thus irradiated. Note that this allows the grouping of an irregular shape of open bixels. The exact shape depends on the translation of the secondary collimator.

jaws-only (JO) decomposition. The VAC would be simpler to construct than the several previously suggested jaws-plus-mask (J+M) combinations. As well as describing a simple VAC for the use with jaws, a design concept of a hybrid VAC was proposed. It was also shown that adding the potential to rotate the simple or hybrid VAC for some components relative to the field to be modulated is advantageous. A DAO algorithm has also been worked out but currently only for a 1D VAC (Webb 2004b).

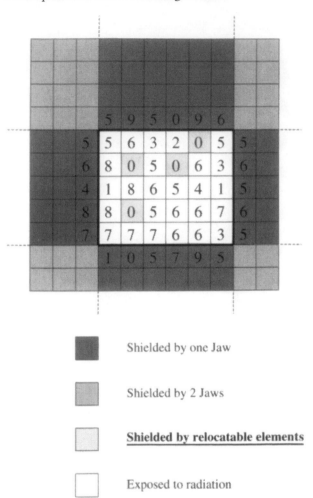

Figure 4.13. The concept of employing SBAs as well as jaws to create field components. The jaws are shown collimating a 5×6 part of a larger 2D IMB (not shown in full). The key indicates the functions of the jaws and the (in this case four) SBAs (relocatable elements). In this example component, 1 MU could be stripped off to create the next residual matrix.

4.5　One-dimensional IMRT

Broggi *et al* (2002) have continued to use the Milan 1D IMRT system using a beam modulator and have developed a rigorous and dedicated QA programme to guarantee its accuracy and reproducibility.

Figure 4.14. Schematic diagram of one form of variable aperture collimator (VAC) showing the six single-bixel attenuators (SBAs) in their parked positions (left) and one possible component (right). Everything outside this is tungsten blocked. The VAC would scan in a double-focused cradle (see figure 4.15). For each scan position, the six attenuators could be moved by springs or push rods.

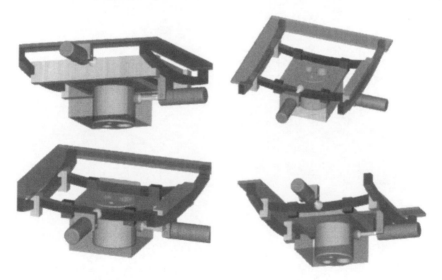

Figure 4.15. Prototype double-cradle arrangement for supporting the VAC. The figure shows the cradle supporting a block of attenuator with three holes which would be replaced by a VAC of the type shown in figure 4.14. The SBAs are double-focused. (See website for colour version.)

4.6 Summary

The main pieces of apparatus for delivering IMRT are the NOMOS MIMiC and MLC. However, the equipment reviewed here whilst playing a very small role, or being only conceptual, represents an attempt at lateral thinking. It removes thought from exploiting the MLC which was never intended for IMRT and thinks *ab initio*. The MIMiC itself was of course the outcome of such lateral thought and, apart from some concerns over matchline dosimetry, was and is very successful. Time will tell the role of the other proposals reviewed here.

Chapter 5

Clinical IMRT—evidence-based medicine?

Whenever a new medical technology is introduced there arises the question of the availability of evidence for its need and efficacy (Bentzen 2004). Then often follows the establishment of a Catch-22 in which the desired evidence cannot rapidly accrue until there is widely diffused technology to gather the said needed evidence. Added to this difficulty, the firm clinical evidence for improved therapeutic ratio may come many years after the trials set-up to gather it. Worse still, phase-3 clinical trials cannot even begin until phases-1 and -2 proof-of-principle trials have been completed. All these features conspire to produce a patchy picture of evidenced-based medicine in the field of CFRT and IMRT. It is this lack of large-scale hard clinical evidence that has fuelled some criticism of the rapid implementation of IMRT and fired discussion (see chapter 1). Conversely, it is this evidence that is exciting and fuelling further clinical IMRT.

The main clinical observations so far have been:

(i) There is lengthy observational evidence from Memorial Sloan Kettering Cancer Institute for the improved normal-tissue sparing of dose-escalated prostate CFRT and IMRT (Ling *et al* 1996) (see detail in section 5.1).

(ii) A positive outcome of the Royal Marsden NHS Foundation Trust phase-3 trial of prostate CFRT showed clear evidence that late rectal toxicity was reduced with improved geometrical shaping (Dearnaley *et al* 1999).

(iii) Chao *et al* (2000) have studied 41 patients with head-and-neck cancers who were enrolled in a prospective salivary-function study to attempt to correlate the functional outcome of parotid-gland sparing with dose received by the parotid during IMRT. It was found that the equivalent uniform dose (EUD) concept suggested that the mean dose to the parotid gland is a reasonable indicator of salivary-function outcome. The outcome of parotid function was assessed objectively by measuring stimulated and unstimulated saliva flow before and at six weeks, three months and six months after the completion of radiation therapy. The study led to an observation that salivary output decreased exponentially as a function of the mean radiation dose with an exponential coefficient of approximately 4% per Gy. Hence, 50% of the baseline saliva flow can be retained if both parotid glands receive a mean dose less than 16 Gy.

Most studies aiming to demonstrate the improvements consequent on the use of IMRT are based on demonstrating improved dose distributions. Conversely, Chao *et al* (2001) have presented some of the first clear clinical evidence for the reduction of late salivary toxicity without compromising tumour control in patients with oropharyngeal carcinoma. Over a 30-year period, up to the end of 1999, 430 patients with carcinoma of the oropharynx were treated at the Mallinckrodt Institute of Radiology. Of these, 14 patients were treated with the NOMOS MIMiC IMRT system post-operatively and 12 patients pre-operatively, definitively without surgery, using the same equipment. Acute and late normal tissue side effects were scored according to Radiation Therapy Oncology Group (RTOG) radiation morbidity criteria with a median follow-up of 3.9 years. The crucial observation was that when IMRT was compared with conventional techniques the dosimetric advantage of IMRT did translate into a significant reduction of late salivary toxicity in patients with oropharyngeal carcinoma. No adverse impact on tumour control and disease-free survival was observed in patients treated with IMRT. Indeed there was statistically insignificant improvement in two-year local regional control.

It was found that, after IMRT treatment, only 17–30% of the patients had late grade-2 xerostomia. Chao *et al* (2001) gave a very powerful presentation of why the blocking of salivary flow is extremely impairing on the quality of life. Whilst the reduction or elimination of salivary output is not life threatening, a severe loss of salivary flow is detrimental to the processing of solid food leading to nutritional deficiency. It leads to a deterioration of dental condition. Saliva facilitates speech and its loss therefore hinders speaking and communication which may in turn cause the patient to withdraw from social interaction. Also the deprivation of salivary flow leads to opportunistic infections in the oral cavity. The prevention of xerostomia is a major goal of head-and-neck IMRT and Chao *et al* 's study demonstrates hard evidence for the reduction in late salivary toxicity (see also Chao and Blanco 2003).

(iv) Maes *et al* (2002) have also produced genuine clinical evidence for decreased xerostomia following IMRT of bilateral elective neck irradiation. In 39 patients treated with conformal parotid-sparing radiotherapy, 78% have no, minor or acceptable complaints of a dry mouth 6–12 months after treatment and no recurrences were seen. Ipsilateral parotid gland sparing is more difficult.

(v) Kwong *et al* (2003) have presented clear clinical evidence for reduced xerostomia and increased alkalinity of saliva for late-response measurement following patients who have received IMRT.

(vi) Patel *et al* (2002), Lee *et al* (2002a) and Münther *et al* (2003) have also given clear clinical evidence of good response of salivary gland function after IMRT for head-and-neck tumours (see also section 5.6.6).

(vii) Claus *et al* (2002) have irradiated 32 patients with sinonasal tumours using IMRT and show that only one suffered dry-eye syndrome, a debilitating consequence of conventional radiation that is, at best, an irritant and, at worst, leads to total loss of vision. Mittal *et al* (2001) have shown that IMRT significantly reduces toxicity in treating advanced head-and-neck cancers and that the patient quality of life was significantly improved.

(viii) Yarnold *et al* (2002) have presented a preliminary analysis of a randomized phase-3 trial between IMRT of the whole breast and conventional two-tangent irradiations. It was found that a change in breast appearance was scored in 52% of cases allocated standard 2D treatment and in only 36% of patients allocated IMRT. These findings suggest that reduction in unwanted dose inhomogeneity impacts on clinically observable late breast changes.

(ix) Mundt *et al* (2001) have presented their clinical experience with intensity-modulated whole-pelvic radiation therapy (see also section 5.5). Butterfly-shaped concave high-dose volumes were created in the pelvis for the treatment of gynaecological malignances and the absence of high-grade toxicity in clinical follow-up was reported. Lujan *et al* (2001b) followed up this study showing that IMRT reduces dose to bone marrow in gynaecological radiotherapy. Mundt *et al* (2002) have designed either five- or seven-field IMRT for whole pelvic treatment using CORVUS and delivered plans using a step-and-shoot delivery on a Varian accelerator (figure 5.1). Forty women have been treated and followed up for reporting of acute toxicity. It was found that there was a significant reduction in both gastrointestinal (GI) and genitourinary (GU) toxicity with no grade-3 acute toxicity recorded at all. It was too soon to assess late toxicity absence but this was expected to follow. One unexpected benefit of IMRT was the reduction of irradiation of bone marrow leading to less severe haematological toxicity. Consequent on this, IMRT provides a means to increase the chemotherapeutic dose to these patients and so gain therapeutic benefit in a concomitant way not involving the radiation.

Against this background and vastly outweighing the presently sparse clinical data, there have been hundreds of papers published showing that CFRT and IMRT generate improved dose distributions with more conformal tumour coverage and sparing OARs. There are so many that no sensible list can be provided. However, chapter 4 of Webb (2000d) summarized the clinical studies up to the year 2000 and there are over 1000 references in this work. Taken with the now wide acceptance that biological models predict that 'better' dose distributions will lead to improved treatment outcome, these studies form the collective evidence for the need for CFRT and IMRT. The studies range through tumour sites within breast,

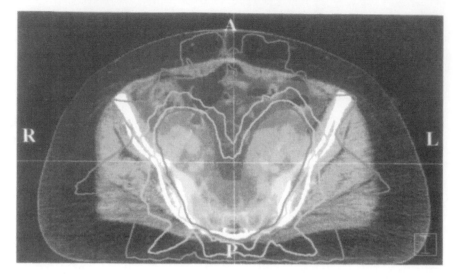

Figure 5.1. Isodose curves from an intensity-modulated whole pelvis radiotherapy plan superimposed on an axial CT slice through the upper pelvis. The small bowel and the PTV are shaded in orange and green, respectively. Highlighted are the 100% (red), 90% (green), 70% (light blue) and 50% (dark blue) isodose curves. The figure shows a high degree of conformity to the shape of the PTV in the upper pelvis minimizing the dose to the small bowel. (Reprinted from Mundt *et al* 2002 with copyright permission from Elsevier.) (See website for colour version.)

prostate, head-and-neck, brain, liver, lung, oesophagus, thyroid and paediatric tumours.

Mohan (2002) has summarized the present state of some future directions of IMRT research, noting that the key missing data concern the clinical outcome in the long run which is subject to debate. Nutting (2003) points also to the future for IMRT.

Schlegel (2003) has reported on the growth of clinical IMRT in Germany. An audit at February 2003 showed that 651 patients had been treated at 11 centres in Germany with DKFZ dominating this number at about 400: 60% of these patients were head-and-neck patients, 26% were prostate patients and 14% made up the rest. Schlegel has made an interesting observation that, if 20% of patients need IMRT, this leads to a prediction of 40 000 patients per year in Germany whereas the current capacity is 400 patients per year. He thus predicts that between 40 and 60 specialized centres will be needed. He points out the early developments in IMRT were made in Stockholm, Heidelberg, Ghent and in London and yet now Europe is behind the United States largely because of the fact that, in the United States IMRT can be billed for at about three times the billing cost for conventional radiotherapy. In both Germany and the UK, there was no extra billing possible for IMRT at the time of his review.

In the following sections, the latest clinical evidence is reviewed.

5.1   IMRT of the prostate showing measurable clinical benefit

One of the definitive implementations of prostate IMRT with the dMLC technique is at the Memorial Sloan Kettering Cancer Center, New York. Between September 1992 and February 1998, 232 patients with histologically proven prostate cancer were treated with CFRT. Of these, 61 were treated with geometrical shaping alone and the other 171 were treated with IMRT. In both sets, the prescription target dose was 81 Gy. The IMRT patients were planned with the DKFZ inverse-planning algorithm of Bortfeld *et al* (1994) modified by Mohan *et al* (1994). This optimized a dose-based cost function with constraints and the possibility to have an inhomogeneous dose distribution to the PTV. The IMRT technique was an isocentric five-field method; in the overlap region of prostate and rectum, an 88% prescription dose was set.

Zelefsky *et al* (2000) have published a report of the improvements consequent on the use of IMRT for this large group of patients (figure 5.2). It was found

(i)   that the dose conformality improved with IMRT (not a surprising finding, agreeing with others but parametrized carefully in the paper);

(ii) that there was a decrease in acute grade-2 and grade-3 GU toxicity but it was not significant;

(iii) that there was a significant decrease in acute grade-1 and-2 rectal toxicity with a concomitant increase in the number of patients having no toxicity;

(iv) that there was no difference in late GU toxicity but;

(v)   that there was a highly significant decrease in late grade-2 rectal toxicity (figure 5.3).

The study provided concrete evidence in support of the routine use of IMRT for these patients. The evidence complemented and extended that found by Dearnaley *et al* (1999) with respect to dose escalation in CFRT.

The study also escalated the dose to 86.4 Gy in an attempt to improve the tumour control probability (TCP) further. The study of Zelefsky *et al* (2000) also gave strong evidence to support the notion that rectal and bladder toxicities exhibit a volume effect. It was also suggested that there may be advantage in identifying specific subvolumes of the prostate which would benefit from a higher dose, something that can be arranged with IMRT dose painting of dominant intraprostatic lesions.

Zelefsky *et al* (2002) provided an update on the early toxicity and biochemical outcome for high-dose IMRT of the prostate. Between April 1996 and January 2001, 772 patients with clinically localized prostate cancer were

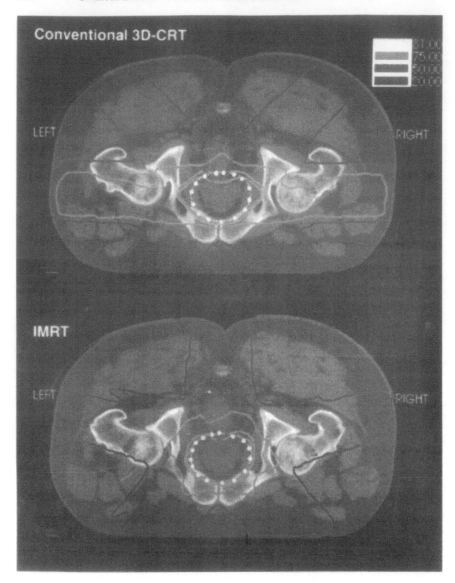

Figure 5.2. Dose distributions of treatment plans designed for a prostate cancer patient. The upper figure shows the outcome of conventional CFRT and the lower figure the IMRT plot. Note the improvement in conformality of the PTV coverage by the 75 and 81 Gy isodose lines in the IMRT plan. Also note that the 50 Gy isodose line avoids the femoral heads in the IMRT plan. (Reprinted from Zelefsky *et al* 2000 with copyright permission from Elsevier.) (See website for colour version.)

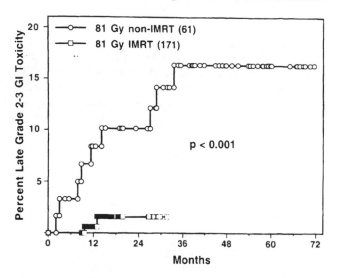

Figure 5.3. Actuarial incidence of the late grade-2 and grade-3 rectal bleeding for patients treated with 81 Gy IMRT compared with 80 Gy conventional 3D-CFRT. There were only two cases of grade-3 rectal bleeding, one in each group. (Reprinted from Zelefsky *et al* 2000 with copyright permission from Elsevier.)

treated with IMRT at Memorial Sloan Kettering Cancer Center: 698 patients were treated to 81 Gy, 74 patients were treated to 86.4 Gy. The median follow-up time was 24 months. Of the patients 4.5% developed acute grade-2 rectal toxicity. No patient experienced acute grade-3 or higher-grade rectal symptoms. Twenty eight percent of the patients developed acute grade-2 urinary symptoms, one patient experienced urinary retention and 1.5% of the patients developed late grade-2 rectal bleeding. 0.1% of the patients experienced grade-3 rectal toxicity. No grade-4 rectal complications were observed. The data demonstrate the feasibility of high-dose IMRT in a large number of patients. Zelefsky (2003) points to there being no decrease in bladder toxicity.

In other centres, the evidence for the efficacy of IMRT of the prostate has been gathering, as follows:

De Meerleer *et al* (2002a) have reported on acute rectal toxicity for 102 patients receiving IMRT for prostate cancer. They showed that acute toxicity was most pronounced during the timespan of fractionated radiotherapy but that grade-3 GI toxicity was not observed. Check-ups at 1 and 3 months after IMRT rarely revealed grade-3 GU toxicity. Three months after treatment any toxicity more than grade-1 was observed in less than 10% of the patients. This was not a randomized controlled trial but is useful information justifying the continuation of IMRT of the prostate.

Fiorino *et al* (2002) showed clear evidence of correlation between rectal dose-volume histograms and late rectal bleeding for planning and clinical data of 510 patients collated from four institutions, An important issue in validating CFRT and IMRT concerns characterizing the dose distribution with dose-space histograms and correlating these with either actual reported or probabilities of complications. This is not easy for the rectum given it is a hollow organ. Fiorino *et al* (2003) have modelled many situations with different geometries of the rectal diameter, wall thickness and varying filling fractions. It was shown that the normalized dose-volume histogram (normalized to equal numbers of calculation points per CT slice) and the true dose-wall histogram correlated well for empty or non-empty rectum. Given the relative ease to compute the former and the difficulty of computing the latter, this is a very useful observation. Conversely, for a very full rectum, the normalized dose-surface histogram (see Hoogeman *et al* 2004) was more representative of the dose-wall histogram.

In multi-institutional trials of new ways to treat prostate cancer patients, CT datasets will be contoured by many different clinicians. Foppiano *et al* (2003) asked 18 observers to contour four patients and then constructed four-and six-field conformal plans for each. The rectal dose was scored in terms of V_{40} through V_{75} to assess the impact of contour variation. This was judged less than 4% after training and it was concluded that the benefits of increased patient numbers from multi-institutional trials outweighed the disadvantages of relying on different observers.

Hoogeman *et al* (2002) followed up 266 patients with prostate cancer who were treated with either conventional or 3D CFRT. It was found that there was no correlation between absolute dose-volume histograms and the incidence of rectal bleeding but there was a volume effect for the relative DVHs of the rectal wall.

Kupelian *et al* (2001) have shown that late radiation damage due to IMRT of the prostate is considerably less than that due to conventional CFRT when the IMRT is delivered at 2.5 Gy per fraction to 70 Gy.

Teh *et al* (2001a) have treated 100 consecutive patients with localized prostate cancer between February 1997 and January 1999 at the Baylor College of Medicine, Houston, TX. They used the NOMOS MIMiC delivery system. They followed up the patients in clinics at six weeks and every four months thereafter and scored data on acute radiation damage using the RTOG acute-radiation-morbidity scoring criteria. It was found that, in general, acute GU toxicity and acute lower GI toxicity were reduced with the use of IMRT. They also found that the toxicities were all grade-2 or below and that the distribution between grades 0, 1 and 2 did not correlate with the mean dose above 65, 70 or above 75 Gy for either bladder or rectum. They interpreted these results as indicating that modest dose escalation can be performed for IMRT with safety. Teh *et al* (2001b) have reported lower late GI and GU toxicity following IMRT for the prostate.

Raben *et al* (2002) have measured acute toxicity and quality of life for patients treated with IMRT for prostate cancer. The overall rate of grade-1 and grade-2 GU morbidity at completion was 41% and 45% and the overall rate of grade-1 and grade-2 GI morbidity at completion was 17% and 4%. The morbidity decreased rapidly with time from the close of treatment.

Hofman *et al* (2002) have shown no major acute rectal complications when delivering 76 Gy to the prostate with a three-field IMRT technique.

Livsey *et al* (2003) have shown that reduced α/β ratios for prostate suggests that optimally selected hypofractionated regimes may produce better tumour control and reduced late toxicity. They reported on the results for the first ten patients studied in which no grade-3 toxicity was seen in a hypofractionated regime. Mott *et al* (2004) described the development of class solutions for hypofractionated prostate cancer treatment.

Clark *et al* (2003b) have conducted a randomized trial to investigate high-dose hypofractionated multisegment CFRT to treat prostate carcinoma. IMRT plans with three fields were created either using a 5 mm-leaf-width MLC or a 1 cm-leaf width MLC and prescriptions were either 74, 70 or 54 Gy to the prostate GTV. Two segments were delivered at each gantry angle. The planning was done by forward techniques.

Now we turn attention to some other clinical implementation of IMRT, concentrating on specific organ irradiation *but reviewing studies where clinical outcomes were not specifically presented (with some exceptions due to difficulty of systematically organizing material here).*

5.2 Comparison of treatment techniques for the prostate

It is a fairly unsatisfactory process attempting to review worldwide experience in implementing a clinical technique, particularly one such as IMRT of the prostate. There are diverse planning systems, diverse IMRT delivery techniques, a plethora of planning variables open to choice and a variety of institution conventions and protocols. There is certainly no agreed protocol for treatment worldwide. As a result any collation such as this is bound to be something of a mixed bag and it is quite hard to draw firm conclusions. Certainly compared with the precision with which one may review physics techniques, this is a more chequered field. As much grouping as possible of similar investigations will follow.

De Meerleer *et al* (2000a) have taken a cohort of 32 consecutive patients with different stages of prostate cancer and performed five planning studies on each. All plans used just three field directions (figure 5.4). They compared:

(i) conformal geometrically-shaped radiotherapy (CFRT) optimized manually,
(ii) CFRT computer-optimized with no limit on PTV inhomogeneity,
(iii) CFRT optimized with a limit of 15–20% on PTV inhomogeneity,
(iv) IMRT without a PTV homogeneity limit and
(v) IMRT with inhomogeneity limit.

The results were averaged and average dose statistics reported. They also reported PTV, rectal, bladder and femoral-head NTCP and the probability of

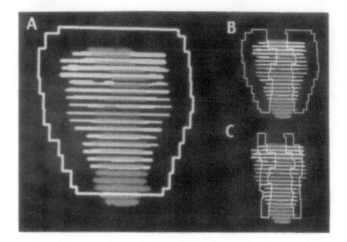

Figure 5.4. This illustrates an example of using segmented beams for IMRT as developed at the University of Ghent. The aperture A spans the whole target volume. Aperture B is the same minus a region in which the rectum is also in the line of sight of the anterior beam. Aperture C is a third segment. The apertures are separately weighted (Reprinted from de Meerleer *et al* 2000a with copyright permission from Elsevier.)

uncomplicated tumour control (P_+) for the series. When calculating TCP, they used some new data on the parameters α and σ_α of the Webb and Nahum (1993) model.

The overall conclusion was that the IMRT technique generated the best TCP for the prostate, generated the lowest rectal NTCP and the highest P_+. Other comparisons were presented in detail along with their significance or lack of it.

De Meerleer *et al* (2000b) further compared IMRT of the prostate using three *versus* five beams . It was found that there were no differences concerning the physical and biological endpoints for the target and that IMRT with three beams better spared the rectum and bladder. The probability P_+ of uncomplicated local control was worse with five beams and, by carefully weighting the anterior beam in the three-field technique, hot spots could be significantly reduced. They concluded that only three modulated beams were needed and were at least as safe as using five IMBs in the treatment of prostate cancer.

Sanchez-Nieto (2000) has shown how the use of TCP modelling can assist treatment-plan optimization. She showed that IMRT with seven fields can reduce the small-bowel irradiation when pelvic nodes are irradiated and that the NTCP is only 11% compared to 34% with three-field conventional therapy and 29% with three-field CFRT. This allows dose escalation to pelvic nodes. Because MRI can show the actual tumour extent within the prostate, it is possible to increase the dose to dominant intraprostatic lesions to 90 Gy delivering 70 Gy to the rest of the prostate. Sanchez-Nieto (2000) analysed this using the ΔTCP model and showed that the TCP increase is entirely due to the dose escalation to the dominant intraprostatic lesion and not due to general increase in dose to the rest of the prostate.

Stasi *et al* (2000) have investigated the MSF technique for prostate cancer using the HELAX treatment-planning system. A simple step-and-shoot technique was performed using five gantry angles and just ten segments and this was compared with a traditional plan created using the CADPLAN treatment-planning system. The dose-distribution changes were remarkable. The dose delivered to 30% of the rectum in the IMRT technique was 10% less than that delivered using a six-field conventional technique and a highly conformal PTV distribution was obtained. The IMRT technique was benchmarked to measurements and found to differ by no more than 1%.

Verellen *et al* (2000) have compared six different IMRT techniques including (i) tomotherapy, (ii) IMRT based on a combination of the CORVUS planning system and the Elekta MLC, (iii) static treatment beams with a miniMLC, (iv) dynamic arc treatment with a mini MLC and (v) IMRT with a mini MLC.

Verellen (2001) has compared five different types of IMRT and their appropriate planning systems. Verellen *et al* (2002) have compared four different treatment strategies for the prostate: (i) sequential (NOMOS) tomotherapy, (ii) dMLC or MSF IMRT with an MLC, (iii) dMLC with a miniMLC, (iv) dynamic field shaping with a miniMLC (dynamic arc). The comparison not only looked at conformality but also at treatment efficiency. The concern was that when all is considered, IMRT may not be the preferred choice. The basis for the comparison was a body of clinical experience in one clinic (AZ-VUB: Brussels). The specific comparison focused on both a convex prostate and a concave prostate.

The NOMOS MIMiC was attached to a Siemens Mevatron KDS2 6 MV accelerator. The dMLC and MSF technique was executed with the Elekta MLCi attached to an SL15 accelerator. The miniMLC dMLC technique was executed with the BrainLAB AG miniMLC attached to a NOVALIS accelerator and the interpreter of Agazaryan *et al* (1999). The dynamic-arc technique also used the BrainLAB miniMLC.

The MIMiC and macroMLC-IMRT techniques were planned with CORVUS. The other two techniques using the microMLC were planned with the BrainSCAN TPS, the dMLC using inverse planning and the dynamic arc using forward planning.

For the convex target, all the treatment modalities met the treatment goals. However, tomotherapy and the conventional MLC-IMRT technique needed a tenfold increase in number of MUs and the miniMLC technique twofold when compared with the dynamic arc method. The same trends were observed for the concave target although then the dynamic arc technique did not meet the desired dose reduction to the rectum.

Ezzell *et al* (2000) have developed a protocol for inverse treatment planning for IMRT of the prostate. They use the CORVUS 3.0 planning system and tested various prescription options including the beam arrangements. Interestingly, they re-planned patients five times, each time with the same set of beam orientations and tissue constraints but accessing beam directions in different random orders to determine the variability in output of the CORVUS inverse-planning system. The following results were found. A set of dose-volume-histogram prescription values for the target and normal structures could be found that reliably satisfied the minimum dose requirements. On average, it was found that a five-field plan with laterals, posterior–anterior (PA) and shallow anterior obliques provided 5% more rectal sparing than other IMRT techniques that included a large number of variations of using six fields or five, seven, nine or fifteen fields in equal angular increments. They also found that, by reducing the allowed dose to the rectum from 70 to 40 Gy, improved rectal sparing could be obtained at the cost of less target-dose uniformity and that the best prescription was 60 Gy to the rectum. Interestingly, it was found that for a given patient and beam arrangement, different computer runs produced quite different solutions. For example, the dose to 30% of the rectum varied by 9% for different computer runs of the optimized five-field plan for one particular patient. However, unequivocally, the IMRT plans were an improvement on conventional treatment planning.

Valinta *et al* (2000) have also used the CORVUS inverse treatment-planning system to optimize a combination of static MLC-shaped beams for IMRT. They used a Siemens PRIMUS accelerator and claim that angle selection is more important than the number of intensity levels in sparing the rectum. They used a limited number of fields and a limited number of levels of intensity. Valinta *et al* (2001b) measured the leaf calibration accuracy to 2 mm and found that doses measured and doses calculated by the treatment-planning system agreed to 2% at the treatment isocentre.

Shentall *et al* (2001) have used the Nucletron PLATO treatment-planning system to create five-field conformal and five-field IMRT treatments. The IMRT plans were produced with two aims. The first was to achieve a PTV dose uniformity of ±5% and the second aimed at sparing rectum as much as possible. The IMRT sequencing was limited to a total of 40 segments. It was found that there was very little difference between the conformal and IMRT plans where the main IMRT aim was PTV homogeneity. However, when the IMRT aim was to spare the rectum, this was achieved with some significant gains. The overall comment was that there is a considerable variation in the clinical results and such comparisons may require to be patient specific.

Mayles *et al* (2002) have also implemented IMRT using the Nucletron PLATO inverse treatment-planning software and delivered with an Elekta linac using the step-and-shoot technique. IMRT was implemented for a prostate boost to 70 Gy to the prostate GTV and 64 Gy to the prostate and seminal vesicles in 32 fractions. QA tests showed that changes were needed to correctly model the collimator and phantom scatter because PLATO uses the collimator scatter value of the largest field rather than allowing for the individual beamlets. When these corrections were made, it was found that measurements in water phantoms and in solid-water phantoms agreed with calculations to generally within 3% when the effects of adding together separate modulated fields were included. Individual modulated fields could differ by more than this 3% from the calculation. The first patient was treated in February 2002 since when the Clatterbridge Centre for Oncology has become one of the sites to implement prostate IMRT using completely commercial solutions (*Wavelength* 2003b).

Young *et al* (2000) have developed a seven-field IMRT technique for radiotherapy of the prostate which is specifically designed to spare the bulb of the penis. The technique involves the use of variable margins to the prostate and was compared with a 3D conformal technique employing seven treatment portals. It was found that the mean dose could be reduced from about 40 Gy to about 25 Gy if a 1 cm target margin was used and that this did not compromise the target homogeneity.

Sethi *et al* (2002, 2003) produced CFRT and IMRT plans for ten patients, the latter plans made using tomographic IMRT and step-and-shoot IMRT. The mean doses to the proximal penile tissue reduced by about 40% depending on prescription dose level to the prostate which was escalated from 73.8 Gy to 90 Gy. The mean doses to the corporal cavernosa and corpus spongiosum (bulb) similarly decreased by above 40%. Buyyounouski *et al* (2004) have also shown the ability of IMRT to reduce dose to the penile bulb and erectile tissues without compromising the target volume. This should help preserve sexual function following high-dose radiotherapy.

Xia *et al* (2001) have studied a case requiring dose escalation to two dominant intraprostatic lesions to 90 Gy whilst simultaneously delivering 75.6 Gy to the remaining prostate. Three techniques were compared. The best dose conformality was achieved by a sequential tomotherapy (MIMiC) plan created using CORVUS. The inverse-planned segmental MLC (SMLC) plan gave the lowest dose to the rectal wall. A forward plan using the UMPLAN system gave the best dose homogeneity to the PTV. It was concluded that all three techniques kept the rectal-and-bladder doses to below the RTOG grade-2 complication rate of 10%.

Kung *et al* (2000a) have studied the use of IMRT to improve rectal sparing in treatment of the prostate for patients who have a metal prosthetic hip replacement. Plans were prepared using nine equally spaced 6 MV coplanar fields each avoiding the prosthesis using the CORVUS inverse-planning system. These showed superior dose sparing compared with geometrically-shaped conformal plans made with the PLUNC treatment-planning system.

Luxton *et al* (2000) have studied the dosimetric advantages of IMRT for treatment of prostate cancer for extended fields that include pelvic lymph nodes. The dose to the lymph nodes, prostate and seminal vesicles was 50 Gy and this was followed by a prescribed IMRT boost to bring the minimum dose to the prostate to 70 Gy. The CORVUS inverse-planning system was used to create the plans for IMRT and the Computerised Medical Systems (CMS) FOCUS treatment-planning system was used to generate a 3D conformal treatment plan using the same data sets. Twenty patients were studied. It was found that

IMRT can be used to provide significant bowel as well as bladder and rectum protection while treating the prostate seminal vesicles and pelvic lymph nodes as compared to conventional CFRT. However, this was achieved at the cost of less dose homogeneity within the target. Luxton *et al* (2004) showed that reduced dose to OARs led to reduced NTCP in prostate IMRT compared with 3D conformal therapy.

Among shorter reports are the following:

- Van der Heide *et al* (2001) have described their micro boost approach to IMRT of the prostate.
- Harnisch *et al* (2000) have shown that short-course IMRT can irradiate the prostate gland to 65 Gy whilst simultaneously irradiating dominant intraprostatic lesions to 80 Gy in 28 fractions. This can be done with rectal-and-bladder doses limited to levels consistent with currently used schedules.
- O'Daniel (2001) has added further planning evidence that IMRT of the prostate leads to clinical advantage compared with fixed-field CFRT.
- Aletti *et al* (2001) have commenced IMRT of the prostate in Nancy, France.
- Mihailidis and Gibbons (2002) have compared five- and nine-field IMRT and shown that the choice between the two is patient dependent and depends on femur and small bowel dose-volume histogram analysis.
- Liu *et al* (2002) are using tissue typing of ultrasound prostate images to stage the cancer for IMRT.
- Zietman (2002) makes compelling arguments for the phase-3 randomized trial of dose escalation of prostate cancer.

5.3   Royal Marsden NHS Foundation Trust pelvic and other IMRT

Dearnaley (2000) has reviewed progress towards the use of intensity-modulated techniques for reducing treatment-related side effects and other morbidities in treating prostate cancer. Specific interests include boosting the dose to dominant intraprostatic lesions and also treating pelvic lymph nodes. Nutting and Dearnaley (2001) have given an update on the past, present and future Royal Marsden NHS Foundation Trust trials on prostate CFRT and IMRT.

The Royal Marsden NHS Foundation Trust and Institute of Cancer Research have implemented IMRT for treating prostate cancer with involved lymph nodes (Adams *et al* 2004). Dose escalation to the prostate whilst simultaneously treating the pelvic nodes with IMRT is also justified by Cavey *et al* (2004). The technique used at the Marsden (Sutton) was the Elekta dMLC method in dynamic (research) mode in the first instance and changed to a Elekta step-and-shoot mode after the first ten patients. This stage of the disease is hard to treat radically due to the presence of the dose-limiting small and large bowel which is enclosed in the horseshoe-shaped PTV. CT scans were acquired at 5 mm intervals and

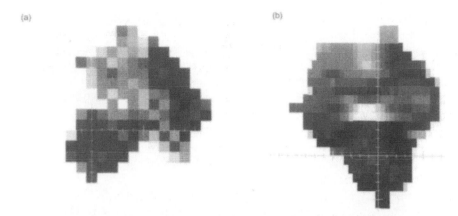

Figure 5.5. Typical beam intensity modulations derived from (a) CORVUS and (b) HELAX TMS for IMRT of the prostate and pelvic nodes. It can be seen that the CORVUS-generated IMB is very noisy and may not be physically realizable without a great deal of careful interpretation using both leaves and jaws. Conversely the HELAX-generated IMB is designed with deliverability checked during the iterative optimization. (Reprinted from Adams *et al* 2004 with copyright permission from Elsevier.)

contoured to specify the bladder, rectum, femoral heads and small- and large-bowel as OARs. A treatment technique was established using five fields planned on CORVUS for the dynamic technique and on HELAX for the step-and-shoot technique (figure 5.5). These were PA, two laterals and two anterior obliques at 30° to the anterior–posterior (AP) line. Doses of 70 Gy to the prostate and 50 Gy to the pelvic nodes were prescribed with the dose to the nodes later escalated to 55 Gy, then 60 Gy as part of a phase-1 study. Planning was tricky but acceptable distributions were obtained (Webb *et al* 2001). The modulations generated by CORVUS were passed into the interpreter developed by Convery and Webb (1998) to derive the delivery sequences (Convery 2001). Those from HELAX generated MLC leaf patterns directly. Before the treatment started, the patient-specific fields were replanned onto an anthropomorphic phantom (see section 3.13.9) at relevant locations and the doses recorded with 80 LiF TLD chips and film between the slices. The TLD doses agreed with calculations to within 2% and the films were digitally scanned and also checked against the planning distributions. Films were also attached to the exit window of the linac head to assess the reliability of the leaf movement (Cosgrove 2001). The first patient was treated in September 2000.

Adams *et al* (2002) have shown that the QA for IMRT can be reduced following observation of successful comparison between predictions and measurements for the first few patients in a phase-1 clinical trial of IMRT of prostate cancer with involved pelvic nodes. Adams *et al* (2003) have further discussed the pruning of the IMRT QA programme that has been possible as the number of patients has increased. Miles *et al* (2003) summarized the median times for performing various aspects of the planning and verification of IMRT and showed that if the individual patient QA was removed, the times were comparable between IMRT and conventional therapy. McNair (2002) has discussed the impact on resources associated with the Royal Marsden NHS Foundation Trust phase-1 clinical trial of IMRT for prostate and pelvic node cancer. By March 2004, some 60 or so patients had been treated at the two geographical sites combined. McNair *et al* (2004) discussed radiographers' perspectives on the clinical implementation but only through a qualitative survey of just this one centre after four months of clinical work. De Langen *et al* (2003) have commented on the continued research that is required once clinical IMRT is established.

IMRT has also been implemented at the Royal Marsden NHS Foundation Trust (Fulham) for radiotherapy of the prostate with involved nodes. The clinical technique was established on a Varian 2100 CD accelerator equipped with a Varian Millennium 120-leaf MLC. Photons of 6 MV energy were used. Planning was performed using the HELIOS system. IMRT was delivered in the dMLC mode. The maximum leaf speed was set to 2.5 cm s^{-1}. The MLC controller monitors and records the position of the leaves every 50 ms. If the leaves are not within a tolerance, then the beam is held off until they are. The HELIOS planning system implements the technique of Spirou and Chui (1998). The resolution is 2.5 mm along the leaf direction and a leaf-motion calculator converts the fluence patterns to leaf patterns taking account of all the physics of transmission, leaf speed, maximum leaf span, rounded leaf-ends (the so-called dosimetric leaf separation [see also section 3.2.3]) and minimum leaf gaps etc. A leaf-position tolerance of 2 mm was used. The leaf transmission was input to CADPLAN as 1.7% for 6 MV photons and 1.8% for 10 MV photons, these being averages of the (variable) measured transmission at different depths. Measurements showed the dosimetric offsets were 1.05 and 1.65 mm for 6 and 10 MV beams respectively, using a measurement technique due to LoSasso *et al* (1998a, b). Planning used five beams at angles of 180°, 270°, 325°, 35° and 100° chosen after planning experimentation. 6 MV was used as the higher 10 MV energy showed no advantages. Planning skill involved sequential adjustments to the relative importance given by the optimization to each organ and a gradual approach to achieving the dose-volume constraints. Malden *et al* (2001) and Clark *et al* (2002a) described these processes together with the verification of IMRT using film measurements with the beams replanned on to a phantom. Measurements were compared with plans for IMRT of the prostate with pelvic-node involvement. This was done by exporting the plan on to a phantom comprising film, delivering IMRT to the phantom and making comparisons between the scanned film and the treatment-planning system (figure 5.6). Dose differences of up to no more than 2% were observed in regions of low dose gradient within the prostate and up to 3% in regions of low dose gradient within the nodes. Portal imaging with an amorphous silicon imager showed agreement between measured fluence patterns and calculated fluence patterns. The whole IMRT process required 23 hr per

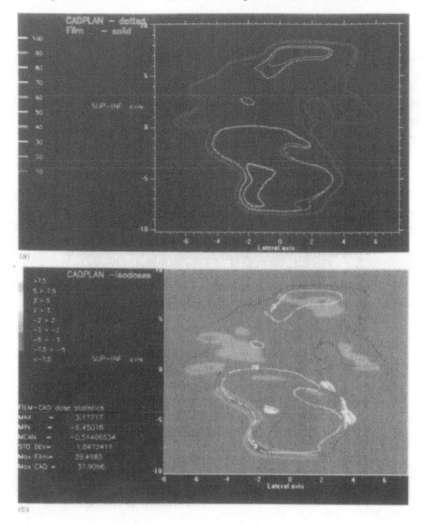

Figure 5.6. An example of how a film measurement is used to check the accuracy of delivery of a modulated beam. The upper panel is an overlay of isodose from film and CADPLAN at 10 cm depth. The lower panel is the difference map of isodoses from film and CADPLAN at 10 cm depth. (From Clark *et al* 2002.) (See website for colour version.)

patient compared with 13 hr for non-IMRT of a prostate only (a detailed table breaks this down into tasks). With experience, the IMRT time was expected to fall. Clark *et al* (2003c) has described this pruning of the QA programme for IMRT at the Royal Marsden NHS Foundation Trust (Fulham).

Staffurth *et al* (2003a) have made a planning comparison of six patients planned with CORVUS version 3.0, HELAX TMS version 6.0 and CADPLAN version 6.3.5 with HELIOS. The CORVUS-planned delivery was simulated as a dMLC delivery from an Elekta accelerator in service mode. The HELAX plan was created as a segmented MLC plan from Elekta accelerators with Precise Desktop version 3 and the CADPLAN was simulated as a dMLC technique from a Varian 2100 CD accelerator. It was found that the PTV coverage was similar across all three treatment-planning systems but that the number of MUs varied dramatically and the shielding of OARs varied between the different planning systems.

Staffurth *et al* (2002) have shown that IMRT of the prostate and involved pelvic nodes has reduced the patient-reported acute GU toxicity compared to a cohort treated with 3D CFRT to the prostate alone. Staffurth *et al* (2003b) have shown through a phase-2 trial of pelvic-node irradiation in prostate cancer that GU toxicity was significantly reduced; GI toxicity did not change. In another centre, Bayouth *et al* (2003a) have similarly reported the ability to dose escalate to the prostate whilst also treating the pelvic nodes with small GI toxicity.

Nutting and Dearnaley (2002) presented an overview of the complete IMRT programme at the Royal Marsden NHS Foundation Trust. A planning study had taken place using 30 patients with head-and-neck, pelvic and thoracic tumours to compare conventional radiotherapy, 3D CFRT and inverse-planned IMRT. The greatest improvements arose for thyroid carcinoma and pelvic-lymph-node treatment because these tumours have a concave PTV. For pelvic tumours nine, seven or five equispaced coplanar IMRT fields gave similar benefits but dose distributions deteriorated with three equally spaced fields. Conversely, for oesophageal and head-and-neck tumours where structures with low radiation tolerance were close to the PTV, it was important to use non-equispaced beams and a special computerized algorithm was designed to calculate the best non-coplanar IMRT beam directions. Multi-field IMRT with only three to four beams was then shown to be quite effective. To quality control the delivery technique, treatment plans were delivered to a humanoid phantom and calculated doses were compared with the doses measured with photographic film, TLD and BANG gel.

Adams *et al* (2001, 2003, 2004) have described the use of the Nordion HELAX treatment-planning system for planning IMRT at the Royal Marsden NHS Foundation Trust. Example applications are treating prostate cancer with pelvic-node involvement, sparing the bladder and rectum. A planning study has also been carried out to look at the complexity of treatment delivery for maxillary sinus tumours attempting to significantly reduce dose to optic nerves and ipsilateral parotid gland. Specific attention has been placed to minimizing the number of field segments because the treatment was delivered with an Elekta accelerator with (at that time) a large intersegment dead time. Plans were exported from HELAX using the DICOM facilities and replanned on to a phantom for experimental verification. TLD measurements have shown that, for a complete five-field treatment, mean discrepancies were about 3%.

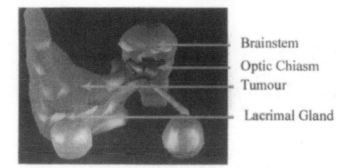

Figure 5.7. Solid geometry representation of tumour mass very close to OARs in the head. Here a meningioma of the right sphenoid bone with retro-orbital extension is adjacent to neighbouring brain stem, optic chiasm and lacrimal gland. (Reprinted from Pirzkall *et al* 2000 with copyright permission from Elsevier.)

5.4   Comparison of IMRT with conformal radiotherapy (CFRT) for complex shaped tumours

Pirzkall *et al* (2000) have studied a comparison of IMRT and geometrically-shaped CFRT for nine patients extracted from the DKFZ Clinic during 1997. The goal of the study was to assess the relative improvement that can be realized in terms of planning quality as well as the impact of such an implementation on a busy department. CFRT plans (those used to actually treat the patient) were created using the VOXELPLAN 3D treatment-planning system using contour data from correlated CT and MRI slices 3 mm thick with 3 mm separation. For six of the targets, a manually driven MLC with a leaf width of 5 mm at isocentre was used and for the other three a commercial Siemens MLC with 10 mm leaf widths was used. Planning limits were set such that preservation of normal function was paramount because many of the lesions were benign and patients were expected to live many decades post treatment. Preserving normal structure was a clinical priority (figures 5.7 and 5.8). IMRT plans were made with CORVUS 3.0 and segmentation was performed separately for CORVUS and for VOXELPLAN implying no contour transfer between systems. Hence, equivalence of segmentation was confirmed by evaluating the slice-by-slice images and post-planning volumetrics statistics.

For each case, three IMRT plans were created: a rotational plan for serial tomotherapy, a seven-field coplanar plan and non-coplanar plan. All plans employed ten intensity levels. A range of evaluation metrics were used based on the RTOG metrics for stereotactic treatments. These included target coverage to desired dose, target coverage to CFRT dose, target conformality, target heterogeneity, integral dose, MUs, treatment time and planning time.

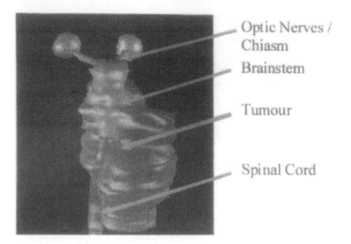

Optic Nerves /
Chiasm

Brainstem

Tumour

Spinal Cord

Figure 5.8. Solid geometry representation of tumour mass very close to OARs in the head. Here a chondrosarcoma of the skull base is adjacent to neighbouring brainstem, optic chiasm and nerves, spinal cord. (Reprinted from Pirzkall *et al* 2000 with copyright permission from Elsevier.)

The key outcome was that all three of the IMRT techniques produced improvement over the CFRT plans in terms of conformality and coverage. They were equivalent in terms of homogeneity, required a greater number of MUs, delivered a greater integral dose and took less time to plan. There were essentially no statistical differences among the three IMRT techniques. The most outstanding observation was the improvement in the parameter target-coverage-to-CFRT-dose, this being the target-volume-greater-than-the-CFRT-desired-dose divided by the total target volume. For benign tumours, this parameter rose from about 0.72 to about 0.98 and for malignant tumours, it rose from 0.81 to about 0.96.

Trade-offs in IMRT included a delivery of higher integral dose outside the target and an increase in MUs between two and eight times the number of MUs per field compared to the CFRT plan. The authors acknowledged that they are not the first to report the benefits of IMRT compared with CFRT nor to stereotactic radiation therapy. However, they claimed that, in all their studies, IMRT was found to be superior overall and this lends weight to the arguments against the criticism of IMRT that it has yet to win its spurs. There is a lengthy discussion of the results centring around what can be done with adjusting the various trade-offs available to the IMRT planner.

Conversely, Vaarkamp *et al* (2003) have shown that some concave target distributions can be created using five MSFs with up to four segments per beam. The shielding of the OARs is not as great as from full IMRT but the plans are considered clinically acceptable.

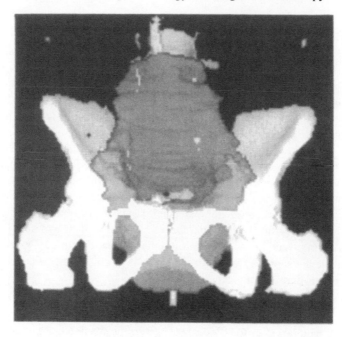

Figure 5.9. Volume rendering of the clinical target volume for a typical patient receiving whole-pelvic IMRT. The CTV contains the upper part of the vagina and parametrial tissues. In women with an intact uterus, the entire uterus was included. The presacral region was included to the bottom of the S3 vertebral body to ensure coverage of the presacral lymph nodes and attachment of the uterosacral ligament. The common internal and external iliac nodal regions were included to the level of the L4-5 interspace. (Reprinted from Roeske *et al* (2000b) with copyright permission from Elsevier.)

5.5 IMRT for whole-pelvic and gynaecological radiotherapy

Gynaecological tumours present another opportunity for radiation therapy improvement with IMRT. Small bowel and rectum are generally included in the irradiated volume of conventional whole pelvic radiotherapy. Roeske *et al* (2000b) have studied the ability of IMRT to reduce the volume of small bowel irradiated in women with gynaecological malignancies who are receiving whole pelvic radiotherapy (figure 5.9). Usually this disease is treated with a standard four-field box with apertures shaped to the PTV in each beam's-eye view, a situation that leads to a significant amount of small bowel being unwantedly irradiated as well as more bladder and rectum irradiation than desirable. Hence, the main acute and chronic toxicity for these patients are small-bowel sequelae including small-bowel obstruction, enteritis and diarrhoea and malabsorption of vitamin B12, bile acids and lactose.

Roeske *et al* (2000b) studied ten consecutive women with cervical or endometrial carcinoma, planning these patients for conventional radiotherapy using the PLUNC 3D treatment-planning system. They then replanned the patients using the CORVUS 3.0 inverse-planning system for nine fields arranged coplanarly and delivering 6 MV photons. This latter choice was made following a preliminary study to determine that, whilst the outcome of IMRT was dependent on the number of fields, the improvement plateaued out after seven to nine fields, consistent with the observations of others. They then showed comparative plan data, comparative dose-volume histograms and comparative dose statistics, extracted for all of these patients, which demonstrated the improved small-bowel sparing of IMRT. A particularly interesting histogram shows the absolute volume of the small bowel irradiated in each patient to a dose of 45 Gy or greater indicating reductions often greater than a half between conventional radiotherapy and IMRT. Similarly, tables showed improved sparing of dose to small bowel and rectum. The price paid was a greater dose inhomogeneity in the PTV and normal tissues but the hot spots were considered not to be of clinical significance because their magnitude and volume were small.

The authors remind us that the Harvard Joint Center for Radiation Therapy originally discussed computer-controlled radiation therapy in the context of pelvic-node irradiation in the 1970s, but, given the available technology at the time, the research was not taken forward to the clinic. The important new work in the current study provides planning proof that IMRT is worthwhile and it remains to be seen whether the benefits translate to reduced treatment sequelae. The patients are being followed closely to answer that question. This work took place at the University of Chicago.

Roeske *et al* (2000a) have shown that, by using nine intensity-modulated fields to the pelvis, cervical cancer can be treated without the requirement for intracavitary brachytherapy. With IMRT, only 10% of the rectal volume received the prescription dose whereas this value was 30% when brachytherapy was combined with conventional whole-pelvic radiotherapy.

Low *et al* (2000a, 2002b) have developed a technique known as applicator-guided IMRT for gynaecological cancer. An intracavitary applicator substitute was placed in the patient's vagina and uterus and 3D images were acquired using MR. An IMRT plan was then computed using the CORVUS planning system to conform to the applicator substitute. Hence, when the dose distribution was aligned with the applicator substitute, it was also aligned with the tumour and critical organs. This technique is expected to replace brachytherapy at this centre.

Wahab *et al* (2004a) have given further details of the technique known as applicator-guided IMRT to deliver IMRT to the cervix. The aim of this study was to define the target volume using positron emission tomography (PET) and then use IMRT instead of the conventional radioisotope high-dose-rate applicator. It was shown that this leads to a more uniform homogeneous dose to the target avoiding the higher low-dose regions that are typical of brachytherapy.

Portelance *et al* (2001) have shown that IMRT can improve the conformality of dose to the cervix with improved rectal and small-bowel sparing. 'Conventional' radiotherapy was planned using the CMS FOCUS system and either two or four fields. IMRT was planned with the NOMOS CORVUS system using either four, seven or nine fields. The volume of small bowel receiving the prescribed dose fell from 35.58% (conventional two-fields) and 34.24% (conventional four-fields) to 11.01% (four IMRT fields), 15.05% (seven IMRT fields) and 13.56% (nine IMRT fields). The volume of rectum receiving more than the prescribed dose fell from 84.01% (conventional two-fields) and 46.37% (conventional four-fields) to 8.55% (four IMRT fields), 6.37% (seven IMRT fields) and 3.34% (nine IMRT fields). The volume of bladder receiving more than the prescribed dose fell from 92.89% (conventional two-fields) and 60.48% (conventional four-fields) to 30.29% (four IMRT fields), 31.66% (seven IMRT fields) and 26.91% (nine IMRT fields). The differences between the IMRT techniques were deemed insignificant, four fields being as good as nine. It was determined that dose escalation to the cervix would be possible with the IMRT techniques. The study was based on the step-and-shoot technique.

Hong *et al* (2002) have assessed the feasibility of inverse planning for whole-abdomen IMRT. They created a PTV comprising the entire peritoneal cavity with a margin. Five 15 MV intensity-modulated beams were created at gantry angles 180°, 105°, 35°, 325° and 255° and IMRT optimization was designed to spare kidneys and bones. Significant dose reduction to the bones and improved PTV coverage were achieved and, because of the very large width of the beams, to minimize field match errors adjacent subfields overlapped by at least 2 cm with intensity feathering in the overlap region.

Vuong *et al* (2002) delivered IMRT for unresectable rectal or rectosigmoid cancer and showed that dose escalation could be achieved with acceptable toxicity.

5.6 Head-and-neck IMRT

So far in this chapter we have tried to separate the statements about the prediction of benefit from those about its actual observation. Clinical IMRT of the head-and-neck was reviewed by Guerrero Urbano and Nutting (2004a,b). Eisbruch (2002) has emphasized that there is a large difference between the expectation of benefit from IMRT and the demonstration of this. He points to IMRT for head-and-neck treatments in which it has recently been conclusively demonstrated that reduced xerostomia arises (Lee *et al* 2002a). It was observed that xerostomia gradually reduced, post treatment, to almost zero after two years. Head-and-neck cancer presents a very real clinical test for IMRT since there are so many potentially non-involved tissues adjacent to tumour including major and minor salivary glands, mandible, pharyngeal musculature, inner and middle ears, temporomandibular joints, temporal brain lobes and optic pathways.

Münter *et al* (2003) have described the treatment technique for stereotactic IMRT of head-and-neck tumours at DKFZ Heidelberg and its inverse treatment planning.   Forty-eight patients with carcinoma of the head-and-neck region were treated between 1999-2002. Planning was performed with the KONRAD treatment-planning system and treatment was through interpretation by the Bortfeld technique to segments delivered by a Siemens accelerator. The technique is very successful with only four patients having xerostomia greater than grade 1. The authors described the now well-known QA techniques implemented at Heidelberg to determine absolute and relative dosimetry (see section 3.13.9).

There is a current European head-and-neck IMRT-*versus*-conventional-therapy protocol and trial (Guerrero Urbano and Nutting 2004a). It is intended to recruit 160 patients. The participating centres are The Netherlands Cancer Institute, University of Ghent, Christie Hospital, Manchester and the Royal Marsden NHS Foundation Trust. The Marsden is coordinating the trial. IMRT can only be routinely implemented by UK NHS hospitals after phase-3 clinical trials have given, what is called, level-one evidence of benefit (Bentzen 2004).

5.6.1   Thyroid IMRT

Parker *et al* (2000) have shown how IMRT can improve the coverage of the PTV for radiotherapy treatment of the thyroid. Four- or five-field directions are used with a few (four to five) MLC subfields for each.

Happersett *et al* (2000) have also developed IMRT for the treatment of thyroid cancer. It was shown that, compared with a conventional technique, IMRT can significantly increase PTV coverage while maintaining similar or better normal structure doses. The normal structures concerned were the lung and spinal cord.

Clark *et al* (2002b) have clinically implemented dynamic IMRT for patients with carcinoma of the thyroid and cervical nodes (figure 5.10). The prescription doses were 60 Gy to the thyroid PTV and 50 Gy to the nodal PTV respectively. Five-field IMRT plans were designed with the aim of reducing unnecessary anterior shoulder irradiation and also including the use of asymmetric fields for improving spinal cord sparing. It was found that the average maximum dose received by the cord could be reduced to below 47 Gy. The dose range in the thyroid PTV was 4.2 Gy and the range was 10.6 Gy in the nodal PTV with all beams covering all of the targets. The dose range in the thyroid PTV was 6.2 Gy and in the nodal PTV was 11.2 Gy, respectively, with asymmetric posterior oblique beams. Comparison of measured doses with calculated doses gave differences of 1% in the thyroid PTV and just 2% in nodal PTV. The conclusion was that IMRT offered improved target homogeneity for patients with thyroid carcinoma as well as reducing the dose to the spinal cord.

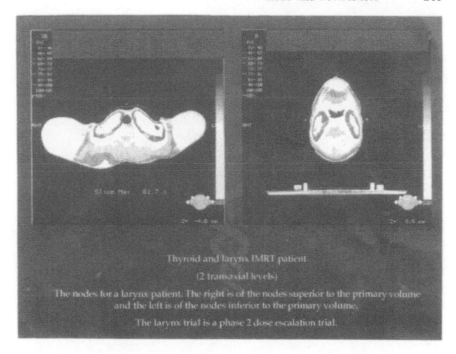

Figure 5.10. Two CT slices showing the nodes of a patient treated for cancer of the thyroid and larynx. The right-hand picture is of the nodes superior to the primary volume and the left-hand one is of the nodes inferior of the primary volume. Notice the highly irregular volume that it is required to irradiate conformally necessitating IMRT. (Courtesy of Dr Catharine Clark.) (See website for colour version.)

5.6.2   Nasopharynx IMRT

Xia *et al* (2000a) have, for one treatment-planning case only, that of a locally advanced nasopharyngeal carcinoma, compared four treatment plans (figure 5.11). These were a conventional treatment, a non-modulated conformal treatment, an IMRT plan generated using CORVUS and deliverable with MSFs and an IMRT plan created using PEACOCKPLAN for the MIMiC. The MSF plan had five gantry angles chosen manually with ten non-zero intensity levels for each beam. The 3D conformal plan had eight gantry angles based on the beam's-eye view. In practice, the plan was delivered using partial transmission blocks and the study was just for technique comparison.

QA was carried out by delivering the plan to a phantom comprising a stack of solid-water slabs with the approximate dimensions of a human head. Relative dose distributions were verified using Kodak XV-film and point-dose distributions were checked with an ionization chamber. Conformality of the plans was ranked such that the best plan was the PEACOCKPLAN, followed by the CORVUS plan

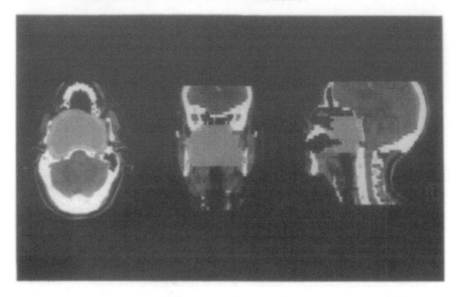

Figure 5.11. Axial, coronal and sagittal displays of GTV in red and CTV in orange for a selected nasopharyngeal case. The complex re-entrant surface and shape of the volumes to be treated is evident and calls for IMRT. (Reprinted from Xia *et al* (2000a) with copyright permission from Elsevier.) (See website for colour version.)

followed by the 3D conformal plan, followed by the conventional bilateral plan. Xia *et al* (2000a) also assessed the dependence of the plans on set-up uncertainty by a technique of shifting the isocentre relative to the plan.

Hsiung *et al* (2002) have compared IMRT with conventional 3D CFRT for boost or salvage treatment of nasopharyngeal carcinoma (NPC). Fourteen NPC patients planned with the ADAC PINNACLE planning system in Taiwan were transferred to the Memorial Sloan-Kettering Cancer Center planning system where new IMRT plans were produced using five or seven radiation fields. Plans were then compared and it was shown that the IMRT plans were superior.

Chu and Suh (2000) have delivered IMRT for NPC using compensating filters. Van Nimwegen *et al* (2001) have also developed IMRT for the nasopharynx. A planning study comprised 20 patients and a five-field technique was used. Lu *et al* (2004) have shown how IMRT can be used for re-treatment of NPCs after failed conventional radiotherapy.

5.6.3 Oropharynx IMRT

Maes *et al* (2000) have developed a class solution using five IMBs for treating oropharyngeal cancer whilst sparing the parotid gland. The technique was based on the HELIOS inverse-planning system from Varian.

Van Asselen *et al* (2001a, 2004) have worked towards a class solution for oropharyngeal cancer. It was found that it was possible to systematically determine a standard set of dose constraints which resulted in a homogeneous dose distribution to the target and acceptable critical organ dose. Seven equi-angular beams were enough to obtain an acceptable dose distribution using between five and ten intensity levels in the sequencing procedure.

5.6.4 Oropharynx and nasopharynx IMRT

Van Dieren *et al* (2000) have shown that intensity modulation for the treament of midline tumours of the head-and-neck can spare parotid gland irradiation. Fifteen patients were studied, each with two modulated portals. Sparing of the submandibular glands was less pronounced for tumours of the oropharynx. For primary tumours of the larynx, all salivary glands could be spared.

The IMBs were generated 2.5 mm apart in the longitudinal direction corresponding to a compensator with this resolution. Four adjacent profiles were averaged to produce the equivalent modulation generatable with an MLC with 1 cm wide leaves. It was found that the clinical outcome was barely affected by this process (see also section 3.9.3).

5.6.5 Larynx IMRT

Levendag *et al* (2000) have performed a multi-institutional evaluation of IMRT. Twenty-two patients with cancer of the larynx without lymph-node metastasis were treated by 3D CFRT to the primary site and to the neck. A CT scan dataset of one (typical) clinical case out of this series was then subsequently selected and with the pre-contoured target and normal tissues was re-planned independently in Amsterdam, Ghent, Brussels and Rotterdam using IMRT techniques. The data provided interesting comparisons between the techniques of the four centres.

Wu *et al* (2001e) and Mohan *et al* (2001b) have developed a simultaneous integrated-boost technique for advanced head-and-neck squamous cell carcinoma using the dMLC technique. Similar work, reported by Dogan *et al* (2003), claimed that the main advantage was greater conformality and the use of a single plan throughout treatment. Conversely, Popple *et al* (2003) have developed planning software that plans both the unboosted phase and the boost phase simultaneously for sequential delivery. Neither of the two phases necessarily respects the treatment plan constraints alone but the sum of the two plans always respects the constraints.

Clark *et al* (2002c, 2003a, 2004) have developed IMRT for six patients treated for larynx carcinoma in a two-phase treatment. They concluded improved target homogeneity was achieved with reduced dose to the spinal cord (figure 5.12). Clark *et al* (2003d) explain the rationale for the use of five, rather than more, fields for some head-and-neck cancers, and also how the use of cut-outs in the immobilization mask can reduce skin dose. Lee and Xia (2003)

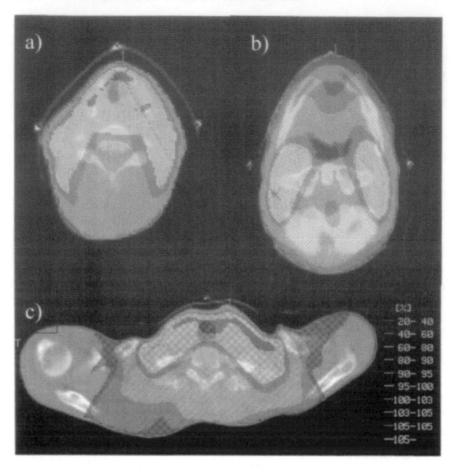

Figure 5.12. Dose distribution for phase-1 treatment of the larynx and nodes to 50 Gy using the IMRT technique. The transverse cross sections are (a) through the central larynx showing the larynx PTV, the nodal PTVs and the spinal cord, (b) the superior nodes showing the nodal PTVs and the spinal cord and (c) the inferior nodes showing the nodal PTVs and the spinal cord. (From Clark *et al* 2004.) (See website for colour version.)

conversely consider that the lack of skin reaction at the Institute of Cancer Research/Royal Marsden Hospital (ICR/RMH) may have more to do with the use of a lower (46 Gy) prescribed dose to nodal regions compared with 60 Gy at University of California at San Francisco (UCSF).

5.6.6   Evidence for parotid sparing

Vineberg *et al* (2000) have shown that the use of optimized beamlet IMRT inside the UMPLAN treatment-planning system in Michigan has led to bilateral parotid sparing. The contralateral parotid dose could be reduced to a mean of 26 Gy while the ipsilateral parotid ranged between 26 and 35 Gy. These results could be improved by increasing the number of beams from four to nine leading to a reduction in parotid mean dose of up to 3–4 Gy. However, inclusion of non-axial beams did not improve parotid sparing.

Bragg *et al* (2001) have also shown that IMRT of the parotid gland can spare the contralateral parotid provided suitable beam directions are chosen.

For head-and-neck planning at The Netherlands Cancer Institute, three to eleven coplanar fields are used, each with two-to-four segments, the goal being to spare parotid glands and to irradiate a concave PTV. Segments were pre-determined; the dose from each segment found and then an iterative looping was used to compute segment weight. It was found that the volume of the parotid raised to more than 25 Gy was reduced from 92% to 48% and the conformity index (the irradiated volume greater than or equal to 95% divided by the PTV) fell from 2.5 to 1.6. It was thus concluded that IMRT was advantageous and also improved PTV homogeneity.

Braaksma *et al* (2002) have shown that some IMRT techniques can increase dose to the minor salivary glands thus tending to offset sparing of the major parotid glands. Braaksma *et al* (2003) have evaluated the parotid sparing of 3D CFRT in a prospective study in node-negative cancer of the larynx. Twenty-six patients were treated and dose distributions of the major salivary glands were correlated with objective stimulated whole saliva flow and subjective (questionnaire: visual analogue scale) salivary-gland function. It was found that the whole-saliva flow reached its nadir six months post-radiotherapy and then improved significantly such that at two years post-treatment measurements were 48% of the pre-treatment values. However, the class solutions for 3D CFRT salivary-gland-sparing technique were inadequate for preserving fully salivary gland function and it was determined through a planning study that further reduction of xerostomia could be achieved using an IMRT technique focused at sparing major and minor salivary glands. This technique was developed through a cooperative effort of four centres (Daniel den Hoed Cancer Centre, Rotterdam; University of Ghent, Belgium; The Netherlands Cancer Institute, Amsterdam; and the University of Michigan, Ann Arbor. These Centres all used different inverse-planning techniques. Following this, all 26 patients receiving 3D CFRT were re-planned with IMRT, showing that the mean parotid gland dose reduced to 23 Gy compared with 28.9 Gy for 3D CFRT. It was also determined that IMRT by the dMLC technique gave very similar results to IMRT performed with the step-and-shoot technique in five patients. The dose-volume histogram findings being superimposable. This study provided important quantitative evidence for the efficacy of IMRT for treating head-and-neck cancer (figure 5.13).

Bel *et al* (2002) have produced treatment plans in which the position of the head-and-neck tumour was sampled randomly from known distributions of head-and-neck tumour position and IMRT plans re-computed. These were compared to the plan created with just a single unmoved PTV. The comparison showed

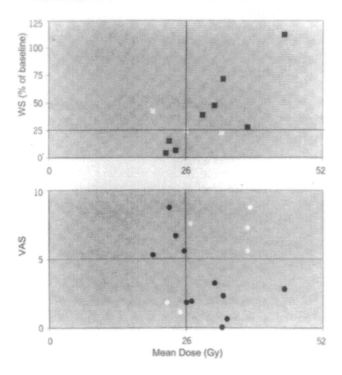

Figure 5.13. The upper panel shows the objective relationship between mean dose (Gy) in the parotid glands (*x*-axis) and residual salivary gland function (expressed by the whole-saliva flow rate, % of baseline, *y*-axis) at 1-year post-irradiation. The lower panel shows the correlation between mean dose in the parotids and subjective salivary-gland function (expressed by visual analogue scores (VAS), *y*-axis). The white squares/dots represent patients in whom a correlation between mean dose and salivary gland function can be depicted (three and six patients); black squares/dots represent patients in whom dose distribution in the parotid glands and outcome on xerostomia do not correlate (8 and 11 patients). Full bold horizontal and vertical lines represent thresholds (Reprinted from Braaksma *et al* (2003) with copyright permission from Elsevier.)

no change with respect to the dose to the clinical target volume and PTV but considerable changes to the NTCP of the parotid glands.

Mazurier *et al* (2002) have implemented IMRT using a HELAX system and a Siemens accelerator. They found that predicted and measured dose distributions agreed to better than 5% and applied the technique to reduce parotid irradiation to avoid xerostomia.

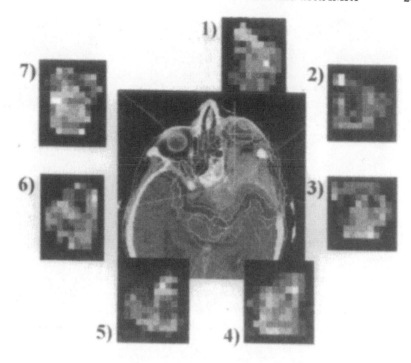

Figure 5.14. This shows the seven co-planar equidistant intensity-modulated fields used to treat a 70-year-old female with extensive sphenoid wing meningioma treated with IMRT. The patient was treated with IMRT for a recurrence after two operations (4 years and 10 months before IMRT). The plan was created with the KONRAD treatment-planning system from Heidelberg. (Reprinted from Pirzkall *et al* 2003 with copyright permission from Elsevier.)

5.6.7 Meningioma IMRT

Debus *et al* (2000) have irradiated 17 patients with large skull-based meningiomas using IMRT in a phase-2 study. The clinical requirement was to treat the complex shape of the meningiomas. Dynamic MLC treatments were delivered with a Siemens Primus accelerator with an integrated MLC and a step-and-shoot IMRT delivery technique. Treatment was quality-assured using film dosimetry and an ionization chamber. Transient acute-treatment side effects were monitored.

Pirzkall *et al* (2003) reported on a follow-up of these patients using MRI and clinical examination showing no radiation-induced peritumoural oedema, no increase in tumour size and no new onset of neurologic deficits. IMRT was considered the treatment of choice (figures 5.14 and 5.15).

Uy *et al* (2002) have reported on 40 patients treated for intracranial meningioma (excluding optic nerve sheath meniongiomas) using the NOMOS

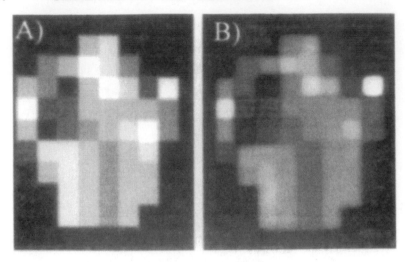

Figure 5.15. The intensity pattern of one IMRT treatment portal: (A) as calculated by the treatment-planning system KONRAD; (B) as acquired during quality assurance and dose verification using a BeamviewPlus portal-imaging system from Siemens Oncology Systems. (Reprinted from Pirzkall *et al* 2003 with copyright permission from Elsevier.)

PEACOCK CKPLAN system between 1994 and 1999. The cumulative five-year local control was 93%.

Some clinical evidence for the benefit of IMRT for treating primary optic nerve sheath meningioma has been presented by Woo *et al* (2002).

5.6.8   Paediatric medulloblastoma IMRT

Papanikolaou *et al* (2002) have shown that IMRT can reduce the dose to the kidneys and the thyroid gland in the treatment of paediatric medulloblastoma and, as a result, reduce the incidence of clinical and subclinical hypothyroidism.

5.6.9   Ethmoid cancer IMRT

Wang *et al* (2002b) have performed a comparative study of IMRT *versus* 3D radiotherapy for treating ethmoid cancer. They found that IMRT gave superior sparing particularly to the lens of the eye. For the Varian 80-leaf MLC, they stated that the physical limitations of the MLC led to a degradation of the IMRT plan.

5.6.10   Other studies reported

Cozzi *et al* (2001a) have compared three different treatment techniques: 3D conformal therapy, intensity-modulated photons and intensity-modulated protons.

Five patients were studied and plans were computed for all treatment modalities. The patients presented with advanced head-and-neck tumours and OARs included the spinal cord and the parotid glands. It was concluded that protons provided significantly the best improved dose homogeneity and also the best sparing of spinal cord. The intensity-modulated photon plans were superior to the non-modulated photon plans but inferior to proton plans. The same results were found for the parotid glands. Cozzi *et al* (2004) further amplified on the benefits of IMRT for treating head-and-neck tumours.

Cozzi *et al* (2001d) further compared six different treatment techniques for advanced head-and-neck cancers. These techniques included conventional mixed electron–photon beams, 3D conformal photon beams with five fields, intensity-modulated photon beams with either nine or five equally-spaced fields and three proton fields either with passive scattering or spot scanning. Plans were evaluated using dose-volume histogram data and predictions of equivalent uniform dose. It was found that all the conventional techniques were inferior to other techniques using IMRT or protons and that, in turn, the proton distributions were superior to those using photon IMRT. Zurlo (2002) comments that IMRT to targets in large parts of the body delivers a higher integral dose than protons to normal tissues. He also calls for international collaboration on trials of proton therapy.

Morgan (2001) used the HELAX TPS to create custom compensation for head-and-neck treatments. Using the Elekta Desktop delivery system and accelerator, the accuracy was verified to better than 2% with isodoses better than 3 mm.

Pawlicki *et al* (2004) have estimated that the lens dose can be an appreciable fraction of the treatment dose in some head-and-neck IMRT treatments. They compared predictions from CORVUS and Monte Carlo calculations. It is important, therefore, to avoid irradiating through the eyes.

Burnet (2003) showed that the use of rotation with a central-axis block could be an alternative to IMRT for spine tumours.

Erridge *et al* (2003) used a simple set of (four) top-up fields added to an 85% open field to perform electronic compensation for head-and-neck tumours to improve PTV uniformity.

5.7   Breast IMRT

Guerrero Urbano and Nutting (2004b) have reviewed clinical IMRT of the breast.

5.7.1   Breast IMRT at William Beaumont Hospital

Kestin *et al* (2000a, b) described the William Beaumont Hospital, Royal Oak, Michigan technique for performing IMRT of the breast (see also *Wavelength* 2001c,e, 2002b). The technique begins by calculating the dose distribution from open tangential fields using tangential beams with correct account of tissue inhomogeneity. Separate MLC segments were then constructed to conform to the beam's-eye-view projections of the 3D isodose surfaces in 5% increments ranging from 100–120% (figure 5.16). Generally, this leads to the generation of between six and eight segments. The beamweights of these segments were then optimized using an inverse-planning technique and by placing 100 reference points uniformly distributed throughout the breast PTV. A special script programme for the ADAC PINNACLE treatment-planning system was then created to optimize the weights of the individual segments by conducting a downhill Simplex search on the 100 randomly placed points in the PTV. It was found that, in general, this technique reduced the number of field segments to between two to four segments per beam direction. Any beam segment that was required to deliver less than four MUs was deleted. A second IMRT plan was then created by the same technique but including lung blocking. It was found that the largest component of dose to the breast was delivered from the open fields (as expected) and that IMRT considerably improved dose homogeneity to the breast compared with the use of standard wedges. It was found that the most dramatic outcome was the reduction from around 10% of the volume receiving 110% of the planning dose down to just 0.1% of the volume when IMRT was constructed with segmented beams with lung blocking. There was also a small decrease in the maximum dose and mean dose. The modest number of segments kept treatment delivery to less than 10 min and the dose homogeneity to the breast improvement was expected to correlate with improved cosmetic outcome. A simplified account of the work at the William Beaumont Hospital appeared in *Wavelength* (2000a) emphasizing the technique of creating MSFs by following isodose lines and reweighting the components. This is a specific form of DAO (Shepard *et al* 2003a) based on the use of isodose lines rather than on projections of target and at-risk structures.

This group considers that this is much more practical than using custom physical compensators but the technique at the William Beaumont Hospital does rely on the use of available high quality CT data. This distinguishes the technique from the early techniques at the Royal Marsden NHS Foundation Trust, which did not use CT data.

Vicini *et al* (2002) have recently reported that the treatment-planning system has been upgraded to permit dose optimization over the entire breast using dose-volume objectives and constraints. Also, whereas formally segments with very low numbers of MUs were rejected, segments with as few as 2 MUs are now routinely used.

By April 2001, 281 patients had been treated with this technique. The number of segments varied depending on the breast size, with more segments being required for patients with larger breasts. Typically between five and eight segments were required and the median treatment time was less than 10 min. The authors presented the dose-volume outcomes and reported that no skin telengiectasias nor fibrosis nor persistent breast pain had been noted in any of the patients. The cosmetic results at 12 months had been rated as excellent or good in 99% of the patients. The authors claimed that widespread implementation of this

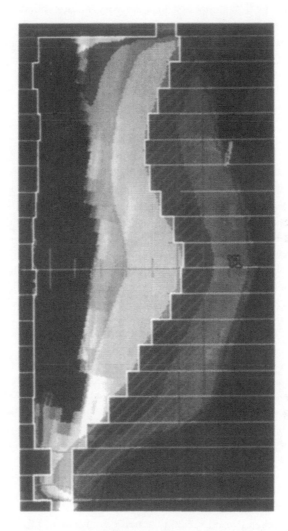

Figure 5.16. Beam's-eye view of an MLC segment with a block corresponding to the 105% open-field isodose surface. The image demonstrates a 2D central-axis reconstruction of the isodose surfaces. Dark blue = 90–94%; light blue = 95–99%; lavender = 100–104%; red = 105–109%; yellow = 110–114%; white = 115–119%. The segments defined in this way then have their weights optimized. (Reprinted from Kestin *et al* 2000a with copyright permission from Elsevier.) (See website for colour version.)

technology could be achieved with minimal imposition on clinical resources and constraints. Specifically, they claimed that their technique does not significantly increase the treatment-planning nor dose-delivery times and they cite rival breast IMRT techniques as being less efficient in these respects.

An editorial (Potters *et al* 2003b) considered the paper of Vicini *et al* (2002). Strom (2002) postulated that the most important impact of the change in planning radiotherapy of the breast using this IMRT technique is on the physician, namely how the physician thinks about the breast and the operative bed as targets. For example, the protocol gives a strict method to tailor the tangential fields to the individual patient's anatomy. Being CT-based, it also allows a boost volume as well as the primary breast tangential fields to be constructed. It is postulated that it largely eliminates differences in dose-prescribing conventions, which are caused by a variety of traditional prescription points because the 100% minimum breast dose must be specified in a tissue region and it also challenges the understanding of what exactly is the clinical target volume.

The treatment technique was verified by delivering the modulated radiation to anthropomorphic phantoms comprising slices of tissue-equivalent material containing, between slices, radiographic film and it was concluded that measurements were within ±3% of treatment plan calculations. The vast majority of the dose (median of 78%) was delivered by the open segments. Finally it was postulated that the technique would probably be less susceptible to the effects of respiratory motion, a hypothesis that is currently being evaluated in this clinic.

5.7.2　Other reports of techniques using small top-up fields

Donovan *et al* (2002b) have made a dose-histogram analysis of 300 patients, half randomized to conventional tangential irradiation and half to IMRT. In standard wedged treatment doses greater than 105% were largely in upper and lower thirds of the breast. Ninety-six percent of the patients had doses greater than 105% in the upper third and 70% had doses greater than 105% in the lower third. Only 4% of patients receiving IMRT had doses greater than 105% in either third. The volume of breast receiving greater than 105% dose decreased by a mean of 10.7% when treatment changed from conventional to IMRT. A goal of the randomized trial was to correlate this improved dosimetry with patient-reported accounts of the location of pain and breast tenderness (figure 5.17).

Van Asselen *et al* (2001b) from Utrecht have also developed a technique to improve the dose uniformity to the breast by using IMRT (figure 5.18). The basis of the technique was to deliver 88% of the dose using two open fields covering the whole treated volume and then the remaining 12% is delivered in four MLC-shaped segments. They found that this decreased the dose inhomogeneity in the PTV from 9% for the conventional technique to 7.6% for the IMRT technique. At the same time, the mean lung dose was reduced for the IMRT technique by approximately 10% compared with the conventional technique.

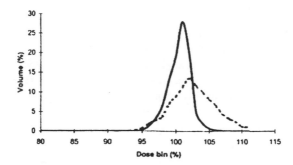

Figure 5.17. An example differential dose-volume histogram plot for a standard and for an IMRT plan. Dotted lines are the standard plan, full line is the IMRT plan. (From Donovan *et al* 2002b.)

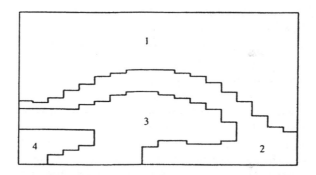

Figure 5.18. The four MLC settings for each intensity level in IMRT of the breast as deduced for the equivalent pathlength map. The first segment includes the areas 1–4, the second segment the areas 2–4, the third segment the areas 3–4 and the fourth segment area 4. (Reprinted from van Asselen *et al* (2001b) with copyright permission from Elsevier.)

In addition to the goal to improve dose homogeneity, the intention was to maintain a vertical match plane between the tangential fields and the supraclavicular and axial fields without using a breast board. The IMRT technique used was a step-and-shoot technique using an Elekta accelerator. Wedges could not be used to optimize the dose distribution because in such an accelerator the wedge direction is perpendicular to the leaf-movement direction. Hence, a MSF IMRT technique had to be developed. For this, 3D CT data for the patient had to be available. Three-dimensional dose distributions were calculated using the Nucletron PLATO RTS version 2.2 planning system and the energy of the beams was 6 MV. The first attempt made was to improve the dose distribution in the central plane of the breast. This was done by calculating equivalent pathlengths taken from ray tracing from the focus of the beam through the CT data set projected on to a plane perpendicular to the beam axes. The translation of such an equivalent-pathlength map into optimal MLC field settings and intensity levels was a complex problem. However, this map was used to obtain the MLC settings. In general, for each of five patients, just four such MLC settings were used. The resulting dose plans were analysed in terms of the volume of tissue with dose lying between particular isodose values. The dose distributions displayed by isodose contours in the central plane showed an improved dose distribution for the IMRT technique compared with the conventional technique. The ipsilateral lung dose decreased with a mean value of 11.8% compared with the conventional technique. Prior to the use of intensity modulation, it was observed that the patient with the largest breast had the worst dose homogeneity. For such patients, the use of more intensity levels or optimized weightings of the segments might result in a further increase in dose homogeneity. It was felt that the ipislateral lung would be spared radiation pneumonitis via the technique. Apart from the use of CT data, the technique has much in common with that at the Royal Marsden NHS Foundation Trust and has similar conclusions.

Remouchamps *et al* (2000) have evaluated whether segmented tissue compensation was superior to the use of physical wedges when treating large-breasted patients. They concluded that, in all respects, it was. Segmented tissue compensation significantly decreased the overdosage in the inframammary fold and axillary region. Ten patients were studied in detail.

Cozzi *et al* (2001c) have shown that IMRT of the intact breast improves on non-IMRT treatments especially with more than two fields.

Li *et al* (2004c) have also developed a breast IMRT technique based on the use of CT data. The conventional tangential field directions were selected (with the option to make small changes) and the bixel intensities were set firstly to optimize the dose at the midplane and then to spare dose to normal structures, lung and heart. Final dose calculations were made by Monte Carlo techniques. As with earlier studies by Donovan *et al* (2002b), improved sparing of dose to normal structures was demonstrated as well as a more homogeneous dose to the target volume.

5.7.3   EPID-based techniques for breast IMRT

Heijmen (2000) has shown that EPIDs can be used to design beam profiles for IMRT (e.g. for treating breast cancer) and that the EPIDs can also be used for verifying the thickness profiles of the compensators so produced (see also Evans *et al* [2000] figure 5.19). The pre-treatment dosimetric verification of beam profiles for the dMLC technique has also been investigated and clinically applied. Methods have also been developed for realtime verification of leaf positions during dMLC treatments. Megavoltage CT has also been developed based on the use of EPIDs.

There have been many techniques suggested for IMRT of the breast. Chui *et al* (2002) have proposed another. In this technique, all that is required is the

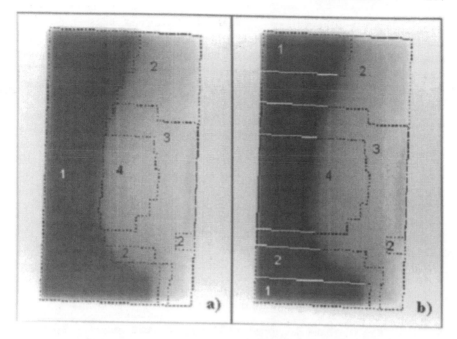

Figure 5.19. Superposition of a patient image from the Royal Marsden NHS Foundation Trust in-house EPID and the four fields to be delivered from this beam direction. A left medial image is shown. This is overlaid with the fields needed (a) for flash method 1 and (b) for flash method 2. The labels inside the images indicate the field numbers. (Reprinted from Evans *et al* (2000) with copyright permission from Elsevier.)

external outline of the breast in the direction of the tangential beams. The beams are then divided into 2 mm × 2 mm bixels and the intensity of each bixel is inversely proportional to the dose delivered to the centre of the line of sight of the line intersecting the breast from that bixel. The aim then is to produce the same dose to all the central positions of such lines. It is important to note that these lines do not necessarily lie in a single plane. The intensity profile is then extended as is usual to produce a skin flash region. Chui *et al* (2002) compared this simplified IMRT technique with a volume-based IMRT technique and a wedged-pair technique for 15 patients with left-sided breast cancer and showed that the dose homogeneity to the breast from the simple technique was considerably better than that from the wedged pair and was similar to the outcome from a full volumetric technique. The main time saving on treatment planning with this simplified technique relative to the volumetric technique comes from eliminating most of the contour delineation that is necessary for the latter. In some ways, this method has similarities with those at the Royal Marsden NHS Foundation Trust from Donovan *et al* (2000) who also attempted to equalize either the mean dose or the maximum dose in intersecting trajectories. However, this technique required the use of a portal-imaging device and other volumetric techniques such as that by Landau *et al* (2001) required the use of CT data and full inverse planning. Chui *et al* (2002) averaged five adjacent profiles of width 2 mm to produce a profile that could be delivered with a 10-mm-wide MLC leaf. They also showed that the simplified IMRT technique provided better sparing to the lung in the high-dose region than the conventional method and lower dose to the ipsilateral lung and contralateral breast. The simplified IMRT technique is practical for large-scale implementation in a busy clinic without requiring significant increase of resources.

5.7.4   Modified wedge technique

Luo *et al* (2000) have produced a breast radiation technique comprising a tangential set-up with a thin medial wedge and thicker lateral wedge. The advantage of this is a reduction of between 5–25% in the dose to the contralateral breast, a matter of particular importance when irradiating young women.

5.7.5   Breast IMRT in combination with use of respiration gating

Dyke *et al* (2001) have developed an IMRT technique for left-sided breast cancer. This technique is combined with a respiratory-gating device developed by Varian Medical Systems which synchronizes the radiation treatment with the patient's breathing pattern (see also sections 6.10.3 and 6.10.7.1). At the time of simulation, two CT scans are performed in immediate succession, the first with the patient holding their breath on exhale and the second on inhale. During treatment planning, the radiation oncologist will then determine from the dose-volume histograms whether to gate the treatments on exhale or inhale. Once this is established, three or four MLC fields are designed to reduce the dose inhomogeneities within the breast.

5.7.6   Comparison of IMRT delivery techniques for the breast

Partridge *et al* (2001a) have compared several IMRT techniques for delivering the top-up modulations needed to create the intensity-modulated fields for tangential breast radiation therapy. The first was the MSF technique in which, for the Elekta accelerator operating at 6 MV, $2N$ control points are required for N equal-fluence fluence increments. The second was the dMLC technique, operated in 'close-in' form, in which $(N + 1)$ control points are required for the same N equal-fluence increments (see figure 5.20). For completeness, a 'sliding-window' delivery was also carried out but with the proviso that this delivered the whole fluence including the baseline constant fluence to which the increments are added in the other two techniques. The basis of all these deliveries is that the ideal fluence modulations are computed first, independent of machine constraints, and then these are 'interpreted'. Measurements were made both with and without

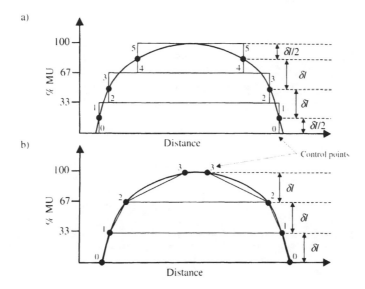

Figure 5.20. Schematic diagram illustrating the numbers of control points generated by dividing a continuous intensity modulation—here with just one peak—into (a) three discrete MSFs and (b) a series of three steps for a dynamic delivery. (From Partridge *et al* 2001a.)

the presence of an additional wedge fluence modulation. The number of control points was varied from 4, 10 to 15.

Measurements were made with film with appropriate build-up and were repeated three times to assess statistical error. Absolute dosimetric measurements were identical for four different delivery options (15-control-point dynamic with and without wedge and 14-increment MSF with and without wedge.) The MU efficiencies of both MSF and dynamic delivery techniques were very similar and decreased slightly as the number of control points increased. In terms of total delivery time, the dynamic delivery techniques were faster than the MSF techniques, the speed decreasing as the number of control points increased due to the several machine checks and deadtime at each control point. The use of the wedge increased the delivery time in all corresponding situations partly due to the need to rotate the head for the wedged portion of the delivery.

Measurements of dose profiles showed very good agreement between MSF, dMLC and ideal dose for 14 fluence increments (15 control points). Use of the wedge only changed the dosimetry in the flash region. The effect of reducing the number of control points (fluence increments) was measured. It was determined that at least 10 are needed for fields without wedges and four for fields with wedges. Close-in also well approximated dynamic delivery. Thus, the use of

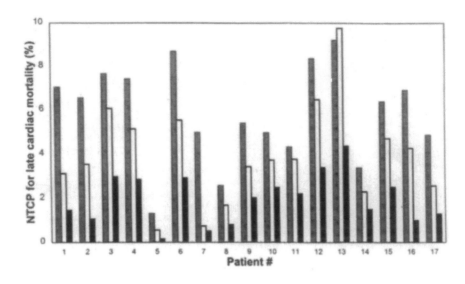

Figure 5.21. Calculated NTCP values for late cardiac mortality for three planning techniques (grey bar = rectangular fields; white open bar = conformal technique; black bar = IMRT. (Reprinted from Hurkmans *et al* (2002) with copyright permission from Elsevier.)

the wedge whilst minimizing the number of top-up fields required, an advantage, conversely led to a disadvantageous increase in treatment time.

Kerambrun *et al* (2003) have compared IMRT of the breast with tangential fields with and without wedges, showing improvement in target homogeneity and lung sparing. The IMRT was verified using a sliced phantom with 25 films and accuracy obtained was 3% in the low-dose region and 3-mm high-dose isodose lines.

5.7.7   Reduced complications observed following breast IMRT

Many studies have already shown IMRT of the breast leads to better dose homogeneity. Hurkmans *et al* (2002) set out to demonstrate that IMRT also reduced radiation pneumonitis and increased dose sparing of the heart so reducing the NTCP for late cardiac mortality (figure 5.21). Importantly, they included in the study the computation of the effects of including geometrical field shaping only as a midway stage between the use of rectangular tangential fields and full IMRT. They found that the NTCP for radiation pneumonitis was 0.3%, 0.4% and 0.5% for IMRT, conformal shaping and rectangular shaping. The NTCP for late cardiac mortality was 2.0%, 4.0% and 5.9%, respectively, for these techniques. Seventeen randomly selected left-breast treatments were included in the study. Non-IMRT planning was on UMPLAN and IMRT planning used KONRAD.

There was concern that the (CT-based) IMRT technique would lead to a much larger clinical burden. It was shown that the cardac NTCP correlated well through a second-order polynomial with the ' maximum heart distance', a geometrical parameter which indicated 'the amount of heart in the field'. This was used to estimate which patients would need IMRT. Cardiac NTCP was estimated using the data from Gagliardi *et al* (1996). This is a planning study and observation of actual clinical outcomes are awaited.

Cho *et al* (2001a) have shown that IMRT of the breast decreases the dose to the heart and both lungs compared to wide tangential-field irradiation. Cho *et al* (2002b, 2004) have shown that a simple class solution for IMRT is possible for IMRT of the left breast and can lead to significant cardiac sparing compared with non-IMRT approaches. Cho *et al* (2002a) have shown that a breast IMRT technique gives comparable results to that of an oblique electron plus photon technique when both breast and internal mammary chain are included in the irradiation. A wide-field photon tangent technique gave worse results. All inverse planning, performed on KONRAD, gave segments which were re-optimized using UMPLAN with a better dose model. Cho (2002) has discussed IMRT class solutions for treating breast cancer reducing the risk of cardiac complications for left-sided breast cancer patients. It was argued that even small relative gains in survival and recurrence-free outcome would translate to large absolute gains.

At NKI Amsterdam, work on IMRT of breast cancer in which the planning system KONRAD was used has shown that IMRT could reduce the normal tissue complication probability of cardiac mortality to 2% whereas it was 4% for conformal therapy and 5.9% for simple rectangular fields. A similar diminution of NTCP for radiation pneumonitis was also observed. This was based on the NTCP model of Gagliardi for heart, showing that NTCP of heart increases with maximum heart distance, being about 6% when the maximum heart distance is 3 cm.

Thilmann *et al* (2002) have developed IMRT for breast cancer when internal mammary and supraclavicular lymph nodes were included in the treatment volume at DKFZ, Heidelberg. It was found that the volume of the ipsilateral lung irradiated with a dose higher than 20 Gy was reduced from 26% to 13% compared with conventional treatment. For patients with left-sided tumours, the heart volume with a dose higher than 20 Gy was reduced from 11% to less than 1% and the technique resulted in improved homogeneity in the target volume. In conclusion, there were substantially improved dose distributions.

Aznar *et al* (2001) showed that the use of additional segmented fields in the treatment of breast cancer led to a higher and narrower differential dose-volume histogram reflecting a homogeneous dose distribution. This effect was not observed in plans created using either compensators or mixed beam energy. A clinical trial is being set-up to evaluate the impact of this improvement on cosmetic outcome and quality of life in patients.

5.7.8   Combination of IMRT with charged-particle irradiation

Fogliata *et al* (2002) have compared a series of different photon plans with and without IMRT to the results obtainable using protons for breast cancer treatments. Not surprisingly the proton treatments were always superior in terms of both the dosimetry of the PTV and OAR sparing. However, a three-field IMRT plan also gave very good conformality.

Li *et al* (2000a) have shown that the dose to normal structures near the breast can be considerably reduced by combining a single electron field with four intensity-modulated photon fields. This was far superior to the use of two tangential fields and comparable to, if not better than, the use of nine IMRT fields without electrons.

5.8   Bladder IMRT at the Christie Hospital

Budgell *et al* (2001b) have described the custom-compensated four-field treatment of carcinoma of the bladder that has been used at the Christie Hospital NHS Trust, Manchester, as a simple test site for the introduction of IMRT. Modulations were converted to MLC leaf-and-jaw settings for dynamic delivery on an Elekta linear accelerator. A full dose calculation was carried out and a test run of the delivery was performed prior to treatment. Absolute dose measurements were made in water or solid-water phantoms. Treatments were verified by *in-vivo* diode measurements and also by electronic portal imaging. At the time of writing, seven patients had been treated using the dMLC technique.

This report brought together the results and observations of a number of earlier technical papers from this group but it was the first to report in detail on the full scope of the treatment technique and its specific clinical implementation. The Christie philosophy is that clinical implementation of IMRT must be approached with the utmost care and therefore the intention was to tackle simple problems prior to full implementation of IMRT. Logue (2000) at the Christie has described how IMRT in the short term is considerably more demanding of resources and more expensive than conventional therapy. It is therefore important to define the clinical needs for IMRT and to define potential benefits. The selection of the appropriate delivery technique needs to include consideration of the complexity, deliverability and efficiency of the technique. It is important to understand that only through attention to all of the stages of the radiation therapy procedure through planning to delivery can the full potential of IMRT be achieved. To this end, although patients received 20 fractions over four weeks, in the initial implementation only four of the fractions were replaced with compensated IMRT fractions, two in the second and two in the third weeks of treatment. The planning process was completely automated and intensity-modulated profiles from each gantry angle were computed to deliver a homogeneous dose within a defined target volume in the isocentric plane normal to the beam axis. The leaf-synchronization technique of Budgell *et al* (1998) was used to ensure that all leaves finished simultaneously and never exceeded their maximum speed. Following this, a full dose calculation was performed based on the calculated leaf trajectories and the use of a macro-pencil-beam algorithm. This takes full account of leaf-and-jaw motions and of the transmissions through leaves and back up jaws. Both head and phantom scatter were modelled at each control point. It was possible to iteratively correct the intensity modulations to recalculate the leaf settings to deliver the altered modulations. Subsequently the dose was calculated for the whole 3D dose volume.

Prior to treating the patient, absolute dose measurements at the isocentre of each beam were made using an ionization chamber in a rectangular water or solid-water phantom and the measured and calculated doses were required to be within 2% overall for treatment to commence.

Treatment was delivered using an Elekta accelerator operating in dynamic mode at 100 MU min^{-1}. The leaf speed was initially restricted to 5 mm s^{-1}. The positions of the leaves was checked 12.5 times per second and a beam-finish facility was available. Additionally, diodes were used for dosimetric verification and a portal imager for real-time geometric verification of leaf positions according to a technique previously described by James *et al* (2000). Budgell *et al* (2001b) showed a typical compensating intensity modulation for a lateral beam for a dMLC bladder treatment showing the increased intensity required at the beam edges to compensate for lack of scatter at the beam edge. It was found that the technique was effective in reducing the high-dose areas found within static plans (figure 5.22). An issue of some concern in all dMLC treatments is the MU inefficiency and it was observed that the inefficiency factor (the ratio of MUs required to peak MUs) lay in the range 1.37–1.75 comparing favourably, with for instance, the 20° wedge factor of 1.51 for the same machine with the same photon energy.

Clinical experience at the Christie began in April 1999 and, up to the end of November 1999, seven patients had been treated. The documented additional time per patient required for planning and delivery using that research system was between 7 and 8 hr. Some of this was due to the necessity to change between clinical and service mode and back again and also to manually change the gantry angle between beams. However, the *in-vivo* dosimetry showed that all beams measured during the pre-treatment dummy run lay within 3% of the calculations. It was determined that a leaf-position accuracy of 1.5 mm was sufficient and that this was achieved by the standard quality-control protocol. It was observed that the Elekta MLC was very precise in dynamic delivery and the overall technique took less time resources than for the production of manual compensators. The group proposed to investigate similar techniques for the breast and for head-and-neck cancer where much higher levels of compensation are required and they concluded that, whilst IMRT of the bladder is not strictly necessary, the experience paves the way for the introduction of more complex and clinically beneficial IMRT techniques. A simplified account of the work at the Christie Hospital appeared in *Wavelength* (2000a) emphasizing improved homogeneity of

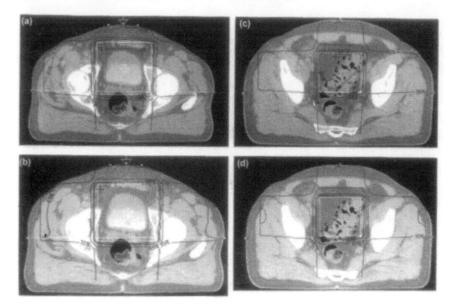

Figure 5.22. A comparison of treatment plans with and without dMLC compensation: (a) isocentric CT slice using static MLC fields (no IMRT); (b) isocentric CT slice using dMLC IMRT technique; (c) most superior CT slice using static MLC fields; (d) most superior CT slice using the dMLC IMRT technique. IMRT improves the target coverage and reduces dose gradients across the target. (Reprinted from Budgell *et al* (2001b) with copyright permission from Elsevier.)

dose to the bladder. Hounsell *et al* (2001) have also described further the Christie Hospital technique for IMRT.

McBain *et al* (2003), at a different centre, used MRI of the pelvis to assess the time-dependent motion and volume of the bladder. Serial MR images were viewed in cine loops. Patients had a characteristic volume-*versus*-time curve which was largely reproducible and linear. Based on this, they developed POLO (Predictive Organ Localization) a planning technique to take account of bladder motion. There have actually been very few reports on clinical IMRT of the bladder.

5.9   Lung cancer IMRT

Guerrero Urbano and Nutting (2004a) have reviewed the limited clinical IMRT of the lung. Dirkx and Heijmen (2000a) have extended their technique to take account of the lateral transport of radiation. They compared modulated dose distributions in lung materials with standard treatments. They investigated the technique for the improvement of the treatment of lung cancer. Compared with the treatment-planning study, the superior and inferior boost fields, used in the beam intensity modulation (BIM) technique for penumbra enhancement, had to have a higher weight and be longer in order to compensate for the lateral transport of electrons in lung tissue. The technique of using reduced-length fields with penumbra enhancement was then robust to this phenomenon. They also showed that calculations made with the CADPLAN system matched closely the measurements taken in model phantoms simulating the treatment of lung cancer (figure 5.23).

Yorke (2001) has compared IMRT using small numbers (4–9) of beams with manually optimized plans using static conformal beams for non-small-cell lung cancer. It was concluded that, independent of the model of lung toxicity that was employed (Lyman or parallel models), IMRT plans were consistently superior for target coverage and normal-tissue protection.

Koelbl *et al* (2002) have compared CFRT and IMRT for ten lung patients. They found that the dose to the ipsilateral lung was significantly reduced with IMRT but dose to the contralateral lung increased within the lower dose levels, probably without clinical significance.

Phillips (2002) has applied inverse planning using projection onto convex sets to improve the radiotherapy of lung cancer.

Sethi *et al* (2001b) have built a phantom to simulate mediastinal irradiation with lung inserts. They then created step-and-shoot and tomographic IMRT plans at 6, 10 and 18 MV for this phantom and made measurements with and without inhomogeneity corrections. It was found that the plan without inhomogeneity corrections delivered overdose to the PTV. However, this overdose decreased with increasing energy.

Conversely, and more controversially, Stevens (2003) argues that it is too early to use IMRT for lung cancer with mobile targets. Superb gated and breath-hold imaging studies have shown the vulnerability of target definition to tumour motion. A study by Liu *et al* (2004) in the same group showed that, in principle, it is possible to use IMRT of the lung to reduce the percentage of lung volume receiving more than 20 Gy as well as the integral lung dose. IMRT also spared dose to oesophagus, heart and spinal cord (Murshed *et al* 2004).

5.10   Scalp IMRT

Locke *et al* (2002) have compared conventional treatment for merkel cell carcinoma of the total scalp with IMRT. They concluded that the IMRT plan provided a more homogeneous dose to the target volume but the critical structure doses were uniformly higher than for the conventional treatment plan and so, for the time being, IMRT is ruled out.

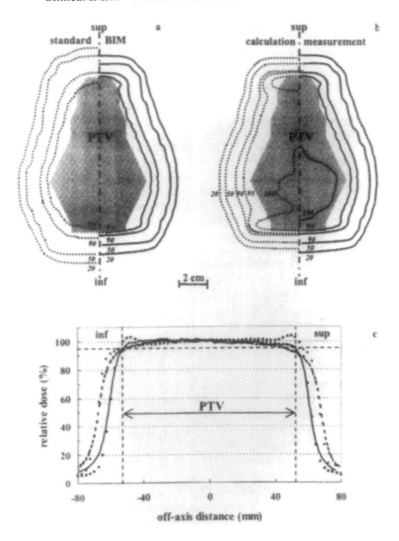

Figure 5.23. (a) Measured isodose distributions in lung-equivalent material at a depth of 12.5 cm for treatment of a PTV (shaded area) with the 10 MV photon beam using the standard field (broken lines) and a 1.4 cm shorter field combined with BIM (full lines). The isodose levels depicted are the 20%, 50%, 90% and 95% levels. (b) A comparison between the measured isodose distribution for the BIM treatment (full lines) and the distribution predicted by the CADPLAN treatment-planning system (broken lines). The 20%, 50%, 90%, 95%, 100% and 105% isodose levels are shown. (c) Measured dose profiles in the superior–inferior direction along the central axis for the standard field (broken line) and BIM (full line) and corresponding dose profiles predicted by the treatment-planning systsem (dots and squares, respectively). (Reprinted from Dirkx and Heijmen (2000a) with copyright permission from Elsevier.)

5.11   Other clinical IMRT reports—various tumour sites

James *et al* (2002a, 2003) have introduced IMRT within a (so-called) small non-academic hospital for eight patients who have been treated with five- or seven-field IMRT to sites within the head-and-neck, pancreas and lung. For head-and-neck tumours, IMRT allowed the construction of irregularly-shaped target volumes avoiding OARs. In the treatment of large inoperable pancreatic cancer, IMRT reduced toxicity by sparing normal small bowel and other sensitive structures and, by using IMRT for paraspinal lung tumours, the high-dose region could be shaped to avoid an unacceptable dose to the cord. The IMRT programme will continue at Suffolk Oncology Centre, Ipswich, with the introduction of further clinical sites. Crosbie *et al* (2002) described the QA of IMRT at this Centre. They found that the calculated cumulative dose at the plan isocentre was consistently greater than the corresponding measured cumulative dose in a perspex block phantom by about 5%, an issue they are continuing to work on. Secondly, the dose from each treatment field was measured separately in a solid-water phantom and compared with the calculated doses. The measurements and the calculations differed by approximately $\pm$ 4.5%.   When calculated fluences, plotted as isodose contours, were compared with measured fluences, the agreement was within $\pm$ 3%. James *et al* (2002b) found that electronic compensation leads to a better dose homogeneity within the breast and this was borne out by phantom measurements in which measured and calculated doses agreed to within 3%. Poynter (2003) also commented that the largest increased workload in developing IMRT at the Suffolk Oncology Centre has been increased time for treatment planning and for QA measurements. At March 2003, seven head-and-neck cases (two neck, two thyroid, two paraspinal and one parotid sparing), six lung cases and two GI cases had been treated. CFRT and IMRT plans were both developed for each patient and patient selection for IMRT was made if the IMRT dose distribution was superior. James *et al* (2003) gave an update on clinical IMRT at the Suffolk Oncology Centre. By the time of this UKRO2 conference (April 2003), 21 patients had been treated (ten head-and-neck cases, nine lung cases and two pancreas cases). The main observation was an increase in the requirement for radiological contouring and about three hours per patient of QA.

Hawliczek *et al* (2000) have used the CORVUS inverse-planning system with the PRIMEVIEW/SIMTEC software at a MEVATRON KD2 Siemens accelerator at the Institute for Radiation Oncology in Vienna. Eleven patients with various indications had been treated and the technical installation dosimetry and integration into clinical service was all performed within two months. They concluded that, using turn-key IMRT systems, the technique was ready for routine use under specific clinical conditions. Schmidt *et al* (2000) have shown that 'plug-and-plan' IMRT was possible in this hospital in Vienna. Schmidt *et al* (2001) are using a Siemens step-and-shoot and a Varian sliding-field and step-and-shoot IMRT system for routine IMRT treatment. They have particularly emphasized the importance of understanding the scatter dose in patients consequent on the use of IMRT.

Lopez Torrecilla (2001) has emphasized the steps of introducing IMRT into the clinic. The pathologies which would benefit from IMRT were studied and the importance of understanding the accuracy of patient repositioning was recognized. In the first phase, a similar dose was used as used in CFRT to acquire confidence in the new technique prior to dose escalation. It was emphasized that different planning systems using different inverse-planning algorithms use their constraints and beam weights differently and so a measure of trial and error must be undertaken jointly by the physicist and radiation oncologist before full understanding is obtained for any given system.

Mazurier *et al* (2001a) are using the HELAX treatment-planning system coupled to a Siemens PRIMUS accelerator for IMRT. They have studied a prostate case and a head-and-neck case and studied the whole chain of IMRT planning and development. After regular calibration of the MLC, it was found that the leaf-positional accuracy was better than 1 mm, that the monitor-chamber linearity was about 3% for 1 MU and that the treatment-planning system had a tendency to overestimate the dose in the penumbra area. A comparison between calculated and measured dose distributions was considered acceptable.

Rosello *et al* (2000) have used the HELAX treatment-planning system with a PRIMUS-Siemens linear accelerator to deliver IMRT in step-and-shoot mode. They discussed the chain of QA tests required to implement the technique.

Henrys *et al* (2003) used an IMRT technique with just a few segments to improve the PTV coverage for colorectal cancer.

Marcié *et al* (2002) have implemented IMRT using the HELAX system and a Siemens Primus accelerator.

Aletti *et al* (2002) have used the CADPLAN inverse-planning system with Varian 23 EX Clinac delivery to treat ten patients in the dynamic mode.

Scielzo *et al* (2002) have described a standard work-up for IMRT using the Millennium-120 MLC.

5.12 Summary

At the time of writing, the situation with regard to clinical implementation is very non-uniform. There are some centres who have developed their own techniques or adopted commercially available ones and who have worked up clinical IMRT for large patient numbers and sometimes (though not often) as part of a clinical trial. These have published full papers at some length and it is possible to obtain a clear picture of their achievements.

With respect to some of the earliest clinical IMRT at Memorial Sloan Kettering Cancer Center (prostate) and Royal Marsden NHS Trust (breast), genuine clinical outcome data are available supporting the thesis that IMRT is needed and advantageous. Some evidence of reduced xerostomia exists too.

However, most implementation is on the basis of *belief*, based on planning studies, that clinical outcome justifies the investment.

Regarding organ site, much attention has been given to IMRT of the prostate, breast, head-and-neck and bladder. Lung IMRT, gynaecological IMRT and other organ IMRT is more patchy. Perhaps not surprisingly, many of the clinical reports are at present just a set of conference abstracts that make it hard to get a clear and full picture and certainly leads to a quite patchily constructed review.

Chapter 6

3D planning for CFRT and IMRT: Developments in imaging for planning and for assisting therapy

6.1 Challenges to IMRT and inverse planning

In this chapter, we shall move away from the techniques to deliver IMRT and its clinical application and now consider some of the bedrocks on which the science of IMRT is based. Specifically these include the application of high-quality 3D medical imaging to assist the planning of IMRT and its verification. Also the use of medical imaging to assist coping with inter- and intrafraction movement will be considered. All of these concepts can be grouped under the generic title image-guided (intensity-modulated) radiation therapy (IGRT or IG(IM)RT). This chapter also groups together reviews of modern inverse-planning and forward-planning techniques.

In particular, it may be important to create inhomogeneous dose distributions within the PTV depending on the proximity of parts of the PTV to OARs or because there is specific information about the non-uniformity of tumour cells and function. IMRT thus requires us to consider how to obtain the required information and the best use of the ability to design these dosimetric inhomogeneities. Pelizzari (2003) has stressed the importance of investigating the dependence of IMRT on new positional information determined through SPECT, PET and MRI. Using those imaging modalities which have higher spatial resolution, it may be necessary to redesign the IMRT delivery, e.g. to use finer leaves in multileaf collimation to produce IMRT with steeper dose gradients. It is also important to consider the effects of patient movement and the complexity of delivery which in turn affects the resource requirements. IMRT also challenges us to consider new ways of evaluating complex treatment plans. The development of IMRT has challenged much conventional thinking in radiotherapy and requires us to reconsider the whole chain of radiotherapeutic processes.

230

6.2   Determination of the GTV, CTV and PTV; the influence of 3D medical imaging

In this section, we discuss the determination of the GTV, CTV and PTV. The growing role of 3D imaging modalities is highlighted. The issue of accounting specifically for tissue movement will be reviewed in section 6.10.

6.2.1   General comments on inhomogeneous dose to the PTV and image-guided planning

Tepper (2000) has emphasized that we are on the brink of a new era in which functional imaging will play a much larger part in treatment planning than in the past. A major task will be to integrate the images in order to effectively use them. At the same time, there has been progress in understanding how volumes can be determined from (more traditionally used) x-ray CT data.

Ling (2001a) has also emphasized the importance of using biological images for planning IMRT. These images should include predictive assays of radiobiological characteristics. Ling (2001b) argues that the planning of IMRT should be driven using images of metabolic, biochemical, physiological functional and molecular information. Images that give information about tumour hypoxia that in turn influences radiosensitivity and treatment outcome can be regarded as radiobiological images. Hypoxia is considered one of the main limiters to radiation response (Chapman *et al* 2003).

Price (2002) has also discussed the introduction of functional and molecular imaging for volume definition particularly in relation to IMRT and CFRT. It was argued that functional imaging is likely to improve the definition of the GTV but not necessarily the CTV due to inadequate spatial resolution. It was also questioned whether the identification of hot spots for dose boost using functional imaging is desirable. Functional imaging can also be used to determine critical areas to avoid. Kilovoltage CT at the time of treatment will enable correlation of the target volume to be treated with the functional images determined pre-treatment (see section 6.11.2).

6.2.2   Interobserver variability in target-volume definition

In the UK, an important clinical trial, known as the Medical Research Council RT-01 trial, is assessing the effect of dose escalation in the prostate from 64 to 72 Gy. This has involved 13 participating centres. An important part of the trial was to ensure that the centres all operated to high and consistent standards. Because the radiotherapy process is a chain, starting with the outlining of targets and normal structures, a study was carried out to determine the inter-clinician variability of outlining (Seddon *et al* 2000). Three clinical datasets of CT images on film were sent to the 13 participating centres together with instructions and also a test set already completed to demonstrate the process. The outlines were returned to the coordinating centre (the Royal Marsden NHS Foundation Trust) and these were analysed as well as being transferred by hand and without modification to the CADPLAN treatment-planning system. Beam's-eye views of the PTV were then created and analysed for their inter-clinician variability. It was also shown, by re-analysing one set of data by one clinician with a long time interval between, that the transfer process was acceptably accurate. The following was observed:

(i)   there was remarkably little inter-observer variability in the GTV central-slice identification;
(ii)   there was good consistency in outlining the inferior bladder and the superior bladder;
(iii) the definition of the superior rectum was very variable;
(iv)   the prostatic apex was very hard to define;
(v)   the definition of the base of the seminal vesicles presented problems.

As a result, at the Royal Marsden NHS Foundation Trust, it is now common practice to include the use of MR images in the work-up towards the planning of prostate cancer. These observations correspond well with the experience of clinicians and it is not really surprising that these differences have been observed.

Moeckli *et al* (2002) have conducted a study investigating how oncologists in 11 Swiss centres delineated target volumes according to ICRU-50 recommendations for a prostate and a head-and-neck case. The results were astonishingly varied between the centres with great discrepancies indicating no clear consensus on interpreting diagnostic images.

6.2.3   Margin definition

Dobbs (2000) and Graham (2000) have emphasized the importance of understanding the determination of the PTV and the associated geometrical uncertainties. Complementing this, MacKay and Amer (2000) have modelled the effects of movement of the target volume with respect to CFRT in terms of the change in TCP and normal tissue complication probability.

McKenzie *et al* (2000) have shown that the margin which should be added to the CTV is given by

$$w = 2.5\Sigma + \beta(\sigma - \sigma_p) \tag{6.1}$$

where Σ is the standard deviation of the systematic set-up error on the CTV, σ_p is the standard deviation of the penumbra function for a single beam and σ is the standard deviation of the random movement errors (figure 6.1). Van Herk *et al* (2000) previously showed that $\beta = 1.64$ is the appropriate value for a single beam irradiation to ensure that the whole of the PTV lies within the 95% dose envelope and McKenzie *et al* (2000) extended this to consider the effect of superposing a number of beams to create the high-dose volume. The value of β was given in these circumstances for up to six beams. The derivation is somewhat empirical for idealized conditions but this is commented upon extensively. McKenzie (2003a)

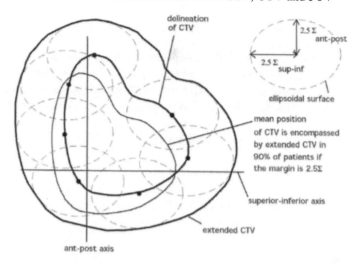

Figure 6.1. The doctor's delineation of the CTV results in a systematic error Σ. In 90% of cases, the mean position of the CTV will be entirely encompassed by drawing a margin 2.5Σ wide around the delineated CTV (From McKenzie *et al* (2000)).

has used this methodology to derive a typical margin for a prostate CTV. As well as applying a margin to the CTV, margins should also be applied to OARs (McKenzie *et al* 2002) via

$$w' = 1.3\Sigma \pm 0.5\sigma. \tag{6.2}$$

This margin is smaller than the CTV-to-PTV margin because movement of the OAR away from the beam is helpful whereas any movement of the CTV is unwanted. McKenzie (2003b) described the new British Institute of Radiology Report on geometric uncertainties in radiotherapy.

Visser *et al* (2000) have shown how systematic and random variations in patient positioning can be included into radiation therapy treatment planning. In general, the margin should be given by approximately twice the standard deviation of the systematic variations plus 0.7 times the standard deviation of the random variations. Other workers have shown similar but not identical numerical relationships.

McKenzie (2000) has provided a mathematical argument which shows that the correct way to include the effects of varying tumour position due to breathing (for example in the upper thorax) is to add the full amplitude of the breathing-related motion to the width of the CTV and only then add the sum of the quadrature errors due to set-up, machine geometry and beam penumbra. The rule of thumb is that the CTV must be kept inside the full breathing-accommodated volume and this is the opposite of expecting that uncertainties, including the breathing motion, should all be added in quadrature.

Often the data for these standard deviations are not available and so resort is made to adding margins determined through experience. Deshpande and Poynter (2000) have shown that increasing the superior–inferior margin of the prostate to 10 mm instead of the usual isotropic 8 mm used in the Medical Research Council RTO1 dose-escalation trial for CFRT of the prostate leads to an increased TCP achieved at the expense of only a small increase in low-grade rectal toxicity.

Jackson (2002) has shown that increased rectal-shielding and posterior-PTV margin sizes of about 0.6 cm reduce rectal complication rates in IMRT and prostate cancer.

Kirby and Shentall (2001) have shown that IMRT is more sensitive to patient movement or set-up inaccuracy than conventional therapy and advise an additional 0.2 cm margin to give the IMRT treatment the same CTV underdose with movement as would be obtained for a non-IMRT treatment with the smaller margin.

Caudrelier *et al* (2000) evaluated the use of fuzzy logic for volume determination in MRI and demonstrated that it improved accuracy and reproducibility. The fuzzy logic method to determine the volumes of interest (VOI) asks observers to give a confidence to the contours being created. The observer specifies a maximum contour outside which a structure definitely does not lie and a minimum contour inside which a structure definitely does lie. The region in between is the fuzzy region and is parametrized by a sigmoid acceptance function (Caudrelier *et al* 2003). This contrasts sharply with the classical method of obtaining a volume by scanning the contour and multiplying by the slice thickness (over all slices). A number of phantoms containing VOI structures of different shapes and sizes were imaged by MR and the VOIs determined both ways. The classical method was often erroneous by up to 70% whereas the fuzzy logic method was accurate to 2%. The former was not robust with respect to slice thickness whereas robustness is a feature of fuzzy logic techniques.

6.2.4   Use of magnetic resonance for treatment planning

There has been wide reporting of the use of MR to assist the definition of target volumes in treatment planning.

6.2.4.1   Distortion correction

Tanner *et al* (2000) have presented details of the Royal Marsden NHS Foundation Trust's technique for removing (or at least correcting for as far as possible) distortions in pelvic MR images which will be used for planning radiotherapy treatment. They observed that the distortions were primarily of two types: those due to the MR equipment and those due to the presence of the patient. The technique presented referred to the former. These were a consequence of the nonlinearities in the field gradients and inherent inhomogeneities in the main

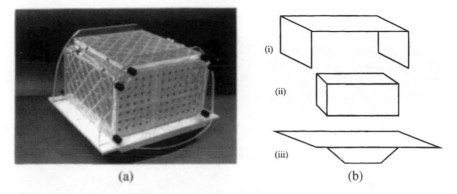

Figure 6.2. (a) Linearity test object mounted within the reference frame phantom, (b) schematic exploded view of components shown in (a): (i) reference frame, (ii) linearity test object, (iii) couch insert. (From Tanner *et al* 2000.)

applied magnetic field B_0. Uncorrected images displayed geometric distortions of up to 16 mm, unacceptable for planning radiotherapy.

The patient was positioned on a flat couch insert to simulate the treatment position. A map of system-generated distortions was created by designing a special linearity test object to measure the difference between the expected (true) positions and the apparent positions (figure 6.2). It was very important to determine a specific set of MRI parameters which remained the same both for the measurements of the test object and for those of the patient. The test object comprised a series of tubes which could be filled with contrast material and which then showed as either points or lines in the images. Additionally, a ring of marker spots was located at the field periphery. The linearity test object was designed in such a way that there were enough tubes for the purpose but that they were adequately spaced so as not to interfere with the measurement.

Software was created to automatically locate the marker positions. This was the centre of gravity of the pixels indicating the markers. Then a distortion-corrected image was created by assigning to each pixel in the new image an interpolated value from the distorted image. Tanner *et al* (2000) developed consistent QA procedures for the process. In particular, before each new patient, the images of the external ring of points, distortion-corrected, was compared with the reference distortion image and this determined the subsequent actions, either acceptance of the correction or a request for a recalibration.

Experiments showed that by applying consistent imaging procedures, stable system distortions could be maintained over many weeks with mean changes of less than 1 mm in the measured positions of the rods. It was also shown that the system corrections (mean shift) were highly dependent on the MRI parameters

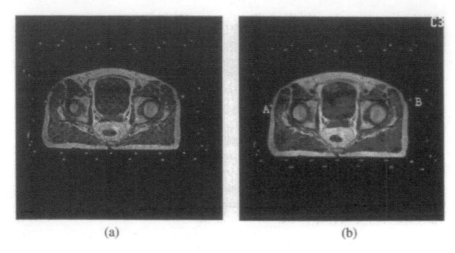

(a)   (b)

Figure 6.3. A pelvic image (a) before and (b) after correction for system distortion. (From Tanner *et al* 2000.)

demanding these stay consistently the same. For this reason, when an MR scanner is upgraded or replaced, these measurements and calibrations must be repeated.

For patient imaging, it was shown that the corrections were not so stable because of the effect of the changing size of the patient (figure 6.3). The largest patient (actually volunteer) led to errors of some 3 mm. It was left to future work to characterize these patient-related distortions.

Wu *et al* (2004) have shown a method to de-distort images of the prostate to obtain a map of citrate to aid the planning of intraprostatic lesions.

6.2.4.2 Use of contrast agents

Barentsz (2000) has shown how contrast-enhanced MRI can assist with planning radiotherapy of the pelvis. This capitalizes on the improved detection and staging that is consequent on the use of contrast-enhanced MR. Gadolinium-enhanced MRI can show the tumour extension. New more-specific lymph-node contrast agents can also show the pelvic metastatic lymph-node spread. This may replace lymphography.

Beavis and Liney (2002a, b) and Beavis (2003a) have used the paramagnetic agent gadolinium-diethylenetriaminepentaacetic acid (DTPA) to produce dynamic contrast enhancement of solid tumours visible with MR. This has enabled them to define boost volumes for simultaneous delivery with IMRT.

6.2.4.3   *Planning based on MR images alone*

Fransson *et al* (2000a) have developed treatment planning based on MR images alone. A 0.2 T Siemens Open Vision MR system generated images which were segmented by ANALYSE into tissue types. Backfrieder *et al* (2000) segmented MR data for converting the images into Hounsfield units for radiation therapy treatment planning. The segmented MR images were compared with co-registered CT images and accurate image characterization was achieved using manually selected training sets. The tissues characterized were air, bone, bone-marrow, adipose tissue, muscle, brain, cerebrospinal fluid and tumour. The segmented tissue regions were then assigned appropriate CT Hounsfield values and the converted MR data were input into a HELAX TMS treatment-planning system for calculation of dose distributions. Results were compared to the corresponding CT-based dose distributions and found to differ by no more than 1%. They concluded that MR can provide the basis for treatment planning. It was not clear whether the same conclusion would have been obtained if the MR contours would have been different from the CT contours as is sometimes the case (see section 6.2.4.4).

Lee *et al* (2003) have shown that T_1-weighted MR images can be segmented and bulk-assigned electron densities can be allocated to each segment, distinguishing air, water and bone (T_1 is the MR longitudinal relaxation time). Plans made on such images gave dose distributions differing by no more than 2% from those created using x-ray CT images.

Chen *et al* (2003) have also shown that MRI alone can be used to plan IMRT for the prostate.

6.2.4.4   *Coregistered CT and MR planning*

Technique

Toner *et al* (2000) have evaluated the accuracy of using external MR- and CT-compatible fiducial markers for image fusion in stereotactic treatment planning. They placed six markers external to a Rando head phantom for both imaging modalities and three markers internally which were used for evaluation. Image fusion was then performed using the BrainLAB software to match the centroids of structures. Spatial errors for the registration were then computed from the displacements of the centroids of the internal markers for three separate fusions using either three, four or six external markers. The error in the registration had a maximum value of 1.6 mm. It was concluded that multimodality markers as the basis of a fusion tool were sufficiently accurate for most stereotactic planning cases.

Fransson *et al* (2000b) have registered MR and CT data using both the BrainLAB 3-point technique and the HELAX eight-anatomical-landmark-point technique. Five anatomical regions were outlined on the MR images and transferred to the coregistered CT data and the registration outcome was evaluated based on a visual inspection of these five regions on the CT images. It was found that the two techniques were rated similarly but that the BrainLAB technique took longer. However, this was compensated by a more user-friendly environment and also the additional manual adjustment of the registration which was allowed by the BrainLAB technique.

Van Herk *et al* (2003) showed how chamfer matching and mutual information techniques can generate merged CT/MRI data for delineating target structure. 'Maps' of the urological structures (like world maps) were constructed to show the differences in delineation of the structures in the two modalities. Matching algorithms are also a good way to quantitate organ motion.

CT and MR definition of the prostate

Most comparative studies have concluded that MR is a very useful adjunct to CT for prostate definition. However, reports do not lead to a consistent statement concerning which volume is greater and whether rectal sparing improves or not with the use of MR images for planning.

A study by Parker *et al* (2003) of comparative CT and MR definition of the prostate showed:

(i)   MR and CT volumes were almost identical;
(ii) interobserver variation of GTV definition was lower for MR than for CT;
(iii) MR-assisted definition of the prostatic apex and
(iv) gold seeds provided a useful registration tool (figure 6.4).

Freedman *et al* (2001) have shown that the use of MRI can indicate the prostate better than CT. MRI data were used for creating IMRT treatment plans and it was shown that, although the MRI prostate volume was larger compared with that from CT, the MRI-based treatment plan did not result in increased dose to the bladder or rectum. Conversely, Lebesque *et al* (2001b) found that MRI generated much smaller target volumes for the prostate, some 30% smaller, just as was found for head-and-neck. Also Steenbakkers *et al* (2002) showed that, for IMRT of the prostate, the dose delivered to the rectum was significantly reduced if the treatment planning was based on an MR delineation of the prostate compared to a CT delineation. Vanregemorter *et al* (2002) have reported the volume of the prostate viewable using CT and MR prior to CFRT.

De Meerleer *et al* (2002b) have used IMRT to deliver a higher dose to an intraprostatic lesion than the dose given to the rest of the prostate. To do this, they used a high-resolution turbo T_2 weighted MR image allowing delineation of the intraprostatic tumour.

CT and MR definition of head-and-neck tumours

Gregoire (2002) indicated that functional MRI and PET imaging could be used to define sub-GTVs, for example highly hypoxic or proliferative areas in which an

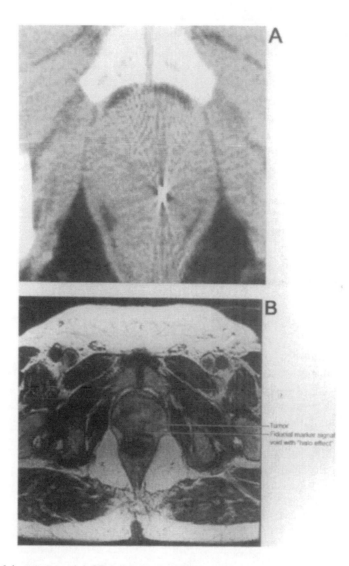

Figure 6.4. (a) An axial CT image through mid-prostate. The fiducial marker is well visualized but the prostate gland is not easily distinguished from the surrounding soft tissues. (b) The corresponding FSE sequence axial MR image from the same patient. The fiducial marker appears as a signal void (arrowed). The prostate outline is clearly demarcated from surrounding tissues. In addition, a dominant intraprostatic lesion is seen in the right peripheral zone. (Reprinted from Parker *et al* (2003) with copyright permission from Elsevier.)

extra boost dose could be sculpted using IMRT, for example, the sparing part of the salivary gland with high secretory activity. A key issue is that images need to be obtained under a very similar immobilization conditions.

Fenton and Lynch (2002) have shown that the use of MRI data, combined and registered with CT data, changed the volume of brain excluded from the tumour volume by more than 50% in six out of eight of the cases studied.

Chytka *et al* (2000) have used parameters extracted from CT and MR to distinguish between low- and high-grade glioma without performing a stereobiopsy. Using a statistical model, the discrimination was found to be 83% accurate when applied to 214 patients.

Aoyama *et al* (2001) have shown how registering MR and CT data can reduce the inter-observer variability in determining the GTV. They also showed that the MR-GTV was significantly lower than the CT-GTV for cerebellum/brain-stem lesions.

Loi *et al* (2002) have imaged high-grade glioma using CT, MR and SPECT. They showed good correlation between the SPECT and MR functional images and an advantage over the CT determination. Possibly even the number of clonogenic cells could be determined to guide IMRT planning.

6.2.4.5   *Increased protection of structures*

Liu *et al* (2000) have incorporated functional MRI into radiation treatment planning. For three patients, two MR datasets were collected. The first was a fairly standard gadolinium-enhanced set of images and the second was a functional MR image corresponding to situations in which the patients perform some task to stimulate the part of the brain required to pursue that task. They describe these as motor paradigms and visual paradigms (figure 6.5). The functional MR data were then registered using Interactive Data Language (IDL) with the gadolinium-enhanced MR data and in turn these were co-registered with x-ray CT data using the Xknife treatment-planning system. Plans were then prepared both using the functional MRI data and ignoring it. The dose distributions to the target and to the so-called eloquent cortex were compared. It was found that, taking an average over the three patients, the dose to the eloquent cortex decreased by 32% from 2.4 to 1.63 Gy whilst the dose to the target remained within acceptable boundaries at a mean of 21.13 Gy.

Liu *et al* (2000) reviewed other work in which functional MRI has been shown to give a different image of the tumour. They pointed out that these previous papers did not investigate the radiation therapy consequences. This is an important development of the use of multi-modality imaging for radiation therapy treatment planning. The application was to stereotactic radiosurgery not IMRT.

Garcia-Alvarez *et al* (2003a, b) have conducted functional MRI to establish the location of the eloquent cortex which was then protected during IMRT planning. The scan was made as the patient performed right-handed finger–thumb

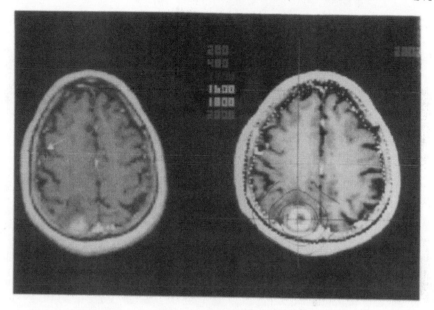

Figure 6.5. The fMRI brain map (left) and dose distribution (right) for a patient conducting a finger-tapping 'motor paradigm'. The activated area is at the motor cortex (pointed by the arrow), which is anterior to the tumour. Right: The prescribed dose distribution from fMRI-aided treatment plan is plotted on and around the tumour. (From Liu *et al* 2000.)

opposition. Beavis (2003a, 2004a) and Beavis *et al* (2003) have also reported on guiding conformal avoidance via functional imaging.

Borrosch *et al* (2000) have incorporated functional MRI in the treatment planning of CFRT for re-irradiation of malignant brain tumours. The aim of the study was to make use of the MR data to provide information on the function of the brain that it was required to preserve. For visual stimulation, opening and closing of the eyes was used as an activation and control condition and for language stimulation silent word generation was alternated with rest. The functional MRI data were then fused with x-ray CT data for treatment planning to irradiate tumours avoiding these functional regions of the brain.

6.2.4.6 *Monitoring the response to radiotherapy via MRI*

Pickett *et al* (2000a) have shown that MR spectroscopic imaging allows differentiation of normal and malignant prostate tissue based on the choline-plus-creatine to citrate ratio. This provides a means of monitoring tumour response for between one and four years after IMRT for prostate cancer. The treatment technique used raised the dose of the dominant intraprostatic lesion above that of the rest of the prostate and 46 patients were treated with forward-planned segmental multileaf collimation and also re-planned with inverse-planned segmental multileaf collimation. The former led to a more homogeneous dose distribution and the latter led to a dose distribution which was conformed better to the prostate and led to high doses within the dominant intraprostatic lesion of 130%. High-dose regions within the prostate appear to be associated with a more dramatic resolution of metabolic activity as demonstrated using MR spectroscopy. Pickett *et al* (2000b) have made use of gold seeds placed in the prostate and ultrasound guidance and an endorectal balloon to align MR images with CT images. Using IMRT of the prostate and seven fields, it was shown that the dose to OARs could be reduced.

6.2.5   Use of functional information from SPECT and PET for treatment planning

6.2.5.1   *Generalities and prostate imaging*

The accurate delineation of the target is crucial for the performance of CFRT. Van Herk (2000) has shown that IMRT requires multimodality image fusion for treatment planning, allowing the integration of MRI, PET and SPECT into the treatment-planning process. PET has proven itself as a sensitive indicator for lung tumours and lymph nodes and it is also useful to differentiate tumour from atelactasis. PET can also measure lung damage. Moya Garcia (2001) has shown how the use of PET can yield PTVs which are more appropriate for the performance of IMRT. In this section are reviewed reports of the use of SPECT and PET to assist target definition. Ciernik *et al* (2002) have shown that, when CT and PET data are fused in the planning of IMRT, the GTV delineation changed in over 53% of the cases. The changes depended on the tumour site and were not always in the same direction. The importance of PET in aiding the planning of IMRT has also been emphasized by Chapman (2003), Kessler *et al* (2003) and Cosgrove *et al* (2003). Van der Heide *et al* (2003) have used PET imaging in oncology to upstage or downstage disease prior to treatment planning. Treatment planning was based on geometrical registration of CT and MRI data and subsequently the matched PET data were added in order to review the contour delineation. Miften *et al* (2004a, b) and Das *et al* (2004) have assumed that the active clonogenic cell density is proportional to [18]fluorodeoxyglucose (FDG)-PET-generated functional images. This information has been used to create functional-dose-volume histograms (see Lu *et al* 1997) for the lung and functional EUD (fEUD). It was proposed that this gives a more realistic basis for plan comparison. The technique could equally be applied for the prostate.

Xing *et al* (2002a) have shown how to obtain inhomogeneous dose distributions within the PTV when the 'abnormality level' (AL) of disease in the PTV can be established through functional medical imaging (figure 6.6). The dose to the PTV is guaranteed a particular minimum value and then the use of an assumed linear relationship between dose and AL determines the whole needed dose distribution. A similar assumed relationship can be

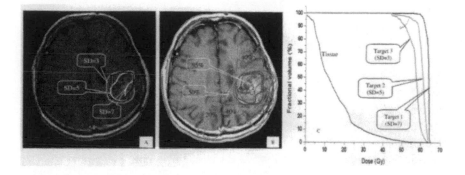

Figure 6.6. IMRT treatment plan for a malignant glioma case. Three action levels are shown in (A). The isodose distribution is shown in (B). The sensitive structure and target DVH for different metabolic action levels are shown in (C). (From Xing *et al* 2002a.)

used to differentially spare normal tissues. The scheme operates on a voxel-specific basis. The main difficulty with this approach is that the required biological information is generally not known. The authors have proceeded with assumptions; but this establishes a framework for the future incorporation of functional imaging. For example, Xing *et al* (2002b) have shown that when PET and MR spectroscopic imaging were incorporated into the planning of IMRT, it was technically feasible to produce deliberately non-uniform doses according to the functional requirements. Xing *et al* (2004) have made functional images of the prostate using MR spectroscopy and then registered these data to CT using thin-plate splines as proposed by Bookstein (1989). This allows the determination of the abnormality level. Miften *et al* (2004a,b) used such functional abnormality data to create functional DVHs.

The introduction of IMRT requires a re-evaluation of the complete chain of radiotherapy processes. The intention at the William Beaumont Hospital, Royal Oak, was for more than 50% of all patients to be treated with IMRT from 2001. The goals of IMRT are to dose escalate, to reduce toxicity and to improve the quality of life. IMRT should be linked to adaptive radiotherapy, accelerated radiotherapy, image guidance and active breathing control. The way they are doing this is to identify a lead physician and physicist and the resources in personnel needed for each target for IMRT. For prostate (see also *Wavelength* 2001c), adaptive radiotherapy and inverse planning will be based on HYPERION and PINNACLE. For lung, active breathing control will be combined with inverse planning and for breast, techniques are being developed for whole breast irradiation, regional node irradiation and quadrant irradiation. At Tubingen also, the link between IMRT and image-guided radiotherapy has been emphasized. Wong (2000) is using PET to image the washout of the ^{15}O positron-emitting isotope when made by *in-situ* activation with 29 MV external photon irradiation. This washout serves as a surrogate of perfusion. Analysis of time-binned images clearly showed regions of rapid and slow washout in the tumour of a canine patient. It was emphasized that the study of tumour biology with PET imaging is very much in its infancy.

Ling (2001c) has emphasized the role of the Memorial Sloan Kettering (MSK) Cancer Center in New York in having treated over 1100 patients with IMRT since October 1995. A decrease of functional tumour volume measured with FDG-PET as the dose built up throughout the course of treatment was shown. For the future, it was suggested that imaging biological information can yield maps of tumour hypoxia, predictive radiosensitivity and genotypic immunology to feed a better tumour definition for CFRT and IMRT. Ling (2001c) offered the 'soundbite': '2001 A Space Odyssey' will become '20–20 Vision'. Both MSK and the Fox Chase Cancer Center (FCCC) observed a dose dependence of rectal grade-2 or -3 toxicity and, as the dose escalated (Price *et al* 2003), the technique was changed to keep this within acceptable values. Ling (2003) has begun to investigate the use of functional images for identifying areas of hypoxic tumours. It was shown that the ratio of choline-plus-creatine to citrate is a good indicator of Gleason score and that regions in which the choline-to-citrate ratio is greater than two indicates hypoxic areas that would benefit from increased dose in an inhomogenous dose distribution, for example for dominant intraprostatic lesions. Hypoxic cell markers which also relate to changes in TCP, for example the markers ^{18}F fluoromisonidazol (FMISO), ^{124}I iodo azomycin galactoside (IAZG) and ^{60}Cu diacetyl-bis-methylthiosemicarbazone (ATSM) were investigated. It is important to validate that regions which showed changed uptake of these hypoxic cell markers are indeed hypoxic and Ling (2003) is using measurements of oxygen in mouse tumours to correlate with the distribution of uptake of these hypoxic cell markers. Somewhat tongue-in-cheek, Ling (2003) has pointed out that the application of imaging to planning radiotherapy is having something of a renaissance since Röntgen, Becquerel and Curie all got together ('figuratively') in the late 1890s to apply the use of x-rays to planning radiotherapy. Lewellen (2001) has also emphasized that molecular imaging using PET can define volumes for boost protocols in radiation therapy.

Dehdashti *et al* (2003) have shown that there is increased uptake of ^{60}Cu ATSM in hypoxic tumours. They found that this was a good indicator of disease-free survival, this being lower for hypoxic tumours due to their increased radioresistance. PET images of ^{60}Cu ATSM could also provide maps of hypoxia allowing the generation of deliberately inhomogenous dose distributions via IMRT.

6.2.5.2   *Head-and-neck imaging*

Nishioka *et al* (2000) have evaluated the impact of ^{18}FDG-PET imaging combined with x-ray CT and MRI for radiotherapy planning of head-and-neck tumours. They found that, among primary tumours, the volume of the primary tumour was changed in two out of 16 patients by making use of the PET data,

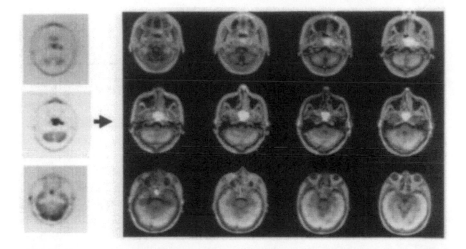

Figure 6.7. An example of [18]FDG-PET/MRI fusion for nasopharyngeal carcinoma. An automatic multi-modality image registration algorithm enables accurate registration between the two modalities using the brain as an internal reference. Anatomic locations of [18]FDG-PET uptakes which were unclear on the original PET images (left), became clear on the fusion images (right). (Reprinted from Nishioka *et al* (2002) with copyright permission from Elsevier.) (See website for colour version.)

50% larger in a case of carcinoma of the base of the tongue and 35% smaller in a case with tongue cancer. Treatment planning was improved for 10 out of the 16 patients by the introduction of the PET data coregistered with CT and MR data.

Nishioka *et al* (2002) have fused [18]FDG-PET and MRI/CT image data for radiotherapy planning for oropharyngeal and nasopharyngeal carcinomas. They studied 12 patients in the former category and nine in the latter. Except for three cases with superficial tumours, all primary tumours were detected by PET. The GTV volumes for primary tumours were not changed by image fusion for 19 cases, were increased for 1 case by 49% and decreased for another case by 45%. Parotid sparing became possible in 15 patients whose upper neck areas near the parotid glands were tumour free by PET (figure 6.7).

Daisne *et al* (2002) have shown that CT, MR and FDG-PET lead to quite different definitions of GTV in head-and-neck squamous cell carcinoma. Klabbers *et al* (2002) have registered PET and CT data for head-and-neck tumours in which the patient was immobilized inside a mask to provide a method of image registration. Paulsen *et al* (2002) have shown that areas of hypoxia in head-and-neck tumours can be identified using the hypoxic marker FMISO. By combining PET and CT data, volumes in which the dose need to be boosted due to hypoxia were identified. Nüsslin *et al* (2003) have used HYPERION to plan for inhomogeneous dose painting using the information from fluorine-misonidazole PET imaging of a head-and-neck tumour.

6.2.5.3 Lung imaging

Ung *et al* (2000) have shown that FDG-PET significantly alters treatment planning in cases where atelectasis or unsuspected node involvement exists. The combination of PET and x-ray CT data was useful for accurate determination of the GTV and minimizing geographical miss and was a fundamental step in treatment planning for lung cancer. Benard *et al* (2000) have shown that FDG-PET imaging for suspected lung cancer can improve the treatment of radiation therapy for the lung. Erdi *et al* (2000) have also shown that the use of FDG-PET in conjunction with x-ray CT improves the treatment planning of non-small-cell lung cancer. Transmission PET images were acquired simultaneously with emission PET images. The PET transmission images were registered with the x-ray CT images in the treatment-planning system using either manual or automatic methods which in turn leads to the consequent registration between x-ray CT and the PET emission images. Six patients were evaluated and, for three of the patients, the PTV was increased about 10% to include nodal disease. In one patient, PET detected a second lesion in the lung. The incorporation of PET data improved the definition of the primary tumour and should therefore improve the chances of achieving local control. Balogh *et al* (2000) also showed that co-registered FDG-PET images with x-ray CT improved the tumour localization for non-small-cell carcinoma of the lung. Schmuecking *et al* (2000) have also shown that whole-body PET imaging can improve the target definition for non-small-cell lung cancer. Findings for 31 patients were recorded. It was found that the use of PET reduced the PTV by between 3–21% compared with conventional x-ray CT imaging, largely due to the exclusion of atelectasis resulting in a smaller PTV. However, in some patients, the PTV increased because lymph node metastases which were inconspicuous in CT were also determined using the PET imaging. Bradley *et al* (2004) found the use of PET-CT changed the treatment plan in over 50% of cases. Christian *et al* (2003b) and Partridge *et al* (2003) have correlated CT scans with SPECT perfusion scans for patients receiving treatment for lung cancer (figure 6.8). They were able to show that beam orientations could be selected which reduced the irradiation of healthy (i.e. perfused) lung tissue. Orientations could be chosen which were passing through already damaged lung. Das *et al* (2003, 2004) have used FDG-PET scans to characterize and define a lung tumour and SPECT to characterize lung function simultaneously. They were thus able to define beams which spared normal lung whilst treating the tumour. Deliberately inhomogeneous PTV dose distributions were created assuming that the PET uptake correlated to tumour proliferation. Simultaneously the SPECT uptake guided avoidance of normal lung with good perfusion. Paulino and Johnstone (2004) expressed concerns about FDG-PET imaging of lung tumours:

(i)   what threshold should be used for identifying a tumour?

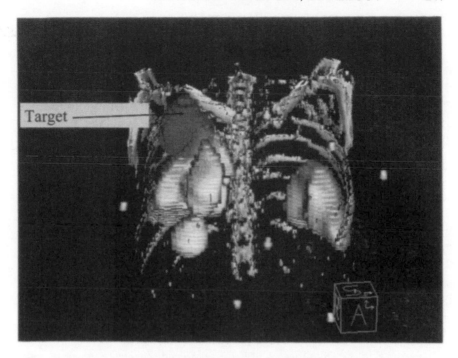

Figure 6.8. A fused image of 3D SPECT perfusion of the lung and 3D CT showing the areas of non perfusion in the upper pole of the left lung. Beams directed at the tumour via this region preferentially may be better than others at sparing normal lung. (Courtesy of Dr M Partridge.) (See website for colour version.)

(ii) What is the effect of poorer spatial resolution than in x-ray CT?

(iii) Who will contour the volumes and with what prior experience?

(iv) will patient respiration not be a bigger problem to solve than concern over tumour definition?

6.2.5.4 *Para-aortic lymph node (PALN) imaging*

Mutic *et al* (2003) and Esthappan *et al* (2004) have used PET to create a GTV for involved para-aortic lymph nodes (PALNs) for four cervical cancer patients (figure 6.9).   FDG is taken up by cancer cells because of the high glucose use of tumours arising from increased blood-flow, increased glucose phosphorylation and decreased dephosphorylation and increased uptake of cell membrane transporters. IMRT for PALNs has been designed to escalate the dose beyond that to the PALN bed defined as CTV. Beitler (2003) has emphasized the importance of having PET images to guide planning of advanced cervical cancer and Miller and Grigsby (2003) indicate its relevance also to brachytherapy planning.

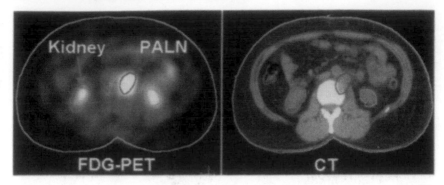

Figure 6.9. Axial PET and CT images showing a positive paraaortic lymph node (PALN) and kidneys. The left-hand image was obtained by PET imaging of fluorodyoxyglucose using a ECAT EXACT scanner from Siemens. The PALN is outlined as well as the uptake in the kidneys. The right-hand image shows the corresponding CT scan with the positive PALN superposed. (Reprinted from Mutic *et al* (2003) with copyright permission from Elsevier.) (See website for colour version.)

6.2.5.5   *Combined PET-CT scanning*

Baumert *et al* (2002) have used a PET-CT combined scanner to image tumours of the head-and-neck, lung, gynaecological or recto-anal carcinoma. They found that, in six out of the 30 patients, the use of the PET data led to significant changes in the dose to the tumour, dose to the lymph nodes and in the extension of the GTV whereas there were no significant differences in the PTV.

6.2.5.6   *Imaging to overcome breathing-motion effects*

Mah *et al* (2000) have investigated the hypothesis that spiral CT may not accurately represent the time-averaged position of thoracic tumours during radiotherapy. Spiral CT can be acquired in about 0.7 s whereas the breathing cycle is about 4 s. In addition, PET with FDG can acquire data over 45 min. They concluded that PET and non-spiral CT may better represent the time-averaged position of the tumour. This study was described in detail by Caldwell *et al* (2003). Measurements of target volume were made by spiral CT and by PET of a phantom with three fillable spheres of known volume. These were measured on an oscillating table. Using CT data, a margin of 15 mm had to be added to ensure no beam geographical miss whereas the volume determined from (lengthy) PET led to good geographical hit and lower dose to normal structures. Van Sörnsen de Koste *et al* (2003a) also showed that the use of a single slow CT scan gave a more faithful PTV than would be obtained from a rapid spiral CT scan. Averaging over six rapid spiral CT scans was an alternative way to get the optimal PTV but much more time consuming. They used the slow-scan CT to create a PTV averaged

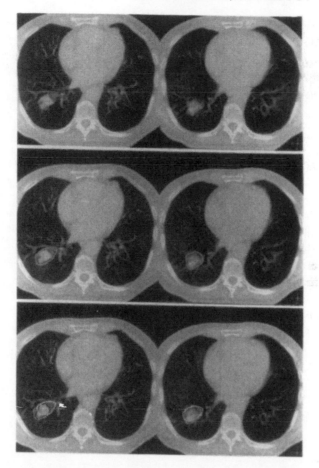

Figure 6.10. The upper panel shows co-registered rapid (left) and slow (right) CT scans of a tumour in the lung. Blurring of the edges of the mobile structure such as vessels in the heart are obvious on the slow scan. The middle panel shows CT scans showing the corresponding contoured GTVs. The lower panel shows the projection of all contoured GTVs on each type of scan. The conclusion is that the slow-scan CT can create an averaged PTV over the breathing cycle more conveniently than averaging over six rapid spiral CT scans. (Reprinted from van Sörnsen de Koste *et al* (2003a) with copyright permission from Elsevier.)

over the breathing cycle for lung tumours and a single fast spiral CT scan for visualizing other mediastinal structures (see also section 6.10.12.2) (figure 6.10).

Low *et al* (2001b) have shown that thoracic movement can significantly blur the distribution of activity in a lung tumour as viewed by a PET scanner. They proposed to use active breathing control (ABC) to minimize the motion contribution and to subsequently allow the patients to be treated with IMRT for upper thoracic tumours.

Humm *et al* (2003) have pointed out that the different data acquisition times for PET and CT can lead to concerns for registration. For this reason, gated PET or respiratory synchronized PET may be the solution.

6.2.6   Use of pathology specimens to compare with GTV and PTV

Teh *et al* (2003) have compared the pathologic prostate volume (PPV) determined from postoperative prostate pathology to the CT-based GTV and PTV for 10 patients selected from a study of 712 patients who underwent prostatectomy. They found that the GTV was on average twice as large as the PPV. The average PTV was 4.1 times larger than the PPV when a 5 mm margin was used. They thus concluded that the extent of prostate disease including extracapsular extension was well covered during CFRT. The rest of this study is a detailed analysis of the specifics of extracapsular extension.

Gregoire *et al* (2003) have also carefully correlated the GTV determined from CT, PET and MR for squamous cell carcinoma of the head-and-neck with pathology excision specimens. They found the very interesting (and worrying?) conclusion that the 3D imaging modalities overestimated the GTV. The PET-defined GTV was closest to pathology reality. Also all imaging modalities missed small microscopic extension.

Nabid *et al* (2002) correlated the size of tumours reported by total-body PET scans and pathology reports for resected tumours but no detailed results were reported.

6.3   New inverse-planning methods for IMRT

Methods for inverse planning for IMRT have been developed now for about 15 continuous years (see reviews in Webb 1993, 1997, 2000d). At least for the first half of this period, the techniques were almost exclusively those based on minimizing a simple dose-based cost function by the analytic technique, gradient descent or the iterative technique, simulated annealing. Since about 1997, dose-volume constraints have been incorporated into cost functions. These are generally of the type 'let no more than $x\%$ of the volume of an OAR receive more than $y\%$ of the target dose'. Lian *et al* (2002) have introduced a probabilistic dose prescription as the basis for IMRT planning. The idea is to specify the preference for a particular dose that a voxel has in terms of a probability function rather than a strict single value. By replacing the rigid dose prescription by a probability function, it was shown that plans can be improved. Bedford (2003) has reviewed optimization and inverse planning for IMRT describing the fundamental concepts and practical factors in its implementation.

6.3.1 Gradient-descent inverse planning

Wu and Mohan (2000) have developed an algorithm for inverse planning which has been made to operate inside the ADAC treatment-planning system. It uses dose-based and dose-volume-based cost functions and a gradient-descent technique. The code allows overlap regions between PTV and OAR as PTV, OAR to be specified or as a separate region. They showed the effect this can have on tailoring the dose distribution although only three clinical examples were shown.

6.3.2 Simulated annealing inverse planning

Sham *et al* (2001a, b) have reformulated inverse planning using volume constraints instead of dose-volume constraints. Instead of the gradient method, the simulated annealing method to minimize a volume-based cost function was also used. They find this advantageous for prostate cases and show that it overcomes difficulties associated with the dose-volume constraint.

6.3.3 Equivalent-uniform-dose-based inverse planning

Wu *et al* (2000a, 2002b) have used the equivalent-uniform-dose (EUD) concept as the basis of a cost function in inverse planning and compared the results with those generated using dose-based and dose-volume-based criteria (figure 6.11). It was argued that dose-volume constraints are simplified surrogates of the underlying biological effects. Mathematically, dose-response functions are highly degenerate functions of dose-volume combinations and therefore of dose distributions. It was argued that the use of the EUD, whilst remaining in the dose domain, makes it easier for the clinician to specify the requirements for plan evaluation. The EUD mimics the reality of dose response more closely than the simple dose-volume parameters. A case of prostate IMRT and another case of head-and-neck IMRT were examined and it was noted that improvement resulted from the use of EUD. It was found that the EUD-based optimization improved the sparing of critical structures but decreased PTV dose homogeneity. The target was included as a normal structure in order to circumvent this and to constrain the maximum dose to the target.

Wu *et al* (2003b) have extended the IMRT optimization technique based upon the generalized equivalent uniform dose (gEUD) concept. The main advantage proposed is that the gEUD technique always leads to a plan which improves on those based on dose-volume constraints. The extension is a two-stage process; if the gEUD-based optimization does not totally meet all the criteria then a subsequent second stage of dose-volume-based optimization takes place. The study considered two planning situations and showed improved dose-volume histograms based on the new hybrid approach.

Bortfeld *et al* (2002c) have also developed a new inverse-planning technique based on evaluating the EUD in an organ and looking at the trade-offs between different entities. Thieke *et al* (2003a–c) have used EUD as the base of the cost

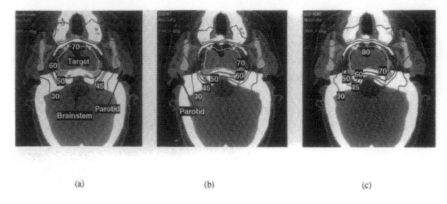

Figure 6.11. Dose distributions in a transverse plane through a head-and-neck tumour for (a) dose-volume-based objectives, (b) EUD-based objectives with dose homogeneity constraint, (c) EUD-based objectives without dose homogeneity constraint. (From Wu *et al* 2002b.)

function of a projection-onto-convex-sets iterative inverse-planning technique inside the KONRAD system. Schwarz *et al* (2003) have made a series of prostate IMRT plans based on EUD and using different values of the EUD-volume parameter 'n' showing that plans vary according to this value.

Lian and Xing (2004) have further developed IMRT optimization based on the EUD model. They have, instead of using a fixed value of the 'a' parameter in the formula for the EUD, included the possibility to vary this parameter according to a probability distribution (in their study a Gaussian distribution). The objective function was then reformulated for each structure in the presence of an uncertainty in this 'a' parameter in terms of joint probabilities and an overall objective function was formulated that takes account of the uncertainty in the parameters. Using two model problems, it was shown that the outcomes are very dependant on the form of the distribution of this EUD parameter 'a'. For two planning cases, four extreme forms of distribution were considered.

Schreibmann *et al* (2004b) have developed beam-orientation optimization for IMRT based on EUD constraints.

6.3.4 Maximum entropy inverse planning

Wu and Zhu (2001b) have introduced a maximum entropy (ME) method for the planning of CFRT. The philosophy behind the technique is that maximizing entropy minimizes information in the treatment plan and this corresponds to the maximum homogeneity of dose and the closest approximation of the calculated dose to the prescription dose. The technique relies on the fact that fluence distributions are positive and additive. The Gull and Skilling (1984) definition of entropy was used and is given by

$$E(X) = \sum_{k=1}^{N_T} \left(D_k(X) - D_k^{PTV} - D_k(X) \log\left(\frac{D_k(X)}{D_k^{PTV}}\right) \right) \qquad (6.3)$$

where D_k^{PTV} is the expected dose delivered at point k in the PTV, N_T is the number of points in the PTV, $D_k(X)$ is the calculated dose distribution at the same point k.

A cost function was constructed which includes a weighted negative least-squares error term for dose constraints on OARs in addition to the entropy term. This second term is given by

$$U(X) = \sum_{k=1}^{N_s} (D_k(X) - D_k^{OAR})^2 \qquad (6.4)$$

where D_k^{OAR} is the maximum allowable dose at point k in the OARs and N_s is the number of constraint points in the OARs. The cost function becomes

$$Q(X) = E(X) - \lambda U(X). \qquad (6.5)$$

A Newton–Raphson algorithm was used to solve the minimization of the constructed cost function. If the quantity λ in the cost function is high, then protection of the OARs is relatively more important than the elimination of the tumour cells in the PTV. Conversely, if the quantity λ is low, the homogeneity of dose distribution in the PTV is much more important than that of the OARs and the entropy in the PTV dominates the cost function. The technique was implemented inside the PLUNC treatment-planning system at the University of North Carolina, Chapel Hill, and its effectiveness was demonstrated compared to manual planning for CFRT of two difficult planning cases.

Arellano *et al* (2000) have developed a variant of the dynamically penalised likelihood inverse-planning algorithm based on the maximum likelihood estimator. This accepts few and only clinically relevant pre-optimization parameters and calculates several alternative optimized solutions for the same clinical case.

6.3.5 Genetic algorithms

One of the main problems in inverse planning is the selection of the importance factors (IFs) which govern the importance of dose conformation and/or dose avoidance and which in turn determine to a large extent the planning outcome. Wu and Zhu (2001a) and Wu *et al* (2000b) have developed a technique to automatically determine both the beamweights (in conventional planning) and the IFs using a genetic algorithm. The chromosomes at the centre of the algorithm have as their components the set of beamweights for the set of fields and the set of

IFs for the PTV and for the OARs. Then the algorithm progresses by first fixing the IFs and iterating genetically for the beamweights followed by a stage with the weights held constant and the iteration being made for the IFs. The process cycles until the cost function based on these IFs and the prescribed dose-volume constraints is satisfied or some termination condition is reached. They showed that this gives better plans than manual plans for three cases. i.e. for satisfied PTV constraints, the doses to OARs are lower.

Li *et al* (2003a–c) have developed a new form of DAO. The method simultaneously determines the number of segments per beam, the segment shapes and their weights. An interim stage involved determining the number of segments per beam based on the IMB complexity determined by a conjugate gradient method and then weights by a conjugated gradient method. The shapes for this set were then optimized using a genetic algorithm in which mutations and crossovers of selected parts of field shapes were interchanged. The goal was to decrease the total number of segments needed for a conformal dose distribution; the result is superior to the use of equally distributing the segments between beams. Roughly 30 segments were determined to be adequate for quite complex planning cases. Li *et al* (2004b) have used genetic algorithms to optimize beam directions.

Cotrutz and Xing (2003a,c) have used a genetic algorithm to directly select and optimize MLC shapes for IMRT (figures 6.12 and 6.13). The three chromosomes representing an aperture coded for the positions of the left and right leaves respectively and the weight of radiation for segments so defined. This algorithm led to highly conformal dose distributions with fewer segments than would have been obtained by segmenting beams obtained from a bixel-based optimization scheme. No topological restrictions were applied other than these imposed by the MLC itself. It was observed, as expected, that conformality increased with increasing the number of prespecified segments, convergence being correspondingly slower. For a prostate case, typically five to seven apertures proved sufficient for acceptable conformality.

6.3.6 Single-step inverse planning

Chuang *et al* (2003) have presented a new technique for inverse planning which determines the beam intensity profile in a single step instead of the more common deterministic or stochastic iterative algorithms. The single-step method utilizes a 'figure of merit'. The figure of merit (figure 6.14) is effectively the ratio of the dose deposited in the tumour to the dose deposited in the normal tissues for each ray and the first step in the process is to create the beam profiles with the bixel values simply set to the figure of merit for each ray. This figure of merit contains a number of factors which allow it to work more effectively. For example, there is a dose offset term which enables the dynamic range of the figure of merit to be limited. If this dose offset is set very high, the beams would automatically become almost uniform. There is a term representing the intersection of each ray with each pixel and other terms control weighting factors applied to the tissue,

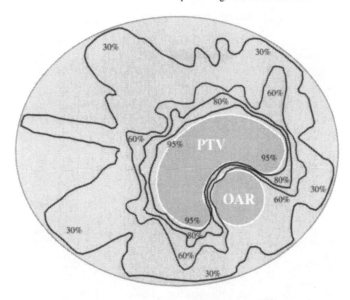

Figure 6.12. Isodose distribution for a C-shaped tumour optimized using seven apertures per beam and nine equispaced 6 MV photon beams and a genetic algorithm for creating segments. (From Cotrutz and Xing 2003c.)

penalty functions for dose to normal tissues and a so called star-pattern correction matrix that corrects for the backprojection star which is created when beams from different directions intersect and which naturally leads to a higher density of intersection close to the isocentre than further away.

The second stage of the algorithm is to maintain the form of the fluence modulation profile and then to weight each whole profile accordingly in order to maximize the uniformity of dose to the tumour. This is an iterative process but is much quicker than changing the individual beamweights of bixels. It was shown that, compared to a commercial algorithm (CADPLAN), the method yielded a lower rectal dose but a greater inhomogeneity in the prostate dose distribution. The smoothness of the intensity profiles was controlled by dose offset in the calculation of figure of merit. The authors proposed an extension to the concept whereby dose-volume constraints and other hard dose constraints will be applied to individual beam elements. The basic idea is that if each individual beam fulfils the constraints, then the linear combination of multiple beams will also satisfy the constraints. This technique appears to have much merit.

6.3.7 Simulated particle dynamics

Hou and Wang (2001) and Hou *et al* (2003a) have developed an optimization algorithm for IMRT based on simulated particle dynamics. The algorithm is

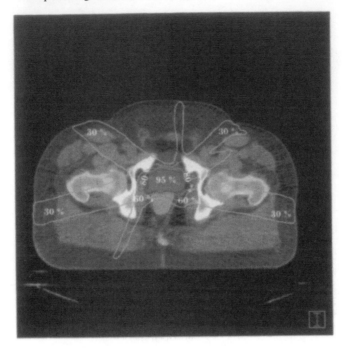

Figure 6.13. Axial slice of the final dose distribution optimized with seven equispaced 6 MV photon beams, each with three segments created using a genetic algorithm. (From Cotrutz and Xing 2003c.)

based on an analogy between the procedure of searching for the optimum intensity profiles in IMRT and the relaxation of a dynamic system of many interacting particles. Hard constraints that require non-negative beamlet intensities can be implemented and also constraints that the dose at any voxel in an OAR cannot exceed a maximum tolerance are implemented as semi-transmittable potential barriers of infinite height. The process takes place without the need to specify IFs.

Hou *et al* (2003) used a genetic algorithm to select beam orientations and a simulated dynamics method for optimization of beam fluence at each chosen set. They confirmed earlier results from Stein *et al* (1997) that some IMBs preferentially traverse OARs. Also fewer beams were better than a large number of beams as candidates for optimization of beam direction. The preferred orientations also depended on the prescription goals. It was also found that, for a complicated clinical case, IMRT with five or seven well-optimized directions gave better results than nine equally spaced beams, a result known from Rowbottom *et al* (2001).

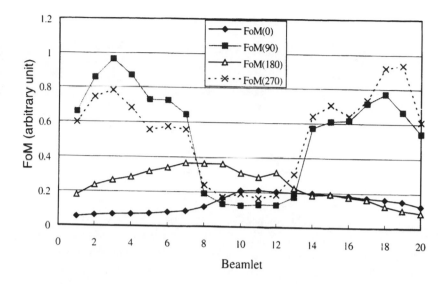

Figure 6.14. An example of a plot of figure of merit for four orthogonal angles of beams irradiating the prostate. The dose offset was set at 100 units and the rectum was weighted by a factor of three with respect to target. (From Chuang *et al* 2003.)

6.3.8   Optimization of surrogate parameters in beam space

Markman *et al* (2002) have presented a very interesting and quite novel technique for inverse planning. Usually 'conventional' IMRT is separated into two distinct problems, that of inverse planning and that of interpreting the inverse-planning-generated IMBs into practical IMRT delivery techniques. The essence of this study was to couple together the problems by ignoring direct bixel optimization and instead optimizing a set of parameters which relate to the bixels but into which can be built delivery constraints.

The first of the techniques investigated was to optimize the fluence at a set of 'anchor points', these being more widely spaced than the conventional bixels. The number of anchor points is thus fewer than the total number of bixels and the bixel intensities are interpolated using functions into which physical properties can be factored. The types of interpolation used were:

(i)   nearest-neighbour interpolation (which is effectively the same as optimizing over the bixel elements directly at a lower bixel resolution);
(ii) linear interpolation which ensures bounded gradients and which should be less sensitive to positioning errors and, finally,
(iii) cubic interpolation which may result in steeper gradients but reduce higher-order derivatives.

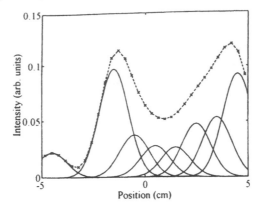

Figure 6.15. An example of determining the fluence profile (1D) for IMRT using radial basis functions. The number of exponential distributions, their centres and widths are pre-defined by the user. The optimization algorithm then determines the height of each function to calculate the fluence profile. The bixel intensities are then sampled from the resulting function shown as a dotted line in the figure. (From Markman *et al* 2002.)

The fluence at the anchor points was directly determined by the optimization method and the remaining bixel fluences were then calculated through interpolation.

A second form of indirect optimization of a parameter set is to expand an IMB as a set of positive radial basis functions (RBFs). The RBFs used were exponentials centred around different positions and of different widths and in the study ten such RBFs were used spaced uniformly in the open field. By optimizing the amplitude of these RBFs, only 10 optimization points per beam entered into the calculation. The final IMB will consequently have been constructed on a much finer grid (figure 6.15). This scheme optimizes a number of parameters which are less than or equal to the number of fluence bixels. It also enables favourable properties of the beam to be incorporated (e.g. direct smoothing) and, in principle, can speed up the optimization by a factor of between two and three. The optimization itself takes place in a fairly traditional way using a quadratic dose-based least-squares objective function with relative importance weights.

Two clinical cases were investigated with five equally spaced 6 MV photon beams for each. Conversely, standard optimization was also performed and comparisons were made between the two newly introduced techniques and the standard method. Results showed typically a factor of two speed-up in run time and the reduced parameter cases preserved the general shape and magnitude of the dose distributions whilst introducing the required smoothing into the beam profiles. It was also shown that, by deliberately misaligning the beam profiles by a millimetre, that the smoother profiles were less sensitive to misalignment.

6.3.9 Comparison of inverse-planning techniques

The effects of three (Dejean *et al* 2000b) and four (Dejean *et al* 2001a) regularization methods for performing IMRT using the singular value decomposition (SVD) technique have been shown. Dejean *et al* (2000c) have compared analytic SVD with the iterative gradient technique. For the gradient technique, the parameter selection is very user dependent whereas SVD is a more natural inverse process. Dejean *et al* (2001b, c) have investigated the use of biological cost functions to compare with the least-squares cost function used in their SVD inverse-planning algorithm. Dejean *et al* (2000a) have used receiver operating characteristic (ROC) analysis to evaluate IMRT. Rousseau *et al* (2002) identified that, of the two inverse-planning techniques embodied in the Dosigray inverse-planning module, the SVD technique gave a better homogeneity in the PTV but worse OAR results when compared with an iterative projected-gradient technique. The application was to a nasopharynx tumour attempting to spare brain stem and cord.

Wu *et al* (2001a) have compared a fast simulated annealing stochastic planning algorithm with a deterministic inverse-planning algorithm and showed that they both generate similar optimal solutions for the cases studied. They studied the effectiveness in optimization, the convergent speed and the sensitivity to the initial guess. It was claimed that this was the first time such a comparison had been explicitly made.

Llacer *et al* (2001) have carried out a very thorough comparison of five inverse-planning algorithms. These were the dynamically penalized likelihood (DPL) algorithm, a variation on this, a new fast adaptive simulated annealing algorithm, a conjugate gradient method (of Spirou and Chui 1998) and a Newton gradient method (of Bortfeld *et al* 1990). These were applied to test phantoms and patient cases. All five can yield similar results but they conclude that the DPL algorithm has advantages (figures 6.16 and 6.17). Filtering of IMBs was specifically discussed as well as the need to avoid 'hot beamlets'.

Crooks and Xing (2001) have developed inverse planning techniques which use linear algebra. They compared the outcomes with the results from the commercial CORVUS system for several clinical cases. They concluded that very similar results were obtained with the new technique but with much less computer time and thus advantage. Linear programming was also developed by Romeijn *et al* (2003). Luo *et al* (2002) have also developed a linear algebraic approach that, instead of minimizing the quadratic difference between the desired and computed doses, minimizes an objective function which is the quadratic summation of negative beamweights. Goldman *et al* (2004) have described an inverse-planning algorithm called FIDO (fast inverse dose optimization), which works out the beamweights for IMRT by solving linear equations in which the conventional quadratic cost functions have added to them a term which forces the beamlets to stay positive. The linear equations so generated are then solved by fast matrix inversion methods. A patent is pending on this planning method.

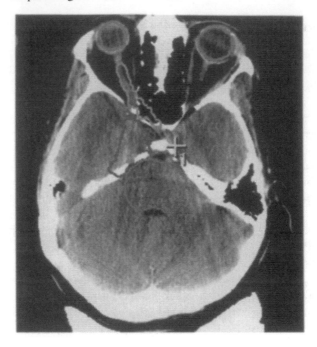

Figure 6.16. Plane containing the isocentre (cross) of a cavernous sinus meningioma case showing two of the specified OARs (left optic nerve and brain stem) and the PTV. (From Llacer *et al* 2001.)

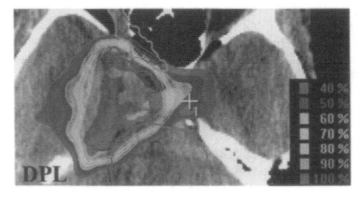

Figure 6.17. Dose distribution resulting from optimization of the cavernous sinus meningioma case by the dynamically penalized likelihood algorithm. Here we see just a part of the plane shown in figure 6.16 in order to focus attention on the PTV and the high-gradient borders with the OARs. Seven isodose levels are shown in patches of different grey from 40% to 100% dose. (From Llacer *et al* 2001.)

Morrill *et al* (2001) compared, as black boxes, two inverse-planning systems for IMRT. They found the two (unnamed) systems comparable when static step-and-shoot fields were delivered with 1 cm-wide leaves.

Richter *et al* (2000) have compared a rotation technique with IMRT and found the latter superior for concave targets. Calculations were performed using the HELAX treatment-planning system.

Yi *et al* (2002) have developed a radiotherapy research tool box for IMRT optimization in which they compared, and showed to be similar, the outcomes of simulated annealing and the gradient-search optimization technique.

6.3.10 Features and comparison of commercial planning algorithms

Cozzi *et al* (2001b) have compared the HELAX treatment-planning system and the Varian CADPLAN-HELIOS treatment-planning system for IMRT. They studied advanced head-and-neck cases and found that comparable dose distributions could be obtained provided the step-and-shoot inverse plans have between seven and eight intensity levels. With less than this, the dose distributions calculated were clinically unacceptable. Fogliata *et al* (2003) extended this study to include a third system, the CMS FOCUS, again finding all three systems gave comparable results provided common planning strategies and constraints were applied.

O'Daniel *et al* (2002) have compared inverse planning using the ADAC PINNACLE system and using CORVUS. They found that, in general, PINNACLE IMRT plans produced a more homogenous target coverage with more efficient treatment delivery than CORVUS but that the CORVUS plans were more conformal and provided more critical-structure sparing.

Dogan *et al* (2001b) have compared IMRT plans created with the NOMOS CORVUS and with the CMS FOCUS treatment-planning systems. Plans created on both systems used seven non-opposed coplanar beams for the following sites: brain, lung, head and neck, nasopharynx and prostate. Critical-structure dose constraints were site-specific and the same for both treatment-planning systems. CORVUS uses simulated-annealing optimization and finite-size pencil-beam dose-calculation algorithms. CMS FOCUS uses simultaneous iterative inverse-planning optimization and superposition dose-calculation algorithms. Both systems produced satisfactory target coverage. CMS FOCUS demonstrated somewhat more inhomogeneous target dose distributions but better sparing of dose to critical structures.

Rosello *et al* (2001) have shown significant discrepancies between the IMRT predictions of a HELAX treatment-planning system and experimental measurements when small fields are placed in the corner of large areas.

Schwarz *et al* (2002) have considered the importance of accurate dose calculation outside the beam edges in IMRT treatment planning. For 10 patients (five prostate and five head-and-neck cases) they showed the UMPLAN planning system tends to overestimate the dose whereas PINNACLE tends to underestimate the dose for prostate plans. This was not considered significant.

Samuelsson and Johansson (2003) have made a systematic study of the Varian HELIOS planning system for planning head-and-neck treatments. They found conformality always increased as the number of beams increased, that varying the collimator angle had very little effect and that to achieve dose-volume specification, lower dose-volume constraints should be set.

Martin and Hachem (2003) have commented on the lack of superposition of dose when multiple field segments are added together to create an intensity-modulated pattern. They note that the HELAX planning system effectively adds together all the components of a field before making a one-shot computation of dose. However, in practice, the dose is built up from individual segments and, because of the TG effect and also because of penumbra phenomena, the delivered dose can differ from the calculated dose in specific regions. This is because internal penumbra is not taken account of in the HELAX planning system. They call this a 'nonlinear accident'.

Johansen *et al* (2003) have compared the TMS and the CORVUS system for inverse planning noting subtle differences in the generation of minimum dose to target, being case dependent.

6.3.11 Dependences of IMRT plans on target geometry

Hunt *et al* (2000) have evaluated the dependence of IMRT on the number of fields (5, 7 or 19), on the dose constraints set for normal-tissue tolerance and on the geometrical separation distance between normal tissue and targets. Not surprisingly, they found that at small target-to-normal-tissue separations, normal-tissue sparing is improved as the number of beams increases and target uniformity is hard to achieve. Their results were used to determine sensible constraints for inverse treatment planning. Ezzell (2003) has also commented that if the planner asks for an impossible plan this will at best lead to a suboptimal plan and at worst may lead to no plan being possible. In short: 'An inverse plan may become a perverse plan'. He provides a recipe for those who wish to commence learning about inverse planning. Experience is built starting with simple planning problems and building to more complex ones.

Hunt *et al* (2002) have further investigated the effects on target homogeneity and coverage in inverse planning for IMRT of varying the radius of curvature of the target, the separation between the target and the normal tissues to be spared and the tolerances which were set for the normal tissues. This study tried to achieve a more systematic understanding of how the dose distributions achieved by inverse planning for IMRT depend on these particular parameters.

The optimizations were designed using a least-squares objective function and a gradient minimization technique developed at the Memorial Sloan Kettering Cancer Center and all constraints in this system are so called soft constraints. To get an understanding of these dependences, a variety of concave targets and normal tissues were simulated, mapping situations commonly encountered in head-and-neck and other disease sites within a cylindrical phantom.

The results are quite complicated and a good understanding of the behaviour can only be had by looking at the many figures in the paper (e.g. figure 6.18). It was found, for example, that relaxing the maximum normal-tissue constraints led to more uniform target doses and less concave dose distributions. Similarly, as the PTV-to-normal-tissue separation became smaller, the uniformity index and the normal tissue dose began to exceed the constraints by increasing amounts. All of these behaviours are completely understandable as far as the direction of behaviour is concerned. The strength of the study is that it leads to a strategy for optimization. In particular, to produce an acceptable plan, the user-specified values for the target- and normal-tissue constraints had to be more stringent than the actual clinically imposed dose limits. Secondly, identical constraints were found to lead to different results depending on patient-specific geometries of target and normal tissue. The optimization results always exhibited a strong dependence on the geometric relationships in the patient. It was therefore decided to operate an iterative inverse-planning technique in which the whole planning process was put in a do-loop over constraints. Initially the dose constraints for targets and normal tissues were set to values approximately 10% lower than the clinical goals and the normal-tissue penalties were equal to or less than the target penalties, Then after the first optimization and re-normalization, the criteria were selectively adjusted with progressively smaller modifications until the changes between successive optimizations were considered clinically insignificant. The study provided useful quantitative information that guides treatment planning for IMRT. Kron *et al* (2004b) similarly showed that the degree of conformality correlated well with the degree of overlap of PTV and OAR for helical tomotherapy.

Kapatoes *et al* (2000) have also pointed to the importance of assessing dose gradients in IMRT treatment planning. The nature of IMRT is such that it is characterized by large dose gradients and therefore misalignments of the patient with the modulated beams can have quite serious consequences in failing to achieve the planned dose distribution. It was found that gradients above 30% per cm were associated with interfaces between the target and OARs but that the gradients were at least half this between target- and normal-tissue boundaries. Kapatoes *et al* (2001b) have developed a strategy for correcting the small number of underdosed and overdosed voxels in a dose distribution that may arise due to optimization limitations. They call this the fluence adjustment strategy.

6.3.12   Multiple local minima and the global minimum in optimization

Many workers have shown that the use of specific cost functions in inverse planning leads to the existence of local minima (independently: Deasy, Spirou and Chui, Llacer, Webb, Rowbottom, Magereras, Mohan and Wu). Stochastic

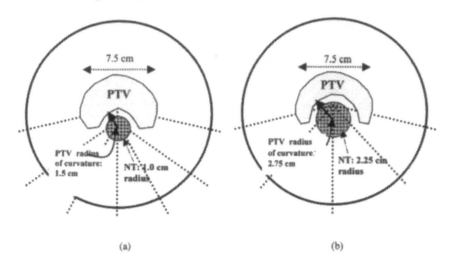

Figure 6.18. Examples of the phantom, target, normal-tissue geometry and beam arrangements used in the study by Hunt *et al* (2002): (a) Seven-field beam arrangement with the 1.5 cm radius of curvature PTV and the 1-cm-radius normal tissue positioned 8 mm from the PTV edge; (b) five-field beam arrangement with the 2.75 cm radius of curvature PTV and the 2.25-cm-radius normal tissue positioned 8 mm from the PTV edge. These are two of the software phantoms designed to study the effect of the radius of curvature and the disposition of the PTV and OARs in the study. (Reprinted from Hunt *et al* 2002 with copyright permission from Elsevier.)

search techniques such as simulated annealing have been used to find a global minimum avoiding local minima.

Wu and Mohan (2001) have, for example, shown that multiple local minima exist in IMRT inverse plans based on dose-volume constraints and dose-based objective functions. A particular configuration was studied with 500 different plans optimized with random initial beamweights and it was shown that the plans clustered in terms of multiple local minima, some plans being significantly better than others. It was thus concluded that avoiding local minima in inverse planning was desirable. However, Wu and Mohan (2002) studied whether in practice the use of some solution corresponding to a local instead of global minimum would have any clinical significance. For selected cases they ran a gradient descent optimization similar to that of Bortfeld *et al* (1990) hundreds of times from different starting conditions thus identifying local minima. They analysed cost, dose-volume histograms, dose-plans and modulation patterns to deduce if there were any clinically detectable differences. For a phantom case they observed multiple local minima in cost and beamweight when only two to four beams were used without intensity modulation. For clinical cases with intensity modulation, the situation changed and no local minima were observed. They postulated that the existence or otherwise of local minima depends on the geometry, number of beams and number of bixels and that the more degrees of freedom there were, the less the likelihood of local minima providing a problem that has clinical significance. Wu *et al* (2003a) have discussed the problem of local minima in IMRT optimization. They have developed a method of intercepts in parameter space to analyse local minima caused by dose-volume constraints and shown that many of these local minima correspond to very similar cost functions. However, occasionally they can correspond to much larger values of the cost and would represent traps which would impede progress towards a global minimum.

Llacer *et al* (2003) have presented an extensive study in which the problem of whether the presence of non-convex spaces affects the optimization of clinical cases in any significant way is addressed. It has always been well known that the use of a pure quadratic dose cost function with no constraints is a convex problem with a single minimum. Work by Deasy and by Bortfeld independently in the 1990s showed that if dose-volume constraints were applied, then the problem becomes non-convex and potentially then contains local minima. As is well known, techniques exist to overcome the problems of local minima and achieve global minima. The study by Llacer *et al* (2003) is seminal because it presents techniques whereby the behaviour of beam-space, dose-space and cost-space very close to global minima are systematically explored with the use of double-precision algebra and very small thresholds for determining the change that leads to convergence. A very large number of clinical cases was examined and it was found that either there was no evidence of multiple minima or that, when there is evidence of multiple minima, the minima are so close together in cost space that they do not affect the resulting dose distributions of the optimization although the resulting fluence maps can be quite different from each other. Fourteen different clinical cases were studied. It was expected that the results would largely be independent of optimization technique and a Newton-gradient technique was used. Zhang *et al* (2004b) came to similar conclusions and also claimed that the Newton technique was the fastest of all inverse-planning methods and robust even in the presence of local minima.

Inverse problems in IMRT are strongly degenerate in cost due to ill conditioning. It is plausible that optimization is attempting to reach a very broad minimum with extremely small gradients in the nearly flat final part of the optimization. Similar discussions have been made by Alber *et al* (2002a,b).

Caccia *et al* (2001) have shown that advantage may be taken of the local minima in inverse-planning algorithms. It was shown that the error due to the implementation through MLCs is comparable with the differences introduced by the rather wide distributions of solution, justifying simple and computionally cheap approaches to the optimization process for typical cases.

Choi and Deasy (2002) have investigated whether inverse planning using the gEUD concept leads to local minima. It was proved that when the only parameter in the EUD , 'a', is equal to or greater than 1 minimizing the EUD on a convex feasibility space leads to a single minimum; if 'a' is less than 1, then maximizing

EUD on a convex feasibility space leads to a single minimum. Jeraj *et al* (2002b) have also shown that, with of variety of other objective functions, there are no local minima in inverse treatment-planning calculations.

6.3.13   Sampling the dose matrix for IMRT optimization speed-up

All IMRT inverse planning requires the specification and use of a matrix which links the dose to each voxel to the (unit) fluence in each bixel. Problems can arise. Because photon scatter leads to dose many centimetres from the 'line-of-sight' of a bixel, it is necessary to compute very large dose matrices which is time consuming. Storing them also requires a very large space (or may even be impossible leading to the requirement to calculate these matrices on the fly many times during inverse planning). Even if they can be stored, looking them up and repeatedly accessing all the elements of the matrix is time consuming. Hence, one technique often adopted has been to truncate the 'effect' of a bixel at a certain lateral distance from the central axis of the line of sight. It is for this reason that many inverse-planning algorithms are presented, in their test phases, using simply the line-of-sight connection and some simple exponential dose model.

Thieke *et al* (2001, 2002) have proposed a technique whereby the large-scale dose matrix is stored in full within a smallish radius (say about 25 mm) of the central axis but beyond this it is *sampled*. To retain the total integral dose, the sampled values are renormalized by the inverse of the sampling probability. This (i) reduces the size of the stored matrix, (ii) speeds up access and therefore the overall computation time and (iii) approximates the use of the full dose matrix. Thieke *et al* (2002) explored the use of three probability sampling schemes. It was then shown that the use of the dose-dependent probability-sampling scheme gave results which are clinically no different from the use of the fully sampled dose matrix, yet at about one-third the computation time (figure 6.19). Conversely, the use of a truncated dose matrix led to significantly different results for some clinical cases. The reason for the success of the probability sampling is that whilst noise is introduced for individual beam elements, the superposition of such elemental contributions smooths out the noise in the final dose distribution.

Cho and Phillips (2001) have presented alternative techniques to reduce the computational dimensionality in inverse planning by using sparse-matrix operations (figure 6.20). In the equation $d = A \times b$, A is an $M \times N$ dose calculation matrix, b is a column vector containing the individual beam element weights and d is a dose vector whose elements are the dose sampling points distributed in the 3D patient space. The matrix A can become very large. However, only elements along principal diagonals relate to primary contribution from individual bixels to individual dose points in the line of sight of those bixels. All the other elements in the matrix A represent scattered radiation. Cho and Phillips (2001) truncated all the elements below 0.03% of the maximum to create an extremely sparse matrix. They then stored this matrix not as an $M \times N$ matrix but by storing the non-zero elements in one array and the locations, column-

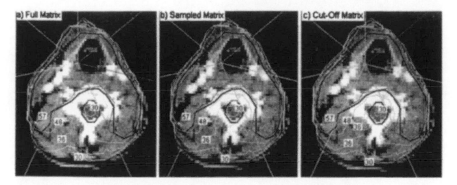

Figure 6.19. Transverse CT slices of the head-and-neck case. The outlines of the horseshoe-formed target and the spinal cord are shown as bold lines. The isodoses of the dose distribution were optimized: (a) by a full matrix, (b) by a sampled matrix and (c) by a cut-off matrix and are labelled with the absolute dose values in Gy. Additionally, the directions of the seven beams are also shown. The 30 Gy isodose inside the spinal cord is almost identical for the full and sampled matrix optimizations but is located more centrally for the cut-off matrix optimizations. The use of a sampled dose-calculation matrix leads to a significant speed gain for the dose calculation. (From Thieke *et al* 2002.)

wise and row-wise of these elements into other arrays. Typically they found that this reduced the storage requirement for matrix A from about 1.4 Gbytes to 94.4 Mbytes, i.e. to about 10% of the original size of the matrix A.

The effect of disregarding these small scatter components was then investigated by repeating the optimization first with the sparse matrix and secondly with the (normal) dense dose-calculation matrix. It was found that the results of the optimization using the dense matrix and the sparse matrix, when viewed as dose-volume histograms, showed that a comparison revealed very little difference in quality of optimization. This observation fits in with the observation of Webb (1991b) who also observed that when scatter was added to inverse-planning code the overall features of the IMBs and the corresponding dose distributions were largely unchanged.

Cho and Phillips (2001) provided the appropriate sparse matrix vector multiplication code. The basis for this work was the observation that if, in order to reduce the storage capability of a computer to store matrix A, an alternative of decreasing the dose sampling resolution was evoked, the inverse planning would deteriorate. The sparse matrix storage scheme used alternatively is the compressed row storage format. All optimizations were performed with the technique of projections onto convex sets (POCS). A hexagonal sampling scheme to replace the more normal rectangular sampling scheme was proposed. Cho *et al* (2003) showed that the use of sparse sampling, variably dependent on an organ, can improve the conditioning of the ill-conditioned inverse problem. Specifically

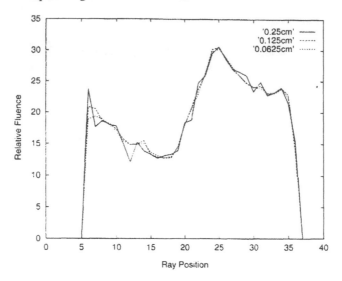

Figure 6.20. A comparison of inverse planning using three different dose-sampling resolutions: 0.0625 cm, 0.125 cm, 0.25 cm. The beam element dimensions were 0.25 cm × 1.0 cm. (From Cho and Phillips 2001.)

the spectral condition number of the matrix linking beam space to dose space is considerably reduced.

Deasy and Wickerhauser (2001) proposed to use wavelet decomposition to decrease the memory required to store 3D pencil-beam dose kernels for IMRT. It was found that 10:1 compression can be achieved with less than 1% error anywhere in the reconstructed dose distribution. Deasy and El Naqa (2003) have proposed a 'gridding' technique whereby points used to control optimization are more densely packed in regions where high dose gradients are required.

The issue of the time taken to perform IMRT optimization involves consideration of the computational time spent evaluating the objective function. Lee *et al* (2002b) have shown that it is possible to use just a fraction of the voxels forming the PTV and OARs and compute the objective function in just this fraction. It turned out that the final dose distributions obtained with a randomly sampled subset of voxels was comparable to that produced when a full set of voxels was sampled but with great savings in computer time.

Otto *et al* (2003) have conversely argued for a very fine spatial resolution for the point-spread kernel based on the use of film dosimetry for the accurate determination of dose in IMRT planning.

6.3.14 Creating a uniform PTV dose in IMRT; cost tuning

It has often been observed that IMRT optimization creates non-uniform dose to the PTV in order to spare the OARs. Vineberg *et al* (2002) set out to study whether and how it is possible to maintain the homogeneity of dose whilst simultaneously sparing OARs. This was done in the context of using the UMPLAN TPS together with its optimization tools for both IMRT and CFRT. The clinical application was optimization of plans for head-and-neck treatment with parotid sparing to reduce xerostomia. Each patient was replanned using nine equispaced fields. Targets were defined with the aid of MRI. All work was carried out on a cluster of COMPAQ Alphastations running VMS. Fast simulated annealing was used for the optimizations.

This planning system makes use of costlets which combined to form a cost function. The costlets symbolize different objectives. Some of these are as follows: a dose-based costlet to force dose to specific values at a point; a mean-dose costlet to force the mean dose in a structure to a specific value; a dose-volume-based costlet to force the volume of a structure greater than or less than some dose value to a specific volume value. Generally each of these costlets involves the difference between 'actual running value' and 'desired value' raised to some power. The power is the key to tuning the optimization, high power giving a costlet more importance. Additionally, weights can be used when combining the costlets to form the overall cost function. The highest priority costlets have the highest power .

Vineberg *et al* (2002) specified a whole range of dose homogeneity values for the PTV and showed how these ranges can be achieved in practice by varying the power of the costlets. For all of 120 tested plans, the objectives were obtained. Specifically, with respect to OAR sparing, it was found that subtle changes to the costlet functions had effects on the trade-offs between PTV dose homogeneity and OAR dose sparing. There was so much subtle variation however that no specific rules of thumb emerged (figure 6.21). It could, however, be definitely said that the PTV dose homogeneity could be kept within the traditional 5% whilst simultaneously achieving all the objectives of sparing OARs. By relaxing PTV dose homogeneity, better OAR sparing resulted. This implies that trade-off discussion will become the dominant theme for the future of plan optimization. Stated differently, the authors refuted the notion that target inhomogeneity is a natural consequence of OAR sparing for head-and-neck cancer.

6.3.15 Importance factors

It should be remembered that IFs are just numbers that modify the effect of contributing terms in a cost function. They have no physical dimensions or meaning and sometimes large changes are required for them to effect small changes in the dose plan (Bortfeld 2003).

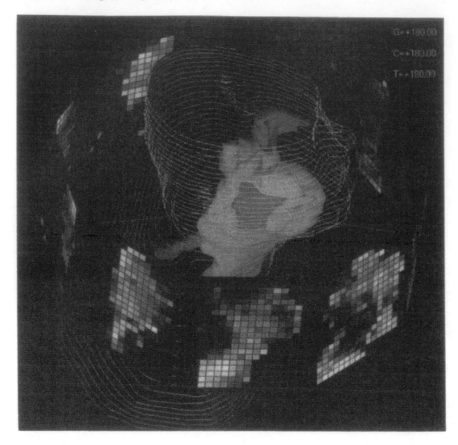

Figure 6.21. A 3D view of the patient, target volume and the parotids and the nine IMBs used for the plan including the intensity modulation for each beam. This is an example plan used by the authors to attempt to obtain uniform PTV dose sparing OARs. It is shown here, however, as a general example of the typical complexity of IMRT planning. (Reprinted from Vineberg *et al* 2002 with copyright permission from Elsevier.)

Xia *et al* (2002b) have created an objective function for inverse planning to include a power of the dose difference between calculation and prescription. Then, by varying the power depending on the organ of interest, better plans with more uniform dose to the target volume and/or more dose reduction to sensitive structures can be obtained when compared with conventional quadratic objective functions. Xia *et al* (2004) have shown that using a power of four instead the conventional power of two led to improved results. The behaviour is not the same as would be achieved by increasing the IFs for organs because increasing the IF for an organ applies the same importance to all dose points whereas higher powers give a greater contribution to the cost function from those dose points with a larger departure from prescription.

Ayyangar *et al* (2003) have made a planning comparison of three systems for inverse planning and found that direct use of the IFs set on one system (CORVUS) to two others (ADAC PINNACLE and CMS FOCUS) did not yield suitable plans. They thus confirmed that the selection of plan parameters is system dependent.

Yan *et al* (2003) used a fuzzy inference system, a form of artificial intelligence, to arrive at a dose prescription and IFs for inverse planning for IMRT. Yin *et al* (2004) have developed a fuzzy logic system for IMRT optimization in which there are three parameters taken through a fuzzifier and a defuzzifier. These parameters are the weighting factors for beams, the dose specification and the dose prescription. The use of this form of artificial-intelligence-guided inverse planning improves treatment outcome.

Phillips *et al* (2004) and Meyer *et al* (2004) have developed Bayesian decision-making algorithms for optimizing IMRT. These provide the ability to make use of an 'influence diagram' to take account of as much information as is known or can be estimated about the patient prior to inverse treatment planning.

Bedford and Webb (2003) have developed an inverse planning algorithm for CFRT generating the most feasible solutions without the need to specify IFs *a priori*. The system is known as AutoPlan.

6.3.15.1   Voxel-dependent IFs

Inverse-planning in its simplest form is driven by selecting some parameter (e.g. the quadratic difference) that characterizes delivered dose compared with prescribed dose. This concept is now well over ten years old. It is usual to assign IFs to each region in space to allow the inversion optimization to 'do better' in one region than in some other region. In the past these IFs have been set globally for the whole PTV and OARs. A disadvantage is that if a dose-volume histogram of a structure fails to meet the objective, it is usually necessary to start all over again, to re-specify the IFs and replan until the needed agreement ensues. Sometimes this turns out also to be impossible to achieve because of the conflicts in the dose specifications.

Xing *et al* (2000a) have shown that a two-stage optimization process is required to arrive at IFs which allow the closest match to be achieved between the expected and arrived-at dose-volume histograms. The optimization iteratively cycles between optimizing a cost function in dose space and optimizing a cost function in dose-volume-histogram space. Cotrutz (2002) and Cotrutz and Xing (2002) have provided a very elegant solution (figure 6.22). At first a solution is constructed based on a cost function in which IFs are globally specified to regions. Then the dose-volume histogram is inspected noting its inadequacies. Then, using a graphics tool, an inspection is made of all those voxels which specifically violate the dose-volume histogram objective. The voxels are displayed graphically and also flagged. Then the IFs of *just these voxels* are selectively adjusted and

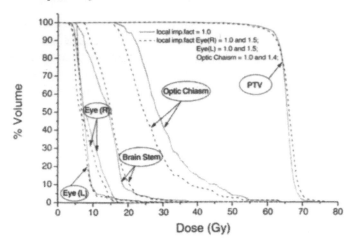

Figure 6.22. Dose-volume histograms for plans optimized with unit value local importance factors (the plain lines) *versus* optimized using higher value of local importance factors for both the eye structures and the optic chiasm (the broken lines). Notice that by switching on local importance factors that are non-uniform the protection of critical structures can be improved. (From Cotrutz and Xing 2002.)

replanning commences. This opens up regions of the cost space for investigation that have previously been unreachable. It was found that after a few cycles of this process the dose-volume histograms are much improved (e.g. for OARs) whereas those for the PTV are minimally disturbed. Essentially the improvement is a consequence of introducing local level dosimetric constraints. Importances are attached to voxels within structures in a differential way. Cotrutz and Xing (2003a) showed, most importantly, that the same effect could not be obtained by simply changing the overall structure-based IFs; it was the differential IF within a structure that actually led to the observed improvement. Thus it was possible to fine tune target and OAR doses flexibly.

Shou and Xing (2004) have observed that, in inverse planning for IMRT, voxels within a structure are generally not equivalent in achieving their dosimetric goals; this is because the spatial location of the voxel relative to the beams and the other patient geometry affects the intrinsic ability of a voxel to meet its dose prescription. This they describe as the 'dosimetric capability' of a voxel and a method is given to establish this parameter on a voxel-by-voxel basis. The optimization strategy is then to assign a higher penalty to those voxels with lower capabilities and this differential penalty scheme, when tried for a hypothetical test case, shows that there is enlarged solution space when this non-uniform importance is allowed. This enables one to break away from the mantra that, in optimization, it is generally true that there is no net gain (i.e. an improvement in one structure is usually accompanied by a dosimetrically adverse affect in another structure). They comment that this form of local IF is another way of dealing with attempting to *a priori* set a non-uniform dose prescription.

Jiang *et al* (2001a) have also highlighted the issue that not only the magnitude of hot spots is important but also their location with respect to OARs. This is because a hot spot close to an organ-at-risk might migrate into the organ-at-risk when patient motion is included. To take account of this effect, points in space were spatially coded according to the distance of the points from the external contour of the volume under consideration. In this way the spatial importance of dose information can be factored into the inverse planning. The proof of principle was demonstrated for a 2D prostate gland with five equally-spaced coplanar fields.

Deasy and Zakarian (2003) also applied 'dose-distance constraints', IFs that varied within a structure according to the distance a voxel is from some defined structure. For example, it was possible to weight those PTV voxels closest to OARs differently from the general PTV weight.

6.3.16 Biological and physical optimization

De Gersem *et al* (2000) have developed a multileaf position optimization algorithm which is based on a biophysical objective function. The tool iteratively changes the individual leaf positions. The dose distributions before and after adaptation are compared using this biophysical objective function. The method has been applied clinically and ten patients were treated. For three prostate cases, a better sparing of the rectum was obtained using leaf optimization. For seven head-and-neck cases, a better sparing of critical organs was achieved as well as a higher minimum target dose.

Holloway and Hoban (2001) have developed an IMRT inverse-planning technique in which the cost function is a physically (dose)-based optimization method but with IFs determined from biological characteristics. They argued that the benefit of this dual scheme is the maintenance of homogeneous dose to the target whilst simultaneously achieving the biological endpoints required.

Jones and Hoban (2002) have developed physically and radiobiologically based optimization for IMRT. The method was developed in two dimensions and applied to a plan with 35 beams, each consisting of 99 pencil beams. A ratio technique was used to develop the fluences for each bixel. They compared a ratio technique in which the ratio was specified using dose-based functions only with two other techniques in which the ratio was determined using biological parameters, EUD and the integral biological effective dose (IBED). The first of these biological methods considered that the tumour comprised only tumour cells and the second method assumed that the tumour comprised tumour and 10% normal cells. The goal of the second biological method was therefore to spare the normal cells within the tumour volume. It was found that the maximum tumour dose using the biological optimization techniques could become extremely large (greater than 100 Gy) particularly for the method which did not include normal cells within the tumour volume. Including normal cells within the tumour volume reduced maximum dose but still to nothing like the usual range of values seen from the outcome of physically based optimization. Overall biological optimization led to very inhomogeneous dose distributions because the EUD and IBED parameters do not vary rapidly with dose. The authors also considered the influence of the variation of the α/β ratio and α itself. Altogether they comment that the biological parameters are very poorly known as is the fraction of normal cells in a tumour volume and, somewhat surprisingly, the last sentence of their paper concludes that one should stick to physically-based optimization techniques. This study represents a bold attempt to improve treatment planning using biological modelling but at a state of knowledge where understanding of the biological parameters is so poor that it is unlikely that workers would trust the outcome of these biological optimizations.

Oelfke *et al* (2002) have created inverse planning within the KONRAD treatment planning system based on EUD to include biological affects in the inverse-planning process. Wiesmeyer and Beavis (2003) have developed a technique to incorporate knowledge of biological function into an EUD-based optimization in which the relative voxel intensity in a functional image is used to provide increased weighting.

Stavrev *et al* (2001) have proposed a hybrid physico-biological approach to treatment-plan optimization based on the constrained minimization of a biological objective function. The concept was to introduce physical constraints that specify the minimum and maximum target doses but define the objective to be minimized in terms of the weighted sum of normal tissue complication probabilities.

Vineberg *et al* (2001) have compared dose-volume and biologically-based cost functions for IMRT plan optimization. They came to the conclusion that the basis of the cost function is not the determining factor for the conformality of the dose distribution. Rather it is the balancing of the power and weights between various components of the cost function that drives the dose distribution to the desired solutions.

Zackrisson *et al* (2000) have shown that, if the time for each fraction increases with IMRT, it may be better to change the dose per fraction. They have studied a radiobiological model of this effect emphasizing the requirement to consider the fraction time when moving from conventional treatment with a small number of beams to fully optimized IMRT.

Nahum (2003) has introduced a reminder that uncertainty in the α and β radiobiological parameters can have significant impact on the calculation of the TCP from inhomogeneous radiation in the PTV. Also is questioned the concept of whether delivering a boost dose to part of the PTV non-simultaneously with the other dose has radiobiological implications. Measurement of hypoxia could guide dose painting. Dasu *et al* (2003) re-emphasized that TCP models require the use of a distribution of radiosensitivity parameters in order to obtain believable values for clonogenic cell density and for radiosensitivity when such models are fitted to clinical data (as discussed by Webb and Nahum 1993).

Wong *et al* (2002b) have described a small-animal radiation research platform. This comprises a rotating ring gantry with a 60 cm aperture and three 225 kVp x-ray beams and a 20 cm × 20 cm flat-panel amorphous silicon imager. The goal is to reproduce as far as possible the conditions for human irradiation, noting that generally when animals are irradiated in the laboratory it has been with uniform radiation. The authors want to study *in-vivo* TCP and normal tissue complication probability, interactions at the tumour and normal-tissue interface and the quantitative study of radiation-induced gene therapy. Pelizzari (2003) has also commented on the possibility to influence gene therapy using highly conformal dose distributions.

6.3.17 Pareto optimal IMRT

Schlegel *et al* (2003), Bortfeld (2003) and Thieke *et al* (2003a, b) have described how KONRAD creates a database of potential treatment plans which cover the space of Pareto-optimal plans (figure 6.23). These are the plans in which an improvement in one organ will inevitably lead to a worse result in at least one of the other organs. Then after the automatic generation of this database, the treatment planner uses an iterative search tool to select the optimal plan for the considered case. Kuefer *et al* (2003) emphasize the advantages of this approach. (Vilfredo Pareto [1848–1923] was a socioeconomist who stated the principle that one cannot make any one person better off without making someone else worse off.)

6.3.18 Combined CFRT and IMRT

Gaede *et al* (2001) addressed the familiar problem that IMRT inverse planning generates fluence modulations that significantly depart from conventional treatment beams. For three-field planning it was decided to constrain one field to be uniform and modulate only the other two. This provides a solution which is closer to conventional experience but nevertheless allows some modulation. They also, somewhat unusually, chose parallel-opposed modulated fields with the modulation identical for each pair of such fields. This contradicts the usual advice not to use parallel-opposed fields for IMRT. Gaede *et al* (2002, 2004) have argued that beam directions for an N-beam plan are likely to be favourable beam directions in the search for a suitable $(N + 1)$-beam plan. So they addressed the issue of how many beams are required for IMRT in this systematic way until the addition of further beams does not further reduce a cost function. At each stage of adding a beam, full inverse plan optimization was performed.

Vervoort *et al* (2001) have included two orthogonal static fields with five IMRT fields for a 3D IMRT treatment plan. It is these two orthogonal static fields which can then be verified using portal imaging on each treatment fraction to ensure accurate patient positioning.

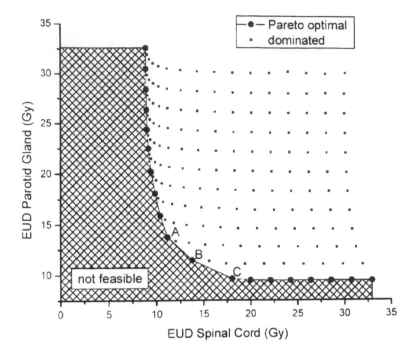

Figure 6.23. The figure shows the exploration of the Pareto front for a head-and-neck case by brute force methods. Each black dot represents one treatment plan and 256 plans are so represented. The fat dots represent the Pareto front for this case, being the set of efficient treatment plans. Pareto optimization involves making an improvement in one particular parameter, for example that shown on the x-axis, which corresponds to a worsening of the parameters shown on the y-axis. (From Bortfeld 2003.)

Al-Ghazi *et al* (2002) have compared IMRT with conformal treatment with a combination of the two. They found that, in some circumstances, the combined therapy was superior to either of the two separate therapies alone.

Coolens *et al* (2003) also showed that the conformality of plans did not dramatically deteriorate when a few beams were constrained unmodulated in a five-field plan.

6.3.19 Split modulation

Grosser *et al* (2002) have developed a procedure for minimizing the number of field components in IMRT using the step-and-shoot approach. Compared with leaf sweep, the minimal field number is up to 50% lower for intensity maps calculated with KONRAD.

For IMRT, Webb (2002c) described a simple technique which splits any $N \times N$ matrix I of fluences into two mathematically equivalent matrices I_A

```
The 2D IMB to be stripped is

  1 110  7  9  8  2  5  2  8  2  5  8  4  6
610  2  5  5  4  7 510  2  6  5  5  4  6
  2 510  3  3  1  1  3  9  2  7  4  2  1  2
  1  2  2  4  7 810  7  7  9  7  6  4  3  9
  1  1  8  7  5  4  3  3  9 310  4  9  5  9
  8  7  5  8  8  8 910  6  7  8  6  4  8  6
  5  1  6  8  2 710  4  7  8  7  3  7  8  8
110  9 310  8  2  2  5  3  5 910  6  5
  9  5  6  3  4  3  2  7  1  2  6  3  3  2  4
710 510  8  2  8  1  9 510  9  4  7  7
  8  2  7  9  3  2  3  3  3  8  9  6  9  9  3
  8  1  8  9  4  8  4  2  4  1  7 110  1  1
  5  2  2  8  8  9  7  4  5  4  2  3  4  3  8
  6  3  3  9  3  9  9  3  3  9  3  4  1  6  8
  6  7  7  3  4  3 510  5  2  2  7  1  1  3

MAX IN IMB IS      10
THE NEW EVEN-SET MAX IS     10
THE HALF-MAX BREAKPOINT IS      5
```

The 2D IMB(A) to be stripped is   The 2D IMB(B) to be stripped is

```
  2 210 810  8  2  6  2  8  2  6  8  4  6     0 010  6  8  8  2  4  2  8  2  4  8  4  6
610  2  6  6  4  8 610  2  6  6  6  4  6    610  2  4  4  4  6 410  2  6  4  4  4  6
  2 610  4  4  2  2 410  2  8  4  2  2  2     2 410  2  2  0  0  2  8  2  6  4  2  0  2
  2  2  2  4  8 810  8 810  8  6  4 410       0  2  2  4  6 810  6  6  8  6  6  4  2  8
  2  2  8  8  6  4  4 410 410 410 610         0  0  8  6  4  4  2  2  8 210  4  8  4  8
  8  8  6  8  8 81010  6  8  8  6  4  8  6     8  6  4  8  8  8 810  6  6  8  6  4  8  6
  6  2  6  8  2 810  4  8  8  8  4  8  8  8    4  0  6  8  2 610  4  6  8  6  2  6  8  8
21010 410  8  2  2  6  4 61010  6  6        010  8 210  8  2  2  4  2  4 810  6  4
 10  6  6  4  4  4  2  8  2  2  6  4  4  2  4  8  4  6  2  4  2  2  6  0  2  6  2  2  2  4
810 610  8  2  8 210 61010  4  8  8        610 410  8  2  8  0  8 410  8  4  6  6
  8  2 810  4  2  4  4  4 810 61010  4       8  2  6  8  2  2  2  2  2  8  8  6  8  8  2
  8  2 810  4  8  4  2  4  2  8 210  2  2     8  0  8  8  4  8  4  2  4  0  6 010  0  0
  6  2  2  8 810  8  4  6  4  2  4  4  4  8   4  2  2  8  8  8  6  4  4  4  2  2  4  2  8
  6  4 410 41010  4 410  4  4  2  6  8       6  2  2  8  2  8  8  2  2  8  2  4  0  6  8
  6  8  8  4  4  4 610  6  2  2  8  2  2  4   6  6  6  2  4  2 410  4  2  2  6  0  0  2
```

The two standard deviations were: 2.800   2.928 sum 5.728

Figure 6.24. The upper panel shows a 15×15 IMB matrix to be split into two. The matrix is random with a peak value of 10. The lower two panels show the result of applying the algorithm of Webb (2002c).

and I_B summing to $2I$, each of which requires considerably fewer segments to deliver IMRT using the jaws-plus-mask technique (Webb 2002b) (figure 6.24). When I_A and I_B are delivered alternately at fractions throughout the treatment the total number of field segments required is between one-half and two-thirds that of delivering matrix I when unadjusted.

Faria *et al* (2002) presented clinical evidence to show that, when a ^{60}Co unit was used to treat a four-field plan but only two fields were treated each alternate day, the results were no different in terms of survival and late side effects to the conventional delivery technique of delivering four fields each day. This is of relevance to the recent suggestions from Grosser (2000), Grosser *et al* (2002) and Webb (2002c).

6.3.20 Summary on inverse-planning techniques

Inverse planning, first proposed by Brahme (1988), has had an explosive development in the last 15 years or so and there seems no end to the flow of papers on this topic. Gradient-descent and simulated-annealing techniques have remained popular. Algorithms based on EUD, genetic algorithms and maximum-entropy algorithms have been developed. Attention has focused on the form and basis of cost functions, both physical and biological. There has been continued controversy over the existence of local minima in such cost functions. Techniques have been developed to speed up inversions through clever sampling schemes. IFs have been defined on ever smaller spatial scales rather than being applied to whole targets or organs. Some studies have addressed the geometrical factors that lead to specific planning outcomes (gradients and dose uniformities). Inverse planning is now standard practice inside commercial planning systems and there have been some comparisons of commercial systems and of planning algorithms in general. Sadly it is next to impossible to make clear consensus statements. Inverse-planning algorithms like cars have their good and bad points, their fans and critics and it may be that 'one size does not fit all'.

6.4   New forward-planning methods for IMRT; direct aperture optimization

Generally IMRT plans are computed by inverse-planning techniques. These include iterative techniques, algebraic techniques and a whole gamut of sophisticated algorithms (see section 6.3) which can include dose-volume constraints and individual weightings for different tissue structures and even voxels within them. They form the basis for most of the IMRT planning taking place in today's commercial treatment-planning systems. These commercial developments go back to the early work done by, for example, Brahme (1988), Bortfeld *et al* (1990) and Webb (1989, 1991a, b, 1992).

   Siochi (2000b) has pointed out that 'simpler IMRT' may be possible and, in many circumstances, effective by performing a field-within-field technique. A few fields with a few intensity levels may lead to sufficient conformation. For example, Vanregemorter *et al* (2000) have used the ADAC PINNACLE[3] treatment-planning system to do forward planning for IMRT. 'Fields' consist of four or five consecutive identical-gantry-angle fields with different apertures. These shapes are determined to conform to different dose levels within the patient. The relative weights of the fields were optimized by the planning system in order to yield equal dose over identified points within the PTV. It was concluded that the methodology of forward IMRT planning was a reasonable alternative to inverse planning. The breast IMRT technique at the William Beaumont Rose Cancer Center is another example of a field-within-field method (see section 5.7.1 and Kestin *et al* 2000a, b). Beaulieu *et al* (2004) have discussed ways to create such apertures automatically.

6.4.1 Segmental inverse planning at Thomas Jefferson University (TJU)

In recent years the group at Thomas Jefferson Medical College in Philadelphia have challenged the conventional wisdom that inverse-planning techniques are always required for IMRT. Instead, they have developed a form of forward treatment planning for, what might be called, 'simplified IMRT' or segmental IMRT. The technique is to determine a beam's-eye-view field for the complete PTV and then to select sub-fields of different and smaller areas, each collimated by a different pattern of the MLC, to irradiate those parts of a PTV not in line with OARs. The Cimmino algorithm can then be used to optimize the weights of these fields having specified dose constraints and IFs for the various tissues. Xiao *et al* (2002a) have presented other mathematics of optimizing the intensities of segments in an aperture-based inverse-planning technique. They showed that the solutions use the 'least intensity' and are very similar to those generated by the well-known Cimmino algorithm. Chen *et al* (2002b) elaborate on aperture-based optimization.

This approach is, of course, controversial because it somewhat kicks against many conventional developments during the last decade and Xiao *et al* (2000a) were careful to point out that the method probably works best when the target structures are moderately well separated from the OARs and/or have a fairly convex shape. They pointed out that the more concave a PTV becomes, the more likely it is that full inverse-planning techniques are required. They are certainly not rejecting these. Indeed, they have used the CORVUS system for routine inverse treatment planning. This technique falls into a class of techniques called segmental IMRT. Similar techniques are being developed at Ann Arbor and in UCSF, particularly for irradiation of dominant intraprostatic lesions. The paper by Webb (1991a) had diagrams of the main concept. Verhey (2001) has also emphasized the importance of studying simple two-segment IMRT compared to full IMRT.

Galvin *et al* (2001) use, for segmental planning, a combination of forward and inverse planning to derive IMRT fields which are ten times more efficient to deliver than those consequent on the use of simulated anealing via the NOMOS CORVUS system. This DAO has been implemented inside a CMS treatment-planning system. They have compared the outcome with an equivalent plan produced using CORVUS which gives very noisy data and have 'credentialled' inverse planning by dose-volume histograms. It was shown that in one case the MUs for a CORVUS MIMiC plan were 3552, for a CORVUS dMLC plan were 864 and for the DAO plus MLC plan were only 296. It was found that the number of segments was 478 for the CORVUS plan and only 49 for the DAO plus MLC plan and thus concluded that small-field dosimetry was avoided and that MU efficiency improved. Xiao *et al* (2000a, b), in this group, compared the outcome of the forward-planning approach with a corresponding plan computed using the NOMOS CORVUS treatment-planning system and their study showed one particular treatment-planning case in which the dose to the rectum could be considerably spared using the forward treatment-planning technique without greatly compromising the dose homogeneity in the target. This Thomas Jefferson University approach was also implemented inside the Elekta PrecisePLAN IMRT system. Simplified accounts of this work appeared in *Wavelength* (2000a, 2003c) emphasizing that this planning is an extension of conventional planning.

Galvin *et al* (2000a) have applied the segmented-field IMRT technique to the breast. Selected isodose curves were projected onto the plane perpendicular to the tangential fields and these were used to define and weight additional field segments to improve the homogeneity of dose to the breast. Galvin *et al* (2000b) have stressed the advantages of the forward-planning technique. These are that it does not require a translator or interpreter, the intensity patterns are simple and quite intuitive, QA is simplified because small fields are avoided and each gantry angle includes at least a segment that conforms to the beam's-eye view of the whole target and standard verification port filming can be used. The method can be applied without the need for any additional hardware or software. The segments are selected based on rules that are a modification of those used for traditional forward planning. Chen and Galvin (2001) have shown that a useful technique for forward-planning IMRT is to add one new field on each iteration, renormalizing the beamweights of the other existing field components, so that the plan is improved. This avoids numerous difficulties encountered in beamlet inverse planning. Chen *et al* (2001b) have refined the automatic-aperture-selection technique for IMRT plan optimization. Segments are sequentially and continually added to the calculation with re-normalization of the other beam weights until an optimum set (usually less than 100) of segments is obtained.

Bednarz *et al* (2002) in this group have created an optimization technique based on predetermined field segments. They considered nine fields with several segments per field. Then they optimized the weights of these using a mixed-integer programming technique to satisfy dose-volume constraints. They compared the outcome to the optimization of field segments based on the Cimmino algorithm and also compared to a 'full' inverse-planned technique using CORVUS. They concluded that the segmental optimization techniques could perform as well as full modulated-bixel IMRT with the additional advantage of simplicity, of easier QA and of stability. Earl *et al* (2003c) provide similar evidence for the reduction in the number of segments.

6.4.2   Aperture-based planning at the University of Ghent

The group at the University of Ghent have pioneered the use of aperture-based IMRT (see review by Webb [2000d] and reports on the developments in IMRT at the University of Ghent to that date [*Wavelength* 2000a]). One patient was treated in 1996, five patients in 1997, 17 patients in 1998, 47 patients in 1999 and 68 patients in 2000. By May 2001 the number of patients stood at 175. The delivery technique gradually evolved. Up until May 1997, the MLC was operated manually. Between then and 2000, the Elekta dynamic prototype equipment was used and by 2001 Elekta RTDesktop was in use. A major application has been to paranasal sinus cancer. Planning is performed using the GRATIS treatment-planning system on a DEC Alpha computer which in 2001 had an interface to two Elekta accelerators, an SL25 and SL18 both then operating the Elekta Javelin operating system with MLC$_i$ and IMRT 3.1.0. Alternatively the Elekta RTDesktop system can be driven. GRATIS is linked to both systems through software based on the DICOM standard (*Wavelength* 2001b). Claus *et al* (2000) have shown that the anatomy-based IMRT approach at the University of Ghent does not necessarily lead to a large increase in MUs compared to conventional radiotherapy.

De Gersem *et al* (2001a) have described in detail the anatomy-based segmentation tool (ABST) This generates superposed segments which, after beamweight optimization, results in IMBs. ABST is a complex algorithm but essentially grows segment shapes that are in accordance with principles elucidated over 20 years ago by Brahme *et al* (1982). For example, the beamweight of radiation needs to increase sharply close to the canyons created by the views of an OAR in order that the PTV be uniformly irradiated. This is implemented algebraically in the technique and a feature of the technique is the automatic generation of multileaf collimated field shapes. These are not created *a posteriori* after intensity-modulated inverse planning. Also the geometric constraints of the MLC can be included. This technique was applied to specific head-and-neck treatments (figure 6.25).

De Gersem *et al* (2001b) have described the details of a tool for adjusting the segment shapes and weights for step-and-shoot IMRT. The tool is called SOWAT (Segment Outline and Weight Adapting Tool). The way this works is as follows. Initially beam segments are created using geometrical views of target and OAR structures. These are then optimized for beamweight to create conformal therapy. The tool SOWAT then starts with the existing set of geometrical shapes and the corresponding dose distribution and adjusts the shapes sequentially and cyclically to attempt to improve the distribution (figure 6.26). It does this by inspecting each shape, first making the adjustment with quite large leaf position changes, cycling around and accepting changes which are beneficial and gradually reducing the leaf-position change at each iterative cycle until a better set of segments is generated. In an example, typically 41 cycles were executed.

The dose-computation engine inside this technique is much simpler, for computation speed, than that which is used to finally compute the dose from geometrical shapes. Once the final set of shapes is found, the ADAC PINNACLE treatment-planning system or the differential scatter-air-ratio method of GRATIS are used for a final dose calculation. The GRATIS system is also linked to the ADAC PINNACLE system for image fusion and contour delineation and to the PEREGRINE dose-planning Monte Carlo module and also linked to the HELAX planning system. CRASH and SOWAT home-made software developments are required because there are no equivalents in commercial systems (*Wavelength* 2001b). De Gersem *et al* (2001b) showed the technique applied to two particular

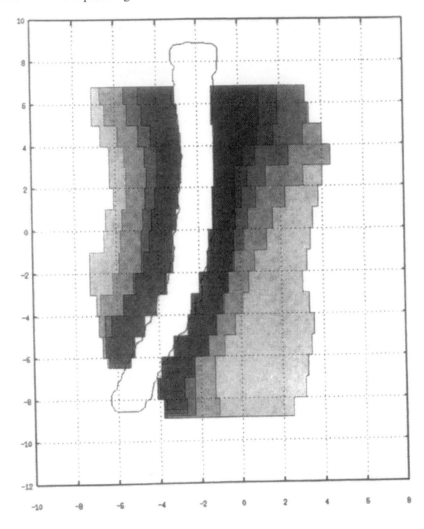

Figure 6.25. An example of the final leaf-and-jaw positions for all segments created by ABST. (Reprinted from de Gersem *et al* 2001a with copyright permission from Elsevier.)

planning problems: a two-phase plan for elective nodal irradiation on both sides of the neck and an ethmoid sinus cancer case.

Claus *et al* (2001) have performed a statistical evaluation of the SOWAT leaf-position optimization tool for IMRT of head-and-neck cancer. This tool evaluates the effects of changing the position of each collimating leaf of all segments on the value of an objective function only retaining changes that improve the value of the objective function. Although clinical implementation of IMRT at the Ghent University Hospital started in 1996, before December 1999 IMRT plans were

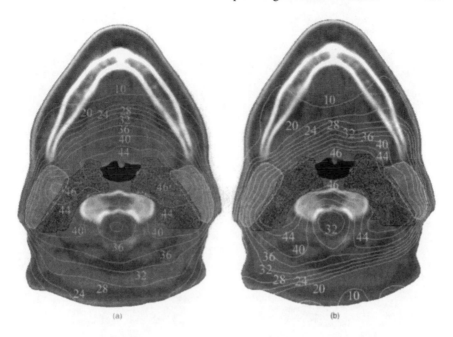

Figure 6.26. Dose distribution of the first phase (0–46 Gy) of the treatment plan in a transverse slice through the body of C2. Speckled areas are enclosed by contours of lymph node regions II left and right, which are part of a CTV1. The light grey areas contoured in white are the left and right parotid glands. (a) Dose distribution before the use of SOWAT, (b) dose distribution after the use of SOWAT. (Reprinted from de Gersem *et al* 2001b with copyright permission from Elsevier.)

generated without the SOWAT tool. After December 1999, SOWAT was in use. Thirty head-and-neck patients were treated with IMRT between December 1999 and January 2002. Two distinct patient groups were distinguished: pharyngeal and laryngeal tumours (17 patients) and sinonasal tumours (30 patients). Dose statistics of the treatment plans with and without SOWAT were analysed.

It was found that, when using SOWAT for the pharyngeal and laryngeal cases, the PTV dose homogeneity increased with a median of 11%, whilst the maximum dose to the spinal cord was decreased for 14 of the 17 patients. In four plans, where parotid function preservation was a goal, the parotid mean dose was lower in one plan without SOWAT and in four plans with SOWAT. For the sinonasal tumours, the PTV dose homogeneity increased with a median of 7% and SOWAT lowered the mean dose to 53 of the 63 optic pathway structures. It was concluded that SOWAT is a powerful tool to perform the final optimization of IMRT plans without increasing the complexity of the plan or the delivery time.

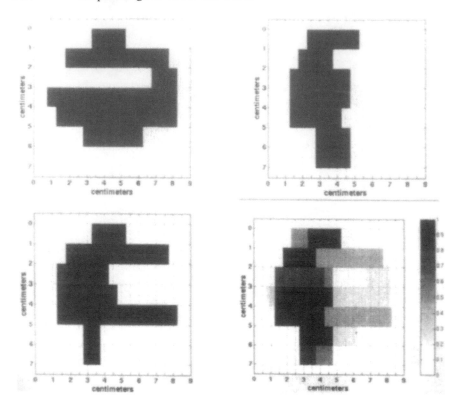

Figure 6.27. Showing how three direct aperture shapes, each of which meets the constraints imposed by the Elekta MLC, combine to give an intensity map for one beam direction in which six non-zero beam intensities occur. (From Shepard *et al* 2002b.)

6.4.3 Direct aperture optimization (DAO) at the University of Maryland

Shepard *et al* (2001, 2002b) have described the DAO technique for step-and-shoot IMRT. At each stage of an iterative optimization, both the shape and beamweight of a segment can be adjusted. Initially all shapes are set to the beam's-eye view of the PTV. Then shapes are adjusted by randomly selecting MLC leaf pairs. By this means the number of segments can be strictly controlled. All MLC leaf constraints can be built into the optimization. If desired, shapes can be constrained to have a minimum size and/or minimum number of MUs thus overcoming concerns about the use of small shapes and low MUs. This approach contrasts vividly with that of first finding IMBs and then *a posteriori* sequencing them. Simulated annealing is used to minimize a quadratic cost function but with dose-volume constraints inbuilt (figure 6.27).

Because of the possibility of overlapping apertures the number of fluence levels per orientation is $N = 2^n - 1$ where n is the number of apertures. Thus very few apertures yield many modulation levels. With conventional two-stage sequencing, it is the other way round. It is claimed that DAO can reduce the requirement to only three to five apertures per beam direction. Jiang *et al* (2004b), for example, showed that, for a seven-field prostate plan, no more than seven apertures were required since the objective function did not improve with more. This corresponds to $2^7 - 1 = 127$ fluence levels, rather more than suggested by earlier workers such as Stein *et al* (1997).

Shepard *et al* (2001, 2002b) performed dose calculation using Monte Carlo EGS4 with pencil beams computed for all needed options. The final configuration was also recomputed. The method was applied to a phantom, a prostate patient and a brain patient and computational runs were made to study the increasing conformality and rate of increase as more beam segments are added. The successful avoidance of cost function local minima was also investigated by rerunning many cases with different random number sets.

Plan parameters were compared with corresponding CORVUS-generated plans which use the two-stage process. It was found that both the number of segments and the number of MUs were greatly (by about a factor of five) reduced for the DAO method (Shepard *et al* 2003a).

The DAO method clearly obviates the whole problem of finding segments *a posteriori* from modulated beams. Yu, at the AAPM Summer School (Palta and Mackie 2003) described this as ' having the solution before one has the problem'. A strength is that it does away with bixel-based optimization; a weakness is that it only does this for MSF (not the dynamic) MLC delivery (Shepard *et al* 2003a). DAO also allows the concept of angular cost to be introduced. This is the cost of omitting a beam angle and the philosophy is that where the angular cost is low, this is a region in space where it is not necessary to develop large numbers of different field shapes. It might be useful to constrain the variation of shapes between angles which are close by in space.

Earl *et al* (2002) have simplified IMRT planning with DAO. Two to five apertures per beam were sufficient to produce treatment plans that were comparable to those produced with other inverse treatment-planning systems. Earl *et al* (2003a) have presented an inverse-planning technique for IMAT based on DAO. The two independent parameters to determine are the field shapes and the value of the beamweight at each field shape. DAO starts by setting each aperture equal to the projection of the PTV and the beamweights to one. Then different apertures are arbitrarily selected and tested for whether they improve the dose distribution. The technique is based on simulated annealing and iteratively cycles until no further improvement is obtained. The width of the distribution of the field size and the beamweights gradually reduces as iteration continues. The technique is computationally demanding and so compression techniques are used to reduce the amount of space required to store the pencil beams by a factor of about 500. At all stages of iteration, the aperture shapes are checked to determine if they satisfy the delivery constraints. The objective functions used were a ranged-based and a dose-volume-histogram-based weighted least-squares function. Several planning cases demonstrated the efficacy of the technique. Earl *et al* (2003a) argue that the absence of a direct robust IMAT inverse-planning technique has held up the clinical implementation of a technique that was first announced in 1995 (Yu 1995). Earl *et al* (2003b) have described the use of the DAO technique to create IMRT using only the jaws of an accelerator.

6.4.4 DAO wobbling the MLC leaf positions

Van Dalen *et al* (2000) observed that when a too mechanistic formula is used for setting the MLC leaves for static therapy (e.g. adding a fixed margin to the beam's-eye view of the PTV), then the 95% contour does not always acceptably sheath the PTV and spare nearby OARs. This is the case when multiple, and possibly non-coplanar, beams superpose. They developed a technique which iteratively adjusts the individual positions of each leaf pair to optimized a cost function representing the conformality. Ma *et al* (2000b) also developed a method to shape radiation fields using dosimetric concepts.

6.4.5 Other forward-planning studies

Bos *et al* (2001) have compared two methods of creating beam segments for static-field IMRT of prostate cancer. They studied one five-field prostate case with a simultaneous boost. The two systems compared were a forward-planning system in which beam segments were designed semi-automatically within the UMPLAN treatment-planning system with subsequent optimization of segment beamweights and the second system was the KONRAD treatment-planning system which automatically generated fluence modulations which, when segmented, again yielded static field components. The overall conclusion was that inverse planning of segments for static IMRT did not improve the 3D dose distribution compared to forward planning of segments. Bos *et al* (2004) extended this study comparing forward plans created by UMPLAN with inverse plans created by HYPERION and for five patients. It was found that inverse planning created a more conformal plan, specifically reducing dose to the rectum but that the inverse plan required typically about three times the number of field segments. This study was a collaboration between the groups at The University of Tübingen and that at the Netherlands Cancer Institute who routinely use segmented irradiation for the prostate (see later).

Bär *et al* (2002, 2003) have compared forward and inverse planning for a particular case where a treatment aimed to spare the function of parotid glands. The forward technique constructed segments which viewed the tumour but excluded views of the OAR and then weighted these with computer optimization. The inverse-planning approach instead produced a fluence-modulated beam that was then interpreted. In general, the resulting number of segments for the inverse-planned IMRT delivery was of the order 35 compared with just 11 for the forward approach even though the same field directions were used. It was found that the inverse planning had the potential to significantly improve dose distributions compared to forward planning for a complex tumour sites in the head and neck.

The two-segments-per-field technique is also in use at the Netherlands Cancer Institute. They also use HYPERION and have compared the forward-planning technique that they have been using with the HYPERION inverse planning. The variation in PTV between CT and MRI has been quantitated and movement of the PTV has also been quantitated through the use of repeat CT scans. The CT volumes were some 40% larger than the MR volumes.

Li *et al* (2002,2003a) have also proposed a DAO technique whereby the beam directions and the number of apertures are defined ahead of a calculation which then uses a genetic algorithm to determine aperture shapes. Cotrutz and Xing (2003b) have also used a genetic algorithm to determine optimum aperture shapes and beamweights by varying the MLC leaf positions (which map the chromosomes). Petersen *et al* (2002) have also used forward planning to create IMRT plans for 23 patients.

Corletto (2001) have compared inverse and forward planning for 1D and 2D IMRT. It was concluded that 2D IMRT of the prostate led to better TCP of the prostate than 1D IMRT which in turn was better than a standard conformal plan.

Lee *et al* (2004) have developed a technique of forward-planned IMRT in which each beam direction has the beam divided into a number of discrete segments whose weights are then optimized. These correspond to projections of the critical structures. Over a period of seven years, 38 patients with head-and-neck cancer had their treatments optimized by this method. The number of segments was considerably lower than for inverse-planned IMRT without much loss of conformality.

6.4.6 Summary on aperture-based IMRT

A review of IMRT planning written five years ago would probably have concluded that IMRT absolutely requires inverse planning without exception. Certainly techniques in which the bixel fluences are very finely spatially modulated *do* require inverse planning for all the reasons stated when inverse planning was first mooted. There are just too many variables to tie down by forward trial-and-error techniques. However, recent years have seen the establishment of IMRT techniques in which each field direction has only a few components. Techniques have been developed to determine these shapes and their radiation beamweights. Whilst this can certainly be done by inverse-planning methods, alternative forward methods are possible. Many studies have addressed the question of how the IMRT conformality changes as the number of such segments is reduced. There is a trade-off between conformality and practicality. Maybe some confusion surrounds this topic. Aperture-based planning can be inverse but does not have to be. Also sometimes the technique is wrongly regarded as the newest approach to IMRT whereas in fact it was one of the earliest proposed before fully-modulated bixel-based IMRT was established (Webb 1991a).

6.5   Smoothing IMBs

6.5.1   Smoothing techniques from the Royal Marsden NHS Foundation Trust

Many inverse-planning algorithms and commercial systems (e.g. CORVUS) generate IMB profiles that have considerable structure. This is the desirable outcome of the quest for high dose-space conformality. Webb *et al* (1998) showed that the application of a median-window filter during the inverse-planning iterative process could reduce the high-amplitude high-frequency bixel-to-bixel intensity fluctuations without adversely damaging the corresponding deliverable dose distributions. This reduced the number of MUs required for delivery with consequent advantages on reduced leakage contribution to dose.

When CORVUS-generated profiles are realized experimentally using the dMLC method of delivery, the MU efficiency can be quite small with unwanted consequences. Also the interpretation of these fields leads to the generation of small field segments, again with undesirable consequences. Webb (2001b) showed that the features of beam space can be user-controlled by incorporating beam smoothness into the optimization cost function to minimize these problems. There is a trade-off between obtaining desirable features in beam space and high conformality in dose space.

6.5.2   Smoothing techniques from the University of Virginia

Mohan *et al* (2000) have also investigated the influence of intensity fluctuations on a number of issues in the delivery of IMRT by the dMLC technique. Firstly the formalism on which the dMLC technique is based was restated, including giving again the equations which determine the leaf patterns from the desired fluence profiles. These closely follow the formalism of Spirou and Chui (1994, 1996). The familiar 'MU-*versus* leaf-position diagram' was constructed in terms of the so-called effective fluence, being the combination of direct primary fluence and an indirect fluence transmitted through, as well as scattered by, the leaves. By assuming a constant leaf transmission, τ, leaf transmission can be accounted for by the method of Spirou and Chui (1996). The so-called 'Stein equation' shows that the total number of MUs is given by the sum of a constant term representing the time taken for the leading and trailing leaves to cross the field at constant maximum speed together with a term representing the sum of the positive-fluence increments along the track. From this follows directly the observation that, the more complex the IMB becomes, the larger the number of MUs required to deliver it. Six examples of increasingly more complex IMBs were shown to illustrate this (figure 6.28).

The solution for the effective fluence cannot be accepted from the first pass of the equations. This is because the position of the leaves changes the penumbra effects and the MLC scattering. So Mohan *et al* (2000) adopted the familiar technique suggested by Stein *et al* (1994) and iterated to convergence. Only a

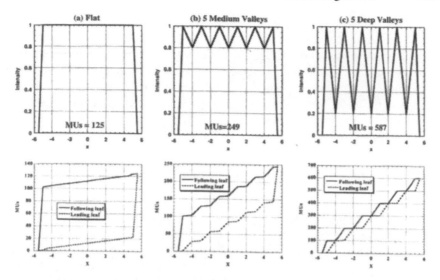

Figure 6.28. A neat demonstration of the dependence of both the leaf trajectory and the window width on the fluctuations of the intensity pattern. The left-hand pair of figures shows the delivery of a flat distribution of fluence. The right-hand figure shows the delivery of five deep valleys and the central figure the delivery of five medium valleys. As can be seen, the number of MUs dramatically rises as the depth of the valleys increases. The figures also show that the width of the open window would also dramatically differ between these three different deliveries. (From Mohan *et al* 2000.)

few iterations were required. This slightly modifies the MU *versus* leaf-position diagram. From these data the head scatter can be convolved (see, e.g. also the work of Convery *et al* [2000]) to obtain the total fluence from which the 3D dose distribution can be computed by folding in a dose-kernel and using the linear attenuation coefficient for transmitted leakage. These calculations were performed within the ADAC PINNACLE planning system.

Mohan *et al* (2000) used a clinical example (a nine-field prostate plan) to show that, when dose-volume constraints on a number of OARs are applied, the IMBs become more structured than they would be if these constraints were not applied. Using some of the six model IMBs referred to earlier it was shown that, the more complex the IMB is, the more likely it is that small field segments will arise. There are a couple of consequences: (i) a large fraction of the intensity at a point may be received indirectly and (ii) small fluence values may be undeliverable due to the summations of leakages and scatter being larger than the required fluence even when there is no direct contribution. Mohan *et al* (2000) showed that by modelling the effects of leaf transmission, penumbra, head scatter and MLC scatter, they can compute dose to within 1% of measured values.

They then suggested a filtering method (the use of a Savitzky–Golay filter) which reduces IMB fluctuations. This has the desired effects of (i) reducing significantly the number of MUs required for the delivery and (ii) creating a better match between the desired fluence profile and the deliverable fluence profile.

Wu *et al* (2002c) have adopted a retro-mapping of IMRT beams to improve delivery efficiency. Instead of filtering the whole intensities uniformly, the proposed scheme adopts a strategy to filter only as much as needed to reduce the MUs needed to deliver the beam.

6.5.3 Smoothing technique from the Memorial Sloan Kettering Cancer Institute

Spirou *et al* (2000, 2001) have implemented techniques to generate IMB profiles that are as smooth as possible. Two approaches were studied. In the first, smoothing takes place at the end of each iteration. In the second, penalty terms for the unsmoothness of the beam profile were included in the objective function used to drive the optimization process. The latter method was preferred. It was more efficient and required fewer iterations because the process of smoothing did not undo the process of optimization as was the case in the former method. If smoothing is introduced after each iteration, then the act of smoothing tends to cancel out the act of conforming at the iterations and, after a while, the whole dual process ceases to make progress. Also there is no inherent mechanism for balancing the removal of genuinely required beam modulation from that introduced due to noise in the iteration process. Conversely, when the smoothing is controlled by the application of a term in the cost function, then it can be user-constrained. The user can balance how much aggression is aimed at conformality *versus* how much importance is attached to smoothing. The form of the smoothing is a Savitzky–Golay filter. Beam element fluences either side of the fluence bixel being filtered are fitted to a polynomial. The order of this polynomial controls the smoothness; the higher the order, the more peaks are preserved; the lower the order the more effective is the removal of high-spatial-frequency modulations. Second-order polynomials were preferred. The number of elements either side can be user-selected. In general, symmetric choices were made. The technique was tailored to the dMLC delivery with bixel size 2 mm along the direction of leaf travel. Sharper dose gradients were obtained when the smoothing was included in the cost function.

6.5.4 Smoothing technique from the University of Tübingen

Alber *et al* (2000) have developed the inverse treatment-planning system HYPERION in which the limitations of the treatment equipment are specifically taken into account for static and dMLC delivery with the result that the optimized fluences are readily deliverable. The concept behind HYPERION (Alber 2001a) is to provide a whole range of potential objectives for optimization. Essentially all the experience of the doctor goes into dose-volume-histogram settings and 'what you want is what you get'. The basis of inverse planning is an optimization engine which makes use of Monte Carlo dose data. Inclusion of biological and practical constraints makes IMRT a more accessible and practical option. The shape of the developed dose-volume-histograms is determined by the choice of constraints and treatment intentions are made explicit. For example, it is possible to restrict the dose variance to the PTV whilst simultaneously restricting the quadratic deviation from the prescribed maximum dose. This is just one example from many. It is possible to control the number of segments and to suppress undesirable field shapes.

Smoothing is also built into the inverse planning and does not greatly effect the dose distribution. Alber and Nüsslin (2001) have described how to create smoother IMRT fields to raise treatment-time efficiency by decreasing the number of segments required under the constraints imposed by commercial MLCs. Bär *et al* (2001c) have developed an inverse-planning algorithm to minimize the number of segments for multisegment IMRT subject to dose constraints. This is achieved by incorporating smoothing constraints into the fluence optimization algorithm and also limiting the segments by a constraint on the size of the minimum size of segment. Additionally, where possible, merging of segments is performed. As a consequence, the complexity of the profile to be transformed and the resulting number of segments is reduced, sometimes by as much as 60%. Alber *et al* (2002a) have shown that the objective function in IMRT is entirely flat in the neighbourhood of the minimizer in most directions. The dimension of the subspace of vanishing curvature thus serves as a measure for the degeneracy of the solution. This high degree of degeneracy is exploited to find solutions which have more practicality (e.g. less complexity/more smoothness) than others without degrading the treatment. They conjectured that it is the conflict between the goals of dose in the PTV and sparing dose to adjacent OARs that leads to high curvature in the objective functions and that once optimization iterations have started to drive cost down this high-curvature part of the function, little is achieved by more exploration of cost space. 'Conflict resolution spectroscopy' is discussed (see also Alber *et al* 2002b). Additional beams are only useful if they aid resolution of these conflicts.

Laub *et al* (2000b) have studied IMRT of the rectum and specifically studying small bowel toxicity. They have compared inverse plans created using a commercial inverse-planning system with those produced by HYPERION. It was found that IMRT could reduce the bowel volume irradiated to a dose of 95% of the prescription dose by about 70% compared with conventional three-field treatment techniques. However, whereas between 120–160 step-and-shoot MLC segments were necessary to deliver the five intensity profiles derived from the commercial system, this number was dramatically reduced to between 20 and 40 segments for the HYPERION plan. The reason this is possible is because HYPERION enables the smoothness of the intensity profiles to be constrained.

6.5.5   Smoothing technique in the Nucletron PLATO TPS

The NUCLETRON PLATO inverse-treatment-planning system is being used at the Christie Hospital, Manchester. The inverse-planning module has a median-window filter which can create smoother beams. It is possible to change the width of the filter and apply it either to peaks or to troughs or both. The IMBs are sequenced using a step-and-shoot interpreter developed in Utrecht (*Wavelength* 2001d). The general trend is that the higher the smoothing the fewer the number of segments. Segments are then replanned onto an anthropomorphic phantom for beam-intensity checks and plans are compared with film measurements. Five-field clinical plans are also replanned onto a Rando phantom. These are preliminary results requiring recalibration of films at right angles to the beam. These class solutions are used for a hypofractionation trial for IMRT of the prostate.

Mott *et al* (2001) have contributed to the debate over the relative merits of inverse- and forward-treatment planning. They have developed smoothed generic dose profiles, class solutions, applicable to specific patient and phantom geometries and investigated how these class solutions then perform for cohorts of patients. It was found that it was possible to apply phantom-created IMBs to patients with additional compensation to get good conformality.

6.5.6   Smoothing technique at the Thomas Jefferson University

The Cimmino algorithm and mixed integer programming for inverse planning have been compared showing that the Cimmino algorithm could produce very smooth fluence patterns. Xiao *et al* (2001) have used the Cimmino algorithm to segment multiple-static MLC fields in such a way that the number of segments is minimized. At the same time this also leads to the generation of smooth intensity patterns that have efficiency advantages. Xiao *et al* (2002b) have applied Cimmino's algorithm for solving the beamlet-based inverse optimization problem for clinical cases and, in particular, have applied special filters in the optimization process to improve the smoothness of the final fluence patterns. It was found that the dose-volume histogram analysis showed similar dosimetric coverage of tumour and sparing of nearby critical structures compared with plans made using CORVUS but that the modulations were much smoother. Xiao *et al* (2003) have shown that a projection method with dose-volume objective can also lead to smoother beams.

6.5.7   Smoothing techniques at University of Maryland

Ma (2002a) has implemented smooth IMB sequences which accounted for the hardware constraints. Large reductions (up to 90%) in the number of leaf segments were found when applying the smoothing operation that minimized the MLC leaf-travel distance, the total beam-on time and the number of leaf segments. Ma (2002b) has extended his matrix decomposition method, which determines leaf trajectories from IMBs to include the possibility to generate smooth beams (figure 6.29). This exploits the multiple valid leaf-sequencing outcomes which each and all correspond to a required IMB. The smooth beams so created require fewer segments with all the advantages this carries. The technique was applied to sets of random-pattern IMBs and also to clinical cases. The reduction in the number of beam segments was greatest for physically unconstrained trajectories. When interdigitation, TG and synchronization requirements were considered the benefit fell. The effectiveness of the smoothing also depends on the complexity of the IMBs. For complex cases with numerous treatment constraints, shallowness of the cost function minimum may be too low to permit much 'room for manoeuver'. Wide-bottomed cost functions lend themselves best to the technique. Li *et al* (2004a) further extended this work to include the possibility to eliminate leaf-end abutments and simultaneously minimize the number of MUs and the number of field segments (see section 3.1.6).

6.5.8   Smoothing techniques at University of California, San Francisco

Sun and Xia (2004) have implemented smoothing in IMRT planning in yet another different way. Firstly they optimized a plan by a conventional simulated annealing approach. Then they applied a smoothing filter to the IMBs. In doing so, only those bixels labelled at projecting to specific structures (controlled by a 'bixel structure index') were included in groups for smoothing. This smoothing process degraded the conformality. So a third step was then taken to re-optimize the weights of the segments from the outcome of the smoothing stage. They showed for several clinical examples that this led to large reductions in the number of components without violating the required conformality.

6.5.9   Smoothing techniques at Sichuan University, China

Hou *et al* (2004) have described a novel way to perform inverse planning that not only improves spatial resolution but also increases beam smoothness leading to the requirement for fewer segments. In this method, several (two and four were tested) matrices of modulated bixels of conventional size 1 cm × 1 cm were produced for each beam angle with the N matrices displaced relative to each other by multiples of $1/N$ cm (e.g. two matrices are spaced by 5 mm apart; 4 matrices are spaced 0.25 mm, 0.5 mm and 0.75 cm apart.) Then the cost function was specified so as to simultaneously optimize all these constituent matrices. The resulting IMB was then the sum of the shifted matrices at each beam location. It was found, using a test case and a clinical case of treating tonsil, that the resulting matrices had higher spatial resolution, greater smoothness and yet simultaneously better conformality was achieved.

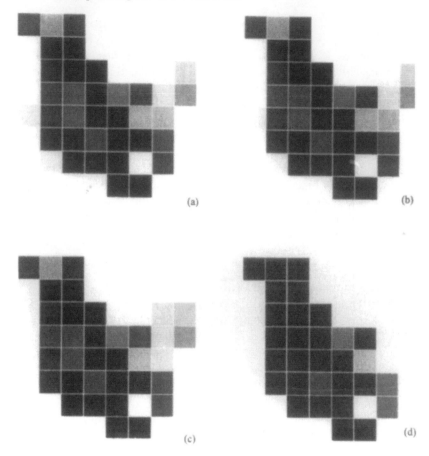

Figure 6.29. Illustration of the algorithm of Ma (2002) for a typical IMB: (a) the tolerance level at 0% that resulted in 42 segments, (b) the tolerance level at 5% that resulted in 34 segments, (c) the tolerance level at 10% that resulted in 25 segments, and (d) the tolerance level at 15% that resulted in 12 segments. Notice the coarsening effect of the intensity maps as the tolerance level increases. (From Ma 2002b.)

6.5.10   Summary on smoothing

When IMRT is delivered by the dMLC technique, the treatment time is directly coupled to the variability of the IMBs through the Stein equations. Hence, smoothing IMBs leads to shorter treatment times. When, alternatively, the MSF technique is used, then smoother beams lead to the requirement for fewer segments. Again this speeds treatment time although possibly now not so much as in early days when the intersegment deadtime was a large fraction of the treatment time. To some extent, this has obviated the need to reduce the number of segments.

6.6 Incorporating MLC equipment constraints in inverse planning

Bortfeld (2000) has reviewed the step-and-shoot IMRT technique, concentrating on the various hardware constraints and putting special emphasis on solutions that reduce TG underdose and matchline effects. Leaf sequences have been developed which keep the number of segments within practical limits. Step-and-shoot IMRT has been compared with the dMLC IMRT technique and the advantages and disadvantages of each method have been presented.

Alber (2001a, b) has also incorporated delivery constraints for static and dMLC delivery into IMRT inverse planning and has discussed both hard (MLC machine constraints) and soft constraints in IMRT treatment planning. Hard constraints are built in to minimize the gap between opposing leaves and to disallow interdigitation. Also some soft constraints are applied to minimize the segment size, the segment width and the segment separation. At the University of Tübingen, the inverse-planning system HYPERION is linked to the ADAC PINNACLE planning system to drive the Elekta RTDesktop delivery system. The aim is for seamless integration between the ADAC PINNACLE planning system and HYPERION. CT data are taken first into ADAC PINNACLE for contouring, then into HYPERION for inverse planning, then the segments back into ADAC PINNACLE for dose verification and finally taken to the Elekta linac for delivery of IMRT. HYPERION delivers step-and-shoot segments. Inverse planning incorporates a step-and-shoot sequencer with variable fluence steps. After optimization and sequencing, there is a re-optimization of segments because the sequencing changes the dose distribution. The result is a set of weights and segments, which do not violate the constraints. The idea of minimizing the segment size is to improve the dosimetry and is very similar to one of the goals of Webb (2001b). It is claimed that the dosimetry of the ADAC PINNACLE system is accurate to 1.5% for 1 cm × 1 cm fields.

Holmes (2001) has developed a technique for inverse planning in which the leakage and head scatter are incorporated during each iteration of the optimization process. The method has been applied to serial tomotherapy in just two dimensions as delivered by the NOMOS MIMiC. This is to overcome the well-known observation that, if the delivery constraints are not included in the inverse-planning stage, then when they are put in *a posteriori* they will deteriorate the dose distribution, generally making it less conformal.

The way the delivery features are built into the inverse-planning system is to identify active bixels as those which are in the beam's-eye view of the PTV. These active bixels may be open or shut for specific fractions of the total radiation time. During the time in which they are closed, whose fraction f will be known, these bixels will contribute to the active dose through leakage and this may be factored in, knowing the transmission of the MIMiC vanes and the associated back-up jaws. Correspondingly, the contribution from inactive bixels may also be computed. The essential difference between this and other methods of inverse planning are that these contributions, including head-scatter effects, are factored in at each stage of the iteration during a steepest-descent method with a quadratic least-squares objective function. The outcome will then be intensity-modulated profiles which are already corrected for these physical effects. In Holmes' study, graphs of beamlet weights illustrate the effect of correcting the active beamlets for leakage and head scatter and show the relative contributions. Dose conformality improves with increased stratification of IMBs. This work has been extended by Seco *et al* (2001) for inverse planning to create distributions which will be delivered with the MSF and the dMLC techniques.

Constraints for segments shapes can be built into planning. The University of Ghent have parametrized 'weirdness' and put this into the objective function and concluded that 'weird shapes' were not needed. In particular, the field patterns developed did not have the 'salt and pepper' noise associated with CORVUS field modulations.

Siebers *et al* (2002b) have incorporated MLC delivery constraints and features (leaf transmission, TG effect) into an inverse-planning optimization code at the Medical College of Virginia (figure 6.30). The algorithm is a gradient-descent technique. It was argued that the resulting plans conform more closely to the plan constraints than plans in which the planning and delivery are considered as separate operations. The integrated approach was compared with the sequential approach for 17 patient cases, mainly but not exclusively in the head and neck. In general, dose to the PTV was adequately realized in both approaches but the integrated approach led to better critical organ sparing. It was observed that if the sequential approach is used the dose-volume constraints have to be overemphasized in order to achieve the actually required constraints. Also much trial and error is required which is eliminated by the integrated approach. Because the leaf sequencer tends to smooth profiles, the integrated approach leads to more MU-efficient deliveries in all cases. This study refered to similar work by Cho and Marks (2000) and by Holmes (2001). Lauterbach *et al* (2001) further reported on the building of this sequencer inside the IMRT optimizer in order to ensure that the fluence maps so created are deliverable. For a head-and-neck case and for a prostate case, the resulting isodose distributions showed a clear improvement for the deliverable optimization technique.

6.7   Beam direction optimization

Macklis *et al* (2000) have provided an editorial on IMRT treatment-planning paradigms. They extended the definition of IMRT, from the simple indication that photon fluence is varied spatially either through direct spatial variation or through temporal variation, to include a new (so-called) 'world-view' that IMRT encompasses not only arranging for extremely high levels of target-structure dose conformality but a very high degree of automated beam selection through

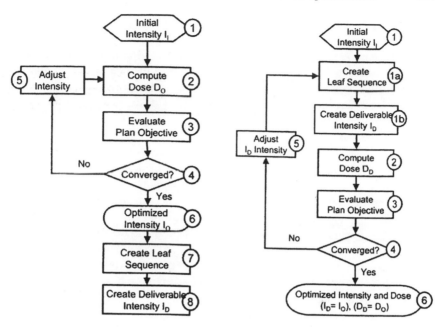

Figure 6.30. Two flow diagrams indicating the difference between traditional IMRT optimization and deliverable-based optimization. On the left is shown the flow diagram for traditional IMRT optimization. Intensities are iteratively updated to improve the plan quality until convergence. Following convergence, the optimized intensity is converted to MLC leaf sequences and deliverable intensity distributions. On the right is shown the flow diagram for the deliverable-based optimization method. Intensities are converted to MLC leaf sequences and deliverable intensities inside of the iterative IMRT optimization loop. The same leaf sequencer was used as in traditional optimization. (From Siebers *et al* 2002b.)

inverse treatment planning and a willingness to question the traditional teachings of conventional radiotherapy.

The problems of determining the optimum beam orientation and the optimum set of IMBs are coupled because the optimum IMBs will change depending on the gantry and couch angles selected. There has been a considerable amount of research trying to unravel the problems.

Holloway and Hoban (2002) have developed an IMRT planning technique that starts with a very large number of beams. The process relies on gradually reducing the number of beams used, based on the average fluence delivered by each beam. Beams with the lowest average fluence are removed at each stage of the iteration and the process is repeated until the desired number of beams is reached. For the situation of 180 initial beams, gradually reduced by three beams at a time, a plan was achieved vastly improved on that using evenly spaced beams after ten iterations.

Gaboriaud *et al* (2000) have shown that use of IMRT with non-coplanar beams improves the treatment of brain tumours. In their study, conformal geometrically-shaped fields were compared with intensity-modulated fields for the same beam directions, the greatest benefit being dose reduction in the OARs especially the brain stem and the temporal lobes.

Woudstra and Storchi (2000) have developed a way to iteratively select both the orientations and the beamweights of a set of beams which, in principle, can be as large as 36 coplanar beams in [0,360°]. The method does not use a plan-based cost function but sequentially optimizes the placements of 'the next beam' by selecting that one which improves the plan most at that stage of selecting beams. This is a 'no way back' method as orientations and beamweights selected early in the process cannot be deselected. However, by including a stage of continuously varying the IFs, the method seems to work. Woudstra *et al* (2002) have developed automatic techniques for selecting beam orientations and IMRT beam profiles for pancreas, oesophagus, head and neck and lung and found that the automatically generated plans, even with a few beam numbers (3–9), compare well with equiangular beam set-ups.

Meedt *et al* (2001) have presented a new algorithm for optimizing beam directions in IMRT. The algorithm starts by creating a standard N-beam treatment plan using fluence-profile optimization. Then, one-by-one, new test rays with the smallest desirable field size are added and optimized and then, following this process, the beam which can best be compensated-for by the remaining beams is removed and the next iteration is started. The process terminates when attempting to add new beams takes the dose distribution no closer to the desired dose distribution. Meedt *et al* (2002) have applied this beam-direction optimization algorithm to head-and-neck IMRT. It was found that a small number (4–6) of non-coplanar beams can sometimes improve upon a larger number (10–20) of coplanar equally spaced beams. The algorithm employs a fast-exhaustive search of about 2000 directions equally spaced on the unit sphere to find a new beam best capable of improving an arrangement and then deleting a corresponding less useful beam. Meedt *et al* (2003) have developed a beam-direction search strategy for IMRT based on selecting beams with 'paths of little resistance' and an exchange scheme to converge on an acceptable outcome.

Webb (1995) introduced the idea of creating a set of beams with maximum geometrical separation to achieve a starting point for CFRT. This was done by a Monte Carlo technique which maximized the sum of the angular separations between beampairs. Wagner *et al* (2001) have capitalised on this as follows. They have started with the previously described distributions. Then they rotate the locked set to an orientation which works best. Then they individually reorient each beam within a preset small angular region. This is done to minimize a score function which has the effect of avoiding couch–gantry collisions, of avoiding 'difficult' entry orientations and of minimizing OAR dose. IFs were used to ensure that avoiding more critical structures outweighed avoiding less critical structures. When the individual beam directions are adjusted, they are done so within a cone of 45° about their starting location. Clearly this removes the precise maximally-spaced-out property of the isotropic beam bouquet. A clinical case bears out the goal of the orientation adjustment showing the tuning of beam directions improves OAR sparing and retains high dose gradients. The original notion from Webb (1995) was never intended to be the complete clinical solution so this advance has brought the notion to the clinical arena.

Levine and Braunstein (2002) have presented a theoretical exposition of the number and directionality of beams required for 3D IMRT. The number of beams predicted vastly exceeds current practicality. The paper is of theoretical interest only.

Meyer *et al* (2001) have investigated an interesting and novel problem. When *N* IMRT fields are equispaced, coplanar, around a PTV and when *N* is fairly large (e.g. nine) some of the field arrangements can lead to intersections of the couch and the beam which is not permitted. They studied this phenomenon for three types of couch and came up with a set of rules to predict such 'collisions'. They predicted the percentage of times this would occur for each type of couch and value of *N* for some model tumour geometries. Then they adjusted the angles of beams so as to avoid the collision without seriously affecting dose conformality.

Meyer (2003) compared an equiangular seven-field plan with a beam-angle-optimized plan on an artificial phantom with two OARs and showed that the OAR dose could be considerably decreased using the non-equispaced beams.

Das *et al* (2001a) have developed a new technique to automatically select beam directions for IMRT when the modulations are relatively simple. The technique comprises alternating optimization and elimination steps until the user-specified number of beams remains together with the appropriate modulation. Das *et al* (2001b) have developed a beam-orientation scheme for IMRT based on target EUD maximization. Essentially optimization oscillates between beam-orientation and beamweight selection and, at alternate stages, those beams which contribute least to maximizing the target EUD are sequentially removed until the desired number of beams remains. Das *et al* (2003) showed that this EUD-based technique yielded a satisfactory dose distribution for just three to five beams for a prostate case compared with that produced by a large number of unselected equispaced orientations.

Li and Yu (2001b) have also developed an IMRT algorithm that simultaneously optimizes beam intensity and beam orientation.

A large number of approaches to beam-orientation optimization has been made by Pugachev and colleagues. Here follows a detailed review of these in date sequence. Pugachev *et al* (2000a) observed that the problems of optimizing beam profiles and beam directions in IMRT are coupled. This problem was solved by using a simulated annealing algorithm to determine the orientations of nine beams in 2D IMRT planning. At each iteration and the angles selected, filtered backprojection was used to compute beam profiles. This relatively fast technique was practicable. It was shown that the resulting dose distributions were improvements on those resulting from optimizing beam profiles at fixed, regularly-spaced, gantry orientations. It was also shown that if simulated annealing were used to also optimize beam profiles then the outcome was much the same but took much longer to achieve. The work is analogous to that of Rowbottom (1998).

Pugachev *et al* (2000c, 2001) have investigated the improvement of conformality of IMRT with introducing non-coplanar beams, with and without optimization of the beam-orientation direction. Baseline plans were created for nine coplanar beams which were equispaced in the gantry-orientation angle. Dose distributions were compared with the results of planning in which the beam orientations were allowed to vary in a single plane and, alternatively, in which beam orientations were constrained to vary non-coplanar inside a fixed domain of gantry and couch angles. For each beam orientation tried, an intensity-modulated plan was created using PLUNC and the so called simultaneous-iterative inverse-treatment-planning algorithm of Xing *et al* (1998). The beam-orientation algorithm has a problem with local minima and so was attacked using the technique of simulated annealing. The algorithm randomly selected beam orientations, computed the beam profiles at such orientations and then accepted or rejected the trial on the basis of a cost function. The simulated annealing was controlled as usual by starting at a high temperature and then gradually reducing the temperature to freeze the gantry and couch locations. The technique is a brute-force technique, quite computationally intensive, needing upwards of a 100 hr on a Silicon Graphics computer but yielded results. Pugachev *et al* (2001) solved three problems: the first was a case of prostate cancer; the second a nasopharynx case and the third a paraspinal case. It was observed that, for the prostate cancer case, very little advantage accrued to optimizing beam directions, whether coplanar or non-coplanar, in fact only a 5% improvement in cost function resulted with very little extra sparing of bladder and rectum. However, for the nasopharyngeal case, the cost function decreased by 27% for the plan with optimized non-coplanar beam orientations and led to a significant sparing of brain stem and optic chiasm. Correspondingly, for the paraspinal case, the cost function decreased by 19% for the plan with optimized non-coplanar beams and there was considerable sparing of the kidney and liver. It was thus concluded that the effect of beam-orientation selection on the quality of the treatment plan varies from site to site.

Pugachev and Xing (2001a) have decoupled the beam-orientation selection from the fluence-modulation calculation for inverse planning for IMRT. The first stage was to recognize and exclude 'bad gantry angles' in which too much radiation is delivered to OARs. The search space then concentrated on the remaining 'good gantry angles' and optimized the fluence profile using simulated annealing. The results showed that the quality of the final treatment plan was not affected by this restriction of the search space.

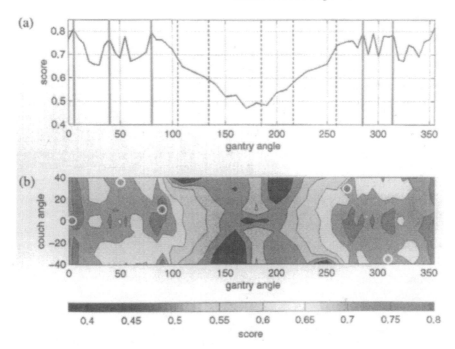

Figure 6.31. A pseudo beam's-eye-view score as a function of beam orientation for (a) a coplanar beam configuration and (b) a non-coplanar beam configuration indicating preferential angles of attack towards the PTV. (Reprinted from Pugachev and Xing (2001c) with copyright permission from Elsevier.)

Pugachev and Xing (2001c) have introduced a concept of pseudo beam's-eye-view (pBEV) to establish a framework for choosing beam orientations in IMRT (figure 6.31). The idea was to introduce a score function for each $1\,cm \times 1\,cm$ beamlet which crosses a target and to assign the maximum intensity to that beamlet that could be used without exceeding the tolerance dose of the OAR. The score function was then constructed in terms of these individual beamlet functions. The idea then was to make a plot of the score function as a function of orientation, decide how many beams to use and choose the directions which have the largest value of score function whilst also ensuring that beams are not antiparallel and are spaced out as widely as possible. It was shown that this technique is especially useful for complicated clinical cases.

The pseudo beam's-eye view is an extension of the classical beam's-eye-view technique which just measures the quality of the beam direction as the binary function based on whether the beam does or does not hit a sensitive structure. The pseudo beam's-eye-view technique accounts for beam modulation. The technique was incorporated inside the PLUNC treatment planning system. Couch angles were discretized in increments of $10°$ and gantry angles in increments of $5°$.

The concept is based on observing that a beam orientation is preferable if it can deliver more dose to the target without exceeding the tolerance of OARs or normal tissue located on the path of the beam. For each beamlet crossing the target the maximum intensity that could be used without exceeding the tolerance of the OARs and generic normal tissue was calculated.

The technique starts assuming that only one beam is going to be used for irradiation. Firstly, the voxels crossed by the beamlet were found. Secondly, the beamlet was assigned an intensity that would deliver a dose equal to or higher than the prescription dose in every target voxel. Thirdly, for each OAR or normal tissue voxel which is crossed by the beamlet, the ratio was calculated by which the beamlet intensity had to be reduced to ensure the tolerance dose is not exceeded. The minimum ratio from that data was then identified. The beamlet intensity was then reduced according to this ratio and the value obtained represents the maximum usable intensity of the beamlet. These steps were then repeated for all the relevant beamlets to obtain the maximum beam-intensity profile in which none of the beamlet intensities can be further increased without violating the structure tolerance. Then an overall score function, a single number, was calculated to reduce these data to a single number.

It is emphasized that this maximum-intensity profile as previously defined is not at all the optimum fluence profile for IMRT. It simply allows us to create a plot of score function as a function of gantry orientation from which beam directions can be chosen for subsequent optimization of IMRT. It was argued that this captures the main features of the planner's judgement of the goodness of a beam direction. It is a trick for significantly reducing the size of the search space.

Once the plots of score function *versus* orientation had been determined, the next task was to select a fixed number of beams and this was done by adding in the separate criterion that beams should be angularly as-well-spaced-out-as-possible and also not antiparallel.

Two clinical cases were investigated showing that the tool is an efficient one for IMRT beam-orientation selection, acting as a filter. It certainly does not provide the final beam configuration for IMRT treatment. This still requires the intervention of a human or inverse planning. The fields so selected were optimized using CORVUS and it was found that, for prostate therapy, five-field techniques could be improved over the use of equally-spaced beam configurations but that nine-field IMRT could not be so improved. However, when performing treatment planning for paraspinal and nasopharangx in which the improvement depended more on the degree of overlap with OARs it was found that, even when nine fields were used, more suitable beam orientations could be selected using the pseudo beam's-eye-view optimization technique. It was argued that, whenever a multidimensional dataset is projected onto a lower dimensional space for decision-making by a human being, the details of the local quantities are often buried.

Pugachev and Xing (2002b) have extended this work by developing a technique that uses beam's-eye-view dosimetrics (BEVD). The goal of the work was to provide an efficient means to speed up the beam-orientation optimization by incorporating *a-priori* geometric and dosimetric knowledge (figure 6.32). The BEVD is a scalar scoring parameter that describes the ability of a particular beam direction to contribute to the PTV dose independent of the contributions from other beam directions. Having set a pre-set number of orientations, the BEVD were computed for these orientations. Those with a very small value were then rejected altogether from consideration. Thus the candidate orientations were pre-screened. The BEVD was then used to weight the number of times a particular beam orientation is sampled in an iterative sampling scheme, the logic being that it is preferable to re-visit those orientations which are considered to be good.

The technique was implemented inside the PLUNC planning system. The dose objective function was a weighted quadratic and the pre-ranked angular search space was sampled using simulated annealing. Whilst, in principle, any sampling density function could have been used, it was found that a sampling proportional to the square of the BEVD score provided an efficient convergence and speeded up the conventional simulated annealing calculation by about a factor of 10. This is because the pre-screening and the weighting reduced the chance for the algorithm to spend valuable computing time on those orientations that were less likely to end up in the final solution. This improvement was compared with the use of linear beam's-eye-view volumetrics and found that the two techniques gave the same gantry angles at the end of the calculation but that the new technique was distinctly faster. Incorporating the prior knowledge also improved the convergence behaviour of the simulated-annealing calculation based on a realistic cooling scheme. However, the technique still took more than two hours on a fast workstation. The whole technique is predicated on a fundamental rule for estimation theory that the use of prior knowledge (Bayesian priors) will lead to a more accurate estimation.

Pugachev and Xing (2001d) have developed a computer-assisted selection of beam energy and orientation in IMRT. A score function was plotted as a function of gantry angle for each energy, this being the largest intensity which could be delivered from that angle without violating the tolerance of the structures located on the path of the beam. An algorithm then selected beam angles and beam energies from the two curves, such that the beam angles are as far apart as possible and the better of the two beam energies at such angles is used. The technique has considerable potential for simplifying IMRT treatment planning.

Pugachev and Xing (2002a) studied eight nasopharynx cases and ten prostate cases to try to decide if class solutions were acceptable. They came to the conclusion that they were not and that individualized beam configurations were needed to maximally utilize the technical capacity of IMRT.

Djajaputra *et al* (2003) proposed a beam-direction optimization algorithm for IMRT based on a fast dose calculation engine (table look-up method) and a fast simulated annealing algorithm for beam-direction selection. This dose-calculation method is justified in that the matrix linking dose space to beam space is very sparse and bixels are planned to contribute only to voxels in their line of

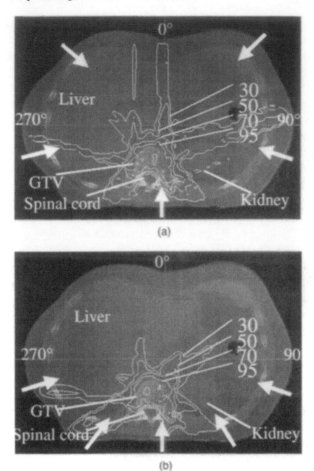

Figure 6.32. (a) The dose distribution of an IMRT para-spinal treatment with five equiangular-spaced beams. (b) The dose distribution of the IMRT para-spinal treatment with five beams obtained using the beam's-eye-view-dosimetrics-guided beam orientation algorithm. The incident beam orientations are indicated by arrows. (Reprinted from Pugachev and Xing (2002b) with copyright permission from Elsevier.)

sight. For each selected set of beams, the fluence was optimized by a gradient descent technique. The number of beams was fixed and not optimized. Several planning cases were studied. For a three- or five- or seven-field prostate case, beam-angle selection led to an improved plan compared with a uniformly spaced set of beams. Rotating the uniform set did not improve these plans as far as a beam-angle-optimized set, although the relative trade-offs in dose sparing to OARs changed. For a uniformly spaced set with a large number of beams, the rotation had less effect. Similar behaviour was observed for a head-and-neck case.

Generally, in IMRT planning, the optimization of beam orientations and beam weights for CFRT become separate problems. Wang *et al* (2003a) have combined the two problems and solved them using mixed integer programming to optimize both the beam orientations and the beamweights simultaneously. The problem was solved using four steps.

(i) First a pool of beam-orientation candidates was set-up with the consideration of avoiding any patient–gantry collisions and avoiding direct irradiation of OARs with low tolerances.
(ii) Each beam-orientation candidate was represented with a binary variable and each beamweight with a continuous variable (in this initial application each beam could comprise one of four wedged dose distributions).
(iii) The optimization problem was set-up according to dose prescriptions and the maximum number of allowed beam orientations.
(iv) The optimization problem was solved with a commercial ready-to-use mixed integer linear programming (MILP) solver. After optimization, the candidates with unity binary variables remained in the final beam configuration.

Wang *et al* (2003a) use PLUNC to generate the dose and anatomical data files and then a home-made program to write the MILP data files in the format of MPS (a standard mathematical programming system). The time for set-up depended on the number of fields and the same was true of solution and typically both can be achieved in under an hour. It was then shown for two cases, the first a prostate and the second a head-and-neck case, that the conformality of radiation improved using this technique. The gains were much greater for the head-and-neck case than they were for the prostate case. The application was anticipated to be applicable to IMRT.

Schreibmann *et al* (2004c) have presented a technique which generates a Pareto set of optimized solutions among which the user can choose.

6.8 Monte Carlo dose calculation

6.8.1 The debate over the usefulness of Monte Carlo dose-calculation techniques

In the *Medical Physics* 'Point and Counterpoint', Mohan *et al* (2001a) have debated whether or not Monte Carlo techniques should replace analytical methods for estimating dose distributions in radiotherapy treatment planning. Mohan believes they should. He states that, were it not for the limitations in the past of computer speed, Monte Carlo techniques would have been used all along. Computers have now become sufficiently fast that the time required to obtain a typical treatment plan has shrunk to a few minutes. He argues that some find the statistical jitter in Monte Carlo results troubling. He also argues that it is hard to assess whether improvements in accuracy are clinically significant and worth additional cost, the reason being that randomized trials in which half of the patients are treated with less accurate methods are not feasible. Mohan's arguments therefore centre on the believe that continually reinventing approximate dose computations and tweaking them to meet every new situation is unacceptable and should be stopped. Broadly speaking, the Monte Carlo techniques are applicable with the same degree of accuracy for photon treatments, electrons and brachytherapy and, therefore, separate models are not required for each of these modalities. The availabity of Monte Carlo calculations leads to a dramatic reduction in the time, effort and data required for commissioning and Monte Carlo allows the calculation of dose distributions in regions of electron disequilibrium. Mohan argues that, because Monte Carlo techniques are affordable and practical, it is not necessary to conduct clinical trials to demonstrate the clinical significance of the improved dose accuracy. In summary, provided Monte Carlo developers and users ensure that the approximations have no significant impact on accuracy, Mohan believes that the stochastic nature of the Monte Carlo approach is no longer an impediment to its use.

Antolak is against the proposition and believes that a fundamental difference between Monte Carlo and analytic techniques is that the latter are deterministic, i.e. independent calculations of the same problem will always give the same answer, whereas Monte Carlo techniques being stochastic give different answers each time that they are used. The error in the answer will also be proportional to the inverse of the volume of the dose voxels and to the inverse square root of the computational resources allocated to the problem. Antolak believes that it is not known how much noise in the dose distributions is acceptable and also believes that Monte Carlo algorithms usually start with a source model that requires trial-and-input data adjustments to match the measured dose data. He also believes that modelling all of the field segments for a complex dMLC fluence pattern would be impractical under normal circumstances, although Siebers *et al* (2001b) and Liu *et al* (2001a, b) have modelled just this situation. Antolak also believes that current Monte Carlo treatment-planning algorithms (those being put forward for clinical use) do introduce approximations that greatly speed up the calculations but may affect the accuracy of the results. In summary, Antolak believes that Monte Carlo methods should be used as an independent verification of dose delivery or to document dose delivery but should not replace analytical methods for estimating dose distributions in radiotherapy treatment planning. Siebers and Mohan (2003) provide a comprehensive overview of this topic. The next paragraphs summarize the progress made in Monte Carlo IMRT planning since 2000 and contribute to this debate.

6.8.2   Determination of photon spectrum and phase space data for Monte Carlo calculations

Monte Carlo calculations need to start from a specification of the materials and geometry of the radiation-delivering equipment and a knowledge of the beam spectrum. Partridge (2000) has used data taken from electronic portal imaging devices to derive transmission measurements in materials which enable a reconstruction of megavoltage photon spectra. The technique relied on placing *a priori* constraints on the Schiff model to avoid the otherwise badly constrained mathematical problem which would be too sensitive to noise in the source data and non-uniqueness.

Fogg *et al* (2000) have developed a Monte Carlo-based phase-space source model to predict IMRT dose distributions. Aaronson *et al* (2001) have also developed a Monte Carlo-based phase space source model for accurate IMRT dose verification. Seco *et al* (2003) have modelled the Elekta SL15/25 accelerator and the Varian 2100 CD accelerator and compared the predictions of Monte Carlo dosimetry favourably with measurement for homogeneous phantoms.

6.8.3   Comparison of Monte Carlo and pencil-beam calculations

Laub *et al* (2001) have compared the predictions from the KONRAD IMRT inverse-planning system, which uses a finite-size pencil-beam (FSPB) algorithm, with the calculations made by Monte Carlo for specific IMRT treatments of a phantom with uniform lung inserts. It might therefore be expected that the KONRAD planning system would produce results that would not be in agreement with the Monte Carlo calculations. The authors, however, showed that this is not the case and quite good agreement was obtained (figure 6.33). This is because, first of all, the IMRT planning process tends to down-regulate the beams which pass through the simulated lung tissue. Secondly, the lung tissue is uniform whereas real lung tissue normally contains large volumes of air leading to electron disequilibrium and, thirdly, selected beam directions tend to avoid the lung. So, all in all, the conclusion is that one should still be interested in Monte Carlo calculations for more accurate IMRT dosimetry even though in this particular experiment they did not seem to show large differences. Laub and Bakai (2001) have compared the dose distributions calculated with a pencil-beam algorithm with the same dose distributions computed with the Monte Carlo code EGS4 and with measurements for an inverse plan of a thorax phantom. They found that all three methods of measuring or computing dose agreed very closely and that no significant overestimations of dose values inside the target volume by the pencil-beam algorithm were found in the thorax phantom. The reason for this as previously stated was that the dose to the low-density region lung is reduced by the use of a non-coplanar beam arrangement and by intensity modulation. However, Laub *et al* (2000a) have developed a hybrid method in which Monte Carlo dose computations were incorporated into the inverse planning. When

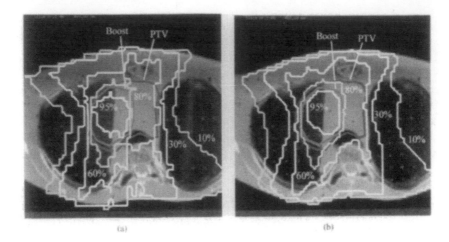

Figure 6.33. Dose distribution of an IMRT treatment plan in the thorax region of a phantom as calculated with a pencil-beam algorithm in KONRAD (left) and using the Monte Carlo code (right). (From Laub *et al* 2001.)

tissue inhomogeneities occur, the computed fluence maps were different from those computed with a pencil-beam dose kernel.

Wang *et al* (2002a) have used Monte Carlo techniques to compute IMRT plans and noted differences between plans computed with Monte Carlo calculations and plans computed with simpler measurement-based pencil-beam algorithms. In general, due to computer limitations, most inverse planning uses approximate dose-calculation algorithms. This study reported on 10 plans created at the Memorial Sloan Kettering Cancer Center using a pencil-beam algorithm as part of an inverse-planning technique. The fluence modulations thus created were then converted to dose distributions once again using this pencil-beam technique. Simultaneously, the same fluence patterns were converted to a different set of dose distributions using a Monte Carlo technique. Five lung patients and five head-and-neck patients were studied. The dose distributions obtained in the two ways were then compared in terms of both tumour coverage and critical organ sparing. It was found that, whilst the IMRT plans produced with the pencil-beam algorithm were all considered clinically acceptable, dose distributions did change with the use of Monte Carlo dose calculations. Specifically, the patient-specific images were used to provide the electron density array from which the particle pathlengths were evaluated and dose scoring was performed. It was found from a detailed study of the comparisons of both isodose distributions and dose-volume histograms that, whilst large differences were observed for the application of single fields, the differences tended to decrease when multiple fields were in use (figure 6.34). The largest differences were seen for IMRT plans in the lung when for one patient there was a particularly low lung density due to emphysema. There was very

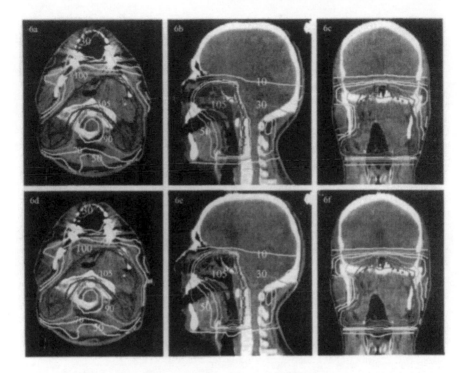

Figure 6.34. Comparison of the treatment plan percent isodose distributions for a seven-field nasopharynx IMRT plan. The images on the upper row are for a standard plan and those on the lower row for the Monte Carlo-generated plan. The PTV is the bold line of dots. The isodose levels are labelled. (From Wang *et al* 2002a.)

little change in spinal cord dose for the head-and-neck plans and the total lung doses differed slightly. In summary, the authors found that the limitations of the pencil-beam algorithm led to varying degrees of differences depending on beam energy, tumour location, the density of the inhomogeneity, the field sizes, the beam directions and even the number of beams. This agrees with Laub *et al* (2001) who also did not find big differences between Monte Carlo calculations and pencil-beam calculations.

Zheng *et al* (2002a) have shown that the predictions of the finite-sized pencil-beam algorithm incorporated within the treatment-planning system CORVUS can differ from the predictions of the Monte Carlo code PEREGRINE by as much as 7% when the patient has metallic implants. However, the difference is never more than 3% even in build-up regions when comparison is made for other tissues.

Holmes *et al* (2000) have developed a three-stage inverse-planning process for IMRT. In the first stage, a pencil-beam optimization scheme was used to generate the intensity maps—this was done with CORVUS. The second stage was a forward-planning stage which used a Monte Carlo calculation to determine dose distributions that were used in the final stage which was the beamweight optimization stage. The whole process was very time consuming taking many days but it increased the accuracy of dose computation to make agreement between Monte Carlo simulation and measurement better than 1% compared with the 3–4% for the pencil-beam model.

Jeraj *et al* (2002a) have studied the effect of using Monte Carlo-generated beamlets in inverse planning for IMRT compared with superposition of pencil-beam calculations. Regarding the plan created using the Monte Carlo beamlets as the 'accurate plan', it was shown that the use of finite-size pencil beams can introduce a very significant (between 5% and 10%) systematic error. The systematic error is the difference between the dose distributions computed with the dose-algorithm and the 'error-free' (MC) dose-calculation algorithm. An additional error studied is the convergence error which is a direct consequence of using the 'wrong' beamlets. Essentially the plan converges to the optimum for the use of the wrong beamlets rather than the error-free (MC) beamlets. The overall conclusion of the study was that inverse-planning systems should be upgraded to use Monte Carlo-generated beamlet doses.

Keall *et al* (2002b) have described how IMRT beams, computed using superposition, the default dose calculation algorithm on a particular planning system, were then recomputed using a Monte Carlo technique and, in 5% of the cases, MC results in MU changes for some of the segments.

Ma *et al* (2000c) have shown that the FSPB algorithm in CORVUS agreed with the predictions of a Monte Carlo computation to within 4% in IMRT calculations where most of the tissue was homogeneous. Discrepancies up to 20% can occur in inhomogeneous regions (figure 6.35). The Monte Carlo calculations themselves were benchmarked to experiment to within 2%. Ma (2000) shows how Monte Carlo is now forming the basis of inverse-planning algorithms and IMRT dose verification.

In conclusion, there does not seem to be a consistent pattern emerging from these studies. Under some circumastances when the dose from modulations generated from inverse planning which uses pencil-beam distributions is recomputed using Monte Carlo methods, the outcome shows little change. However, the use of Monte Carlo kernels actually in the inverse planning does seem to change the outcome.

6.8.4   MCDOSE

Some of the earliest Monte Carlo work done with a 'package' was performed with the EGS4/DOSXYZ code developed at Stanford. More recently, MCDOSE has been developed. Ma *et al* (2002a) have described its operation in some detail. It has been shown to be accurate to about 2% when compared with experiments. The main feature of MCDOSE is that all the dose-modifying structures such as jaws, physical and dynamic wedges, compensators, blocks etc are included into

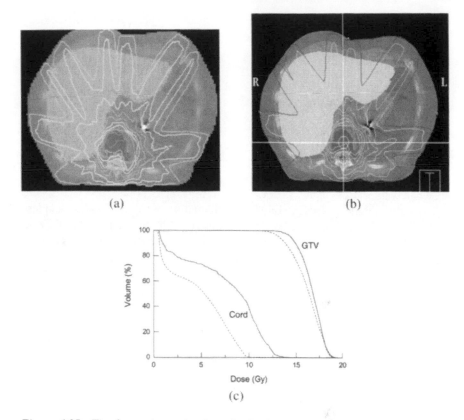

(a)   (b)

(c)

Figure 6.35. The figure shows the dose distributions for the treatment of the vertebra calculated by both Monte Carlo simulation and using the CORVUS system. The plan was generated using the CORVUS system for 15 MV photon beams with nine coplanar gantry angles (20°, 55°, 90°, 140°, 180°, 220°, 260°, 300°, 340°). The intensity was modulated using a Varian dynamic MLC with 80 leaves and the prescribed target dose was 18 Gy. The isodose lines shown are 17.6, 15.6, 13.7, 11.7, 9.8, 7.8, 5.9, 3.9 and 2.0 Gy respectively in each figure. (a) shows the distribution calculated by Monte Carlo; (b) the distribution computed by CORVUS; (c) shows the dose-volume histogram as calculated by Monte Carlo (full curves) and CORVUS (broken curves) for the target and the spinal cord. The Monte Carlo dose distribution shows slightly better target coverage than the CORVUS dose distribution. (From Ma *et al* 2000c.)

the code. Variance-reduction techniques lead to this code operating some 30 times faster than EGS4/DOSXYZ. It was shown how the beam parameters are 'tuned'. Essentially the free parameters are adjusted until the experimental measurements match the predictions for specific beam geometries.

Ma *et al* (2002b) and Price *et al* (2003) have described how Monte Carlo treatment planning for IMRT has been implemented at the Fox Chase Cancer

Center. It has been used to investigate the beam delivery accuracy for microMLC-based radiotherapy and also to study the dose-calculation accuracy for MRI-based treatment simulation as well as other techniques including image-guided robotic radiotherapy.

6.8.5   Speeding up Monte Carlo dose calculations

Deasy *et al* (2002) have developed a prototype Monte Carlo-based IMRT treatment-planning system that uses beamlet dose distributions which are stored in wavelet-compressed format.

Siebers *et al* (2000) have developed a method to speed up the Monte Carlo dose calculation for dynamic IMRT plans. In this, incident photon weights were calculated using the probability of photon attenuation in the moving leaf and leaf tip and incident electron weights were modified by the probability that they do not intercept the leaf. TG effects were also accounted for but secondary electrons and photon scatter from the MLC was ignored. Provided the dMLC efficiency was greater than 0.3, this approximate Monte Carlo method was not different by more than 1% from a full Monte Carlo simulation.

Siebers *et al* (2001a) have commented that accurate convolution/superposition or Monte Carlo dose methods are currently considered too time consuming to take part in iterative IMRT dose calculations. They thus proposed a voxel-by-voxel dose-correction ratio matrix which can be applied to the pencil-beam dose results during the optimization which then aims to converge both in terms of the inverse-plan cost function and also converge such that there is no difference between the ratio-matrix results and the pure convolution/superposition results.

Siebers *et al* (2002a) presented two methods of what they call hybrid planning. Inverse-planning proceeded using a fast pencil-beam algorithm for dose deposition. However, every so often, the dose matrices were recomputed using the convolution/superposition algorithm which is known to be more accurate. The dose matrices were corrected either multiplicatively or additively for the differences between the outcomes of using the pencil-beam and the convolution/superposition techniques. It was shown that this leads to a very fast inverse-planning algorithm without sacrificing dose-calculation accuracy.

Siebers *et al* (2001b) compared a number of different schema for IMRT, specifically studying issues of speed *versus* accuracy. Four techniques were:

(1) use of a pencil-beam algorithm throughout (fast but maybe in error);
(2) use of a convolution/superposition (CS) algorithm throughout (much [100 times] slower but said to be more accurate);
(3) recalculation by CS of the dose distribution using the fluences found in (1);
(4) as (1) but followed by a full CS optimization.

They inspected five prostate patients and concluded that

(i)   techniques (1) and (2) gave quite different plans but;

(ii) technique 3 gave a result quite close to technique 2.

It appears the a hybrid method can generate (quicker) a solution very similar to the full lengthy CS calculation.

(iii) Large differences in dose distributions for single fields tend to be washed out by multiple fields.

6.8.6 Monte Carlo calculation accuracy and error

Jeraj and Keall (2000) have evaluated the effect of statistical uncertainty on inverse treatment planning based on Monte Carlo dose calculation. They pointed out that if inverse plans are created using Monte Carlo-generated dose kernels (which are themselves subject to statistical noise), the dose plan will change if the dose distribution is recalculated using noise-free data. Thus, the use of Monte Carlo kernels actually introduces an error in the inverse planning. They call this error the 'noise convergence error'.

6.8.7 Application to the dMLC technique

Keall *et al* (2001a) have developed code to compute the dose from IMRT delivered via the dMLC technique. The code assumed that dMLC treatments may be modelled as a set of linked static MLC-shaped treatments. The method accounted for intraleaf thickness variations, interleaf leakage and leaf-tip shape in the particle transport. The MLC model modified the weight of the incident particles by their probability of transmission which was determined from the leaf-sequencing file. The contribution through the open apertures, leaves and leaf tips were calculated separately and then added. The model included the effects of Compton-scattered photons. This was claimed to be better than direct Monte Carlo calculation which would require the number of histories to scale in inverse proportion to the (low) treatment efficiency. The low efficiency requires careful calculation of leakage contributions due to to the larger number of MUs needed for dMLC treatments. Keall *et al* (2001a) compared the results of the calculations with measurements made using film for model dMLC treatments created using the Varian 80-leaf and 120-leaf MLC operating at two energies. They showed the agreement was better than that between collapsed-cone calculation and measurement. The code was very effective at predicting the leakage through the MLC leaves (figure 6.36).

Shih *et al* (2001) have used the EGS4/BEAM code to parametrize the phase space of the fluence from a 6 or 18 MV beam from a Varian 2100 C accelerator and, thence, to predict the distribution of dose from a dynamic wedge. Monte Carlo calculations agreed with measurements to within 2% or 2 mm in the high dose gradient.

Verhaegen and Liu (2001) and Liu *et al* (2001a, b) have simulated the generation of a universal dynamic wedge for a Varian accelerator using Monte

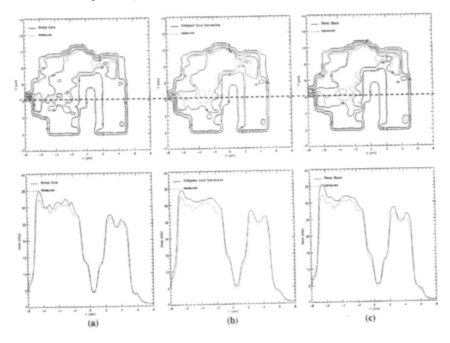

Figure 6.36. This figure shows how closely measured intensity-modulated dose distributions are to the calculations from three different planning algorithms. The isodose and dose profile comparisons are with measurement (film) at 5 cm depth in a water phantom: (a) Monte Carlo, (b) superposition and (c) pencil beam. The broken line on the isodose plot indicates where the doses profile was taken. As can be seen the dosimetric agreement between calculation and measurement was closer for the Monte Carlo calculations than it was for the other two calculation techniques. (From Keall *et al* 2001a.)

Carlo calculations. They have done this in two ways. In the first way, known as the position probability sampling (PPS) method, a cumulative probability distribution function was computed for the collimator position, which was then sampled during a single simulation to create a single phase-space distribution of fluence. In the second method, known as the static-component-simulation method, a dynamic field was approximated by MSFs in the step-and-shoot fashion requiring the computation and storage of a large number of phase-space fluence distributions corresponding to positions of the jaws at different fractions of the total delivery time. It was shown that the two methods gave comparable results and required comparable storage but that the first method had the elegance of avoiding simulating MSFs and having to store many phase-space files. Corrections were made for the monitor chamber backscatter effect. The overall effect is to generate the number of particle histories simulated for each segment in proportion to the MUs for that segment. Calculated and experimentally-measured dose distributions were also compared at depth d_{max} to exclude scatter effects, for a sinc function in 2D delivered by either step-and-shoot or dynamic delivery with a Varian accelerator and an 80-leaf MLC. The same was done for a field created by CORVUS. The error on the Monte Carlo simulation was estimated as 2.5% and the experiment was 2%. The two agreed to better than 2% for most data points which justified the approach.

Fix *et al* (2001a, b) have developed a Monte Carlo simulation using a multiple-source model They modelled two step-and-shoot cases and also two clinical treatments delivered with a dMLC. The latter were a head-and-neck and a bronchus carcinoma. The dose profiles for all applications studied showed good agreement between calculated and measured dose distributions. Fix *et al* (2002) have commented on the fact that the delivery of IMRT with small beam segments can lead to a change in the spectral characteristics of the beam. Using a Monte Carlo method, they found that sweeping a small moving slit of 2.5 mm width across the field led to an increase of the mean photon energy of up to 16% at 6 MV.

6.8.8 Monte Carlo calculations in tomotherapy

Jeraj *et al* (2001) have used Monte Carlo (MCNP) code to simulate the performance of the tomotherapy prototype at the University of Wisconsin. This was necessary because the Siemens accelerator and target used are mostly non-standard. There is no flattening filter; instead there is a thinner target, an electron stopper and a primary compact collimator. Good agreement was observed between the Monte Carlo simulations and the measurements for all dosimetric characteristics for a variety of different leaf pattern openings. The TG effect was studied and, in addition, it was calculated that the change in the spectrum off-axis is very minor.

Lee *et al* (2001a) have used the MCDOSE EGS4 user code to compute the dose distributions arising from the NOMOS MIMiC collimator using Monte Carlo techniques. These were compared with the CORVUS implementation of the finite-sized pencil-beam algorithm and agreed to within 4% or 3 mm lateral shift in isodose lines. This is a significant advance on the work done previously by Webb and Oldham (1996).

6.8.9 Monte Carlo calculations of IMAT

Li *et al* (2000d, 2001c) have used Monte Carlo treatment-planning code to verify IMAT. It was found that Monte Carlo calculations agreed with measurements in a Rando phantom to better than 2% for absolute dose and within 3% for relative dose distributions. However, up to 8% discrepancies were found between the Monte Carlo calculations and the RENDERPLAN treatment-planning results, largely due to the improper calculation of the effects of head scatter and tissue inhomogeneities.

6.8.10   Other reports on Monte Carlo dosimetry

Martens *et al* (2001c) have shown good agreement between Monte Carlo dose calculations incorporated into a treatment-planning system and measurements of dose near air cavities in the mucosa.

Alber *et al* (2001a) have described a two-stage IMRT planning algorithm. Inside their planning system HYPERION, a pencil-beam dose calculation is used to create the basic segment locations of multileaves and then a Monte Carlo dose calculation is used to make a more accurate dose calculation.

Lewis *et al* (2001) have developed a Monte Carlo-based system for predicting the complete chain of events when a Varian Millennium MLC, attached to a Varian 2100 CD Linac, is used for IMRT. The detailed model included modelling the linear accelerator, the multileaves and incorporated CT data as well as modelling the portal imaging detector.

Francescon *et al* (2001) have compared dose distributions produced by the ADAC PINNACLE[3] IMRT planning system with those produced by EGS4/BEAM and found differences were always less than 3% for the total treatment.

Sánchez-Doblado *et al* (2001) have used the BEAM/DOSXYZ code to model the segments in an IMRT treatment and compared these with the predictions of inverse treatment-planning systems and measurements using film. Only when these dose maps agree can the treatment be delivered to the patient.

Hartmann *et al* (2001, 2002a) have developed a new inverse Monte Carlo code for IMRT optimization and have shown the improvements in dose calculations compared with the non-Monte Carlo plan from the Nordion HELAX TMS planning system.

Pawlicki *et al* (2000a, b) have investigated the role of set-up uncertainty in IMRT. The target volume and beam orientations were defined and then two beamlet optimizations were done, one with and the other without the uncertainty incorporated into the planning stage. In both cases, the beamlet weights were optimized. Then a final dose calculation was made with the set-up uncertainty included in each plan and the dose calculations performed by Monte Carlo techniques. The overall conclusion was that including the set-up uncertainty in the pre-optimized beamlet dose distributions ensured the optimal target coverage.

Li *et al* (2000c) have developed a water-beam imaging system for verifying IMRT dose distributions delivered with the dMLC technique. Reconstructed dose distributions were compared with Monte Carlo simulation results.

Papanikolaou *et al* (2001) have studied a large number of inhomogenous dose-calculation algorithms applied to fluence modulations for IMRT and showed that significant differences arose, indicating the need to always compute dose distributions using inhomogeneity corrections.

6.9 Energy in IMRT

Pirzkall *et al* (2000a, 2002) have studied the effect of beam energy and the number of fields on photon-based IMRT for deep-seated targets. Ten patients with prostate cancer were selected and IMRT plans were produced using the CORVUS treatment-planning system. Plans were created for 6, 10 and 18 MV photons using either four, six, nine or 11 coplanar non-opposed fields. The 12 plans produced for each patient were then compared. It was observed that the target and sensitive structure metrics were the same for all plans, thus confirming earlier work by Sternick *et al* (1997) and the conventional wisdom that IMRT is independent of energy. However, it was found that plans with 6 MV photons that use less than nine fields might result in an increase in dose in regions distant from the target volume, for example near the skin surface. This is because these regions are not included as constraints in the inverse planning and the finding is in line with observations that such regions can occur in conventional treatment with low-energy photons. All plans were created with five intensity levels and plans were evaluated using standard conformality indices or metrics (figure 3.37). It was argued that although the study was done with a particular planning system and for a particular set of patients, the conclusions had generic applicability and the issue of the selection of beam energy in IMRT seems to have more to do with decreasing the cost of accelerators than improving treatment plans. Lu *et al* (2000) have also demonstrated the equivalence of IMRT at 6 and 15 MV.

Papanikolaou (2001) has linked the ADAC PINNACLE treatment-planning system for IMRT to a Varian 2100 EX linac operating in step-and-shoot mode. Calculations were performed at 6 and 18 MV in three modes: assuming tissues were homogeneous, 2.5 dimensional (convolution) and a full 3D scatter transport (superposition) calculation. Differences as high as 15% in the absolute dose were found between algorithms, particularly to OARs and it was argued that the choice of energy appears to be of lesser importance. A low-energy beam was recommended for practical reasons. This differs from the recommendation of some other authors.

O'Brien *et al* (2002) have looked again at the issue of energy in IMRT and found that plans at 6 and 18 MV are essentially equivalent. Dong *et al* (2003) have done the same and shown equivalence of plans. High energy is not required for IMRT.

Wierzbicki and Blackmore (2003) have compared four IMRT delivery systems, MIMiC on MDX Siemens 6 MV beam, Varian Clinac 6 MV beam with Millennium MLC, Siemens Primus 6 MV beam with MLC and Siemens Primus 18 MV beam with MLC. Four clinical cases were planned for each using CORVUS. The conclusion was that the systems all performed much the same and that high energy was not necessary.

Subramanian *et al* (2002) have argued for the construction of special linacs for IMRT.

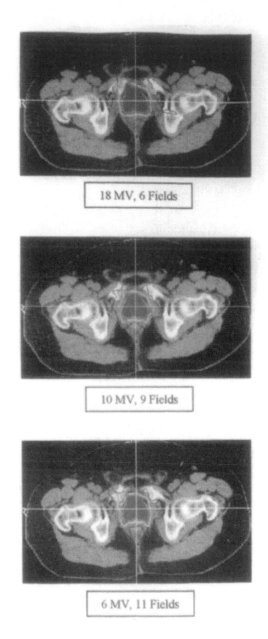

Figure 6.37. Plan equivalence by dose distribution. Equivalent plans in terms of the 55–95% isodose lines. (Reprinted from Pirzkall *et al* 2002 with copyright permission from Elsevier.)

6.10 Measuring and accounting for patient/tumour movement

Most of the development of IMRT has taken place assuming that the organs of a patient do not move from fraction to fraction and are well represented by their positions determined from some pre-planning 3D imaging study, be it x-ray CT, MR or functional imaging. As the ability to conform to the target so delineated has reached near perfection, attention is now turning to not accepting this limitation and attempting to quantitate organ movement and to account for it in IMRT planning and delivery. A nice soundbite could be that 'IMRT of the moving patient is like completing a jigsaw on a jelly'. A thought-provoking commentary on the importance of this problem has been written by Goitein (2004) specifically with respect to historical aspects of considering motion and also segregating the different types of motion and their importance and methods to cope with them.

We shall see later in this section that it has been shown that if the goal is to create a dose distribution conforming to the PTV then provided the PTV is defined to adequately cope with all motion of the CTV within it, the IMRT dose distribution to the PTV after all fractions have been delivered even with patient movement is faithful to what it would have been if there had been no movement. This is a consequence of the central limit theorem. Pursuing the analogy, it is that none of the individual 'jigsaw on a jelly' pieces really fit at each fraction but when all the pieces are taken together over all fractions, the summed jigsaw fits. Pieces from the jigsaw from one day fit with other pieces from the jigsaw from some other day. However, this methodology considers delivering the conformal dose to the PTV. It also makes assumptions that each fractional delivery is individually dephased with respect to the tissue movement. To deliver a conformal dose distribution to the CTV, other methodology is needed to either 'freeze' movement or track the tumour.

6.10.1 General review

Langen and Jones (2001) have put together a detailed critical review of organ motion and its management. They have studied three types of motion. The first is position-related organ motion which can be minimized if the patient's planning scan is performed while the patient is immobilized and in the treatment position. The second kind of motion is interfraction organ motion, i.e. motion that occurs when the target volume position changes from day to day. This is mainly a problem for organs that are part of, or adjacent to, the digestive/excretory system. This part of the review is a detailed collation of the work of all those authors who have studied this type of motion and is classified under various headings: gynaecological tumours, prostate (the largest group) and bladder and rectum. Their overall conclusion for prostate is that comparisons between the many studies are hampered by the different protocols and study end-points used.

The third type of motion is intrafraction organ motion, generally due to respiratory and cardiac functions which disturb other organs. The review collates the data for liver, diaphragm, kidneys, pancreas, lung tumours and prostate. Specific tables indicate the studies, the number of patients, the conditions and the average movements recorded. A number of techniques aim to minimize intrafraction motion: e.g. synchronization of the diagnostic and treatment procedures with breathing including voluntary breath holding, synchronized equipment gating and forced breath holding. There are other techniques that continually track the organ position. These are reviewed later in detail. This very detailed paper from Langen and Jones (2001) contains a wealth of data from some 66 clinical studies. However, it is very hard to summarize the outcome in simple terms due to the many different conditions holding. The tables themselves must be studied for the detail.

Several techniques have been proposed to manage interfraction motion (Wong 2003). One is to acquire CT images just prior to treatment. The second is the use of radio-opaque markers and realignment of the patient prior to each fraction. The third is the use of uretheral catheters and the fourth is a kind of adaptive process as used in the William Beaumont Hospital. These will be reviewed later.

Image-guided CFRT and IMRT have been recognized as growth areas (*Wavelength* 2001a). Image-guided radiotherapy (IGRT) attempts to image the moving tumour online in the treatment position, during treatment or at least before each fraction and to take account of the variation in the position of the target. Flat-panel imagers, such as that made by Elekta, and cone-beam CT are tools being explored to take forward IGRT (see section 6.11.2).

There is much debate about whether every patient receiving 3D CFRT or IMRT needs to receive image-based target localization prior to turning on the beam for each fraction. In a *Medical Physics* 'Point and Counterpoint', Herman *et al* (2003) debate this proposition. One view is that escalating doses or reducing margins should not be routinely practised until evidence of the benefit of target localization for each fraction can be produced. Another view suggests that daily imaging introduces substantial operational costs that are only justified for some tumour sites and should not be blindly applied to all.

6.10.2 Some observations of the effects of movement

Chuang *et al* (2000) have investigated the effect of uncertainties in patient positioning and patient motion in IMRT. They found that random patient motion and set-up error led to quite small inaccuracies but consistent set-up error such as the chin-up motion introduced 20–35% dose changes to critical structures such as the brainstem and optic chiasm.

Verhey (2000) has compared the different delivery techniques for IMRT including sequential tomotherapy. The techniques vary in their sensitivity to patient motion and set-up error.

Keall *et al* (2002a) have proposed 4D IMRT in which respiratory gating takes care of the movement of mobile tumours such as those in lung and breast. Keall *et al* (2001b) have discussed synchronization of breathing motion with IMRT delivered using the dMLC technique and shown that it is very important to cater for breathing patterns. Such patterns can be established from analysis of gated CT scans (see, e.g., Wahab *et al* 2003).

Sohn *et al* (2001a) have studied the effect of breathing motion on IMRT of the breast. For each patient, an average of eight images per field were acquired during treatment so that each patient had approximately 240 images for tangential fields. This was done with the Portalvision imaging system. The motion of the breast was measured and dosimetric consequences were determined.

Samuelsson *et al* (2003) have shown that, provided International Commission on Radiation Units and Measurements (ICRU) margins are applied to CTVs for head-and-neck cancers, IMRT is not particularly sensitive to systematic set-up errors in patient position.

6.10.3 Optical imaging for movement correction

There have been many attempts to couple optical imaging of body-external markers to patient treatment. Most, but not all, optical systems use either active or passive infrared markers attached to the skin surface. Their motion may or may not correlate with the movement of underlying organs and this is an issue to consider in their use. The techniques are reviewed here.

Tomé *et al* (2000) have used a bite-plate connected to the patient's maxillary dentition and attached to which is an array of passive infrared markers. The movement of these markers was then detected and fed to an optically-guided patient localization system which can correct for any variability in the patient positioning from day to day. Tomé *et al* (2001) have developed IMRT for stereotactic fractionated radiotherapy based on this optically-guided system for 3D CFRT using multiple non-coplanar fields. Based on observations for four patient cases, it was found that IMRT improved conformality indices.

Several reports have appeared on the use of the BrainLAB ExacTrac optical navigation-and-guidance system. This is a real-time infrared tracking device for monitoring movement. Alheit *et al* (2000) confirmed that the system improves patient set-up accuracy for prostate treatment. Hagekyriakou *et al* (2000) used the BrainLAB ExacTrac system to locate the prostate at successive fractions and have found that this is accurate to 1 mm. Verellen *et al* (2001) have also used a NOVALIS linac, a BrainLAB integrated miniMLC and the ExacTrac system for stereotactic radiosurgery. The same system was also tested for irradiating extracranial targets such as the prostate. Kim *et al* (2004b) showed that, for head-and-neck patients, the ExacTrac system showed that 95% of the intrafraction patient positions were within 1.5 mm of their planned position.

Wagman *et al* (2001) have evaluated the Varian real-time position-monitor (RTM) respiration-gated system. This is a passive-marker system using an infrared-sensitive camera to track the motion of reflective markers mounted on the abdomen. The system can be used for both acquisition of CT images for treatment planning and also, in the treatment mode, the system allows the linear accelerator to be gated. For liver patients, it was found that motion could be decreased from some 2.3 cm down to just 5 mm with gating.

Baroni *et al* (2000) have used a stereophotogrammetric technique to observe set-up errors, breathing movement and changes in breast volume in external-beam breast radiotherapy. The system used was the ELITE automatic optoelectronic motion analyser. Localization errors of about 4 mm and median errors due to patient breathing of about 8 mm were found.

Lyatskaya *et al* (2002) have used skin-mounted infrared reflective markers tracked by infrared cameras with submillimetre accuracy to monitor the position of the breast during fractions of radiotherapy. The goal was to reduce the rate of long-term cardiac complications by irradiating the patient at deep inspiration.

Chen *et al* (2001a) have tracked the movement of tumours in the thorax using fluoroscopy. They showed that there can be phase changes between the tumour movement and the movement of surface skin markers.

Macpherson *et al* (2002) have designed a photogrammetric system for measuring the intrafraction movement of the patient's skin surface using two CCD cameras. The skin surface motion was correlated to internal organ motion captured through fluoroscopy. It was claimed that respiratory motion can be reduced from more than 2 cm to less than 2 mm.

Ozhasoglu *et al* (2001) have used optical surface monitoring of chest and abdomen, strain gauge transducers and spirometry to study the breathing patterns of healthy subjects. They found that over an interval of 10–20 min, the average person's breathing pattern was stable enough to allow real-time detection and correction for tumour motion. However, any successful respiratory compensation system must detect and accommodate some breathing that is not strictly periodic and stable.

Moore and Graham (2000) have introduced the concept of the ' virtual shell'. The 3D surface of the patient is derived from CT data and regarded as the 'correct' surface to which the patient should conform for subsequent treatment fractions. An optoelectronic stereophotogrammetric device then measures the contour of the patient at each treatment fraction by analysing the interference patterns of structured light projected on to the patient's skin. Then the virtual shell and the patient surface contours are computer docked by applying small translations and rotations to the patient position until some cost function representing the goodness of fit is minimized, using first gross manipulations and subsequently simulated annealing to avoid trapping in local minima. The technique was applied to 20 patients with prostate cancer who received weekly CT scans. The advantage of the method was its non-invasiveness. It relied on earlier observations that the position of the prostate could be correlated to the external surface of the patient (MacKay *et al* 2000). The technique was verified by creating a shell for the RANDO-man body phantom and deliberately adding surface imperfections to the shell.

6.10.3.1 Breast movement

George *et al* (2003) have quantified the effect of intrafraction motion during breast IMRT planning and dose delivery. The effects of breathing were studied when this was parametrized through a trace obtained using the Varian RTM system. This trace was then adjusted to create the simulations for shallow and heavy breathing. Using these traces, it was possible to compute the plans that corresponded to IMRT of the breast when the patient was in four breathing conditions: no breathing (artificial of course), shallow breathing, normal breathing and heavy breathing. The results of this comparison showed that the PTV dose heterogeneity increased as respiratory motion increased. The heart-and-lung dose was also increased with respiratory motion. Conversely, the CTV dose heterogeneity decreased as the respiratory motion increased. Due to the interplay between respiratory motion and MLC motion during IMRT delivery, the planned and expected doses are different. However, because of the random phase between leaf motion and breathing, no statistically significant differences between planned and expected doses were obtained when the planned distribution was compared with the breathing-adjusted expected distribution averaged over 25 treatment fractions.

Korreman *et al* (2003) have shown that deep-inspiration breath-hold has led to increased lung-and-heart sparing for treatment of breast cancer.

6.10.4 X-ray imaging of position

Jaffray (2003) has reviewed the field of using external x-rays to guide IMRT. Murphy *et al* (2003) have treated 300 patients over seven years with the Cyberknife at Stanford. By using the amorphous silicon x-ray detectors, a comprehensive picture was built up of the pattern of intrafraction tumour movement for 250 cranial, 23 spinal, nine lung and three pancreas patients. They reported that the average variation in treatment site position was 0.45 mm (cranium), 0.53 mm (cervical spine), 1.06 mm (breathhold lung) and 1.50 mm pancreas. Cumulative distributions of movement were shown. However, these histograms disguise the often observed gradual drifts in target position, drifts which can be corrected using Accutrak (see section 6.10.8) and also the occasional and unpredictable large movements.

6.10.4.1 Intrafraction and interfraction prostate movement measurements

Balter *et al* (2000) have determined the extent of prostate movement due to breathing during radiotherapy by implanting 0.9-mm-diameter gold fiducial markers for four patients with the markers implanted at the periphery of the prostate and observing these with fluoroscopy. It was found, for example, that, in the prone position, the maximum craniocaudal prostate ventilatory motion was between 5.5 and 10.3 mm. However, this reduced to between 2.0 and 7.3 mm in the supine position and as low as 0.5 mm with the use of a false table top that allowed the pelvis to fall posterially. Lateral motion of the prostate was negligible and anterior-posterior motion was the next most significant movement. This potential for the prostate to move out of the high-dose volume may decrease the efficiency of prostatic IMRT.

Kitamura *et al* (2002) have studied the intrafraction movement of the prostate for 10 patients, each patient being observed five times (once a week) creating 50 datasets for analysis. A stereoscopic television system was used to observe the internal movement of implanted gold markers within a specified region of the prostate. Measurements were taken 30 times per second and real-time pattern recognition was used to determine the trajectory of the gold seeds. It was observed that there were no clear differences in movement found between the individuals. Like Balter *et al* (2000), they observed that the amplitude of 3D movement was much lower (0.1–2.7 mm) in the supine position than in the prone position (0.4–24 mm). The large value of 24 mm was due to bowel movement displacing the prostate. The prostate movement is affected by the respiratory cycle and is influenced by bowel movement in the prone position. Internal organ motion is much less frequent in the supine position than in the prone position. The study showed interesting 3D plots of the trajectory of the prostate gland over 2 min from which the magnitude of the problem may be assessed (figures 6.38 and 6.39).

Nederveen *et al* (2002) have used a flat-panel imager to record the position of gold markers in the prostate during a radiotherapy fraction. The markers were small (5 mm long) but the imaging technique was capable of seeing them. Images were taken, using just 2 MU, about every 400 ms and data were recorded for ten patients with 251 fractions in total. Portal images were made of both anatomical position and lateral beams to assess the motion. Individual frames in the movie were subtracted from each other and a template-matching algorithm located the markers. Traces were plotted of the marker positions as a function of time. Within a time window of 2 to 3 min, the prostate moved on average 0.3 ± 0.5 mm in the AP direction and -0.4 ± 0.7 mm in the cranialcaudal direction. Some individual displacements were quite large of the order 9 mm. The authors commented on differences between their observations and those at the Royal Marsden NHS Foundation Trust made using MR movie loops.

Spitters-Post *et al* (2002) and Visser *et al* (2002) have implanted four fine gold markers in the prostate and shown that, relative to bony landmarks, there is intrafraction movement of the prostate. The markers are visible with MR, CT and electronic portal imaging. The movement in the left–right direction had a systematic deviation of 0.6 mm and the movement in the craniocaudal and ventro-dorsal directions had a standard deviation of 0.6 and 2.4 mm, respectively. The random variations ranged from 0.9 to 1.7 mm.

Dehnad (2001) has implanted markers in the prostate and used them to correct IMRT treatments for random and systematic motions. Van der Heide (2001) also implanted markers and corrected patient position using the images of the widest component of an IMRT field. The planning was performed by the KONRAD module within the PLATO TPS.

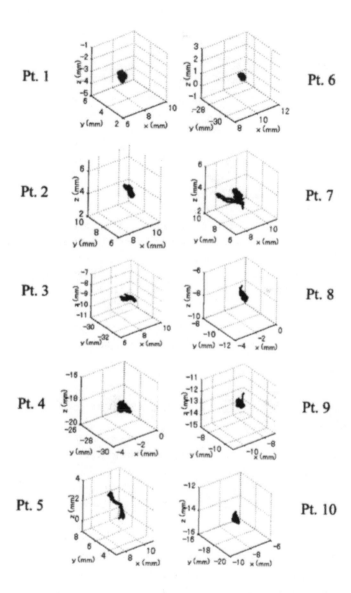

Figure 6.38. 3D trajectory of the prostate gland for a 2 min measurement period, data recorded every 0.033 s for a patient in the supine position. X represents left/right, y represents cranio-caudal and z represents anterior/posterior. Black dots represent the position of the marker implanted in the apex of the prostate gland. Data for 10 patients are separately shown. (Reprinted from Kitamura (2002) with copyright permission from Elsevier.)

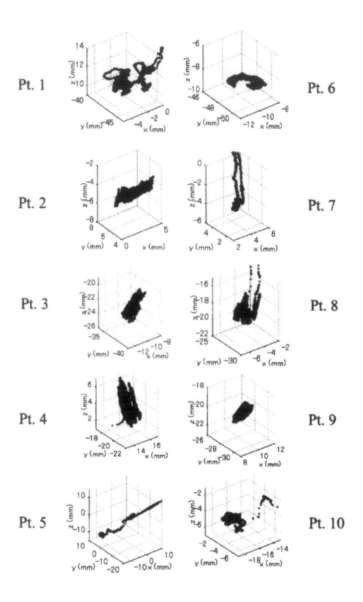

Figure 6.39. 3D trajectory of the prostate gland for a 2 min measurement period, data recorded every 0.033 s for a patient in the prone position. X represents left/right, y represents cranio-caudal and z represents anterior/posterior. Black dots represent the position of the marker implanted in the apex of the prostate gland. Data for 10 patients are separately shown. (Reprinted from Kitamura (2002) with copyright permission from Elsevier.)

Internal target-movement monitoring almost always requires the use of internal markers. However, Berbeco *et al* (2004) have produced a method for tracking in the absence of markers.

6.10.4.2 *Intrafraction and interfraction lung movement measurements*

Shimizu *et al* (2000, 2001) have implanted 2 mm gold radio-opaque marker seeds into a lung tumour and used a real-time tumour-imaging system to track the movement of the lung with normal breathing. The seeds were viewed using two diagnostic x-ray tubes and the coordinates of the markers were stored 30 times per second. The movement was anisotropic for ungated therapy with a maximum excursion to 16 mm. The linear accelerator was then gated to irradiate only when the markers were within some specified range. Repeat imaging then showed the marker movement was restricted to about 5 mm. This is the same system as used for prostate measurements by Kitamura *et al* (2002) (section 6.10.4.1).

Seppenwoolde *et al* (2001, 2002) have used a real-time tumour-tracking system to observe the movement of lung tumours in which gold markers were located. When the tumours were in the upper thorax, it was found that the amplitude of the motion could be as large as 12 mm in craniocaudal direction and motion of both lungs and heart affected the movement of the tumour (figure 6.40). The time-averaged tumour position was closer to the exhale position because the tumour spends more time in the exhalation than inhalation phase. A linear accelerator was gated to irradiate only when the markers were in a small range of positions.

Kini *et al* (2001) have investigated the correlation between the movement of internal markers and the movement of external markers. One-hundred-and-fifty fluoroscopic movies were analysed for six patients for whom there was also simultaneous monitoring of the chest wall through camera-detected infrared markers. A simple mathematical function was found linking the external-marker trace to the amplitude of movement but with a phase shift. This was then used *a-posteriori* for further motion traces in which the internal motion was predicted from the external motion and then compared with the actual measurement of internal motion using x-rays. The variability with good coaching was about 3 mm between prediction and actual motion. Tsunashima *et al* (2003) have also shown a correlation between the movement of external optical markers and internal x-ray markers but with a phase shift. Gulliford *et al* (2004) have developed an artificial neural network (ANN) to learn and predict lung movement up to 10 s ahead of real time in order to guide accelerator gating.

Vedam *et al* (2003a, b) have quantified the predictability of diaphragm motion during respiration by using a non-invasive external marker. For five patients and 63 fluoroscopic lung procedures, the internal position of the diaphragm was monitored using x-ray fluoroscopy with images obtained at 10 Hz. The position of a skin mark was simultaneously acquired using the Varian real-time position management system at 30 Hz. This was done with the patient

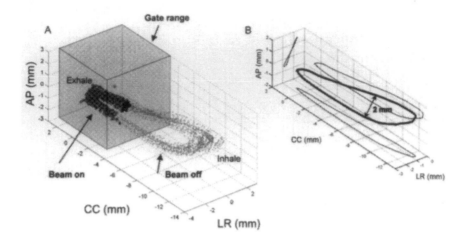

Figure 6.40. (A: left) The 3D path of the tumour for one patient during one treatment portal: the grey dots represent the tumour position throughout the treatment, and the black dots represent the tumour position as the tumour is detected to be inside the real-time tumour-tracking radiotherapy system (transparent box). The asterisk represents the planned zero position. (B: right) the average 3D curve of the tumour: the projections on the coronal, sagittal and axial planes are drawn in thin black lines. Note that the tumour follows a different path during inhalation than during exhalation because of hysteresis. (Reprinted from Seppenwoolde *et al* 2002 with copyright permission from Elsevier.)

either freely breathing or under a variety of instructed breathing patterns. It was shown that the internal diaphragm position correlated well with the respiration signal and the statistical significance of the correlation was high. This correlation was then used to predict diaphragm motion from one session to another and it was shown that the diaphragm motion was highly predictable to within 1 mm standard deviation independent of the method of breathing training. A limitation of the work was that only the motion of one anatomical structure and only one respiration signal were correlated. From this, it was deduced that the margins needing to be added for gated radiotherapy could be significantly reduced for lung tumours.

Blackall *et al* (2003) have instead created MR images at various phases throughout the breathing cycle by gating. A high-quality reference image was also obtained at end-exhalation and a model developed to link this vectorially to the other images thus creating a model of lung movement. Hence, knowing the time in the breathing cycle, the position of lung and lung tumours could be predicted from a single MR scan. Rohlfing *et al* (2004) made a similar gated MR study for liver and showed that the elastically deformable model of liver was needed; non-rigid deformations gave the wrong location for one patient by up to

3 cm. Balter *et al* (2003a) have also developed morphological models of lung movement to account for it in IMRT planning. Zhang *et al* (2003a) have used a deformable model of the lung to create modulations tailored to lung deformation. Lotz *et al* (2003) have built a similar model for characterizing bladder volume change between scan time and treatment time.

Patients breathe asymmetrically and a good mathematical representation is

$$z(t) = z_0 - b \cos^{2n}\left(\frac{\pi t}{\tau}\right) \tag{6.6}$$

for a motion in the z-direction where t = time, z_0 = mean position, b = amplitude, τ is the period of motion and n parametrizes the asymmetry (figure 6.41). The larger n becomes, the more time is spent at end expiration. Values of b, n and τ can be determined from fluoroscopy. Then dose distributions for static patients (e.g. based on a single scan at end expiration) can be convolved with such a function to give a measure of the dose averaged over the breathing cycle. Lujan *et al* (2003) have shown that it is necessary to make many fluoroscopic measurements to adequately estimate the parameters. Also, if the parameters predict a motion that is very different from that assumed at planning, dose errors arise. The effect of part of the fraction time following an irregular or different breathing pattern was also studied. This study was in the context of CFRT of the liver at the University of Michigan.

Szanto *et al* (2002) have implanted gold markers in 1000 patients since 1994. The patients were re-positioned by imaging these markers at the time of treatment and average deviations from a reference position were usually less than 1.5 mm; occasionally shifts as large as 1 cm have been observed.

Unkelbach and Oelfke (2003, 2004) have developed an inverse-planning technique to take account of interfraction organ motion by the use of probability distributions in the cost function to be minimized in inverse planning. They specifically made use of information from a small finite number of CT scans of the patient and addressed two intrinsic problems, that of incomplete knowledge of the distribution of patient geometries and the sampling of this distribution by only a finite number of fractions. They developed a technique which will create the most suitable fluence profile corresponding to a distribution of positions for the target geometry. They applied this technique to the classic Brahme *et al* (1982) problem of a circularly symmetric uniform dose surrounded by a circularly symmetric annulus the OAR.

6.10.5   Ultrasound measurement of position

Bamber *et al* (1996) have suggested that ultrasound images can be used to measure tissue motion and indicate corrections needed for radiotherapy. At that time of writing, the technique was not operating in 3D and not yet fast enough for the suggested purpose. Subsequently, there have been many developments (Meeks *et al* 2003).

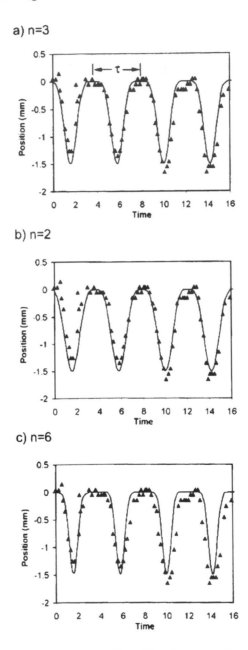

Figure 6.41. Measured time course of breathing for a sample patient based on the diaphragm motion (away from the expiration position) as observed under fluoroscopy (triangles) compared with the model given in equation (6.6) using (a) $n = 3$, (b) $n = 2$ and (c) $n = 6$. (From Lujan *et al* 2003.)

6.10.5.1 The NOMOS BAT

The main piece of commercial apparatus available is the Beam Acquisition and Targeting (BAT) device from the NOMOS Corporation. This takes a 2D slice of the patient and allows an organ to be contoured. The extraction of the prostate contour from ultrasound data is not easy. Hu *et al* (2003) have developed an automated technique. Then this contour is registered with that from a planning stage and any misregistration is corrected by adjusting the position of the couch. In principle, this corrects for interfraction motion.

The technique of using the BAT device commences by docking the probe on the collimator so that the ultasound system 'knows' where the isocentre is (to which the CT data are referenced). Contours of the prostate are then displayed in transverse and sagittal planes and moved manually to lock with the CT data, thus giving the required three orthogonal components of movement. Anterior–posterior (AP) and lateral movements are determined from the transaxial plane and superior/inferior (SI) movements on the sagittal images. There has been a considerable number of studies reported using the BAT. These are now reviewed.

Lattanzi (2000) has used the BAT to measure the accuracy of position of the prostate at different treatment fractions (figure 6.42). To establish the accuracy of the BAT, ultrasound and CT at the same time were compared with good correlation (better than 2 mm). Lattanzi *et al* 1999a, b) also showed that the ultrasound-derived PTV correlated well with the CT-derived PTV *taken at the same time*. Dong and Court (2003) also showed that the BAT measurements closely reflect the anatomical position as determined from in-treatment-room CT.

Lattanzi *et al* (2000) have reported on the clinical experience of using the BAT transabdominal ultrasound device to localize the coned-down part of the prostate irradiated to a boosted dose (2–4 fractions). The technique was initiated at the Fox Chase Cancer Center, Philadelphia in March 1999 and the study reported on 54 patients. Prior to the commencement of the boost phase, the patients received a new CT scan. Prostate-only fields were then determined for a PTV with no margin whatever. Daily ultrasound localization of the prostate was then registered to the PTV from this scan.

The averages and standard deviations of displacements found were -3.0 ± 8.3 mm (AP), 1.86 ± 5.7 mm (lateral) and -2.6 ± 6.5 mm (SI). These might be regarded as quite small, even acceptable, mean errors. However, of more note, was the observation that 21% of AP shifts, 7% of lateral shifts and 12% of SI shifts were greater than 10 mm. These shifts were due to the changed filling of rectum and bladder. This emphasized that mean displacements are of little use in characterizing actual displacements, the extrema of which can be quite unacceptably large. By re-positioning the patient, making use of the BAT, it was possible to have a reduction in planning treatment margins allowing safe dose escalation. The effectiveness of IMRT would have been reduced if re-positioning had not taken place.

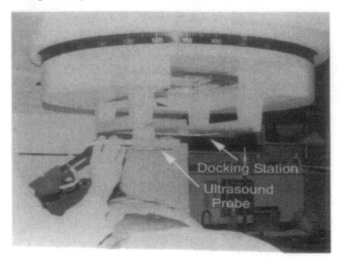

Figure 6.42. The figure illustrates how the ultrasound probe known as the BAT is docked to the linac machine head for the purpose of providing a 3D reference for the probe relative to the machine isocentre. The docking plate is an easily replaceable acrylic sheet which is fitted into the block tray. (From Lattanzi *et al* 2000.)

Beyer *et al* (2000) have also used the BAT system to study the movement of the prostate. Twenty-four consecutive patients were monitored. Adjustments of the isocentre were required on all patients daily and the average daily motion was typically less than 1.5 mm. On any given day, however, more significant adjustments were required, up to 6 mm and, indeed, 3% of readings required adjustments of more than 10 mm, re-emphasizing the observations of Lattanzi *et al* (2000). It was concluded that the BAT system is an effective means to target IMRT in the treatment of prostate cancer.

Willoughby *et al* (2000) have reported on transabdominal ultrasound localization of the prostate with the BAT system for 100 patients. It was found that analysis of alignments performed on the first 50 patients compared quite well to that of the last 50 indicating that the overall average shift is not changed.

Trichter and Ennis (2001) showed that the use of the BAT system detected significant prostate motion unrelated to the position of bony anatomy that was detected using EPID images. It was concluded that the BAT system provided improved positional information for IMRT delivery.

Falco *et al* (2001) have used the BAT system to determine that, over a course of fractionated treatment to patients with prostate cancer, the prostate can shift by as much as 1 cm from its reference position in any direction. Daily displacements were then incorporated into the treatment-planning process to assess changes in the target dose coverage.

Chandra *et al* (2001, 2003) have reported their clinical experience of ultrasound-based daily prostate localization at the MD Anderson Cancer Center in Houston. For 147 consecutive patients, 3509 BAT alignment procedures were made and reported. Ultrasound images were scored depending on their image quality. The ultrasound image quality was judged to be poor or unacceptable in 5.1% of the cases. Of the remaining BAT images with high-quality imaging scores, alignments were unacceptable (> 5 mm error as judged by the reviewing physician) in 3% of cases. The mean shift in each direction of the prostate averaged over all patients was quite small, 0.5–0.7 mm. However, the distribution of shifts was a near random Gaussian in all three major axes and the standard deviation of the spreads was found to be 2.8 mm in the right/left direction, 4.9 mm in the AP direction and 4.4 mm in the SI direction. The percentage of BAT procedures in which the shift was > 5 mm was 28.6% in AP, 23% in SI and 9% in RL directions. The BAT ultrasound system corrects for both set-up uncertainty and internal organ motion, thus compensating for inadequate initial set-up using skin marks. The use of the BAT procedure added an extra 5 min to each treatment slot. By plotting regression plots, it was possible to separate the positioning errors into random and systematic errors. Interestingly, the couch shifts did not vary significantly as the study progressed, showing that there were no significant changes over time. It was hypothisized that, after extensive use of the BAT alignment procedure, the therapist may have paid less attention to the initial set-up using skin marks and relied more directly on the BAT for final position adjustment thus leading to large recorded shifts. It was concluded that, conversely, intrafraction prostate movement was not of grave concern.

Serago *et al* (2002) have also investigated the use of transabdominal ultrasound to localize the prostate and its immediately surrounding anatomy and guide the positioning of patients for the treatment of prostate cancer. They used the BAT system and evaluated the same-day repeat positioning by the same ultrasound operator and also the variation in the measurements made by two different operators. Self-verification tests of ultrasound positioning indicated a shift which was < 3 mm in more than 95% of cases. Interoperator tests indicated shifts of < 3 mm in 80–90% of the cases. A comparison was also made with conventional localization based on CT and skin markers and the mean difference in patient positioning between conventional and ultrasound localization for lateral shifts was 0.3 mm, vertical shifts 1.3 mm and longitudinal shifts 1 mm. However, on a single day, the differences could be > 10 mm for 1.5% of lateral shifts, 7% of longitudinal shifts and 7% of vertical shifts. This rather emphasizes that stating the mean variations is not as adequate as stating the range and maximum variations. It was also found that the pressure of the ultrasound probe displaced the prostate in seven out of 16 patients by an average distance of 3.1 mm. However, interestingly 56% of the patients showed no displacement. The quality of this work is very dependent on the quality of the ultrasound images.

Morr *et al* (2000, 2002) have used the BAT system for prostate localization and alignment with IMRT fields at each treatment fraction. Twenty-three patients were studied, 19 of whom were successfully imaged supine with a full bladder. The BAT system was then used to create ultrasound images of the prostate and these were registered to contours previously determined from CT data. After training, this could be done in about 6 min. The average right–left, up–down and in–out adjustments were 2.6, 4.7 and 4.2 mm, respectively. Some positional adjustments larger than 10 mm occasionally occurred. On average, the time required for verification and position changes was less than 10 min with an average of 5.56 min. Positional adjustments > 10 mm were very rare and related mainly to a misidentification of the target structures on the ultrasound image or patient movement.

Héon *et al* (2002) have used the BAT system to monitor the positional information for 22 prostate cancer patients and 504 ultrasound scans were obtained and analysed. The prostate displacement data showed that the prostate can shift from its intended treatment-planning position by median values of 3 ± 3, 5 ± 4 and 5 ± 3 mm, in the axial, sagittal and coronal axes, respectively.

Little *et al* (2003) have used the BAT system to show the mean and range of prostate motions and related this to the margins required on the CTV to ensure tumour coverage for all fractions. The prostate organ motion shift was defined as the difference between the BAT shift and the set-up error shift defined on portal films taken at the same time as BAT measurements. Significant numbers of patient treatment fractions showed the prostate located outside the PTV and so requiring post-BAT intervention (repositioning) to restore irradiation fidelity.

Van den Heuvel *et al* (2003) showed that the use of the BAT system does *not* improve on positioning. They implanted five gold seeds in the prostate and imaged them each fraction for 15 patients. BAT repositioning was performed on alternate fractions only, allowing assessment of the positioning accuracy with and without ultrasound repositioning. Whilst there were differences, it was concluded that the residual errors did not greatly change the positioning accuracy. The only significant improvement was in the craniocaudal direction. Since this study is negative to the promise of the BAT system a detailed study of its conclusions is recommended. Langen *et al* (2004) also found that the BAT did not reduce the errors in the SI and lateral directions but only in the AP direction. They found that the BAT did not agree with x-ray images of implanted markers.

All these studies specifically concern *interfraction* prostate motion. It probably makes no sense to provide here averages for interfraction prostate movement detected with the BAT because the experiences are varied. Possibly the recorded values reflect local practices. There certainly appears to be some effect of learning, not all of it beneficial. The reader can make up their own mind from the statistical data given here for the several centres that have used the BAT system.

Huang *et al* (2002a) have used the BAT system to quantify the *'intrafraction'* motion of the prostate (quote marks because this was not a continuous measurement during the treatment). This was done by taking ultrasound snapshots before and after each fraction for a total of 400 BAT procedures. It was found that the mean intrafraction movement was negligible, i.e. 0.01 ± 0.4 mm in the left direction, 0.2 ± 1.3 mm in the anterior direction and 0.1 ± 1.0 mm in the superior direction. There were some individual movements much larger than this. Moreover, the movements did not correlate with interfraction movement. It was commented that others, using fluoroscopy over a period of time, had seen larger movements and this may be a weakness of the 'before-and-after' snapshot ultrasound study.

Mah *et al* (2001) have deduced from MRI cine video loops of patients with prostate cancer that no significant respiratory motion could be deduced in terms of prostate displacement. The prostate appears to 'dance' as a result of peristaltic motion of the rectum but only with an amplitude of less than 2 mm which is usually of the order of the set-up error. These findings were confirmed by Huang *et al* (2001b) who used the BAT system to measure intrafraction prostate motion during 100 IMRT treatments for prostate cancer. They noted that the movement was always less than 5 mm. Kitamura *et al* (2001) come to the reverse conclusion that intrafractional movement of the prostate, measured using a real-time tumour-tracking radiation therapy system imaging implanted gold markers, actually shows that intrafraction movement can be as large as 8 mm.

6.10.5.2 *Other ultrasound systems developed*

Bouchet *et al* (2000) have developed a 3D ultrasound technique for high-precision guidance of radiation therapy. A Voluson 530D 3D imaging system was used to create ultrasound images and the ultrasound probe position was tracked using a CCD camera which is focused on four infrared light-emitting diodes attached to the ultrasound probe. This then allows the ultrasound images to be precisely locked onto x-ray, CT or MR scans and displacements at the time of treatment can be corrected using this technique. The claimed accuracy is better than 1 mm in AP, lateral and axial directions.

Sawada *et al* (2002, 2004) have used real-time ultrasound measurement during treatment to correlate images with equivalent ultrasound images created at the planning stage. A CT scanner and linac are in the same room and an ultrasound reference image is recorded at the time of CT scanning. Then the patient is rotated by 180° to the linac. The ultrasound probe is on a firmly anchored robotic arm and then proceeds to record images of the patient on the linac (figure 6.43). The images are correlated with the reference image and a trigger pulse to the linear accelerator is generated when the image correlation index exceeds a predetermined threshold level. This is a form of gating based on the use of ultrasound.

Artignan *et al* (2002b, 2004) have studied the effect of the pressure of an ultrasound probe making measurements of the position of the prostate but unwantedly disturbing the prostate during the measurement. When the probe moved from 0 to 35 mm, the prostate moved from 0 to 10 mm so the abdominal pressure actually *creates* prostate displacement (see also Serago *et al* (2002).

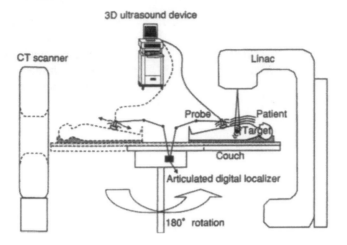

Figure 6.43. Diagram of a system comprising a CT scanner, a 3D ultrasound device, an articulated digital localizer and a linac. The CT scanner and the linac are located in the same treatment room so that a tumour position detected by the CT scanner coincides with the isocentre of the linac when the couch is rotated around the couch support by 180°. The probe of the ultrasound device is attached to the end of the localizer arm. The system allows accurate repositioning of the patient at the treatment time and the recording of ultrasound images to correlate with those at the CT scanner for gating treatment. (From Sawada *et al* 2004.)

Ultrasound images were made of ten healthy volunteers using a transabdominal ultrasound probe with a precise method of indexing the probe relative to the skin surface to quantify 'pressure'. The probe was set to point inferiorly avoiding too much shadow from pubic symphysis and to obtain a clear view of the prostate. The position of the prostate corresponding to light pressure was determined as reference, even though it was hard to view the prostate (the reason more pressure is needed). Then the distance of the probe from the prostate was shortened in carefully controlled 5 mm intervals. The volunteers did not all behave the same way but in all volunteers the greatest prostate movement was in the posterior direction. On average, there was 3 mm movement for every 1 cm of applied 'pressure'. The prostate motion was less than 5 mm in 100% of the volunteers after 1 cm of 'pressure', in 80% after 1.5 cm, in 40% after 2 cm, in 10% after 2.5 cm and in 0% after 3 cm of 'pressure'. No correlations were found with bladder volume or prostate-to-probe distance. There was some motion inferiorly but random motion left–right.

Conversely, the work of McNeeley *et al* (2003) disputes this. They made MR images of the prostate with and without simulation of the pressure of an ultrasound probe and showed the prostate moved only about 1 mm on application of the probe.

6.10.6   Magnetic monitoring of position

Overcoming the problem of the moving tumour is the next challenge in IMRT. Seiler *et al* (2000) and Muench *et al* (2001) have proposed a novel technique to track the real time position of the tumour with respect to the radiation beam (figure 6.44). It relies on the principle of coupling a source of magnetic field to a position-sensitive sensor. This sensor could be attached to the surface of the patient if the tumour were superficial or, alternatively, embedded surgically to be close to a deep-seated tumour. Organ motion generated by breathing, circulation, peristalsis etc can then be tracked. Experiment showed an acuracy of 1–2 mm or 0.5–1° of rotation. A phantom which was moving was then irradiated using this technique to gate a proton irradiation and the resulting dose distribution was very similar to that found for irradiating a stationary phantom (but quite different from that irradiating the movement-untracked phantom). Balter (2003) has also described a technique for monitoring the position of an implanted transponder using electromagnetic fields with reported accuracy to 2 mm at 10 Hz frame rate. Balter *et al* (2003b) have also made magnetic measurements of position using the apparatus from Calypso Medical technologies of Seattle and shown that absolute errors were of the order 1 mm or less.

6.10.7   Gating

One of the potential dangers with IMRT is that dose distributions which have been tightly conformed to a target may be inappropriately placed if the target is moving. The use of target margins is one traditional way to accommodate this but is not a way to effectively overcome the problem. We should remember that margins are an artificial construct invented by physicists to solve specific problems. In an ideal world we should find a way to do without them. There is a growing interest in the use of gated IMRT. The technique works like this. The breathing motion of the patient is monitored and converted into a binary on/off signal. For example, this signal could be set at a particular state of inspiration or expiration. The signal so generated then gates the radiation either on or off with the net result that the target is in almost the same place for each of the gated sequences of radiation. This, unfortunately, leads to a very small duty cycle for radiotherapy.

6.10.7.1   *Gating based on optical measurements of surface markers*

Solberg *et al* (2000) have investigated the feasibility of gated IMRT. Respiration was monitored in real time using cameras which track infrared-reflecting markers placed on the patient's surface and gating occurs for a period around exhalation. Radiation was delivered using a Novalis accelerator and integrated microMLC. The advantage of this system is that the beam can be started and stopped within 0.01 MUs.

    Kini *et al* (2000) have described a gating apparatus (a passive marker video-tracking motion-monitoring system) to improve the treatment of thoracic

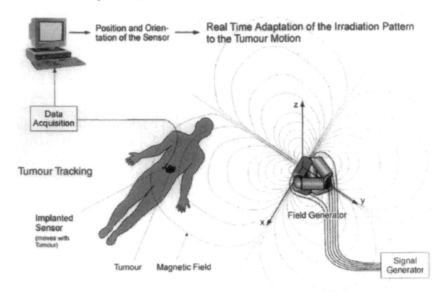

Figure 6.44. A schematic diagram showing how the tumour-tracking technique using miniaturized sensors with the help of magnetic fields takes place. The patient carries an implanted sensor which moves with the tumour. A magnetic field is generated via a six-coil tetrahedron-shaped assembly and the field sensor is a miniaturized induction coil. The alternating magnetic field created by the field generator induces an alternating voltage in the sensor which is measured by the connected data acquisition electronics. (From Seiler *et al* 2000.)

malignancies such as lung and oesophageal carcinoma. Phantom studies have shown that, using respiratory gating, the PTV could be decreased substantially.

Ramsey *et al* (2000) have shown that respiratory gating can improve the treatment for lung cancer patients. A prototype respiratory-gating system was used which correlates the signal from a video-based respiration-monitoring system with the internal location of the target volume. Upper and lower gating thresholds were determined using free breathing by the patients and then those patients whose observed target motion under fluoroscopy was greater than 0.5 cm were assigned to gated treatment paths. It was found that the average beam-on time for respiratory-gated treatment increased by a factor of 2.8.

Vedam *et al* (2001) have determined parameters for respiration gated radiotherapy. Their method determines the amplitude of motion and accounts for any difference in phase between the internal tumour motion and the external marker motion. An infrared camera tracks movements of external markers and a fluoroscopic system creates images of internal markers (figure 6.45). The movements are then tracked to determine amplitude and phase differences. The

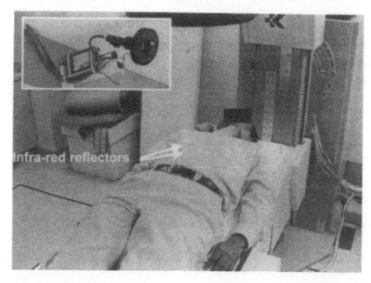

Figure 6.45. A volunteer in a simulation position showing the infrared reflective markers placed mid-way between the umbilicus and the xyphoid process. The camera with an infrared illuminator surrounding the lens is inset. (Note that for patients the marker block is attached directly to the skin rather than to the clothing.) (From Vedam *et al* 2001.)

choice of gating techniques is then whether to gate at inhale or exhale and then the second decision is whether to track using amplitude or phase. The third decision is to determine the cost benefit of how long one wishes to extend the delivery time for the sake of decreased tumour motion. All gating techniques assume reproducibility from breathing cycle to breathing cycle. It was observed that gating during exhalation was more reproducible than gating during inhalation.

Ford *et al* (2002a) have evaluated a respiration gating system using film and electronic portal imaging. The gating system measures respiration from the position of a reflected marker on the patient's chest. Simultaneously fluoroscopic movies were recorded. It was found that the use of the gating system reduced the localization of dose by about 5 mm.

6.10.7.2 Gating based on x-ray fluoroscopic measurements

Shirato *et al* (2000a, b, 2004) have described a technique for real-time tumour tracking and gating of a linear accelerator. The technique relies on four sets of diagnostic x-ray television systems which offer an unobstructed view of the patient (figures 6.46 and 6.47). The system measures the location of a 2 mm gold marker in the body 30 times a second. The marker is inserted in or near the tumour using image-guided implantation. This system allows the tracking of markers placed in the prostate, in lung tumours and in liver tumours. The linear

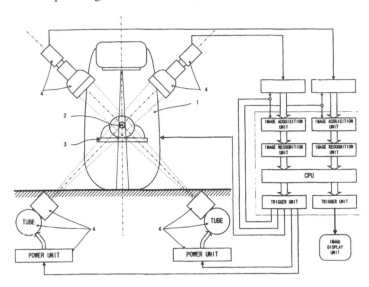

Figure 6.46. The configuration of a real-time tumour-tracking system synchronized to a linear accelerator: (1) linear accelerator; (2) internal marker in the patient's tumour; (3) the patient couch; (4) high-voltage power units, x-ray tubes, image intensifiers. (Reprinted from Shirato *et al* (2000a) with copyright permission from Elsevier.)

accelerator is gated so that the treatment beam is never on at the same time as the diagnostic x-ray units are pulsed and the gating of the linac is performed by means of a grid on the electron gun. The system allows the determination of the 3D coordinates of the tumour marker and the linac was energized only during that period in which the detected location of the tumour marker was within the accepted volume as defined by the allowed displacement value.

A phantom with a moving marker was used to establish that the accuracy of determining the displacement was better than 1.5 mm provided the tracked marker moved at constant velocity. From measurements made with patients, it was determined that the range of the coordinates of the tumour marker during irradiation was 2.5–5.3 mm whereas it would have been 9.6–38.4 mm without tracking. It was established that the dose due to the diagnostic x-ray monitoring was less than 1% of the target dose and was thus an acceptable additional radiation burden. Shirato *et al* (2000a) claimed that their technique has some advantages over others. For example, active breathing control requires a measure of cooperation of the patient; optical trackers can only look at surface markers which may or may not correlate with the position of deep-seated implanted markers. Magnetic tracking of a single marker coil requires somewhat invasive wiring. The main limitation of their technique was the potential for migration of the implanted marker. Another issue is that responding to image data with actions affecting the radiation has inherent delays which have been characterized

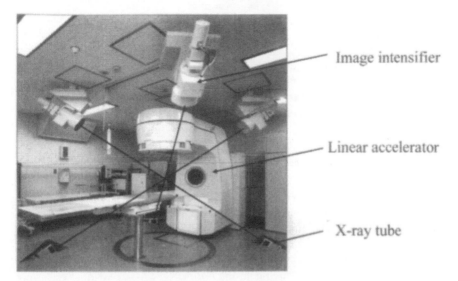

Figure 6.47. The motion-gated linear accelerator system and fluoroscopic real-time tumour-tracking system. Three of the four fluoroscopic systems are shown. (Reprinted from Shirato *et al* (2000b) with copyright permission from Elsevier).

by Sharp *et al* (2003). Provided the imaging is at greater than 10 Hz, there was no advantage to using prediction techniques. At slower frame rates, computer prediction was required.

6.10.7.3 Imaging and therapy gated by respiration monitor

Mageras (2001) and Mageras and Yorke (2004) have used respiratory gating (RG) for both imaging and radiation therapy. For CT and electronic portal imaging, acquisition was triggered by the RG system, whereas, for PET and fluoroscopy, image data were tagged or time stamped with phase from the respiration waveform and then selectively viewed according to phase. Mageras (2002) created respiration-correlated spiral CT images to achieve 4D imaging. The x-ray-on signal from the scanner and the respiration waveform from a position-sensitive respiration monitor placed on the patient were recorded synchronously during acquisition and used to correlate the CT slice with the respiration phase. Then, by selecting reconstructed slices that occured at the same phase in the respiration waveform, 3D images were created at typically 8–10 different phases and these give information on the anatomical variation with phase of the breathing cycle. These systems allowed gated IMRT delivered with either the dynamic or step-and-shoot multileaf collimation. Fluoroscopy showed that external monitor movement correlated well with that of the diaphragm in most patients.

Wong (2002) has highlighted two techniques for compensating for breathing motion during the delivery of IMRT. The first is respiratory-gated radiation treatment (RGRT). Typically, the relatively quiescent segment near end exhalation is chosen with a duty cycle of about 25%. The disadvantage of this technique is this low duty cycle and the need to still apply margins for the movement during the radiation-on period. The second technique is active breathing control (see section 6.10.11) which, in principle, should result in a smaller margin. However, at present, it has not been determined how much margin reduction can be safely achieved and this remains a research priority.

Jiang and Doppke (2001b) have studied the effects of respiratory motion on the treatment of breast cancer with tangential fields. A spirometer was used during CT scanning so that CT scans could be reconstructed at three instances of respiration status. The first was during normal breathing and this was used for treatment planning. The other two sets of data were acquired by breath-holding at inspiration and expiration. When the tangential fields were shifted relative to the CT data sets to simulate the effects of this motion, it was found that the results suggested that breast motion during normal breathing may not be a severe problem in tangential treatment.

Giraud *et al* (2000) have evaluated intrathoracic organ motion during the physiological breathing cycle in order to optimize the 3D CFRT of lung tumours. The patient breaths through a mouthpiece connected to a spirometer. Significant differences were noticed in CT scans taken at different parts of the breathing cycle and it was suggested that a spirometer-gated irradiation would be able to overcome the problems posed by intrathoracic organ motion during treatment.

The signal from a spirometer is directly correlated with respiratory motion and is, therefore, ideal for target respiratory motion tracking. However, it is susceptible to signal drift which deters its application in radiotherapy. Zhang *et al* (2003b) have developed calibration techniques to enable the spirometer to provide a long-term drift-free breathing signal and it was concluded that spirometer-based monitoring is most suitable for deep-inspiration breath hold and less important for free-breathing gating techniques.

Van Herk *et al* (2002a) have created respiration-correlated CT scans for lung cancer patients. A small thermometer is placed under the nose and the temperature difference between inhaled and exhaled air was used to detect the breathing phase. The thermometer signal was correlated with the CT scanner by simultaneously digitizing the x-ray-on signal. This allows a set of 200–300 raw CT scans acquired in 3–5 min to then be grouped, with each group containing just those 40 slices with irregular slice spacing corresponding to a selected breathing phase. The volumetric images are then interpolated and are useful for showing the resulting tumour motion curves demonstrating the precise 3D path of the tumour.

Van Herk (2003) has described two methods of image registration for assisting radiotherapy planning. The first is a pixel-by-pixel registration in which the comparison is based on a so-called cost function describing the goodness of fit between the two scans. The second is a volume registration algorithm which is based on registering previously defined contours, for examples bony anatomy (chamfer matching).

6.10.7.4   *Measurements using oscillating phantoms*

Kubo and Wang (2000) have investigated whether IMRT dose distributions are suitably maintained when the additional constraint of linear accelerator gating is brought into play. Square pulses were used to mimic breathing at regular intervals and also pulses which were generated from the spirometric curve characterizing a patient's real breathing. Experiments were carried out (figure 6.48) to irradiate a film using either an enhanced dynamic wedge or an intensity-modulated fluence created with a dMLC technique. A reference measurement was made with the film stationary with respect to the moving leaves and jaws. Secondly, an irradiation was made with the film oscillating as a result of breathing motion with respect to the same irradiation conditions and, thirdly, a measurement was made with the film moving under breathing conditions but with the radiation gated using the signals previously described. The measurements were done at different accelerator pulse-repetition rates—80, 240 and 320 MU min^{-1} and for two energies—6 and 18 MV. Motions of the film were either in the direction of the leaf motion or at right angles to this. Kubo and Wang (2000) showed that the gating results (the third set) were almost identical to those with no gating and no motion, whereas there were large differences between the results from this third set of irradiation conditions and the dosimetry measured with no gating. The results were largely independent of beam energy, the output rate of the accelerator or the direction of movement of the target. It was therefore concluded that linear-accelerator-gated operations maintain the enhanced dynamic wedge and IMRT dose delivery.

Hugo *et al* (2001) have compared dose distributions for IMRT delivered to a stationary target with IMRT delivered to film in an oscillating target with the oscillating motion removed by gating the accelerator using respiratory triggering based on infrared photogrammetry. No significant dose differences were found between stationary non-gated films and oscillated gating films. Hugo *et al* (2002, 2003a, b) showed that the addition of gating reduced the maximum distance to agreement and dose difference from 4.6 to 0.61 mm and from 15.1% to 4.45% respectively.

Hugo *et al* (2002) have evaluated the use of a BrainLAB ExacTrac commercial patient positioning system for gating IMRT (figure 6.49). An experimental arrangement was constructed whereby a phantom could be oscillated to simulate the breathing function that was derived from the motion of a liver. IMRT was delivered either with single or multiple fields, both with the MSF and dMLC techniques, to the stationary and the moving phantom. IMRT of the moving phantom was gated with a variety of window sizes. It was found that with the smallest window size the error as measured by the γ factor was lowest

Figure 6.48. The experimental set-up for film motion using a respirator: (A) 6-cm-thick solid water, (B) 1.5-mm-thick spacer, (C) film, (D) lead blocks to restrict the film-holding arm from drifting sideways, (E) reflective marker in its holder, (F) diaphragm which moves the film-holding arm in the direction indicated by a double arrow and (G) air inlet from an respirator. (From Kubo and Wang 2000.)

Figure 6.49. The experimental set-up for single field studies involving motion of the phantom with respect to the beam delivery. Two slab phantoms (RSD dry water) were placed on the positioning stage to provide build-up and localization. Film was placed between these phantoms and markers were attached to the build-up material. The phantom could then be oscillated with respect to the beam. (From Hugo *et al* 2002.)

and although the duty cycle was quite small (7%) a 1 mm window placed centrally on the exhalation phase was recommended.

Goddu *et al* (2002) have used an oscillating phantom to show the difference between dose distributions with and without breathing delivered with IMRT using a Varian accelerator.

Suh *et al* (2003, 2004) transferred the patient movement information to the MLC fields for IMRT so that the leaf patterns follow the tumour breathing. Yi (2004) and Suh *et al* (2004) have developed a method to compel a particular breathing pattern using a ventilator. Then it was shown that if an established breathing pattern could be applied to MLC motion, the beams effectively tracked the tumour and the irradiation was as accurate as applying unmoving beams to an unmoving target. This was shown to be effective for both simple unmodulated fields and also for step-and-shoot IMRT components. The direction of MLC leaf movement was arranged to coincide with the direction of the movement of the phantom. It was also assumed that the breathing motion was regular. For patient implementation, the breathing motion was compelled to a fixed rhythm set by the patient themself and also corrections were made for superior–inferior breathing motions by arranging for the MLC leaves to track in this orientation (see also Keall *et al* 2003).

Some other experiments with oscillating phantoms are presented in section 6.10.12.3.

6.10.7.5 Evidence against the need for gating

Zygmanski *et al* (2001a) took fluence distributions from the HELIOS inverse-planning system and converted them to effective incident fluence taking into account the expected motion of tissues due to patient respiration. These effective incident fluences were replanned onto the patient to demonstrate the effects of IMRT treatment with and without respiratory gating. It was found, surprisingly, that respiratory gating did not significantly improve the treatment. Zygmanski *et al* (2001c) studied the problem of IMRT delivered with a dLMC system in the presence of organ motion. They observed that gating effectively eliminates organ motion but extensively prolongs the delivery time, whereas tumour tracking, an alternative solution, requires real-time organ imaging. The dose error was studied for a periodically moving target resulting from IMRT delivered without gating or tumour tracking. IMRT plans for the pancreas and lung were made with HELIOS and transferred to a phantom which was firstly stationary and secondly resting on top of a motorized table executing periodic motion simulating organ motion with an amplitude of 1 cm and a periodicity of between 5–15 s. Although individual IMRT fields showed dose differences between 10% and 30%, composite five-field plans exhibited dose differences of between 5% and 15% indicating the possibility that dMLC IMRT treatments can be used with caution without gating under conditions of normal breathing. Mechalakos *et al* (2004) also observed that, with suitable margins, the majority of lung tumours are covered during conventional lung radiotherapy despite movement although it was conceded that large respiration-induced motion requires further strategies for correction.

6.10.8   Robotic feedback

The development of CFRT and IMRT has provided the specific challenge of the problem of irradiating the moving tumour. There is a considerable history of research into techniques to monitor tumour movement. Many of the ways to improve therapy concentrate on either gating radiation therapy to specific periods in which the tumour is known to be in the same position or treating the envelope of tumour movement. The latter approach involves the use of margins which undoubtedly over-irradiates normal tissue.

A dramatic new development is robotic radiation therapy in which, instead of the use of a conventional linear accelerator and couch, the in-line high-fluence linear accelerator is carried on a industrial robotic arm and can point with six degrees of freedom towards the tumour (see section 4.1).

It has been pointed out (Webb 1999, 2000e) that this particular form of radiation therapy is particularly amenable to coping with the problem of tumour movement. If the movements can be tracked, they can be fed back to the robotic arm to effectively track the moving tumour. Schweikard *et al* (2000) have performed some pioneering experiments to correlate the position of a tumour as determined by stereoscopic x-rays with the position of the external skin surface as determined by infrared sensors. The rationale behind this work is to recognize that, ideally, one would like to monitor the internal position of tumour markers using x-rays throughout the irradiation but that this is not possible for reasons of unwanted radiation dose. Instead the 3D position of tumour markers is correlated with the 3D position of external surface markers to generate a mathematical relationship between the two. Then, throughout the treatment, the position of the external markers is monitored and this relationship used to re-compute high-frame-rate locations of the internal markers and feed this movement data back to the robotic arm for compensation. The technique is even cleverer than this. It can continuously update the relationship between internal and external markers to take account of changes which may gradually take place during the treatment. It can also distinguish between changes in tumour position that are due to breathing and internal movements from those which are due to overall global shifts in the patient's position due to coughing, twitching or other involuntary movements. This method of robotic radiation therapy has been developed for the Accuray treatment-delivery system at Stanford, CA (Accutrak).

Hough and Jones (2002) have developed a robotically controlled treatment couch and chair for patient positioning for proton therapy. They indicate that this would also be a suitable alternative to the current couch technology for photon technology.

Suh *et al* (2002) have alternatively observed that there is a strong correlation between the movement of the skin surface and the target motion for tumours in the lung. Therefore, they consider it is possible to predict the exact target location from the skin motion which will be useful to aid gated radiation therapy. Ten patients were imaged lying down on the simulator with radio-opaque markers on their skin and the fluoroscopic images were recorded simultaneously.

Yan *et al* (2004) correlated the motion of infrared passive skin markers detected with the BrainLAB ExacTrac system and fluoroscopically-detected motion of internal markers. Provided the phase shift was found, there was a good cross-covariance.

6.10.9   Held-breath self-gating

Another technique to control tumour movement in the upper thorax is to irradiate only at one specific phase of the breathing cycle with the patient themselves controlling this. Ideally, this also requires 3D imaging of the patient at this same breathing phase. Forster *et al* (2003) have shown considerable differences in lung position determined from free-breathing CT scans and from gated CT scans. Rietzel *et al* (2003) have also shown that gated CT ('4DCT') leads to clearer definition of specific breathing phases.

Mah *et al* (2000) have evaluated technical aspects of the deep-inspiration breath-hold technique in the treatment of thoracic cancer. In this technique, the patient was verbally coached through a modified slow vital-capacity manoeuvre and brought to a reproducible deep inspiration breath-hold level. The goal was to immobilize the tumour and also to expand normal lung out of the high-dose region. It was inferred from data on the first seven patients that the displacement of the centroid of GTV from its position in the planning scan over some 350 breath-holds was only 0.02 ± 0.14 cm. In this particular technique, the patient held their breath for up to 10 s during a high-dose-rate irradiation. The manoeuvre for deep-inspiration breath-hold as practised at Memorial Sloan Kettering Cancer Center cannot, however, be performed by 60% of patients with lung cancer (Mageras 2004).

Kim *et al* (2001) have evaluated the feasibility of a held-breath self-gating technique in radiotherapy of lung cancer. Sixteen consecutive patients with non-small-cell lung cancer were accrued for the study and the patients were asked to hold their breath at four points in the breathing cycle after a standardized training. These were: maximum and end tidal, inspiration and expiration. While under fluoroscopic visualization, it was found that maximal inspiration and expiration tended to provide the best positional reliability and the standard deviation of diaphragmatic position ranged from 0.13 to 2.57 mm with an average of 0.97 mm. The day-to-day variation of diaphragmatic position was less than 5 mm and the held-breath self-gating technique resulted in a reduction of diaphragmatic movement by an average of 11.9 mm when compared to that seen with the tidal breathing.

Barnes *et al* (2001) have studied 10 patients undergoing radiotherapy for stage 1-3B NSCLC with and without deep-inspiration held breath. This technique of 'immobilizing' the tumour is self-gating with respect to the radiation. With appropriate margin selection, the use of deep-inspiration breath-hold reduced the percent of lung receiving more than 20 Gy from 12.8% to 8.8%. Patients showed a remarkable variability in the duration of breath-hold (19–52s). Della Biancia *et al* (2003) have shown that IMRT at end inspiration generated lower normal lung dose than IMRT at end expiration.

Onishi *et al* (2001) have taken a similar approach to controlling respiratory motion. They asked the patient to hold their breath at deep inspiration and provided the patient themselves with a switch which directly controlled the output of the linac. This was argued as having very high time-efficiency because the patient only operated the switch for the period in which they could hold their breath.

Shimoga *et al* (2001) have developed SmartGate. This is a standard stethoscope placed on the neck with a radio transmitter coupled to a receiver outside the treatment room. The equipment is capable of distinguishing between different parts of the breathing cycle and, in particular, permits gating-out times in which the patient is breathing abnormally or when the patient's breathing is interrupted by nasal urges such as coughing and sneezing. The system uses an expert rule base and human-like fuzzy-decision capability.

Ozhasoglu *et al* (2003) have shown using the Cyberknife's imaging system that natural breath hold can yield lung tumour positions reproducible to within 2 mm. Ma (2003) has also commented on the positioning accuracy of the Cyberknife.

6.10.10   Intervention for immobilization

Lofroth *et al* (2000) have used a urethral catheter to localize the prostate with high precision during treatment. The technique has been used with more than 250 patients and allows margins to be reduced without external technical support.

Soubra *et al* (2001a, b) treat the prostate with the patient prone, lying in a vacuum bean bag, inside a specially designed wooden cradle. The prostate movement was also reduced by inserting a balloon into the rectum to a specified location, inflated to 50 cm^3 of air and patients were treated with the bladder full. It is claimed that this leads to a reproducibility of patient position to better than 5 mm.

6.10.11   Active breathing control

The first active breathing control (ABC) device built at the William Beaumont Hospital, Royal Oak, was based on a modified Siemens servo ventilator 900C (Wong 2003). The second device was somewhat simpler and smaller with a single-valve system, a single flow path for inhalation and exhalation and a single flow monitor. Patients undergo a training session to determine the individual tolerance to breath-hold and the length of time for which they can comfortably

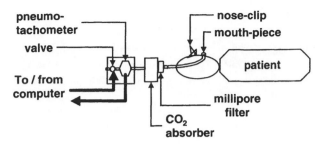

Figure 6.50. A schematic representation of the active breathing control (ABC) apparatus showing a single-valve system with a single flow path interfaced to a personal computer. (Reprinted from Stromberg *et al* 2000 with copyright permission from Elsevier.)

hold their breath. They also view a real-time display of the changing lung volume as well as the intended breath-hold level. Once this is individually established, the ABC device is used to halt the patient's breathing for reproducibly fixed periods in the breathing cycle.

The technique of ABC has been used to study the reduction in normal tissue complications with deep inspiration when treating patients for Hodgkin's disease (Stromberg *et al* 2000) (figure 6.50). The clinical goal was to attempt to reduce the dose to the normal lung and to the heart without compromising the treatment of the tumour. This is desirable given that patients generally survive Hodgin's disease but can then progress to suffer from late normal-tissue radiation complications including secondary cancers, cardiovasular disease and pulmonary dysfunction. ABC may also reduce cardiac overlap by moving the spleen inferiorly with deep inspiration.

Five patients underwent CT scanning of the chest and abdomen using the ABC apparatus. Scans were taken at normal inspiration (NI), normal expiration (NE) and deep inspiration (DI) in the supine position. DI scans were also repeated during and at subsequent weeks to assess the intrafraction and interfraction reproducibility. Stromberg *et al* (2000) described the CTVs and PTVs defined at NI, NE and DI. A composite PTV was also defined based on the range of NI and NE CTVs. Planning on an ADAC PINNACLE system used conventional AP and PA beams.

The assessment of normal-tissue damage was made using dose-volume histograms for the heart and dose-*mass* histograms for the lung, the latter to remove the effect of changing air volume at different stages of the breathing cycle (which is irrelevant in calculating tissue damage). These dose-volume/mass histograms were compared at the several respiratory phases.

ABC was found to be well tolerated with DI breath-holds ranging from 34 to 45s. Dose-mass histograms for all five patients showed a median reduction of 12% lung mass irradiated at DI compared with free-breathing. Advantages were also found for breath-hold using ABC at NI and NE compared with free breathing but these were less significant. All patients showed a decrease in heart–spleen overlap with DI of at least 1 cm. Cardiac DVHs showed the mean volume of heart irradiated at the 30 Gy dose level decreased from 26% to 5% with DI compared with free breathing. Analysis showed intra- and inter-session variabilities of only 5%. It was concluded that all ABC-controlled treatments led to improvements with DI providing the greatest benefits.

Aznar *et al* (2000) have used the ABC device in conjunction with IMRT for improving the sparing of heart in left-breast radiotherapy. Remouchamps *et al* (2002, 2003) have also shown how the use of deep-inspiration breath-hold ABC can lead to lower cardiac doses for left-breast-irradiated patients.

Sixel *et al* (2001) have investigated the use of deep-inspiration breath-hold during tangential breast radiation therapy as a means to reduce cardiac irradiated volume. The ABC device was used for five left-sided breast cancer patients. It was found that the use of the ABC device reduced the dose to the heart for some patients when deep-inspiration breath-hold was applied. The magnitude of the impact of the breath-hold technique depended on the patient anatomy, lung capacity and pulmonary function. This was determined by creating plans based on CT scans which were acquired with and without the breath-hold and subsequent virtual simulation for regular tangent and for wide tangent techniques.

Donovan *et al* (2002a, 2003) has used the ABC device for four patients receiving tangential pair radiotherapy to the left breast. Patients are able to achieve a 15–20 s breath-hold for 6–8 repeat breath-holds with 75% of the maximum lung capacity and the heart was excluded from the treatment fields in all cases. The OSIRIS system was used to acquire outline information with and without the ABC to assess shape changes and reproducibility of position. At the same centre, McNair *et al* (2003a) have measured the mean breathold time as 20 s (range 10–25 s) for ten patients with lung cancer. The mean breath-hold levels were 72% of the maximum inspiration volume. Christian *et al* (2003a) reported good tolerance of the device. A review of the application of ABC at this centre is in *Wavelength* (2003d).

Wilson *et al* (2001) have used the ABC device for treating non-small cell lung cancer with CHARTWEL. Five patients with locally advanced NSCLC underwent three consecutive planning CT scans, one breathing, and the second and third using ABC. ABC was found to be tolerated for breath-holds of between 20–30 s and the CT lung volumes were reproducible. PTVs could be made smaller using ABC with the median reduction of approximately 50 cm^3. Wilson *et al* (2004) reported on a longer study and showed that the use of the ABC device also led to increased lung and spinal cord sparing.

Dawson *et al* (2000) have assessed the use of the ABC device for improving the treatment of liver cancer. Over 100 fractions of radiation have been delivered using ABC in four patients. The average time for each breath hold was 25 s in a range 10–45 s. The motion of the diaphragm and hepatic markers were detected on fluoroscopy during the ABC breathhold and found to be reproducible to within 3 mm over the short term. The average long-term reproducibility of the diaphragm position increased to about 6 mm indicating a drift of breath-hold liver position during the course of treatment. The technique relied on imaging liver micro coils, which have been inserted during hepatic artery catheter placement.

A company perspective on the ABC (C for 'control' has become C for 'coordinator') device has been written (*Wavelength* 2002a). A description of the clinical implementation in two centres (William Beaumont Hospital and Mount Vernon Hospital) appeared in *Wavelength* (2002b) and at Thomas Jefferson University (TJU) and the Royal Marsden NHS Trust in *Wavelength* (2003d). Wong *et al* (2001) have shown how a combination of ABC and respiratory gating can offer an improved approach to target volume immobilization during treatment using an Elekta machine. The TJU DAO method is incorporated in the Elekta PrecisePLAN treatment-planning system (*Wavelength* 2002c). The use of the ABC device for guiding extracranial stereotactic radioablation is described in *Wavelength* (2002d).

6.10.12   Calculating the effect of tissue movement

6.10.12.1   Incorporating movement knowledge into the inverse planning itself

Li and Xing (2000a, b) have discussed the problem of accounting for random organ movement in IMRT. Firstly, they proposed a method to assess the outcome of movement applied to a plan computed without accounting for tissue movement. Call this plan $D_f(n)$ where n labels patient voxel and f represents fixed anatomy during inverse planning. If voxel n has probability $P(n, n')$ of moving to voxel n', then the dose received by voxel n is

$$D(n) = \sum_{n'} P(n, n')D_f(n').$$ (6.7)

The dose-volume histogram computed from $D(n)$ then gives the outcome of incorporating organ movement into the plan computed without organ movement.

A better approach, however, is also suggested in which the objective function includes the effects of movement, i.e.

$$obj = \frac{1}{N} \sum_{n=1}^{N} r_s[D(n) - D_0(n)]^2$$ (6.8)

where $D_0(n)$ is the prescribed dose and r_s the IF for structure 's' and N is the total number of dose-calculation points. Minimizing this cost function minimizes the difference between the motion-smoothed dose distribution (from equation (6.7)) and the prescription. The algorithm was incorporated within the PLUNC treatment-planning system and the simultaneous iterative treatment planning (SITP) iterative algorithm was used for cost-function minimization. Li and Xing (2000a, b) applied this to two clinical cases describing the motion using a 3D Gaussian distribution function. It was shown that the second approach gave better critical-organ sparing than the first for much the same PTV coverage (figure 6.51).

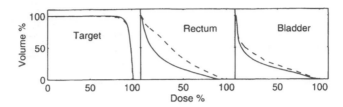

Figure 6.51. The CTV, rectum and bladder DVHs in two different cases: case 1, the DVHs of the treatment plan taking into account the spatial uncertainties of the structures during the optimization with the method described (full lines); case 2, the DVHs of the treatment plan by requiring a uniform dose to the PTV (CTV plus 1 cm margin) and zero dose elsewhere (broken lines). (From Li and Xing 2000a.)

Xing *et al* (2000b) have investigated the effect of respiratory motion in the design of IMRT treatment for breast cancer. The PLUNC treatment-planning system was used. It was found that if the effects of motion of the breast were not included in the static plan and then this plan were applied to geometry in which the breast motion was 'switched on', the dose received by the breast differed significantly from what had been planned in the static situation. For example, the minimum target dose to the CTV decreased by between 6–9% for all six cases studied and the volume receiving lower doses increased. As a result, the average dose in the clinical target volume was decreased significantly. However, if the motion were included in the inverse treatment planning, then the plans obtained with that approach improved the situation greatly and represented the truly optimal solution in the presence of the respiratory movement of the breast.

Loof *et al* (2001) have stressed the importance of including geometrical uncertainties in the optimization process for IMRT. These were simulated by convolving a treatment plan with the probability function representing expected movement. An alternative approach was to create IMRT plans with the HELIOS treatment-planning system in which the pencil-beam kernels had already been convolved with a movement kernel. The two approaches were expected to be similar. The authors then went on to point out that by using the second of these techniques *ab initio* the inverse planning could automatically take account of the motion of tissues.

6.10.12.2   *Use of multiple CT datasets and adaptive IMRT*

Using multiple CT scans to assess a plan made from a day-1 scan

Xu *et al* (2000) have performed multiple daily CT scans for 30 prostate patients with an average of 17 scans per patient. These scans have been registered using the bony anatomy. The plan produced for IMRT of the prostate on day 1 was then reapplied to the geometry obtained from the subsequent CT scanning occasions and, in particular, the change of dose to the rectum was evaluated. It was found that large dose-rectal-wall volume variations existed between patients whereas, by contrast, the dose-rectal-wall volume variation for any one patient was somewhat smaller. It was found that making use of the first few consecutive CT images improved the quality of treatment planning for rectal-toxicity-based dose escalation.

Bignardi *et al* (2000) have studied the effect of variability in the PTV over a period of fractionated treatment of the prostate. Eighteen cases were studied, each with a four-field box technique for CFRT. The CT scans were performed initially and at three and six weeks of treatment and the PTV was re-computed at three and six weeks from these subsequent scans. The beam parameters from the plan using the initial CT scan were then reapplied to the data obtained at three and six weeks. The complication probabilities for the target volume and for the rectum and bladder were remarkably constant over the six-week period.

Large *et al* (2001) have studied the effects of interfraction prostate motion on IMRT delivery. A series of CT scans was taken in the supine position immediately before and after the first three fractions. These scans were then registered. The targets were redrawn and the targets shifts were assessed. The dosimetric effects were seen to be as large as 10%.

Plasswilm *et al* (2002) made multiple CT scans of patients receiving IMRT for prostate cancer and showed that in two out of nine patients there would have been significant underdosage of the CTV if the plan from just the first CT scan rather than the plan from some averaged CT scan had been used.

Happersett *et al* (2003) have performed a study of the effects of internal organ motion on dose escalation in conformal prostate treatments. This study was performed for 20 patients. For each of the patients, a planning CT dataset was created and plans were made on this dataset. Three types of plan were made; the first was a conventional six-field plan with a prescribed dose of 75.6 Gy; the second was the same six-field plan escalated to 72 Gy followed by a boost to 81 Gy and the third was a five-field plan with IMRT delivering 81 Gy. For each of these patients and for each of these three types of plan, three more subsequent CT scans called 'treatment CT scans' were made. The scans were made on the first day of treatment, at the fourth week and at the completion of the treatment course. Each of the three treatment CT scans was then spatially registered in 3D to the planning scan using a chamfer-matching algorithm and the contours from the treatment scans were transferred to the planning scan. It was then possible to evaluate the planned dose distribution on contours representing organs in the treatment scan. No attempt was made to separate organ motion into systematic and random components and in the dose calculation, for what is known as the treatment results, rather than planned results, the dose from each treatment scan was assumed to contribute one-third of the total dose for the entire treatment (this may not be too representative). The study showed that for the planning scans, the TCP for plan types 2 and 3 was 9.8% greater than for plantype 1. Conversely for the treatment TCP, the increases were 9% between plans 2 and 1 and 8.1%

between plans 3 and 1. It was thus observed that the IMRT plan was more susceptible to patient movement. However, given that the TCP for the planned IMRT treatment was already higher than that for the other two treatments, dose escalation with IMRT was deemed to be possible. This is a good example of the use of multiple CT scans to assess the robustness of specific treatment plans to organ motion.

Hoogeman *et al* (2003) have used multiple CT scans (a planning CT scan and 11 repeat CT scans) to determine interfraction prostate motion and, from this, have been able to reduce the systematic error in IMRT of the prostate.

Schaly *et al* (2004) have developed a technique to actually track the dose to the whole 3D volume when the movement of individual voxels is computed from a series of CT scans. The William Beaumont CT datasets were used (15 CT scans for the same patient) and a thin-plate spline method was developed to create a map of the movement of all the tissue voxels. The application was to conformal (not IM) radiotherapy of the prostate. It was found that large (typically 30%) changes to the dose to the rectum occurred for the first fraction and that even after summing over all 15 fractions the rectal dose was still some 20% different from that planned. The prostate dose did not vary much. This is a very significant study since it is one of the first to actually track the voxel motion and compute effects.

Using multiple CT scans to create a composite target volume

McShan *et al* (2001, 2002) have developed the multiple instance geometry approximation (MIGA) which includes defining two or more instances of the patient geometry and optimizing the plan for all instances concurrently. This is a way of taking account of, at present, rigid body translations with respect to one particular planning geometry. Errors due to intrafraction movement are not included. The MIGA solutions are worse than the 'no motion' inverse planning but better than the latter convolved with motion. Fraass *et al* (2002) have used MIGA to improve IMRT for head-and-neck CTVs which are unusually near the skin surface.

Martinez *et al* (2001) have described an adaptive radiotherapy technique (ART) to customise margins of expansion from CTV to PTV for the individual patient. CT and portal images were recorded throughout week 1 of treatment and used to define a 'confidence-limited PTV' in which anisotropic margins were applied. The confidence-limited PTV, PTV_{CL}, was some 24% smaller on average than the conventional PTV. The achievable level of dose escalation is patient specific and, on average, it was found to be 7.5% for IMRT.

Pavel *et al* (2001), and Pavel-Mititean *et al* (2004) have studied the movement of the prostate and rectum during a course of treatment using nine sets of CT scans taken at intervals throughout the treatment for a series of prostate cases. The volumes of target, bladder and rectum were calculated and compared with the planning-scan volumes and, in particular, dose-volume histograms and dose-surface histograms for the mobile rectum were calculated and compared for plans made on each CT scan. A 'total' dose-surface histogram was then compared with the daily dose-surface histogram. It was found that the target volume changed much less than the changed volume of bladder and rectum. The largest translations were seen in the craniocaudal direction. When comparing dose-surface histograms from individual fractions with the planning case, the percentage of the OAR (rectum) surface receiving 60–80% of the target dose increased from 60–68% for conventional plans and from between 30–45% to 40–55% for IMRT plans.

Regarding time trends in prostate movement, Mechalakos *et al* (2002) have reported on a series of 50 patients each of whom had four CT scans during their course of treatment. The study investigated the time course of a large number of physical parameters associated with the planning. It was found that only two parameters showed significant changes, namely the total bladder volume (despite controls) and the increasing separation between bowel and PTV. No trends were observed for the prostate, seminal vesicles or rectum. This study had all patients prone. Its conclusions imply that repeat scanning of patients is not needed, somewhat at odds with other studies.

Using multiple CT scans to change the inverse-planning technique

Errors in the set-up of patients for radiotherapy, like all errors, can be both systematic and random. To estimate the systematic error, one has to take several measurements during the first few fractions of (say a 30 fraction) treatment. Bortfeld *et al* (2002d) have mathematically analysed and solved the problem of determining at which fraction an intervention should take place to minimize the overall systematic error. The idea is that for the first n treatment fractions the error is measured. At the $(n + 1)$th fraction, a correction is made, being the difference between the actual position of the target and the mean of the errors so far and this correction is then applied for all further fractions up to the total number of fractions. The question is what is the value of n? The authors wrote down expressions for the expectation value of the overall quadratic set-up error and then minimized that function. Making the assumption that both the actual and the individual positions are parts of distributions which can be arbitrary, with two different variances, they found that for 30 fractions and typical values of these variances the curve reached a minimum at about $n = 4$. The curve was very flat-bottomed indicating that the exact fraction of intervention was probably not critical.

Wu *et al* (2002a) have proposed a novel technique in which errors introduced into the delivery of the up to nth fraction can be corrected by re-optimizing the $(n + 1)$th fraction (and so on for all n). The method relies on repeated measurement of patient position and dose reconstruction at each fraction. Essentially the $(n + 1)$th optimization is of the total dose prescription minus the cumulative dose to the $(n + 1)$th fraction. Several alternative strategies were proposed. However, the technique assumes no deformable patient. The most appropriate apparatus capable of implementing this is the University of Wisconsin Tomotherapy Unit. Wu *et al* (2004) propose that the Cimmino algorithm can be used to rapidly compute the dose distribution for a warped geometry using the starting parameters from the unwarped geometry.

Birkner *et al* (2001, 2002) have incorporated image feedback into inverse planning. Patients with prostate cancer were sequentially CT scanned during their treatment and a model was made of daily set-up errors and organ motion and deformation. The information from CT scans over the first five days was fed back into the inverse treatment-planning process. Birkner *et al* (2001) incorporated organ motion and deformation into inverse planning (4D IGRT). The rectum is known to move during a course of treatment and motion-averaged dose-volume histograms were developed using a series of CT scans taken through the treatment time. These were then used for optimization. It was claimed that this reduces the high dose to the rectum. The future intention is to provide a seamless integration of HYPERION to ADAC PINNACLE.

Birkner *et al* (2003) have described on 'offline' method to perform adaptive radiotherapy to account for variations in set-up and internal organ motion throughout the treatment fractions. Planning was performed on an initial CT dataset regarded as a reference situation for the geometry. Then $(m - 1)$ further CT datasets are acquired. These m datasets were used to estimate the positions of all voxels in all structures as a fraction of time represented by a probability fraction for the tissue location and this plan was then applied for the remaining $(n - m)$ fractions. The final outcome depends on the value of m but Birkner *et al* (2003) showed that $m = 5$ gave a more conformal overall result (than $m = 1$) and that continuous adaptation beyond $m = 5$ made marginal difference.

Didinger *et al* (2001) have shown the advantages of using adaptive inverse planning during IMRT of prostate cancer. Weekly CT scans were used to replan the patients.

Daily repositioning based on daily CT scanning in the treatment room

Kneschaurek *et al* (2000) have generated daily CT images prior to irradiation of patients with prostatic cancer. These images were then transferred to the BrainSCAN treatment-planning system which automatically generated optimal leaf positions for the BrainSCAN microMLC. By registering the daily CT data sets with the first CT investigation, it was possible to generate a composite dose-volume histogram for the whole treatment, which took account of the movement and different filling of the rectum.

Ruchala *et al* (2002a) have presented a different way to reposition the patient on any given fraction that does not use external fiducial or bony anatomy repositioning. This is called contoured anatomy dose repositioning (CADR). Each time a new CT scan is created, the relevant targets and OARs are contoured. These contours are then compared with those for the planning CT images and patient repositioning is based on the use of these contours. Ruchala *et al* (2002a)

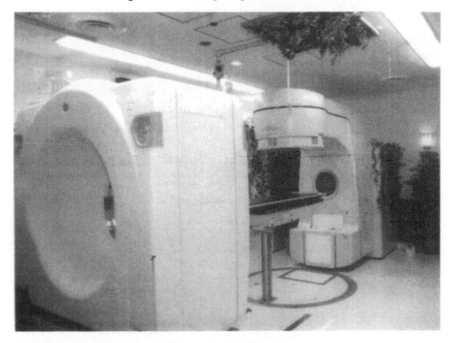

Figure 6.52. A panoramic view of the integrated CT–Linac irradiation system. A linac gantry is placed on the opposite side of the CT gantry. Between these gantries, the common treatment couch is placed. (Reprinted from Kuriyama *et al* 2003 with copyright permission from Elsevier.)

describe the advent of onboard 3D imaging for radiotherapy, which promises to vastly increase knowledge of patient anatomy at the time of treatment as an example of a conundrum called Cassandra's complex of Greek mythology. Cassandra was granted knowledge of the future yet was powerless to change it. The CADR technique is a first step towards optimizing this problem.

Kuriyama *et al* (2003) have designed an integrated facility with the CT scanner in the same room as the linac. The CT scanner moves on rails relative to the patient on the couch rather than the more usual *vice versa* (figure 6.52). The centre third rail carries a magnetic strip with position information ensuring the patient position is correct to better than 0.4 mm for stereotatic radiotherapy (figure 6.53). The arrangement also permits verification of the position of targets for CFRT. Paskalev *et al* (2003, 2004) have designed a technique to correlate such in-treatment-room CT images with planning CT images and to generate the required shifts to align the two on a daily basis. The correlation is based on target information only; no contours need to be defined. It was shown that the method is robust to noise and as good as a method based on image fusion. Reported shifts in the position of the prostate were up to 9 mm.

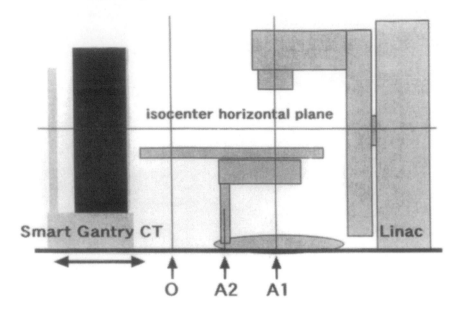

Figure 6.53. Diagram of the integrated CT-Linac irradiation system. The treatment couch has two rotation axes: A1 is for isocentric rotation to make non-coplanar arcs; and A2 is for the rotation between CT and linac. The gantry axis of the linac is coaxial with that of the CT scanner. A special self-driving device is equipped on the bottom of the conventional CT gantry so that the gantry, not the couch, moves when scanning. (Reprinted from Kuriyama *et al* 2003 with copyright permission from Elsevier.)

Dong *et al* (2004) have used the Varian ExaCT CT-on-rails system coupled to a Varian 2100 EX accelerator for in-treatment-room repositioning of the patient at each fraction. By deliberately creating shifts, it was found that the system could correct them to better than 1 mm. The system was used to show, and correct for, GTV shrinkage through a course of treatment as well as to correct interfraction variations in position.

Progress in MVCT and kVCT will be reviewed in section 6.11.

Adapting the delivery of IMBs to take account of interfraction motion

Ruchala *et al* (2002c) have introduced multi-margin optimization with daily selection for image-guided radiotherapy. The study was conducted retrospectively using 17 daily CT scans acquired during a prostate treatment . PTVs were identified with a variety of margins and IMRT were generated for these PTVs (four of them). For each fraction, the most appropriate plan was chosen corresponding to the daily position of the prostate. It was shown that this led to decreased rectal-and-bladder dose.

6.10.12.3 *Modelling the effect of intrafraction movement*

CT scans at different acquisition speeds and at different phases of the breathing cycle

Because lung tumours move during radiotherapy with respect to the beams applied, the determination of lung mobility is a key research area. Van Sörnsen de Koste *et al* (2003b) have made a study of this subject by collecting, for each patient with non-small-cell lung cancer, three rapid and three slow planning CT scans. They then computed a so-called optimum CTV as the envelope of the CTVs determined for all of these six scans. The six scans were co-registered using contour-matching software ACQSIM. The relationship was studied between the CTVs determined from the slow scans, the rapid scans and the optimum scan and, not surprisingly, it was found that this ratio was less than one. The ratio had a mean of 0.68 for the rapid-to-optimal CTV and 0.78 for the slow-to-optimal CTV, thus indicating that the position of the CTV as determined from the mean of the slow CT scans was closer to that regarded as optimal. The key observation was that by applying a standard margin of 5 mm symmetrically to the slow-scan CTV, the optimal CTV would be covered. There was no correlation, however, between tumour mobility and tumour location. They thus concluded that individualized procedures were required for each patient.

Allen *et al* (2004) performed a similar study. Three CT scans were made at end expiration, end inspiration and free breathing. GTVs were defined on each scan. It was found that the application of a margin to the GTV on the free-breathing scan did not adequately cover the tumour position as determined from the end tidal scans. In some cases, areas of tumour were geographically missed and, in other cases, too much normal lung was included.

Keall *et al* (2004) have coupled the Varian RTM infrared system to a 16-slice helical CT scanner to organize projections according to phase in the breathing cycle. Since the CT scanner was 'expecting' a cardiac pulse for triggering, the pulses from the infrared system needed to be shortened for this gating procedure. Then images were reconstructed of different phases. This demonstrated the system worked well for a moving phantom and also for a patient undergoing audio-visual breathing coaching during scan acquisition. Images of the moving lung tissue were presented. However, when the breathing pattern was irregular, missing projection data led to reconstruction artefacts. It was also demonstrated that 'slow CT scans' could be reconstructed by ignoring the phase tag and that the slow scan was equivalent to the scan at any individual phase of a stationary phantom.

Kim *et al* (2003) performed experiments to show that the definition of target was too small when the total time of the CT scan was less than one breath time. The target was correctly defined with a CT scan of exactly one breath period and did not change if the CT scan time thereafter increased.

Frazier *et al* (2000, 2004) have evaluated the impact of respiration on whole-breast radiotherapy using either wedged pairs or IMRT at the William Beaumont

Hospital. CT scans were taken with the patient breathing normally, a so-called free breathing (FB) scan and, at the same time, the ABC apparatus was used to obtain two additional CT scans with the breath held at the end of normal inspiration (NI) and normal exhalation (NE). The plan produced using the FB scan was then copied and applied to the NI and NE scans. Somewhat surprisingly, the dose distribution to the whole breast was observed to change very little between the FB, NI and NE scans for both the wedge technique and IMRT indicating that the dose distributions delivered using multiple-static MLC segments are relatively insensitive to the effects of breast motion during normal respiration. However, this is not true when considering the nodes which may be designed to be excluded during the FB irradiation but turn out to receive significant dose when the same plan is applied to the CT scans taken at normal inspiration and exhalation.

Studies of the effect of movement for a single fraction

Holmberg *et al* (2000) have studied the effect of target movement in step-and-shoot IMRT via numerical simulation. The intrafraction position of a target element in the thorax region was modelled as a function of time through an asymmetric cosine function with a bias towards the exhale position. It was found that the maximum simulated exposure time in a moving target was as high as 170% of the predicted exposure time in a static target when the short-time segment covers the inhalation position and the long-time segment covers the exhalation position. With opposite positions, the calculated maximum reduced to 130%. However, this demonstrates that intrafraction movement can seriously affect IMRT of the thorax for a single fraction of IMRT.

Kung *et al* (2000c) and Zygmanski *et al* (2001d) have developed a technique to predict the dose error in IMRT treatment of lung cancer due to lack of account of respiratory movement. The IMBs from an inverse-planning system were converted into leaf positions for the dMLC device at each portal. Then, for the beam's-eye view of each portal, a 2D fluence grid was created and applied to the moving target volume. This enabled the time-dependent sub-fields to be ray-projected to create an effective incident fluence. This effective incident fluence was then returned to the Varian HELIOS treatment-planning system to produce the dose distribution that is actually received by the moving target at a single fraction. It was found that, without respiratory gating, target volumes could receive dose errors of up to 10% or greater over 15% of the target volume pointing to the need for respiratory gating in thoracic IMRT.

Ramsey *et al* (2001b) have considered that the dosimetric consequences of intrafraction prostate motion in IMRT could be 'disastrous'. They argued that IMRT only works by lining up very precisely different modulated fields and, if the prostate moves between these fields, then the dosimetry will go wrong. Ramsey (2002) has experimentally simulated the effects of respiration motion by driving the patient couch in a simulated respiratory motion relative to the delivery of IMRT using the dMLC technique. Using film dosimetry, they found differences between the dose distribution under these circumstances compared with that when the couch was static (i.e. simulating no motion). With respect to the studies of Bortfeld *et al* (2002a, b), it must be noted that this is simulating only a single fraction delivery and the effects should be less when averaged over a set of fractions (see next section).

Sohn *et al* (2001b) have studied the effect of motion on lung-cancer IMRT. Often the effect of motion on a dose distribution is modelled by convolving the static-patient dose distribution with the mathematical representation of the movement. Sohn *et al* (2001b) did this for a specific modulated field created by CORVUS. They then instead created an experiment in which they measured the dose distribution created by a modulated field when the detector was on an oscillating frame simulating motion. Interestingly, they found that this measurement differed from the convolution calculation by up to 15%. The discrepancy was nearly as large as the motion-induced artifact itself. It was thus concluded that motion convolution models inadequately predict the breathing-motion-induced dose fluctuations from IMRT and a more sophisticated model of the artefacts caused by breathing as a function of the radiation dose delivery sequence is warranted to enable development of motion-robust delivery algorithms. This observation does not disagree with the conclusions from Bortfeld *et al* (2002a, b) because it is a 'one-fraction' experiment and Bortfeld *et al* (2002a, b) explained that individual fraction dose distributions can differ from calculation. The dose distribution *averaged over many fractions* should not (see next section).

Pemler *et al* (2001) have studied the influence of respiration-induced organ motion on dose distributions in treatments using enhanced dynamic wedges. They studied the particular situation in which craniocaudal movement of a tumour began to beat with similar craniocaudal movement of a dynamic wedge for a Varian accelerator. In worse cases studied, the maximum deviation in MUs was 16% for a 60° wedge. It was commented that similar disagreements might result for IMRT with the dMLC technique.

Schaefer *et al* (2004) delivered five modulated fields, corresponding to a lung irradiation, to a moving phantom with the oscillation period of either 12 cycles min^{-1} or 16 cycles min^{-1} and an effective amplitude of 16 mm. The dose was measured at 18 points and compared with that received by a static phantom. The delivery technique was the static step-and-shoot (MSF) method. It was found that, even for a single fraction, the dose varied by no more than 5% when motion was occurring. It was therefore concluded that the damage due to motion would be even less with fractionated radiation therapy (see next section).

Studies of the effect of movement for the whole radiotherapy course

Bortfeld *et al* (2002a, b) have investigated the effects of intrafraction movement for IMRT delivered with a compensator, with an MLC and with a scanning beam (figure 6.54). A sinusoidal motion was modelled with an amplitude of 5 mm and *fractionation was considered*. To calculate the effect of movement, for the compensator the dose matrix was simply moved with respect to the compensator. For the MLC IMRT technique, the individual field patterns were delivered to the moving dose grid and similarly for the scanning beam. It was found that the expected delivered dose distribution was the convolution of the movement pattern with the distribution without motion, i.e. exactly the same as observed for conventional radiotherapy. So the effect of movement could be accommodated by the use of margins, as with conventional therapy. Additional effects specific to the IMRT delivery technique were small. This conclusion is largely as a result of the averaging of individual daily movement patterns over the complete course of (say) 30 fractions. Errors due to organ movement on a given day may be quite large but these will be averaged out over the course of the fractions and were not expected to have biological consequences. It was shown from theoretical considerations that the probability function for the relative deviation of dose from the expected value quickly becomes Gaussian as the number of fractions increases. The width of the Gaussian depends on the IMRT technique. However, this does not mean that attempts to reduce the effects of motion are wasted since margins are never a wanted phenomenon. They compromise conformality to CTV and any attempt to reduce them by the use of gating, tracking etc would be useful. The study was made assuming the dose distributions do not deform with breathing but just relocate relative to the modulation (see also extensions discussed by Bortfeld *et al* [2004]). Chang *et al* (2004b) disagree with the conclusions of Bortfeld *et al* (2002) and argue that deformable organ movement will cause a greater interplay between time-dependent delivery of IMRT to moving tissues and their movement.

Jiang *et al* (2002, 2003c) conducted experiments irradiating a motor-driven platform containing a phantom by different IMRT techniques. The platform oscillated with an amplitude of 2 cm and a period of 4 s to mimic the motion of a lung tumour to first order as a sinusoid. Measurements were made at eight positions in the breathing cycle, at two dose rates, for five fields from two patient plans and for a dMLC delivery and MSF delivery with either 10 or 20 intensity levels. The phase of the movement with respect to the start of delivery was varied. Hence, the experiment observed an 'interplay effect'.

Not surprisingly large dose variations (as large as 30%) were found for delivery of single fractions, but the FWHM of dose dramatically narrowed (to about 2%) when the average was taken over 30 treatment fractions. In the study of Jiang *et al* (2003c), the direction of movement was always at right angles to that of the leaf movement whereas in the study of Jiang *et al* (2002) four angles between MLC and phantom motions were simulated. No differences were observed between delivery techniques. There was no apparent dependence of interplay between the dMLC delivery mode or the angle between MLC motion and tumour motion. This is 'good news' for conventional lung IMRT but somewhat perversely may be 'bad news' for gated IMRT in which the delivery and the movement of the tumour is highly correlated. Also, despite the conclusions, it was recognized

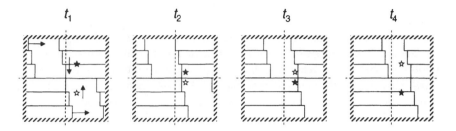

Figure 6.54. Illustration of the interplay between organ motion and leaf motion for the delivery of IMRT with an MLC. The leaves move from left to right. The star symbolizes a point in an organ that moves up and down. The two different versions of the star represent two different phases of the motion. (Let us say that the filled star represents exhalation and the open star represents inhalation.) Depending on the phase relative to the leaf motion, the point can receive very different dose values. In the phase shown by the filled star, the point dose not receive any primary dose between the time t_1 and t_4 and it may, in fact, receive no primary dose at all. In the phase symbolized by the open star, in which the point moves up between t_1 and t_4, it is treated with the full primary dose at all times. This example, of course, represents the extreme limit of possible dose variation and such extremes are certainly relatively improbable. (From Bortfeld *et al* 2002a.)

that modelling is limited and that tumour tracking will be preferable for CTV conformality.

Duan *et al* (2002) have delivered IMRT to a phantom which was set into linear sinusoidal motion with 1.5 cm longitudinal and 0.5 cm transverse amplitude to simulate respiratory motion. Dose-volume histograms show marked differences after one fraction, maxima and minima varying up to 20% from what was required. However, when five fractions were delivered with de-phasing between the fractions, the absolute errors decreased to about 5%. These results demonstrate that although respiratory motion may introduce substantial dose errors in a single treatment, multi-fraction treatments can significantly reduce these errors.

Kim *et al* (2003, 2004a) have computed the time-averaged dose-volume histogram (TDVH) for a treatment by convolving the dose distribution with the probability function for movement of targets and OARs. For a head-and-neck case, they showed that, with a standard deviation of movement of only a millimetre or so, the time-averaged DVH is not significantly different from the non-time-averaged DVH for the PTV although there are differences for the OAR.

Chui *et al* (2003a, b) explained how intrafraction organ motion interferes with the delivery of IMRT by the dMLC or shuttling MLC technique. If the motion is along the direction of leaf travel, then the primary fluence will not be calculated by a vertical-line intersection in the time–distance diagram of leading and trailing leaf patterns but instead by the intersection of the motion line with this diagram. If the motion is perpendicular to the leaf motion, the irradiation points may sample modulations from the adjacent leaves (figure 6.54). Chui *et al* (2003b) modelled the movement of targets using a function whose amplitude varied from ± 3.5 to ± 10 mm and whose period varied from 4 to 8 s. IMRT planning for three breast patients and four lung patients was considered. Calculations were made for each case with 30 randomly chosen phases and fluence maps were then calculated as the sum of these overall 30 phases to simulate the effect over the whole fractionated radiotherapy. Dose distributions were computed with these 'whole course averaged' fluence patterns. This whole procedure was repeated 10 times to estimate uncertainties. For breast patients (movement parallel to leaf movement), the statistics showed small variations about the mean attributed to the fact that the delivery averages over a very large number of periods. The high-dose portion of the DVH does not vary by more than 2% from the result with no motion. The low-dose portion is more affected. This implies the margins were sufficiently large that the PTV coverage was not unduly affected by motion. Conventional treatments (non-IMRT) behaved the same. Very similar results were obtained for the lung patients (movement-perpendicular to leaf movement) except that the low-dose parts of the PTV were more greatly affected. However, the CTV statistics showed almost identical coverage to the calculations without movement. If the margin cannot be increased, then some other intervention such as breath-hold or ABC must be used.

Rosu *et al* (2003) have estimated the effect on liver target and normal-tissue dose due to set-up error and organ motion relative to a planned dose distribution. Forty patients were studied who had had pre-treatment CT scans taken with voluntary breath-hold at normal exhalation. The effects of two forms of motion were computed using convolution techniques. The first form of motion was interfraction set-up variation which they represented using an anisotropic Gaussian kernel to describe the set-up uncertainties. The static dose distribution was convolved with the appropriate kernel to result in a new dose distribution that directly incorporates the effect of geometric uncertainties. Then this blurred dose distribution was further convolved with a 1D (superior-inferior direction only) patient-specific probability distribution function to describe breathing in which more time is spent at exhalation than at inhalation as described by the probability function

$$p(z) = \frac{1}{nb\pi \left(\frac{z_0 - z}{b}\right)^{\frac{2n-1}{2n}} \left[1 - \left(\frac{z_0 - z}{b}\right)^{1/n}\right]^{0.5}} \tag{6.9}$$

where terms are defined after equation (6.6). For the double-convolved dose plans, dose-volume histograms and recomputed normal tissue complication probabilities at the prescription dose were compiled. Also the prescription dose in the static plan was re-calculated at a nominal 20% NTCP level and then the dose required in the convolved plan to restore the NTCP level to 20% was re-evaluated.

As expected, the convolution led to blurred dose distributions creating penumbra regions at beam edges. It was found that the corrections due to breathing predominated over the set-up variations and the main result was that, as a result of breathing, hot spots emerged superior to the structures of interest and cold spots emerged inferior to the target volume compared with the static dose distribution. The minimum doses to the CTVs met or exceeded the minimum PTV doses from the static plans in all but one case. Detailed dose-difference distributions and tables of NTCP differences in Rosu *et al* (2003) need to be studied to understand the detailed changes in dose and complication probability. Finally, it was commented that convolution assumes the shape of the dose distribution is unaffected under small translations of the patient geometry and also assumes no change in the external contours or deformation of the organs. All of these assumptions are strictly untrue and work is continuing to develop a model of tissue deformation.

Modifying the IMBs to account for motion

Deng *et al* (2001c) have incorporated breast organ motion into the MLC leaf sequencing for IMRT. The leaf sequencer reads in the breathing pattern as a function of leaf-pair positions as if the breathing pattern were fed into the MLC controller. The MLC leaf trajectories were then adjusted based on the breathing motion and MC dose calculations were performed using MCDOSE at different phases of the breathing cycle. The inclusion of motion into the adaptation of the IMRT delivery effectively delivered a plan equivalent to that without organ-motion correction.

Deng *et al* (2002) have investigated the importance of breathing motion on IMRT treatment. They incorporated the effects of organ motion into the MLC leaf sequencing. When these components were applied to the moving patient, the dose distribution was more accurate than when the corresponding sequence, not accounting for movement, was applied to the moving patient.

Neicu *et al* (2002, 2003) have developed gated motion adaptive therapy. The original IMRT leaf sequence is modified to compensate for tumour motion based on the average tumour trajectory. The average tumour trajectory was derived by averaging tumour position at each phase of breathing. The radiation beam moves along the average tumour trajectory while the tumour moves according to the measured data.

Gierga and Jiang (2002) and Gierga *et al* (2003) have evaluated the effect of target motion in IMRT. Measured tumour trajectory data for lung patients were used and IMRT plans were created with a commercial IMRT treatment-planning system. A sampling algorithm was then used to generate the photon fluence maps that incorporate tumour motion using the original leaf sequence files and knowledge of the tumour trajectory. These modified fluence maps were then input to a MC dose calculation to calculate the 3D dose distribution. Different initial phases in the breathing cycle were simulated for 20 fluence maps and the corresponding dose distributions for 20 equally spaced initial phases were computed. The expectation value and variation of point doses and dose-volume histograms were then estimated and found to be different for each of the phases. Individual dose distributions could differ widely from those planned. The overall dose distribution was calculated by sampling randomly from the initial dose distributions for each phase of all fields. A distribution averaged over the phases differed much less from that planned. Gierga *et al* (2004) performed a similar study for liver tumours. Fluoroscopy of clips established the trajectories and IMBs for static plans were then re-applied to the moving targets. Perhaps surprisingly, it was found that liver motion did not greatly affect the distributions for many of the patients studied. Jiang *et al* (2003a, b) showed how to make a SMART (Synchronized moving aperture radiation therapy) plan for IMRT. Four-dimensional CT data (Low *et al* 2004, Wahab *et al* 2004b, Keall *et al* 2004) were used to optimize the dose delivery for each phase of the breathing cycle. Duan *et al* (2003a) showed that the errors introduced by motion did not depend on the number of fields making up the IMRT treatment. Jiang *et al* (2003b) also showed how to incorporate knowledge of the movement of tumour into the actual dynamic leaf pattern for IMRT.

The gating signal for 4D reconstruction can be obtained by the following methods:

(i)   the phase from an optical respiration monitor,
(ii) the displacement of an external fiducial marker or
(iii) spirometry-based tidal volume.

The main methods to use gated 4D CT data to create maps of individual voxel movement are:

(i)   use of thin plate splines (Rietzel *et al* 2004, Hartkens *et al* 2002, Malsch *et al* 2004),
(ii) use of optical flow equations (El Naqa *et al* 2004, Zhang *et al* 2004a),
(iii) use of viscous fluid flow equations (Mageras *et al* 2004b) or
(iv) finite element analysis (Brock *et al* 2004).

Rietzel *et al* (2004) have made 4D CT scans to show the patient geometry at 10 different respiratory phases. The Varian RTM system was coupled to a GE Lightspeed CT Scanner operating in axial cine mode. Using these data, Jiang *et al* (2004a) produced a plan that is optimized for each phase (but not overall) when there is no map of voxel displacements available. Then the set of 10 IMBs for any fixed gantry angle for the 10 phases can be sequenced by the same algorithm as used for IMAT because of the analogy between gantry angle and breathing phase, both being from 0° to 360°. This technique is called Synchronized Moving Aperture Radiation Therapy (SMART) (Jiang 2004). Alternatively, if a voxel displacement map can be made, e.g. by optical flow mapping (Malsch *et al* 2004, El Naqa *et al* 2004, Zhang *et al* 2004a, Huang *et al* 2004, Horn and Schunk 1981), then the objective function summed over all phases can be optimized to produce an IMB set. For any particular phase, the plan is then *not* optimal. Deformable alignment has also been done by biomechanical modelling and finite element analysis (Brock *et al* 2004) for liver and breast. Zhang *et al* (2004c) have created a similar technique to correct for breathing motion during tomotherapy. Again a voxel displacement map was computed from 4D CT data and the bixel intensities corresponding to the different breathing phases were computed. Then the breathing motion was monitored during the Tomotherapy and the appropriate bixels selected to 'track' the motion.

Kung *et al* (2003) folded a hypothetical sinusoidal tumour movement into the leaf patterns of a dMLC treatment for lung cancer and showed acceptable dose-volume histogram degradations even with one phase of motion. When phase averaging over fractions was included, the differences were even smaller. The method relied on the transformation whereby the patient (tumour) was considered stationary and its movement was instead transferred to the fluence delivery space of the moving segments.

Chen *et al* (2001c) have studied the movement of upper thoracic tumours and their deformation due to breathing and implications for IMRT. Jones and Hoban (2000) studied the influence of random movement of IMB positions relative to the patient and showed that a decrease in the EUD ensued.

6.10.12.4 *Modelling set-up inaccuracy*

Lomax (2001) has studied the effects of inter- and intrafraction positional errors in radiotherapy for both proton therapy and IMRT. Dose-volume histograms were computed for specific offsets between planning and image position.

Mock *et al* (2000) have computed the effect on prostate therapy of accounting for treatment set-up inaccuracy. Twenty patients were studied and evaluated using portal images. In all, 186 portal images were analysed. By feeding the known systematic and random deviations into the HELAX treatment-planning system and recalculating the dose distribution, it was found that the TCP averaged over all patients decreased by 19.7% from the anticipated (no movements included) value of 73.6%. The normal tissue complication probabilities for bladder and rectum, however, stayed almost unchanged at below 1% and around 5% respectively.

Booth and Zavgorodni (2001) have constructed a Monte Carlo model to simulate organ translations at the CT imaging stage and to evaluate the effects of this uncertainty on the dose distribution. The goal was to understand the effect of a mobile organ not being at its exact mean position at the time of imaging causing treatment to be planned with an organ offset from its assumed mean position. The study was limited to 1D movements and the study of single beams.

Baird *et al* (2001) have studied the effects of set-up uncertainty on dose distributions for intact breast-treatment techniques. They compared three types of treatment planning: (1) standard, paired tangential fields with wedges; (2) electronic compensation and (3) two-field IMRT. Beam isocentres were shifted to represent the average and the maximum set-up uncertainty and the dose was recalculated for each patient using the optimized fluence patterns. It was found that both the mean and the maximum patient movement noticeably affected the dose delivery for the IMRT plans but had no significant impact on the standard wedged plan or electronic compensation (see also Hector 2001).

Jolly *et al* (2001) also simulated the effects of organ motion with respect to tangential breast treatments and concluded that the heterogeneity of dose during the thoracic movement could lead to a greater probability of tissue complication due to overdosing the lungs.

Samuelsson *et al* (2001) have used the University of Michigan treatment-planning system to show the effects on IMRT of patient-position uncertainty during the treatment. The degrees of conformality and sharpness of the dose gradients had a large influence on the sensitivity of IMRT plans to set-up variations. Inclusion of set-up uncertainties into the optimization process can, in principle, correct dose distributions that would otherwise be degraded by the set-up errors.

Manning *et al* (2001) have investigated the effect of set-up uncertainty on normal tissue sparing with IMRT for head-and-neck cancer. The first approach they made was to investigate the effect of adding a planning OAR volume for the contralateral parotid gland and it was shown that this reduced the radiation dose to the contralateral parotid gland but slightly worsened the conformality of dose to the target volume. Then the worse-case scenario was investigated of systematically translating set-up uncertainties of 5 mm in six orthogonal directions without altering the optimized beam profiles. It was found that the plans without the planning OAR volume (PRV) were more susceptible to isocentre movement than the plans with PRV.

6.10.12.5   *Modelling the movement of OARs*

Manning *et al* (2000) have described the concept of PRV in which the effects of internal motion and set-up margin were added, not only to the PTV but also to the OAR. It was then shown that optimized nine-beam IMRT plans for three head-and-neck patients were unable to achieve their desired outcomes if the motions of OARs were not taken into account but were able to do so if the PRV was instead used. This is an interesting study in which the effects of movement are translated into changes in the dosimetry to OARs.

6.11   Megavoltage CT (MVCT) and kilovoltage CT (kVCT) for position verification

6.11.1   MVCT

Nakagawa *et al* (1994) have described how the detector constructed for megavoltage computed tomography (MVCT) can also be used for real-time beam

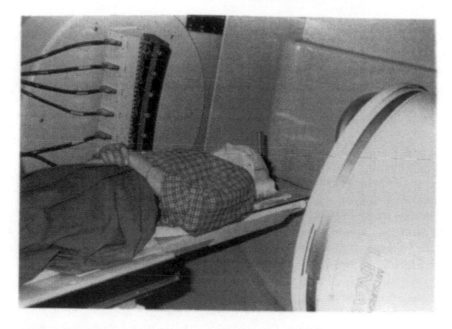

Figure 6.55. Image of the megavoltage CT system developed in Tokyo. (From Nakagawa *et al* 1994)

monitoring in dynamic conformation therapy. The MVCT system developed at the University of Tokyo in about 1991 (figure 6.55) was a fairly direct copy of the systems put together by Swindell and colleagues at the University of Arizona (Swindell *et al* 1983) and later at the Royal Marsden NHS Foundation Trust, Sutton. MVCT gives a picture of the patient in the treatment position just prior to treatment and this takes 38 s using the Tokyo system.

Dynamic conformation is a technique particularly popular in Japan in which, as the gantry rotates, the leaves of the MLC change their positions to define a new aperture shape at each 1° of gantry orientation. This does not involve intensity modulation but is a direct extension of some very early work by Takahashi in the 1950s (see figure 1.1). However, one of the problems with dynamic conformation therapy is that it is only possible to monitor mispositions in the leaf position with respect to planning and not with respect to the patient. The idea of Nakagawa *et al* (1994) was to use the same 1D detector as is used for MVCT to measure the field width for the centre pair of the MLC leaves. This was done by measuring the 50% fluence location to define the field width. This field width was then compared with that at planning and any mismatch would indicate an error. Deviations between actual and detected leaf openings have been found to be less than 2 mm for typical treatment fields. Some disadvantages of this system are that it only measures the field width for one pair of leaves and, therefore, is not a true 3D verification and, secondly, the quality of megavoltage imaging is not sufficient to delineate tumour and organ contours. Therefore, the beam positions can only be viewed with respect to planned beam positions rather than with respect to tumours.

Nakagawa *et al* (2000a) have shown how MVCT can be used to overcome treatment inaccuracy during stereotactic radiosurgery for thoracic tumours (figure 6.56). The MVCT unit comprises a detector array that is mounted on a linear accelerator at a distance of 160 cm from the beam source. A scan is performed during a half rotational irradiation taking 35 s. Spatial resolution is about 3.5 mm at the isocentre level. The slice thickness is 2 cm and the dose to isocentre is 2.8 cGy per scan. The MVCT is done after patient set-up on the accelerator couch and used to reposition the patient if necessary. The MVCT image is then re-used during treatment to superpose the beam, monitored using a real-time measurement on the detector, on to the MVCT image to show an accurate direction of the beam at the tumour. Nakagawa *et al* (2000a) demonstrate that this improves the reliability of conformal stereotactic radiosurgery and claim that the use of this MVCT-assisted stereotatic radiosurgery has improved the clinical outcome for thoracic neoplasms.

Ford *et al* (2002b) have used an amorphous silicon electronic portal imaging device to create MVCT images using doses between 10 and 70 cGy. Using just 28 cGy, contrast greater than 5% was distinguishable from background water in phantom measurements. This is equivalent to the use of just 35 MU.

Pouliot *et al* (2002) and Ghelmansarai *et al* (2003) have made similar studies and shown that MVCT is possible with just 15 MU showing visualization of bony anatomy structures adequate for patient positioning. Pouliot *et al* (2003) then argued that such MVCT images can be correlated to planning CT images to create the required relocation parameters for each fraction.

Guan (2002) has performed MVCT using 50 full projections and 50 truncated projections each with 1 MU and claims that, with an amorphous silicon 500 imaging device, it is possible to produce 1.5% contrast detectability. Nearly double the number of MUs is required for a CCD portal-imaging system also performing MVCT.

Hajdok *et al* (2002) have made MVCT images using a ^{60}Co source. They claim 3 mm high-contrast spatial resolution and 2.8% low-contrast sensitivity, linearity between MVCT number and electron density and an inherent absence of beam-hardening artefacts.

Partridge and Hesse (2002) have constructed MVCT images at the time of treatment based on the use of a portal-imaging device and then backprojected the primary fluence from the measured EPID signal through the MVCT image to construct the input fluence maps for IMRT which may be directly compared with the predicted fluence maps (see section 3.13.1).

Munro *et al* (2002) and Seppi *et al* (2003) have developed yet another piece of apparatus for MVCT (figure 6.57). This is a flat-panel imager comprising a scintillator of individual CsI crystals 8 mm thick by 0.38 mm × 0.38 mm pitch with five sides of each crystal coated with a reflecting powder/epoxy mixture and

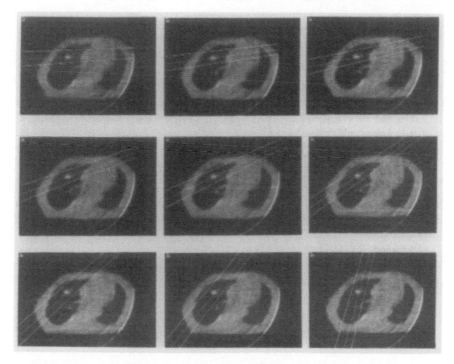

Figure 6.56. The figure shows a series of displays of a megavoltage computed tomography image of the thorax of a patient. Superimposed on this image are displays of the beam direction and shape displaying in real-time monitoring of the MLC for the rotational conformation technique. (Reprinted from Nakagawa *et al* (2000a) with copyright permission from Elsevier.)

the sixth side in contact with the flat-panel sensor. The flat-panel imaging system enables projections to be taken in just two radiation pulses which corresponds to 0.046 cGy dose. By collecting 360 projections, an MVCT image could be created by reconstruction using just 16 cGy of dose and reconstructing with the Feldkamp conebeam CT reconstruction algorithm. Experiments showed that the contrast resolution was about 1% for large objects and high-contrast structures such as air holes and embedded seeds could be seen with a spatial size of about 1.2–2.4 mm in size (figure 6.58). Images can be made with 1 MU per projection. Images with just a total dose of 16 cGy show soft-tissue structure such as the heart, lung, kidneys and liver in reconstructed images of an anthropomorphic phantom. This system bears great resemblance to that constructed by Mosleh Shirazi *et al* (1998). The authors commented on the lengthy 20-year history of the development of MVCT and they also suggest that, if their detector could be improved, it would be possible to image soft tissue such as the prostate. At the moment, this is not possible and also they comment that 16 cGy dose is probably still too large to be used on a daily basis if one is imaging the entire patient cross section. Compromises might include the performance of 'local tomography'.

Mageras *et al* (2004a) made MVCT images with the Varian 4030 HE EPID in cone beam geometry obtaining images with a density resolution of 2%. This detector has an 8-mm-thick pixellated CsI scintillation layer on top of the amorphous silicon photodiodes and a detective quantum efficiency (DQE) some 10 times that of copper-plus-Gadox systems. By gating with the Varian RTM imaging system, 4D CT is also possible.

6.11.2   Flat-panel imaging for kVCT

Jaffray (2002, 2003) and Jaffray *et al* (2002) have developed an imaging system that generates high-resolution soft-tissue images of the patient at the time of treatment with the purpose of guiding radiotherapy and reducing uncertainties (figures 6.59 and 6.60). It uses a flat-panel x-ray detector mounted opposite a kV tube on a retractable arm at 90° to the treatment source. A filtered-backprojection algorithm is used to perform cone-beam reconstructions and yield images 0.1 cm thick with an imaging exposure of 1.2 röntgens and giving a contrast of 47 Hounsfield units. These images, in principle, enable patient positioning at the time of treatment to be adjusted to correct for errors of alignment.

This was the basis of the Elekta iViewGT release 2 which supported IMRT via step-and-shoot. This is a 41 cm × 41 cm amorphous silicon detector projecting back to 26.2 cm × 26.2 cm at isocentre and with a 15 cm offset. It has a motorised retractable arm. Release 3 supports the flat-panel imager with a retractable arm and very good image quality. Image registration using the William Beaumont template mapping is possible. Chamfer matching with The Netherlands Cancer Institute software and seed recognition with the University of Utrecht software (see also *Wavelength* 2001d) is possible and it also supports cone-beam CT (see *Wavelength* 2003a).

The commercial product from Elekta is called the Elekta Synergy system (*Wavelength* 2003e, 2004) and has been developed and evaluated through a consortium of four centres:

(i)   William Beaumont Hospital, Royal Oak;
(ii) The Netherlands Cancer Institute, Amsterdam (van Herk *et al* 2002b);
(iii) Princess Margaret Hospital, Toronto;
(iv) Christie Hospital, Manchester (figure 6.61) (Williams 2003a).

Van Herk *et al* (2002b) have used kVCT on a treatment machine to localize the prostate prior to radiotherapy. To do this, they developed automatic contour-extraction methods and registration techniques so that the patient could be moved to the planning position. Artignan *et al* (2002a) give more details. Van Herk *et al* (2004) used kVCT images to obtain interfraction shifts to realign the patient.

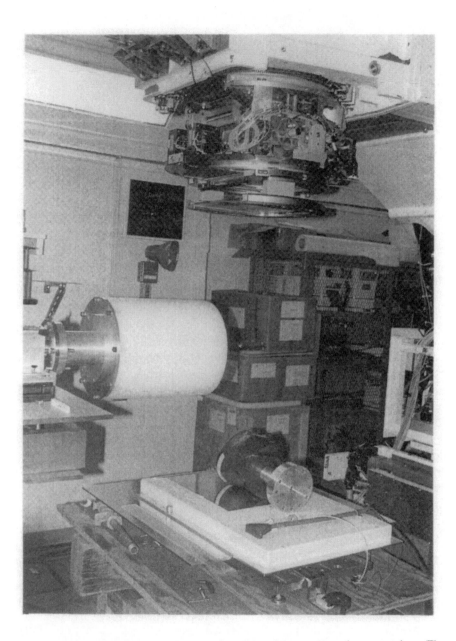

Figure 6.57. The experimental set-up for megavoltage computed tomography. The phantom was rotated with the x-ray source and the image receptor stationary. A cylindrical normalization (polyethylene) phantom is shown mounted on the rotating stage. (Reprinted from Seppi *et al* (2003) with copyright permission from Elsevier.)

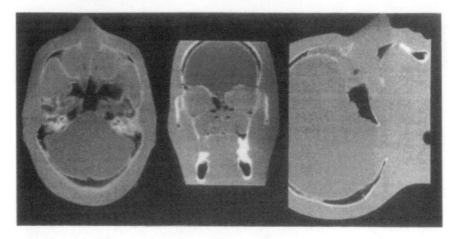

Figure 6.58. Axial, coronal and sagittal CT slices (5 mm thick) of the head phantom acquired at 6 MV using a 16 cGy irradiation. (Reprinted from Seppi *et al* (2003) with copyright permission from Elsevier.)

Zijp *et al* (2004) showed how to create the 'Amsterdam Shroud' from the kVCT cone-beam images which can be image processed to yield the diaphragm motion *without any markers* and so obtain the image phase respiratory signal.

Létourneau *et al* (2002) have shown that the Elekta/Jaffray kilovoltage portal-imaging system can be used for setup verification for radiotherapy.

Nüsslin *et al* (2003) have introduced image-guided radiotherapy into their HYPERION treatment-planning system. They have tools to adjust the contours within the planning system and to evaluate the effects of such adjustments on treatment outcome. These adjustments might be signalled if it were possible to re-image the patient between each treatment fraction. The use of an amorphous silicon flat-panel imager could assist such fraction-by-fraction adjustments.

6.12   MRI and IMRT simultaneously

Lagendijk and Bakker (2000), Langendijk *et al* (2002) and Raaymakers *et al* (2003) have developed the concept of a low-field intensity MRI unit integrated with a low-energy accelerator both with the same isocentre. The MRI unit is a 0.2–0.3 T machine. This enables MRI images to be taken of the patient in the treatment position and used as the basis for treatment planning and also MRI images to provide the basis for on-line position verification which is claimed to be far superior to megavoltage position verification. Raaymakers *et al* (2004) have completed the design study. Active magnetic shielding will uncouple the MRI system and the accelerator. Modifications to the coil windings reduce the attenuating path of the rays from the external accelerator to 10 cm of

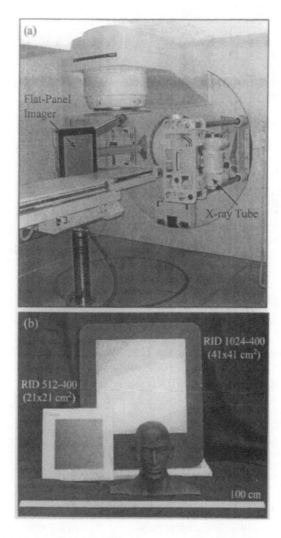

Figure 6.59. (a) A medical linear accelerator has been modified for kV cone-beam computed tomography. A kV x-ray tube has been mounted on a retractable arm at 90° with respect to the treatment source. A large-area (41 cm × 41 cm) flat-panel imager is mounted opposite the kV x-ray tube on a fixed mount. (b) A photograph of the large-area flat-panel detector employed in this investigation (black) in comparison to a smaller detector employed in previous investigations and an anthropomorphic head phantom. Labels indicate the pixel format and the pixel pitch of the imagers as well as the approximate sensitive area. The pitch is quoted in micrometres and the pixels are the number of pixels. (Reprinted from Jaffray *et al* (2002) with copyright permission from Elsevier.)

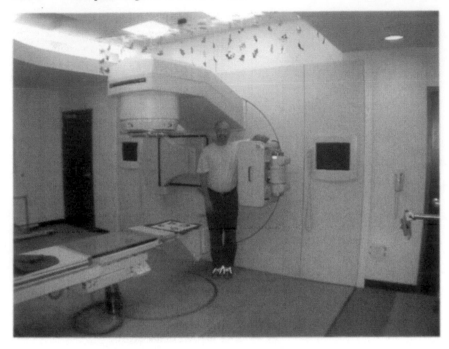

Figure 6.60. A small Hitchcockian cameo as the author props himself up next to the prototype kVCT machine at the William Beaumont Hospital, Royal Oak in October 2003. (Photograph by Dr Mark Oldham.)

aluminium. The prototype is being considered by a manufacturer but is not yet under construction.

6.13   IMRT using mixed photons and electrons

Svatos *et al* (2000) have demonstrated the advantages of using mixed beams of 21 MeV electrons and several configurations of intensity-modulated 6 MV photon beams. The addition of a single electron beam to the target reduced the integral dose two-fold.

Dogan *et al* (2001a) have studied how to match IMRT photon plans with electron beams using the CMS FOCUS treatment-planning system. It was found that, in order to avoid the production of either hot spots or cold spots, a very precise matching of the fields was required and the precise gap between the fields was specifically important. For the case studied, the idealized gap was 3 mm.

Dogan *et al* (2002a) noted that when IMRT photon fields and electron fields are matched, hot spots of up to 145% arose due to the bulging of the electron penumbra. They designed techniques to improve the field matching. These required the PTV to be extended into the region previously covered by

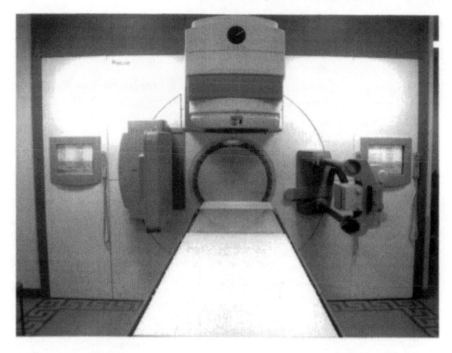

Figure 6.61. A view of the Christie Hospital's research kVCT machine showing the kV source mounted at 90° to the treatment head. The kV source is shown in the extended position and in line with the amorphous silicon flat panel. This is the system of Jaffray *et al* (2002) commercialized by Elekta Oncology Systems. (From *Wavelength* 2002.)

the electrons. The high isodose lines were then designed to bend away from the bulging electron penumbra. The electron penumbra was also further modified by using an MLC. Experimental measurements were performed abutting both tomographic IMRT delivery, step-and-shoot delivery (Varian accelerator) with the electron fields. The film calibration was performed using 'mixed photon and electron' irradiation to properly account for the different spectral effects. It was reported that, with no error in the field matching, the hot spots dropped to 105% in a test case and were still at only 115% when a 1.5 mm matchline error was deliberately introduced.

Korevaar *et al* (2002) have made comparative planning investigations to decide whether the additional use of high-energy electrons in photon IMRT can improve on predicted outcome. Studies showed that largely this was not the case. IMRT plans with and without electrons were very similar and it was recommended that high-energy electrons should not be used.

Ma *et al* (2003c) have compared tangential photon breast radiotherapy, IMRT and modulated electron radiotherapy (MERT) for breast cancer treatment. It was concluded that photon IMRT provided superior target homogeneity compared with conventional tangents but that MERT offered significant improvement over tangents for the plans inspected. The breathing margin could be significantly reduced because electron beams do not require direct tangential fields.

Xing *et al* (2003) have developed a technique to generate uniform doses from matched photon and electron fields. First, the electron plan was created and then the IMRT photon plan was created to dovetail to it along a matchline.

Epilogue

Consider IMRT as a train and the development of IMRT as like a long train journey. The train started in a cold siding a long way from its destination. This siding is where radiotherapy was in, say, the start of the 1980s. The destination is a station at which IMRT is at a state of development whereby it can be wisely but widely applied to all patients who need it, with quality assured and with due notice taken of all the potential impediments. The professionals developing IMRT certainly would like the train to be at this destination right here and now. So would the patients. They need to have a technology providing either a cure or, if this is unachievable, a high quality for their remaining life. Those who manage our health care—hospital managers, funders, fundraisers and politicians—also would like the train now in this station. However, the train is still some way from the destination.

En route, the IMRT train has stopped in several stations with mileposts. The first had the basic concepts of planning worked out (about 1990). The second had the fundamental concepts of IMRT delivery established but only in a few research centres (1994). The train got an extra locomotive in the late 1990s as companies began to market the needed equipment. So the train went faster and, by 2000, IMRT was a technology whose constituent elements one could largely purchase rather than create in-house (company-sponsored courses aided education and provided some interesting IMRT souvenirs [figure E.1]). At other stations, 3D medical imaging was coupled to the train. Also en route at stations the coaches which will carry the passengers have been upgraded. Some new coaches, like spiral tomotherapy, which weren't there at the start of the journey have been hooked up. Quality assurance, that at the start of the journey needed a person to tap the wheels of every coach before proceeding (time consuming), has been speeded up and now fewer checks need to be made per stop. However, from time to time at stations the train itself needs a larger inspection and overhaul. In short, the train that it is hoped will arrive at the destination is almost unrecognizably different from the one at the start of the journey. The trainspotting manuals (this series of books) have evolved to describe these landmarks.

The railway officials are the doctors, physicists, radiographers, engineers, all those professionals whose skills combine to move the IMRT train safely. Like railway officials of old, they work together, the only sensible way to ensure the

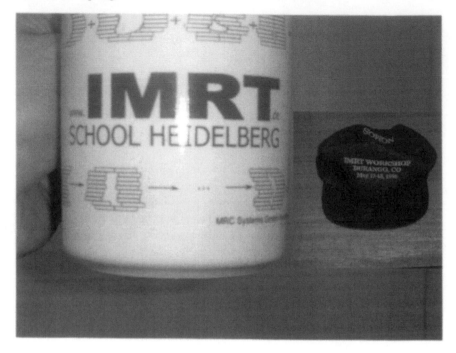

Figure E.1. Two examples of IMRT souvenirs produced by companies sponsoring IMRT teaching courses. The mug on the left was given to all the students at the MRC-DKFZ Heidelberg Schools. The hat on the right was worn by delegates at the first-ever IMRT course in Durango, CO in May 1996.

passage of the train. We should avoid any possibility that those who drive the trains separate from those who run the track. Research and development need to go hand in hand with clinical service. No one professional group owns the IMRT railway. Successful ongoing progress depends on the collaboration.

But there are 'leaves on the line' (the UK readers will appreciate the analogy) and the safety of the train has been questioned. We are quite adept at delivering highly conformal dose distributions to immobile phantoms and also to some patients in certain well-controlled situations. But patient movement makes up some of the leaves on the line. So does our growing awareness that, whereas once diagnostic identification of disease was superior to the ability to treat it, maybe now the reverse is true. The functional status of tissues needs to be established. We would like to know the distribution of clonogenic cells. We would like to establish the relationship between the disease which shows as changed x-ray attenuation and that which shows as changed function. These are more leaves on the line. But the train must not stop. Instead, with wet leaves, it must proceed cautiously whilst ways to sweep away the leaves or ride safely over them are considered.

Analogies pushed too far usually begin to falter and some will have seen some flaws in this one (since the journey is through time, passengers (patients) have been joining this train at different states of its development). So enough of trains. However, the message is clear. IMRT has come a long way from the days, not much more that ten years ago when the first patient was yet to be treated and IMRT was (as I have heard it unfortunately described) a 'physicists' toy'. The challenge is now to solve those final problems, then to roll out the technology as widely as possible (a) so as many benefit as require it and (b) to establish objectively through phase-3 randomized trials the benefit of IMRT. There will always be arguments of cost–benefit in a society with competing needs for so many different forms of healthcare. I believe IMRT will meet a clear need, lead to objectively assessable outcomes and justify the optimism of its developers and supporters. As I am no prophet and sadly also do have a pessimistic side, I expect it to throw up new questions and reach forms that we have yet not begun to imagine.

References

Aaronson R, DeMarco J, Chetty I and Solberg T 2001 A Monte Carlo model for quality assurance of intensity modulated radiotherapy incorporating specific leaf constraints *Med. Phys.* **28** 1268

Achterberg N and Mueller R G 2001 Multifocal static tomotherapy with multileaf collimation and table movement *Radiother. Oncol.* **61** S58

Adams E J, Convery D J, Cosgrove V P, McNair H A, Staffurth J N, Vaarkamp J, Nutting C M, Warrington A P, Webb S, Balyckyi J and Dearnaley D P 2004 Clinical implementation of dynamic and step-and-shoot IMRT to treat prostate cancer with high risk of pelvic lymph node involvement *Radiother. Oncol.* **70** 1–10

Adams E J, Convery D J, Warrington A P and Webb S 2001 IMRT planning at the Royal Marsden NHS Trust using TMS *Insights* **3** 1–2

Adams E, Vaarkamp J, Convery D J, Partridge M, Warrington A P and Webb S 2003 Refinement of a quality assurance programme for IMRT *Clinical Oncol.* **15** (Suppl. 2) S7

Adams E J, Vaarkamp J, Convery D J, Warrington A P, McNair H and Webb S 2002 Quality assurance for IMRT: a diminishing burden *Radiother. Oncol.* **64** (Suppl. 1) S125

Agazaryan N, Llacer J, Ullrich W, Promberger C, Solberg T, Arellano A and Paul T 2001 Deliverability scoring and dosimetric investigation of leaf sequencing in IMRT *Med. Phys.* **28** 1253

Agazaryan N and Solberg T D 2003 Segmental and dynamic intensity modulated radiotherapy delivery techniques for micro multileaf collimator *Med. Phys.* **30** 1758–67

Agazaryan N, Solberg T, Arellano A R and Paul T 1999 Leaf sequencing for fluence modulated radiation therapy *Med. Phys.* **26** 1139

——2000 Three-dimensional verification for dynamic multileaf collimated IMRT *Proc. World Congress of Medical Physics and the AAPM Annual Congress, August 2000* paper MO-BBR-02

Agazaryan N, Solberg T and Llacer J 2002 An investigation of deliverability complexity in intensity modulated radiotherapy *Med. Phys.* **29** 1197

Alaei P, Higgins P, Weaver R and Nguyen N 2002 Comparison of dynamic and step-and-shoot intensity-modulated radiation therapy planning and delivery using the Helios system *Med. Phys.* **29** 1270

Alber M 2001a Enhanced biological planning and expedited delivery of IMRT *Radiother. Oncol.* **61** (Suppl. 1) S75

——2001b Optimisation of IMRT under contraints for static and dynamic MLC delivery *Med. Phys.* **28** 1205-6

Alber M, Birkner M, Laub W and Nüsslin F 2000 Optimisation of IMRT under biological and practical constraints using Monte Carlo dose computation *Radiother. Oncol.* **56** (Suppl. 1) S79

Alber M, Fippel M and Nüsslin F 2001a IMRT optimisation with Monte Carlo dose computation including the MLC *Radiother. Oncol.* **61** (Suppl. 1) S64

Alber M, Meedt G, Birkner M and Nüsslin F 2002b Universal properties of the IMRT optimisation problem and consequences for the optimum number of beams and MLC field segments: conflict spectroscopy *Radiother. Oncol.* **64** (Suppl. 1) S96

Alber M, Meedt G, Nüsslin F and Reemtsen R 2002a On the degeneracy of the IMRT optimisation problem *Med. Phys.* **29** 2584-9

Alber M and Nüsslin F 2001 Optimization of intensity modulated radiotherapy under constraints for static and dynamic MLC delivery *Phys. Med. Biol.* **46** 3229–39

Aletti P, Marchesi V, Bey P, Lapeyre M, Beckendorf V, Marchal C and Noel A 2002 Clinical implementation of the IMRT at the A. Vautrin Hospital (Nancy—France) *Radiother. Oncol.* **64** (Suppl. 1) S216

Aletti P, Metayer Y, Bey P, Beckendorf V, Buchheit I, Noel A and Marchal C 2001 Comparison of IMRT and standard conformal therapy for the treatment of the cancer of the prostate *Radiother. Oncol.* (Suppl. 1) **61** S67

Al-Ghazi M, Dunia R, Sanford R and Radany E 2002 When is a combination of IMRT/conformal radiation therapy superior to either IMRT or conformal therapy alone? *Med. Phys.* **29** 1214

Alheit H, Dornfeld S, Winkler C, Blank H and Geyer P 2000 Stereotactic guided irradiation in prostatic cancer using the ExacTrac-System (BrainLab) *Radiother. Oncol.* **56** (Suppl. 1) S107

Allen A M, Siracuse K M, Hayman J A and Balter J M 2004 Evaluation of the influence of breathing on the movement and modeling of lung tumors *Int. J. Radiat. Oncol. Biol. Phys.* **58** 1251–7

Amols H 2003 Snake oil: Who's buying and who's selling? in regard to Glatstein: 'The return of the snake oil salesmen' *Int. J. Radiat. Oncol. Biol. Phys.* **56** 1507

——2004 In response to Drs. Nutting and Harrington (re: Glatstein/Amols, Snake Oil) *Int. J. Radiat. Oncol. Biol. Phys.* **58** 1317

André L and Haas B 2000 A comparison between intensity modulated treatment fields generated with sliding window algorithm and with the step and shoot algorithm *Proc. World Congress of Medical Physics and the AAPM Annual Congress, August 2000* paper WE-E313-05

Aoki Y and Yoda K 2003 A multi-portal compensator system for IMRT delivery *Med. Phys.* **30** 1347-8

Aoyama H, Shirato H, Nishioka T, Hashimoto S, Tsuchiya K, Kagel K, Onimaru R, Watanabe Y and Miyasaka K 2001 Magnetic resonance imaging system for three-dimensional conformal radiotherapy and its impact on gross tumor volume delineation of central nervous system tumors *Int. J. Radiat. Oncol. Biol. Phys.* **50** 821–7

Arellano A, Solberg T and Llacer J 2000 Clinically oriented inverse planning implementation *Proc. World Congress of Medical Physics and the AAPM Annual Congress, August 2000* paper WE-E313-02

Arnfield M R, Siebers J V, Kim J O, Wu Q, Keall P J and Mohan R 2000 A method for determining multileaf collimator transmission and scatter for dynamic intensity modulated radiotherapy *Med. Phys.* **27** 2231–40

Artignan X, Smitsmans M H P, de Bois J, Jaffray D, Ghilezan M, Fournerel P, Steenbakkers R, Lebesque J V, van Herk M and Bartelink H 2002a On-line image guided radiotherapy for prostate cancer: an accelerated method for prostate localization *Radiother. Oncol.* **64** (Suppl. 1) S288–9

Artignan X, Smitsmans M H P, de Bois J, Lebesque J V, van Herk M and Bartelink H 2002b On-line ultrasound-image guidance for radiotherapy of prostate cancer: the impact of image acquisition on prostate displacement *Radiother. Oncol.* **64** (Suppl. 1) S279

Artignan X, Smitsmans M H P, Lebesque J V, Jaffray D A, van Herk M and Bartelink H 2004 Online ultrasound image guidance for radiotherapy of prostate cancer: impact of image acquisition on prostate displacement *Int. J. Radiat. Oncol. Biol. Phys.* **59** 595–601

Aspradakis M M, Walker C P, Steele A and Lambert G D 2002 Accelerator commissioning for IMRT: small MU segments and small fields *Radiother. Oncol.* **64** (Suppl. 1) S62

Ayyangar K M, Li S, Saw C B, Yoe-Sein M, Pillai S, Chivukula V S and Enke C 2003 Comparison of a complex prostate plan using three IMRT planning systems *Med. Phys.* **30** 1487

Azcona J D, Siochi R A and Azinovic I 2002 Quality assurance in IMRT: Importance of the transmission through the jaws for an accurate calculation of absolute doses and relative distributions *Med.Phys.* **29** 269–74

Aznar M C, Pignol J P, Sixel K E and Benk V A 2001 Improving dose homogeneity in breast irradiation: a dose volume histogram study *Med. Phys.* **28** 1988

Aznar M C, Sixel K E and Ung Y C 2000 Feasibility of deep inspiration breath hold combined with intensity modulated radiation treatment delivery for left sided breast cancer *Proc. 42nd Annual ASTRO Meeting. Int. J. Radiat. Oncol. Biol. Phys.* **48** (Suppl.) 297

Backfrieder W, Fransson-Andreo A, Lorang T, Nowotny R, Dieckmann K and Poetter R 2000 Tissue classification of MR images for use in radiotherapy planning *Radiother. Oncol.* **56** (Suppl. 1) S207

BäckSamuelsson A and Johansson K A 2002 Head scatter on and off axis in small MLC defined beams applied for IMRT treatments with dynamic MLC *Radiother. Oncol.* **64** (Suppl. 1) S37

Baird C, Turian J, Myrianthopoulos L and Hamilton R 2001 Effects of setup uncertainty on dose distribution from intact breast treatment techniques *Med. Phys.* **28** 1292

Bakai A, Alber M and Nüsslin F 2002 Fast quality assurance for highly modulated IMRT beams employing multiple point measurement acceptance criteria *Radiother. Oncol.* **64** (Suppl. 1) S211

Baker C, Wolff T and Fenwick J 2002 Correction of monitor units for head scatter contributions in asymmetric IMRT fields from a commercial treatment planning system *Radiother. Oncol.* **64** (Suppl. 1) S36

Baldock C, Murry P and Kron T 1999 Uncertainty analysis in polymer gel dosimetry *Phys. Med. Biol.* **44** N243–6

Balog J B, Caldwell C B, Ung Y C, Mah K, Danjoux C E, Ganguli S N and Ehrlich L E 2000 Interobserver variation in contouring gross tumour volume in carcinoma of the lung associated with pneumonitis and atelectasis: The impact of 18FDG-hybrid PET fusion *Proc. 42nd Annual ASTRO Meeting. Int. J. Radiat. Oncol. Biol. Phys.* **48** (Suppl.) 128

Balog J, Olivera G and Kapatoes J 2003 Clinical helical tomotherapy commissioning dosimetry *Med. Phys.* **30** 3097–106

Balter J 2003 Demonstration of accurate localization and continuous tracking of implantable wireless electromagnetic transponders *Med. Phys.* **30** 1382

Balter J, Brock K, Kashani R, Coselmon M, Kessler M and McShan D 2003a The evolution of a dynamic lung model for treatment planning *Radiother. Oncol.* **68** (Suppl. 1) S41

Balter J M, Litzenberg D W, Brock K K, Sanda M, Sullivan M, Sandler H M and Dawson L A 2000 Ventilatory movement of the prostate during radiotherapy *Proc. 42nd Annual ASTRO Meeting. Int. J. Radiat. Oncol. Biol. Phys.* **48** (Suppl.) 167

Balter J, Wright N, Dimmer S, Friemel B, Newell J, Cheng Y and Mate T 2003b Demonstration of accurate localisation and continuous tracking of implantable wires *Int J. Radiat. Oncol. Biol. Phys.* **57** S264

Balycyki J, Cattell A, Bloomfield D and Powley S 2001 Commissioning radiotherapy equipment—a multidisciplinary approach *RAD Magazine* **27** (233) 21–3

Bamber J C, Verwey J A A, Eckersley R J, Hill C R and ter Haar G 1996 Potential for tissue movement compensation in conformal cancer therapy *Acoustical imaging* vol 22, ed P Tortoli and L Masotti (New York: Plenum) pp 239–44

Bär W, Alber M, Birkner M, Mondry A, Bakai A and Nüsslin F 2001a Implementation of a step and shoot IMRT concept under consideration of clinical and dosimetric aspects *Med. Phys.* **28** 1202

Bär W, Alber M and Nüsslin 2001b A variable fluence step clustering and segmentation algorithm for step and shoot IMRT *Phys. Med. Biol.* **46** 1997–2007

——2001c A method to reduce the number of beam segments for step and shoot IMRT *Radiother. Oncol.* **61** (Suppl. 1) S13

Bär W, Schwarz M, Alber M, Bos L J, Mijnheer B J, Rasch C, Schneider C, Nüsslin F and Damen E M F 2003 A comparison of forward and inverse treatment planning for intensity-modulated radiotherapy of head and neck cancer *Radiother. Oncol.* **69** 251–8

Bär W, Schwarz M, Rasch C, Alber M, Schneider C, Bos L, Mijnheer B, Lebesque J, Damen E and Nüsslin F 2002 A treatment planning comparison for intensity-modulated radiotherapy of head and neck cancer *Radiother. Oncol.* **64** (Suppl. 1) S124

Barentsz J 2000 Prospects of contrast enhanced MR imaging for radiotherapy in the pelvis *Radiother. Oncol.* **56** (Suppl. 1) S7

Barker J, Patrocinio H and Podgorsak E B 2001 A comparison study of multileaf and micro-multileaf collimators *Med. Phys.* **28** 1986

Barnes E A, Murray B R, Robinson D M, Underwood L J, Hanson J and Roa W H Y 2001 Dosimetric evaluation of lung tumour immobilization using breath hold at deep inspiration *Int. J. Radiat. Oncol. Biol. Phys.* **50** 1091–8

Baroni G, Ferrigno G, Orecchia R and Pedotti A 2000 Real-time opto-electronic verification of patient position in breast cancer radiotherapy *Computer Aided Surgery* **5** 296–306

Barthold S, Echner G, Hartmann G, Pastyr O and Schlegel W 2002 CoRA—A new cobalt radiotherapy arrangement with multiple sources—a feasibility study *Phys. Medica.* **19** 13–26

Bassalow R and Sidhu N 2003 Dark current radiation produced in step-and-shoot IMRT on clinac 21EX linear accelerator *Med. Phys.* **30** 1493

Bate T, de Meerleer G, Bakker M, Villeirs G and de Neve W 2002 IMRT for prostate cancer: the consequences on the workload *Radiother. Oncol.* **64** (Suppl. 1) S88–9

Bate M T, Vakaet L and de Neve W 2000 Quality control and error detection in the treatment process *Radiother. Oncol.* **56** (Suppl. 1) S89

Battista J and Baumann J S 2003 The future of IMRT *IMRT—the State of the Art: AAPM Medical Physics Monograph Number 29* ed J R Palta and T R Mackie (Madison, WI: Medical Physics Publishing) pp 875–82

Baumert B G, Ciernik I, Dizendorf E, Reiner B, Davis J B, VonSchulthess G K, Lutolf U M and Steinert H C 2002 Implication of combined PET-CT on decision making in radiation oncology *Radiother. Oncol.* **64** (Suppl. 1) S209

Bayouth J, Pena J, Culp L, Brack C and Sanguineti G 2003a Feasibility of IMRT to cover pelvic nodes while escalating the dose to the prostate gland: dosimetric and acute toxicity data on 24 consecutive patients *Radiother. Oncol.* **68** (Suppl. 1) S77

Bayouth J E and Morrill S M 2003c MLC dosimetric characteristics for small field and IMRT applications *Med. Phys.* **30** 2545–52

Bayouth J E, Wendt D and Morrill S M 2003b MLC quality assurance techniques for IMRT applications *Med. Phys.* **30** 743–50

Beaulieu F, Beaulieu L, Tremblay D, Lechance B and Roy R 2004 Automatic generation of anatomy-based MLC fields in aperture-based IMRT *Med. Phys.* **31** 1539–45

Beavis A 2003a Dynamic contrast enhanced MRI in the inverse planning of IMRT: Prostate case study *Clinical Oncol.* **15** (Suppl. 2) S5–6

——2003b IMRT in our real NHS world? *RAD Magazine* **29** (340) 33–4

——2004a The need for more intelligence in radiotherapy *RAD Magazine* **30** (346) 31–2

——2004b Is tomotherapy the future of IMRT? *Brit. J. Radiol.* **77** 285–95

Beavis A, Bubula E and Whitton V 2002 Pre-treatment quality assurance for IMRT: the Princess Royal Hospital protocol *Radiother. Oncol.* **64** (Suppl. 1) S85

Beavis A, Ganney P, Whitton V and Xing L 2000a Optimisation of MLC orientation to improve accuracy in the static field delivery of IMRT *Proc. World Congress of Medical Physics and the AAPM Annual Congress, August 2000* paper FR-B309-02

——2000b Slide and shoot: a new method for MLC delivery of IMRT. *Proc. 13th Int. Conf. on the Use of Computers in Radiation Therapy (Heidelberg, May)* ed W Schlegel and T Bortfeld (Heidelberg: Springer) pp 182–4

——2001 Optimization of the step-and-shoot leaf sequence for delivery of intensity modulated radiation therapy using a variable division scheme *Phys. Med. Biol.* **46** 2457–65

Beavis A W, Garcia-Alvarez R and Liney G P 2003 The use of FMRI in IMRT treatment planning: A case study *Med. Phys.* **30** 1384–5

Beavis A and Liney G 2002a Delineation of IMRT boost volumes for prostate treatments using advanced magnetic resonance (DCE) techniques *Med. Phys.* **29** 1257

——2002b The application of an advanced magnetic resonance technique for planning boost delivery with IMRT *Radiother. Oncol.* **64** (Suppl. 1) S210–11

Beavis A W, Whitton V I and Xing L 1999 A combination delivery mode for intensity modulated radiation therapy based on specific patient cases. *Int. J. Radiat. Oncol. Biol. Phys.* **45** (Suppl. 1) 164

Bedford J L 2003 Optimisation and inverse planning for intensity modulated radiotherapy: Fundamental concepts and practical factors *Recent Res. Dev. Radiol.* **1** 23–38

Bedford J L and Webb S 2003 Elimination of importance factors for clinically accurate selection of beam orientations, beam weights and wedge angles in conformal radiation therapy *Med. Phys.* **30** 1788–804

Bednarz G, Michalski D, Houser C, Huq M S, Xiao Y, Anne P R and Galvin J M 2002 The use of mixed-integer programming for inverse treatment planning with pre-defined segments *Phys. Med. Biol.* **47** 2235–45

Beitler J J 2003 Let's accelerate the PET learning curve *Int. J. Radiat. Oncol. Biol. Phys.* **55** 281

Bel A, Astreinidoi E, Dehnad H and Raaijmaakers C P J 2002 Evaluation of the effect of geometric uncertainties on IMRT plans for head and neck treatments *Radiother. Oncol.* **64** (Suppl. 1) S213

Benard F, Boucher L, Boisvert C, Boileau R, Laberge G and Nabid A 2000 The use of FDG-PET imaging for the management of lung cancer patients *Radiother. Oncol.* **56** (Suppl. 1) S40

Benedict S H, Cardinale R M, Wu Q, Zwicker R D, Broaddus W C and Mohan R 2001 Intensity-Modulated stereotactic radiosurgery using dynamic micro-multileaf collimation *Int. J. Radiat. Oncol. Biol. Phys.* **50** 751–8

Bentzen S M 2004 High-tech radiation oncology: should there be a ceiling? *Int. J. Radiat. Oncol. Biol. Phys.* **58** 320–30

Berbeco R I, Mostafavi H, Sharp G C and Jiang S B 2004 Tumor tracking in the absence of radiopaque markers *Proc. 14th Int. Conf. on the Use of Computers in Radiation Therapy (Seoul, May)* pp 433–6

Beyer D, Quiet C, Rogers K and Gurgoze E 2000 Prostate motion during a course of IMRT: the value of daily ultrasound to guide portal localization and correct for random and systematic errors *Radiother. Oncol.* **56** (Suppl. 1) S105

Bieda M, Szeglin S and Das I 2002 Microanalysis and temporal characterization of small MU segments in IMRT *Med. Phys.* **29** 1304

Bignardi M, Galelli M, Spiazzi L, Bonetti M and Magrini S M 2000 Conformal radiotherapy for prostate carcinoma: differences in dose distribution between the original plan and the same plan recalculated (in the same patient) after 3 and 6 weeks of treatment *Radiother. Oncol.* **56** (Suppl. 1) S108

Birkner M, Alber M, Yan D and Nüsslin 2002 Image guided adaptive IMRT of the prostate based on a probabilistic patient geometry *Radiother. Oncol.* **64** (Suppl. 1) S282

Birkner M, Yan D, Alber M, Jiang L and Nüsslin F 2003 Adapting inverse planning to patient and organ geometrical variation: algorithm and implementation *Med. Phys.* **30** 2822–31

Birkner M, Yan D and Nüsslin F 2001 Incoporating organ motion and patient setup into inverse planning using patient specific image feedback *Radiother. Oncol.* **61** (Suppl. 1) S13

Blackall J, Landau D, Ahmad S, Crum W, McLeish K and Hawkes D 2003 Image registration based modelling of respiratory motion for optimisation of lung cancer radiotherapy *Proc. World Congress in Medical Physics (Sydney)* CDROM and website only

Bliss P and Wilks R J 2002 The use of a compensator library to reduce dose inhomogeneity in tangential radiotherapy of the breast *Radiother. Oncol.* **64** (Suppl. 1) S134

Boehmer D, Bohsung J, Eichwurzel I, Moys A and Budach V 2004 Clinical and physical quality assurance for intensity modulated radiotherapy of prostate cancer *Radiother. Oncol.* **71** 319–25

Bohsung J, Groll J, Kurth C, Pfaender M, Grebe G, Jahn U, Würm R E, Stuschke M and Budach V 2000 Clinical implementation of the VARIAN intensity modulated radiotherapy (IMRT) system *Radiother. Oncol.* **56** (Suppl. 1) S195

Boman E, Lyyra-Laitinen T, Kolmonen P, Jaatinen K and Tervo J 2003 Simulations for inverse radiation therapy treatment planning using a dynamic MLC algorithm *Phys. Med. Biol.* **48** 925–42

Bookstein F 1989 Principal Warps: Thin plate splines and the decomposition of deformations *IEE Trans. Patt. Anal. Mach. Intell.* **11** 567–85

Booth J T and Zavgorodni S F 2001 Modelling the dosimetric consequences of organ motion at CT imaging on radiotherapy treatment planning *Phys. Med. Biol.* **46** 1369–77

Borrosch D, Micke O, Papke K, Schaefer U, Schuck A, Schueller P and Willich N 2000 Functional MRI in the treatment planning of conformal radiotherapy for reirradiation of malignant brain tumors *Proc. 42nd Annual ASTRO Meeting. Int. J. Radiat. Oncol. Biol. Phys.* **48** (Suppl.) 258 (Handout material of the AAPM Meeting in 1999)

Bortfeld T 2000 Step and shoot IMRT *Radiother. Oncol.* **56** (Suppl. 1) S83

——2001 Current status of IMRT: physical and technological aspects *Radiother. Oncol.* **61** (Suppl. 1) S11

——2003 Physical optimisation *IMRT—the State of the Art: AAPM Medical Physics Monograph Number 29* ed J R Palta and T R Mackie (Madison, WI: Medical Physics Publishing) pp 51–75

Bortfeld T R, Boyer A L, Schlegel W, Kahler D L and Waldron T J 1994 Realisation and verification of three-dimensional conformal radiotherapy with modulated fields *Int. J. Radiat. Oncol. Biol. Phys.* **30** 899–908

Bortfeld T, Burkelbach J, Boesecke R and Schlegel W 1990 Methods of image reconstruction from projections applied to conformation therapy *Phys. Med. Biol.* **35** 1423–34

Bortfeld T, Jiang S B and Rietzel E 2004 Effects of motion on the total dose distribution *Seminars in Radiat. Oncol.* **14** 41–51

Bortfeld T, Jokivarsi K, Goitein M and Jiang S 2002b Effects of intra-fraction motion on IMRT delivery with MLC, compensators and scanning *Med. Phys.* **29** 1347

Bortfeld T, Jokivarsi K, Goitein M, Kung J and Jiang S B 2002a Effects of intra-fraction motion on IMRT dose delivery: statistical analysis and simulation *Phys. Med. Biol.* **47** 2203–20

Bortfeld T, Kufer K H, Monz M, Scherrer A, Thieke C and Trinkaus H 2002c Radiation therapy planning—a multicriteria approach *Med. Phys.* **29** 1258

Bortfeld T, Oelfke U and Nill S 2000 What is the optimum leaf width of a multileaf collimator? *Med. Phys.* **27** 2494–502

Bortfeld T, Schlegel W, Höver K-H and Schulz-Ertner D 1999 Mini and micro multileaf collimators—handout material; 41st AAPM meeting, Nashville, TN, July 1999

Bortfeld T, van Herk M and Jiang S B 2002d When should systematic patient positioning errors in radiotherapy be corrected? *Phys. Med. Biol.* **47** N297–302

Bos L J, de Boer R W, Mijnheer B J, Lebesque J V and Damen E M F 2001 Forward versus inverse planning of static IMRT of prostate cancer *Radiother. Oncol.* **61** (Suppl. 1) S32

Bos L J, Schwarz M, Bär W, Alber M, Mijnheer B J, Lebesque J V and Damen E M F 2004 Comparison between manual and automatic segment generation in step-and-shoot IMRT of prostate cancer *Med. Phys.* **31** 122–30

Bouchet L G, Meeks S L, Bova F J, Goodchild G, Buatti J M and Friedman W A 2000 Three-dimensional ultrasound optic guidance for high precision radiation therapy *Proc. 42nd Annual ASTRO Meeting. Int. J. Radiat. Oncol. Biol. Phys.* **48** (Suppl.) 195

Boyer A 2001 Clinical implementation of IMRT *Med. Phys.* **28** 1222

——2003 Static MLC IMRT (Step and shoot) *IMRT—the State of the Art: AAPM Medical Physics Monograph Number 29* ed J R Palta and T R Mackie (Madison, WI: Medical Physics Publishing) pp 285–317

Boyer A, Song Y, Yong D and Xing L 2002a Sweeping window arc therapy (SWAT) *Med. Phys.* **29** 1265

Boyer A, Xing L, Luxton G and Ma C 2000 IMRT with dynamic multileaf collimation *Radiother. Oncol.* **56** (Suppl. 1) S84

Boyer A, Xing L, Yu C, Xia P and Verhey L 2002b Linear accelerator quality assurance for IMRT *Radiother. Oncol.* **64** (Suppl. 1) S85

Braaksma M, Levendag P, Muller K, Van de Est H, Van Sornsen de Koste J and Nowak P 2002 The role of IMRT in advanced cancers of the head and neck. Some limitations *Radiother. Oncol.* **64** (Suppl. 1) S78–9

Braaksma M M J, Wijers O B, van Sörnsen de Koste J R, van der Est H, Schmitz P I M, Nowak P J C M and Levendag P C 2003 Optimisation of conformal radiation therapy by intensity modulation: cancer of the larynx and salivary gland function *Radiother. Oncol.* **66** 291–302

Bradley J, Thorstad W L, Mutic S, Miller T R, Dehdashti F, Siegel B A, Bosch W and Bertrand R J 2004 Impact of FDG-PET on radiation therapy volume delineation in non-small-cell lung cancer *Int. J. Radiat. Oncol. Biol. Phys.* **59** 78–86

Bragg C M, Conway J and Robinson M H 2001 Parotid gland tumours—A comparison of intensity modulated radiotherapy and 3-D conformal radiotherapy *Int. J. Radiat. Oncol. Biol. Phys.* **51** (Suppl. 1) 342–3

Brahme A 1988 Optimisation of stationary and moving beam radiation therapy techniques *Radiother. Oncol.* **12** 129–40

Brahme A, Roos J E and Lax I 1982 Solution of an integral equation encountered in rotation therapy *Phys. Med. Biol.* **27** 1221–9

Brock K K, Sharpe M B and Jaffray D A 2004 Biomechanical modelling of anatomical deformation for image-guided registration of soft-tissue surrogates *Proc. 14th Int. Conf. on the Use of Computers in Radiation Therapy (Seoul, May)* pp 467–71

Broggi S, Fiorino C, Castellone P, Bertelli E, Cattaneo G M and Calandrino R 2002 Mechanical testing and dosimetric characterization of a 1D IMRT delivering system for clinical applications *Radiother. Oncol.* **64** (Suppl. 1) S204–5

Bruinvis A D and Damen E M F 2001 Quality assurance of a treatment planning system for IMRT applications *Radiother. Oncol.* **61** (Suppl. 1) S45

Bucciolini M, Buonamici F B and Casati M 2004 Verification of IMRT fields by film dosimetry *Med. Phys.* **31** 161–8

Budgell G 1998 Delivery of intensity modulated radiation therapy by dynamic multileaf collimation: resolution requirements for specifying leaf trajectories. *Med. Phys.* **25** A202

Budgell G 2002 Verification in intensity-modulated radiotherapy *RAD Magazine* **29** (323) 23–4

Budgell G J, Martens C and Claus F 2001a Improved delivery efficiency for step and shoot intensity modulated radiotherapy using a fast-tuning magnetron *Phys. Med. Biol.* **46** N253–61

Budgell G J, Mott J H L, Logue J P and Hounsell A R 2001b Clinical implementation of dynamic multileaf collimation for compensated bladder treatments *Radiother. Oncol.* **59** 31–8

Budgell G J, Sykes J R and Wilkinson J M 1998 Rectangular edge synchronization for intensity modulated radiation therapy with dynamic multileaf collimation *Phys. Med. Biol.* **43** 2769–84

Burnet N 2003 Conformal therapy + rotation and central axis block is an alternative to IMRT for spine tumours *Clinical Oncol.* **15** (Suppl. 2) S23

Buyyounouski M K, Horwitz E M, Price R A, Hanlon A L, Uzzo R G and Pollack A 2004 Intensity-modulated radiotherapy with MRI simulation to reduce doses received by erectile tissue during prostate cancer treatment *Int. J. Radiat. Oncol. Biol. Phys.* **58** 743–9

Caccia B, Del Giudice P, Mattia M, Benassi M and Marzi S 2001 An evaluation of the overall sensitivity of the optimisation process in the IMRT technique *Radiother. Oncol.* **61** (Suppl. 1) S83

Cadman P, Bassalow R, Sidhu N P S, Ibbott G and Nelson A 2002 Dosimetric considerations for validation of a sequential IMRT process with a commercial treatment planning system *Phys. Med. Biol.* **47** 3001–10

Caldwell C B, Mah K, Skinner M and Danjouz C E 2003 Can PET provide the 3D extent of tumor motion for individualized internal target volumes? A phantom study of the limitations of CT and the promise of PET *Int. J. Radiat. Oncol. Biol. Phys.* **55** 1381–93

Cardarelli G, Zheng Z, Hussain K, Cintron O, Shearer D, DiPetrillo T, Tsai J, Engler M, Mohiuddin and Wazer D 2001 Evaluation of a new commercial QA phantom for intensity modulated radiation therapy (IMRT) plan verification *Med. Phys.* **28** 1209

Caudrelier J, Vial S, Gibon D, Bourel P, Vasseur C and Rousseau J. 2000 Evaluation of a fuzzy logic method for volume determination in MR imaging *Radiother. Oncol.* **56** (Suppl. 1) S17

Caudrelier J-M, Vial S, Gibon D, Kulik C, Fournier C, Castelain B, Coche-Dequeant B and Rousseau J 2003 MRI definition of target volumes using fuzzy logic method for three-dimensional conformal radiation therapy *Int. J. Radiat. Oncol. Biol. Phys.* **55** 225–33

Cavey M L, Bayouth J E, Colman M, Endres E J and Sanguineti G 2004 Is dose escalation to the prostate feasible while treating the pelvic nodes with IMRT? *Proc. 14th Int. Conf. on the Use of Computers in Radiation Therapy (Seoul, May)* pp 124–7

Chandra A, Dong L, Huang E, Kuban D A, O'Neill L J, Rosen I I and Pollack A 2001 Evaluation of ultrasound-based daily prostate localization during IMRT For prostate cancer *Int. J. Radiat. Oncol. Biol. Phys.* **51** (Suppl. 1) 167–8

——2003 Experience of ultrasound-based daily prostate localization *Int. J. Radiat. Oncol. Biol. Phys.* **56** 436–47

Chang J 2001 IMRT (intensity modulated radiotherapy) verification using Varian AS500 electronic portal imaging device (EPID) *Med. Phys.* **28** 1282–3

Chang J, Mageras G K, Chui C S, Ling C C and Lutz W 2000a Relative profile and dose verification of intensity-modulated radiation therapy *Int. J. Radiat. Oncol. Biol. Phys.* **47** 231–40

Chang S X, Cullip T J and Deschesne K M 2000b Intensity modulation delivery techniques: 'step and shoot' MLC auto-sequence versus the use of a modulator *Med. Phys.* **27** 948–59

Chang S, Cullip T, Deschesne K, Miller E and Rosenman J 2002 Clinical experience of an alternative IMRT delivery technique intensity modulated via compensators *Med. Phys.* **29** 1305

——2004a Compensators: An alternative IMRT delivery technique *Proc. 14th Int. Conf. on the Use of Computers in Radiation Therapy (Seoul, May)* pp 207–11

Chang S, Cullip T, Deschesne K and Rosenman J 2003 Clinical experience of a compensator-based intensity modulated treatment technique *Proc. World Congress in Medical Physics (Sydney)* CDROM and website only

Chang S, Potter L and Cullip T 2001 Investigation of a camera-type EPID as an IMRT QA tool *Med. Phys.* **28** 1293

——2003 High intensity modulation resolution: the advantage of a 4-bank micro-multileaf collimator system *Proc. World Congress in Medical Physics (Sydney)* CDROM and website only

Chang S, Schulman D, Cullip T, Keall P, Mageras G and Joshi S 2004b The effect of organ motion on cumulative dosimetry: a comparison of different IMRT delivery techniques *Proc. 14th Int. Conf. on the Use of Computers in Radiation Therapy (Seoul, May)* pp 615–17

Chao C and Blanco A L 2003 IMRT for head and neck cancer *IMRT—the State of the Art: AAPM Medical Physics Monograph Number 29* ed J R Palta and T R Mackie (Madison, WI: Medical Physics Publishing) pp 631–44

Chao C, Deasy J O, Markman J, Haynie J, Perez C A, Purdy J A and Low D A 2000 Functional outcome of parotid gland sparing in patients with head and neck (H and N) cancers receiving intensity-modulated (IMRT) or 3-D radiation therapy *Proc. 42nd Annual ASTRO Meeting. Int. J. Radiat. Oncol. Biol. Phys.* **48** (Suppl.) 174

Chao K S C, Majhail N, Huang C-J, Simpson J R, Perez C A, Haughey B and Spector G 2001 Intensity-modulated radiation therapy reduces late salivary toxicity without compromising tumor control in patients with oropharyngeal carcinoma: a comparison with conventional techniques *Radiother. Oncol.* **61** 275–80

Chapman J D 2003 How can PET contribute to our current RT treatment planning? *Radiother. Oncol.* **68** (Suppl. 1) S15

Chapman J D. Movsas B and Nahum A E 2003 Fifty years later: how many current treatment failures are due to hypoxia? *Radiother. Oncol.* **68** (Suppl. 1) S32

Chauvet I, Papatheodorou S, Gaboriaud G, Ponvert D and Rosenwald J C 2001 Preliminary study of optimization parameters for the HELIOS inverse planning software *Phys. Medica.* **17** 115

Chen G T, Jiang S B, Kung J, Doppke K P and Willett C G 2001b Abdominal organ motion and deformation: Implications for IMRT *Int. J. Radiat. Oncol. Biol. Phys.* **51** (Suppl. 1) 210

Chen H, Yang J, Stapleton P and Murphy S 2002a Calibration of MLC leaf positions with an electronic portal imaging device for IMRT delivery *Med. Phys.* **29** 1262

Chen J Z and Wong E 2004 MLC leaf optimisation for intensity modulated arc therapy *Proc. 14th Int. Conf. on the Use of Computers in Radiation Therapy (Seoul, May)* pp 132–4

Chen L, Price R, Li J, Wang L, Qin L, Ma C and Pollack A 2003 MRI-based treatment optimisation for prostate IMRT *Med. Phys.* **30** 1335

Chen Q-S, Weinhouse M S, Deibel F C, Ciezki J P and Macklis R M 2001a Fluoroscopic study of tumour motion due to breathing: facilitating precise radiation therapy for lung cancer patients *Med. Phys.* **28** 1850–6

Chen Y, Boyer A L and Ma C-M 2000a Calculation of x-ray transmission through a multileaf collimator *Med. Phys.* **27** 1717–26

Chen Y and Galvin J 2001 Optimization of conformal radiotherapy by beam weight renormalization *Med. Phys.* **28** 1262

Chen Y, Hou Q and Galvin J M 2004 A graph-searching method for MLC leaf sequencing under constraints *Med. Phys.* **31** 1504–11

Chen Y, Michalski D, Houser C and Galvin J M 2002b A deterministic iterative least-squares algorithm for beam weight optimization in conformal radiotherapy *Phys. Med. Biol.* **47** 1647–58

Chen Y, Michalski D, Xiao Y and Galvin J M 2001c Automatic aperture selection and IMRT plan optimization by beam weight renormalization *Int. J. Radiat. Oncol. Biol. Phys.* **51** (Suppl. 1) 74–5

Chen Y, Xing L, Luxton G, Li J L and Boyer A L 2000b Automated quality asurance process for MLC-based IMRT using CT virtual simulation *Proc. 42nd Annual ASTRO Meeting. Int. J. Radiat. Oncol. Biol. Phys.* **48** (Suppl.) 194

Cheng C-W and Das I J 2002 Comparison of beam characteristics in intensity modulated radiation therapy (IMRT) and those under normal treatment condition. *Med. Phys.* **29** 226–30

Cheng C W, Das I J and Huq M S 2003 Lateral loss and dose discrepancies of multileaf collimator segments in intensity modulated radiation therapy *Med. Phys.* **30** 2959–68

Cheng C, Das I, Xiao Y and Galvin J 2002 Beam characteristics in intensity modulated radiation therapy (IMRT) of modern accelerators *Med. Phys.* **29** 1266

Cheng C W, Wong J R, Ndlovu A M, Das I J, Schiff P and Uematzu M 2000 Dosimetric study of virtual mini-multileaf and its clinical application *Proc. 42nd Annual ASTRO Meeting. Int. J. Radiat. Oncol. Biol. Phys.* **48** (Suppl.) 342

Cho B C J, Hurkmans C W Damen E M F and Mijnheer B J 2001a A planning study comparing intensity modulated versus non-intensity modulated radiotherapy in treating breast cancer *Radiother. Oncol.* **61** (Suppl. 1) S44

Cho B C J, Hurkmans C W, Damen E M F, Zijp L J and Mijnheer B J 2002a Intensity modulated versus non-intensity modulated radiotherapy in the treatment of the left breast and upper internal mammary lymph node chain: a comparative study *Radiother. Oncol.* **62** 127–36

Cho B, Schwarz M, Mijnheer B and Bartelink H 2002b A simple intensity modulated radiotherapy class solution using predefined segments to reduce cardiac complications in the treatment of left-sided breast cancer *Med. Phys.* **29** 1284

——2004 Simplified intensity modulated radiotherapy class solution using pre-defined segments to reduce cardiac complications in left-sided breast cancer *Radiother. Oncol.* **70** 231–41

Cho J 2002 Intensity modulated radiotherapy in breast cancer *Radiother. Oncol.* **64** (Suppl. 1) S8–9

Cho P S and Marks II R J 2000 Hardware-sensitive optimisation for intensity modulated radiotherapy. *Phys. Med. Biol.* **45** 429–40

Cho P, Meyer J, Yee D, Phillips M and Parsai H 2003 Numerical instability and conditioning in IMRT optimization *Med. Phys.* **30** 1489

Cho P S, Parsaei H and Phillips M H 2001b Influence of random field perturbations on dynamic intensity modulated beams *Med. Phys.* **28** 1203–4

Cho P S and Phillips M H 2001 Reduction of computational dimensionality in inverse radiotherapy planning using sparse matrix operations *Phys. Med. Biol.* **46** N117–25

Choi B and Deasy J O 2002 The generalized equivalent uniform dose function as a basis for intensity-modulated treatment planning *Phys. Med. Biol.* **47** 3579–89

Christian J A, McNair H A, Symonds-Tayler J R N, Knowles C, Bedford J L and Brada M 2003a Reduction in lung tumour motion using active breathing control *Radiother. Oncol.* **68** (Suppl. 1) S115

Christian J A, Partridge M, Cook G, McNair H A, Cronin B, Courbon F, Bedford J L and Brada M 2003b Optimisation of lung radiotherapy planning using accurate co-registration of planning CT and SPECT perfusion images *Radiother. Oncol.* **68** (Suppl. 1) S54

Chu S and Suh C 2000 Dose planning of intensity modulated radiotherapy for nasopharyngeal cancer using compensating filters *Proc. World Congress of Medical Physics and the AAPM Annual Congress, August 2000* paper WE-E313-09

Chuang C F, Verhey L J and Xia P 2002 Investigation of the use of MOSFET for clinical IMRT dosimetric verification *Med. Phys.* **29** 1109–115

Chuang C, Xia P, Nguyen-Tan F, Fu K and Verhey L 2000 Investigation of the uncertainties in patient positioning and patient motion in IMRT treatment *Proc. World Congress of Medical Physics and the AAPM Annual Congress, August 2000* paper WE-FXH-45

Chuang K-S, Chen T-J, Kuo S-C, Jan M-L, Hwang I-M, Chen S, Ling Y-C and Wu J 2003 Determination of beam intensity in a single step for IMRT inverse planning *Phys. Med. Biol.* **48** 293–306

Chui C 2000 Clinical implementation of IMRT *Proc. World Congress of Medical Physics and the AAPM Annual Congress, August 2000* paper TU-ABR-02

Chui C S, Chan M F and Ling C C 2000b Delivery of intensity-modulated radiation therapy with a multileaf collimator: Comparison of step-and-shoot and dynamic leaf motion methods *Proc. World Congress of Medical Physics and the AAPM Annual Congress, August 2000* paper MO-EBR-06

Chui C-S, Chan M F, Yorke E, Spirou S and Ling C C 2001a Delivery of intensity-modulated radiation therapy with a conventional multileaf collimator: Comparison of dynamic and segmental methods *Med. Phys.* **28** 2441–9

Chui C-S, Hong L and Hunt M 2002 A simplified intensity modulated radiation therapy technique for the breast *Med. Phys.* **29** 522–9

Chui C, Hong L, Yang J and Spirou S 2001b Splitting of large intensity-modulated fields *Med. Phys.* **28** 1252

Chui C, LoSasso T and Palm A 2003a Computational and empirical verification of intensity modulated radiation therapy delivered with a multileaf collimator *Proc. World Congress in Medical Physics (Sydney)* CDROM and website only

Chui C-S, Spirou S and LoSasso T 1996 Testing of dynamic multileaf collimation *Med. Phys.* **23** 635–41

Chui C-S, Yorke E and Hong 2003b The effects of intra-fraction organ motion on the delivery of intensity-modulated field with a multileaf collimator *Med. Phys.* **30** 1736–46

Chytka T, Novotny J Jr, Vymazal J, Liscak R, Marek J and Vladyka V 2000 Glioma malignancy prediction model based on patient's clinical data and CT and MRI imaging *Radiother. Oncol.* **56** (Suppl. 1) S37

Ciernik I F, Dizendorf E, Reiner B, Burger C, Khan S, Lütolf U M, Steinert H, VonSchuthess G K and Baumert B G 2002 Treatment planning of 3D-conformal radiation therapy based on the integrated computer-assisted Positron Emission Tomography (PET/CT)imaging *Radiother. Oncol.* **64** (Suppl. 1) S92

Clark C, Bidmead M, Mubata C, Harrington K, Evans P R and Nutting C 2003a Does IMRT provide an improved treatment for carcinoma of the larynx and cervical nodes *Clinical Oncol.* **15** (Suppl. 2) S5

Clark C H, Bidmead M A, Mubata C D, Harrington K J and Nutting C M 2004 Intensity-modulated radiotherapy improves target coverage, spinal cord sparing and allows dose escalation in patients with locally advanced cancer of the larynx *Radiother. Oncol.* **70** 189–98

Clark C, Bidmead M, Mubata C, Harrington K, Rhys Evans P and Nutting C 2002c Planning and quality assurance issues for clinical implementation of intensity modulated radiotherapy (IMRT) for treatment of carcinoma of the larynx and cervical nodes *Radiother. Oncol.* **64** (Suppl. 1) S245–6

Clark C, Hansen V N, Miles E, McNair H, Bidmead M, Webb S and Dearnaley D 2003c Optimization of forward planned IMRT for hypofractionated treatment of the prostate *Clinical Oncol.* **15** (Suppl. 2) S24

Clark C H, Miles E A, Bidmead M A, Mubata C D, Harrington K J and Nutting C M 2003d (in regard to Lee *et al* 2002 *Int. J. Radiat. Oncol. Biol. Phys.* **53** 630–7) *Int. J. Radiat. Oncol. Biol. Phys.* **55** 1150

Clark C, Miles E, Mubata C, Meehan C, Alonso S and Bidmead M 2003b Optimization of a patient specific quality assurance procedure for dynamic IMRT *Clinical Oncol.* **15** (Suppl. 2) S7

Clark C H. Mubata C D, Bidmead A M, Humphreys M, McNair H, Harrington K J and Nutting C M 2002b Clinical implementation of dynamic intensity modulated radiotherapy: Planning and quality assurance issues for treatment of patients with carcinoma of the thyroid and cervical nodes *Proc. IPEM Biennial Meeting (Southampton, 16–17 July)* pp 9–10

Clark C H, Mubata C D, Meehan C A, Bidmead A M, Staffurth J, Humphreys M E and Dearnaley D P 2002a IMRT clinical implementation: prostate and pelvic node irradiation using HELIOS and a 120-leaf MLC *J. Appl. Clinical Med. Phys.* **3** 273–84

Claus F, Boterberg T, Ost P and de Neve W 2002 Short term toxicity profile for 32 sinonasal cancer patients treated with IMRT. Can we avoid dry eye syndrome? *Radiother. Oncol.* **64** 205–8

Claus F, de Gersem W, Vanhoutte I, Duthoy W, Remouchamps V, de Wagter C and de Neve W 2001 Evaluation of a leaf position optimization tool for intensity modulated radiation therapy of head and neck cancer *Radiother. Oncol.* **61** 281–6

Claus F, de Wagter C, Remouchamps V, de Gersem W and de Neve W 2000 Time efficiency and monitor unit efficiency of anatomy-based IMRT *Radiother. Oncol.* **56** (Suppl. 1) S195

Convery D J 2001 Implementation of inverse-planned IMRT with dynamic MLC delivery at the Royal Marsden NHS Trust, Sutton *Compendium of the 3rd IMRT Winter School, Heidelberg (London, Jan 26/27)* pp D2.1–D2.7

Convery D J, Cosgrove V P and Webb S 2000 Improving dosimetric accuracy of a dynamic MLC technique *Proc. 13th Int. Conf. on the Use of Computers in Radiation Therapy (Heidelberg, May)* ed W Schlegel and T Bortfeld (Heidelberg: Springer) pp 277–9

Convery D J and Webb S 1997 Calculation of the distribution of head-scattered radiation in dynamically collimated MLC fields *Proc. 12th Int. Conf. on the Use of Computers in Radiation Therapy (Salt Lake City, UT, May)* ed D D Leavitt and G Starkshall (Madison, WI: Medical Physics Publishing) pp 350–3

——1998 Generation of discrete beam-intensity modulation by dynamic multileaf collimation under minimum leaf separatation constraints *Phys. Med. Biol.* **43** 2521–38

Coolens C, Webb S, Evans P M and Seco J 2003 Combinational use of conformal and intensity modulated beams in radiotherapy planning *Phys. Med. Biol.* **48** 1795–807

Cooper P, MacKay R I and Williams P C 2001a Dosimetric evaluation of a resolution-enhancing tertiary collimator for radiotherapy *Radiother. Oncol.* **61** (Suppl. 1) S57

——2001b A resolution-enhancing tertiary collimator for radiotherapy: artefacts generated using the protytype and a strategy to avoid them *Radiother. Oncol.* **61** (Suppl. 1) S81

——2001c Optimizing the indexing of a resolution-enhancing tertiary collimator for radiotherapy *Phys. Med. Biol.* **46** N263–8

——2002 Indexing artefacts using a tertiary collimator and a method to avoid them *Phys. Med. Biol.* **47** N191–201

Corletto D 2001 Inverse and forward optimisation by 1D and 2D IMRT for prostate cancer *Radiother. Oncol.* **61** (Suppl. 1) S12

Cosgrove V 2001 Verification and QA of IMRT by DMLC technique *Compendium of the 3rd IMRT Winter School, Heidelberg (London, Jan 26/27)* pp V1.1–V1.8

Cosgrove V P, Carson K J, Hounsell A R, Grattan M W D, Stewart D P, Eakin R L, Fleming V A L and Jarritt P H 2003 Direct use of an integrated PET/CT imaging system for radiotherapy treatment planning in non-small-cell lung cancer *Radiother. Oncol.* **68** (Suppl. 1) S29

Cosgrove V P, Convery D J, McNair H A, Vaarkamp J and, Webb S 2001 Quality assurance in a phase 1 prostate and pelvic node IMRT trial *Phys. Medica.* **17** 160–1

Cosgrove V P, Convery D J, Murphy P S, Nutting C M and Webb S 2000 Dynamic MLC delivered IMRT: verification using polyacrylamide gel dosimetry *Proc. 13th Int. Conf. on the Use of Computers in Radiation Therapy (Heidelberg, May)* ed W Schlegel and T Bortfeld (Heidelberg: Springer) pp 311–13

Cotrutz C 2002 Inverse treatment planning with interactively variable voxel-dependent importance factors *Med. Phys.* **29** 1336

Cotrutz C, Kappas C and Webb S 2000 Intensity modulated arc therapy (IMAT) with centrally blocked rotational fields *Phys. Med. Biol.* **45** 2185–206

Cotrutz C and Xing L 2002 Using voxel-dependent importance factors for interactive DVH based optimization *Phys. Med. Biol.* **47** 1659–69

——2003a IMRT dose shaping with regionally variable penalty scheme *Med. Phys.* **30** 544–51

——2003b Segment-based IMRT inverse planning using a genetic algorithm *Med. Phys.* **30** 1372

——2003c Segment-based dose optimisation using a genetic algorithm *Phys. Med. Biol.* **48** 2987–98

Cozzi L and Fogliata A 2001 Dosimetric validation of a step and shoot intensity modulation technique using ionisation chambers, films and a commercial electronic portal-imaging device *Radiother. Oncol.* **61** (Suppl. 1) S107

Cozzi L, Fogliata A and Bolsi A 2001a A treatment planning comparison of 3D conformal therapy, intensity modulated photon therapy and proton therapy for treatment of advanced head and neck tumours *Radiother. Oncol.* **61** (Suppl. 1) S32

——2001b Comparative analysis of IMRT step and shoot and sliding window techniques on two commercial treatment planning modules *Radiother. Oncol.* **61** (Suppl. 1) S65

——2001c Intensity modulation for intact breast cancer treatment. A first approach *Radiother. Oncol.* **61** (Suppl. 1) S83

Cozzi L, Fogliata A , Bolsi A, Nicolini G and Bernier J 2004 Three-dimensional conformal vs intensity-modulated radiotherapy in head-and-neck cancer patients: comparative analysis of dosimetric and technical parameters *Int. J. Radiat. Oncol. Biol. Phys.* **58** 617–24

Cozzi L, Fogliata A, Lomax A and Bolsi A 2001d A treatment planning comparison of 3D conformal therapy, intensity modulated photon therapy and proton therapy for treatment of advanced head and neck tumours *Radiother. Oncol.* **61** 287–97

Crooks S M, McAven L F, Robinson D R and Xing L 2000 Treatment time reduction through multileaf-collimator leaf-sequencing in static IMRT *Proc. 42nd Annual ASTRO Meeting. Int. J. Radiat. Oncol. Biol. Phys.* **48** (Suppl.) 340

——2002 Minimizing delivery time and monitor units in static IMRT by leaf-sequencing *Phys. Med. Biol.* **47** 3105–16

Crooks S M, Wu X, Takita C, Watzich M and Xing L 2003 Aperture modulated arc therapy *Phys. Med. Biol.* **48** 1333–44

Crooks S M and Xing L 2001 Linear algebraic methods applied to intensity modulated radiation therapy *Phys. Med. Biol.* **46** 2587–606

Crosbie J C, Poynter A J, James H V, Scrase C D, Morgan J S, Bulusu V R and LeVay J 2002 Quality assurance program for effective delivery of dynamic IMRT *Proc. IPEM Biennial Meeting (Southampton, 16–17 July)* p 36

Cumberlin R L, Deye J and Coleman C N 2002 In response to Drs Schulz and Kagan *Int. J. Radiat. Oncol. Biol. Phys.* **52** 274–5

Curran B 2003 IMRT delivery using serial tomotherapy *IMRT—the State of the Art: AAPM Medical Physics Monograph Number 29* ed J R Palta and T R Mackie (Madison, WI: Medical Physics Publishing) pp 221–45

Dai J-R and Hu Y-M 1999 Intensity-modulation radiotherapy using independent collimators: an algorithm study *Med. Phys.* **26** 2562–70

Dai J and Zhu Y 2001a Minimizing the number of segments in a delivery sequence for intensity-modulated radiation therapy with a multileaf collimator *Med. Phys.* **28** 1251–2

——2001b Minimizing the number of segments in a delivery sequence for intensity-modulated radiation therapy with a multileaf collimator *Med. Phys.* **28** 2113–20

Daisne J F, Duprez T, Weyant B, Sibomana M, Reychler H, Hamoir M, Lonneux M, Cosnard G and Grégoire V 2002 Impact of computed tomography (CT), magnetic resonance (MR) and positron emission tomography with fluorodeoxyglucose (FDG-PET) image coregistration on GTV delineation in head and neck squamous cell carcinoma (HNSCC) *Radiother. Oncol.* **64** (Suppl. 1) S25–6

Danciu C and Proimos B 2003 Gravity oriented absorbers for conformal therapy of small and irregular targets with protection of their vital neighbours *Proc. World Congress in Medical Physics (Sydney)* CDROM and website only

Das S, Bell M, Zhou S, Miften M, Munley M, Whiddon C, Craciunescu O, Baydush A, Wong T, Rosenman J, Dewhirst M and Marks L 2003 Feasibility of functional image-guided radiotherapy evaluated using a novel heuristic fluence modulation algorithm *Med. Phys.* **30** 1335

Das S, Cullip T, Tracton G, Chang S, Marks L, Anscher M and Rosenman J 2003 Beam orientation selection for intensity-modulated radiation therapy based on target equivalent uniform dose maximization *Int. J. Radiat. Oncol. Biol. Phys.* **55** 215–24

Das S, Halvorsen P, Cullip T, Chang S, Tracton G, Marks L and Rosenman J 2001a Equivalent uniform dose based automated beam selection for intensity modulated treatment planning *Med. Phys.* **28** 1307

Das S K, Halvorsen P, Cullip T, Tracton G, Chang S, Marks L, Anscher M and Rosenman J 2001b Beam orientation selection for intensity modulated radiation therapy based on target equivalent uniform dose maximization *Int. J. Radiat. Oncol. Biol. Phys.* **51** 74

Das S, Miften M, Zhou S, Bell M, Munley M, Whiddon C, Craciunescu O, Baydush A, Wong T, Rosenman J, Dewhirst M and Marks L 2004 Feasibility of optimising the dose distribution in lung tumours using fluorine-18-fluorodeoxyglucose positron emmission tomography and single photon emission computed tomography guided dose prescriptions *Med. Phys.* **31** 1452–61

Dasu A, Toma-Dasu I and Fowler J F 2003 Should single or distributed parameters be used to explain the steepness of tumour control probability curves? *Phys. Med. Biol.* **48** 387–97

Dawson L A, Brock K, Litzenberg D, Kazanjian S, McGinn C G, Lawrence T S, Ten Haken R K and Balter J 2000 The reproducibility of organ position using active breathing control (ABC) during liver radiotherapy *Proc. 42nd Annual ASTRO Meeting. Int. J. Radiat. Oncol. Biol. Phys.* **4** (Suppl.) 1675

Dearnaley D P 2000 Optimisation of high dose treatment for prostate cancer from conformal to IMRT *Radiother. Oncol.* **56** (Suppl. 1) S123

——2004 (in regard to 2003 The return of the snake oil salesman *Int. J. Radiat. Oncol. Biol. Phys.* **55** 561–2 and **56** 1507–8) *Int. J. Radiat. Oncol. Biol. Phys.* **58** 1644

Dearnaley D P, Khoo V S, Norman A R, Meyer L, Nahum A, Tait D, Yarnold J and Horwich A 1999 Comparison of radiation side effects of conformal and conventional radiotherapy in prostate cancer: a randomised trial *Lancet* **353** 267–72

Deasy J and El Naqa I 2003 Adaptive gridding for IMRT dose calculations *Med. Phys.* **3** 1487

Deasy J, Lee E, Kawrakow I and Zakarian C 2002 A prototype Monte Carlo-based IMRT treatment planning research system *Med. Phys.* **29** 1254

Deasy J and Wickerhauser M 2001 IMRT optimization based on stored pencil beam dose distributions: Compression, denoising and dose calculations using wavelets *Med. Phys.* **28** 1261

Deasy J and Zakarian C 2003 Dose-distance constraints: A new method for controlling intensity modulated treatment planning dose distributions *Med. Phys.* **30** 1334

De Brabandere M, van Esche A, Kutcher G J and Huyskens D 2002 Quality assurance in intensity modulated radiotherapy by identifying standards and patterns in treatment preparation: a feasibility study on prostate patients *Radiother. Oncol.* **62** 283–91

Debus J, Pirzkall A, Thilmann C, Grosser K H, Rhein B, Haering P, Hoess A and Wannenmacher M 2000 IMRT in the treatment of skull base tumours *Radiother. Oncol.* **56** (Suppl. 1) S86

De Deene Y, de Wagter C and Baldock C 2003 Comment on 'A systematic review of the precision and accuracy of dose measurements in photon radiotherapy using polymer and Fricke MRI gel dosimetry' *Phys. Med. Biol.* **48** L15–18

De Deene Y, de Wagter C and de Neve W 2000 Monomer/polymer gel dosimetry *Radiother. Oncol.* **56** (Suppl. 1) S116

De Gersem W, Claus F, de Wagter C and de Neve W 2001a An anatomy-based beam segmentation tool for intensity-modulated radiation therapy and its application to head-and-neck cancer *Int. J. Radiat. Oncol. Biol. Phys.* **51** 849–59

De Gersem W, Claus F, de Wagter C, Van Duse B and de Neve W 2001b Leaf position optimization for step-and-shoot IMRT *Int. J. Radiat. Oncol. Biol. Phys.* **51** 1371–88

De Gersem W, Vakaet L, Claus F, Remouchamps V, Van Duyse B and de Neve W 2000 MLC leaf position optimization using a biophysical objective function *Radiother. Oncol.* **56** (Suppl. 1) S96

Dehdashti F, Grigsby P W, Mintun M A, Lewis J S, Siegel B A and Welch M J 2003 Assessing tumor hypoxia in cervical cancer by positron emission tomography with ^{60}Cu-ATSM: relationship to therapeutic response-a preliminary report *Int. J. Radiat. Oncol. Biol. Phys.* **55** 1233–8

Dehnad H 2001 Clinical use of IMRT for prostate treatment with marker detection *Abstracts of the Elekta IMRT User Meeting 2001 in NKI Amsterdam* pp 18–20

Dejean C, Grandjean P, Lefkopoulos D and Foulquier J 2000b Application of regularisation methods to the IMRT inverse problem *Radiother. Oncol.* **56** (Suppl. 1) S211

Dejean C, Lefkopoulos D, Grandjean P, Berre F and Touboul E 2000c Comparison of two inversion techniques: the analytic singular value decomposition and the iterative gradient with constraints *Radiother. Oncol.* **56** (Suppl. 1) S212

Dejean C, Lefkopoulos D, Platoni K, Keraudy K and Julia F 2000a ROC analysis: a useful evaluation tool to define the most conformal prescription isodose for IMRT plans *Radiother. Oncol.* **56** (Suppl. 1) S115

Dejean C, Lemosquet A, Grandjean P, Lefkopoulos D and Touboul E 2001a Influence of IMRT inverse problem regularisation on biological indices *Radiother. Oncol.* **61** (Suppl. 1) S84

Dejean C, Lemosquet A, Lefkopoulos D and Touboul E 2001b Comparison of biological assessment functions in intensity modulated radiotherapy: two-dimensional study *Radiother. Oncol.* **61** (Suppl. 1) S84

——2001c Comparison of biological assessment functions in intensity modulated radiotherapy: A two-dimensional study *Phys. Medica.* **17** 115

De Langen M, Essers M, Dirkx M L P, Heijimen B J M 2003 Bringing IMRT into clinical practice: The complete trajectory *Phys. Medica.* **19** 59

Della Biancia C, Hunt M and Amols H 2002 A comparison of the integral dose from 3D conformal and IMRT techniques in the treatment of prostate cancer *Med. Phys.* **29** 1216

Della Biancia C, Yorke E, Mageras G, Giraud P, Rosenzweig K, Amols H and Ling C 2003 Comparison of two breathing levels for gated intensity modulated radiation therapy (IMRT) of lung cancer *Med. Phys.* **30** 2340–1

Deloar H M, Kunieda E, Kawase T, Saitoh H, Tohyama N, Ozaki M, Kimura H, Fujisaki T and Kubo A 2004 Tomotherapy with kilo-voltage X-ray energy: Preliminary trials with Monte Carlo simulation *Proc. 14th Int. Conf. on the Use of Computers in Radiation Therapy (Seoul, May)* pp 194–7

De Meerleer G, de Neve W, Verbaeys A, Goethals C, Villeirs G, Oosterlinck W and Vakaet L 2002a Intensity modulated radiotherapy (IMRT) for prostate cancer: acute toxicity of 102 patients treated *Radiother. Oncol.* **64** (Suppl. 1) S56

De Meerleer G O, Vakaet L A M L, de Gersem W R T, de Wagter C, de Naeyer B and de Neve W 2000a Radiotherapy of prostate cancer with or without intensity modulated beams: a planning comparison *Int. J. Radiat. Oncol. Biol. Phys.* **47** 639–48

De Meerleer G, Vakaet L and de Neve W 2000b Three versus five beams in the IMRT treatment for prostate cancer: a planning comparison *Radiother. Oncol.* **56** (Suppl. 1) S107

De Meerleer G, Villeirs G, Bral S, de Gersem W and de Neve W 2002b Simultaneous boost to the intraprostatic lesions: IMRT planning for step and shoot delivery by direct segment aperture optimization *Radiother. Oncol.* **64** (Suppl. 1) S286

Dempsey J F, Ahuja R K, Kamath S, Kumar A, Li J G, Palta J R, Romeijn H E, Ranka S and Sahni S 2003 The leaf sequencer: an underestimated problem? *Radiother. Oncol.* **68** (Suppl. 1) S3

Deng J, Guerrero T, Ding M S, Jolly J, Pawlicki T and Ma C M 2001c Incorporate organ motion into MLC leaf sequencing for intensity modulated radiation therapy *Int. J. Radiat. Oncol. Biol. Phys.* **51** (Suppl. 1) 92–3

Deng J, Guerrero T, Hai J, Luxton G and Ma C-M 2001b Source modelling and beam commissioning for a Cyberknife system *Med. Phys.* **28** 1303

Deng J, Lee M, Ding M and Ma C 2002 Effect of breathing motion on IMRT treatment of breast cancer: a Monte Carlo study *Med. Phys.* **29** 1263

Deng J, Pawlicki T, Chen Y, Li J, Jiang S and Ma C 2000 The MLC tongue-and-groove effect on IMRT dose distributions *Proc. World Congress of Medical Physics and the AAPM Annual Congress, August 2000* paper WE-E313-06

——2001a The MLC tongue-and-groove effect on IMRT dose distributions *Phys. Med. Biol.* **46** 1039–60

Depuydt T, Van Esche A and Huyskens D P 2002 A quantitative evaluation of IMRT dose distributions: refinement and clinical assessment of the gamma evaluation *Radiother. Oncol.* **62** 309–19

Deshpande N and Poynter A 2000 Effect of differential field margins in conformal therapy of the prostate *Radiother. Oncol.* **56** (Suppl. 1) S107

De Wagter C, de Deene Y, Vergote K, Martens C, de Gersem W, Van Duyse B and de Neve W 2001 The entire IMRT delivery process *Radiother. Oncol.* **61** (Suppl. 1) S16

Didinger B H, Nikoghosyan A, Schulz-Ertner D, Zierhut D, Schlegel W, Wannenmacher M and Debus J 2001 Benefit of adaptive inverse planning during intensity modulated radiotherapy of prostate cancer—a comparative DVH-analysis *Int. J. Radiat. Oncol. Biol. Phys.* **51** (Suppl. 1) 300

Dirkx M L P, de Langen M and Heijmen B J M 2000 Comparison of two algorithms for leaf trajectory calculation for dynamic multileaf collimation (DMLC) *Radiother. Oncol.* **56** (Suppl. 1) S96

Dirkx M L P and Heijmen B J M 2000a Beam intensity modulation for penumbra enhancement and field length reduction in lung cancer treatments: a dosimetric study *Radiother. Oncol.* **56** 181–8

——2000b Testing of the stability of intensity modulated beams generated with dynamic multileaf collimation applied to the M50 Racetrack Microtron *Med. Phys.* **27** 2701–7

Djajaputra D, Wu Q, Wu Y and Mohan R 2003 Algorithm and performance of a clinical IMRT beam-angle optimisation system *Phys. Med. Biol.* **48** 3191–212

Dobbs H J 2000 ICRU-62: A clinical perspective *Radiother. Oncol.* **56** (Suppl. 1) S51

Dogan N, King S and Emami B 2003 Sequential versus simultaneous integrated boost IMRT for treatment of head and neck cancers *Radiother. Oncol.* **68** (Suppl. 1) S74

Dogan N, Leybovich L B, King S and Sethi A 2001b Comparison of static step-and-shoot IMRT plans developed by two different commercial treatment planning systems: NOMOS-CORVUS vs CMS-FOCUS *Radiother. Oncol.* **61** (Suppl. 1) S82

Dogan N, Leybovich L B, King S, Sethi A and Emami B 2000c Comparison of treatment plans developed for concave-shaped head and neck tumors: IMRT versus 3-D conformal versus 2-D *Proc. 42nd Annual ASTRO Meeting. Int. J. Radiat. Oncol. Biol. Phys.* **48** (Suppl. 1) 176

Dogan N, Leybovich L B and Sethi A 2001a Investigation of dose distributions in adjacent static step-and-shoot IMRT and electron fields *Radiother. Oncol.* **61** (Suppl. 1) S66

——2002b Investigation of split IMRT fields for step-and-shoot-IMRT delivery *Radiother. Oncol.* **64** (Suppl. 1) S211

——2003b An automatic feathering technique for split IMRT fields *Med. Phys.* **30** 1335–6

Dogan N, Leybovich L, Sethi A and Emami B 2000b Comparison of IMRT plans for treatment of head and neck cancers using modified and unmodified tomographic and MLC-based static step-and-shoot techniques *Radiother. Oncol.* **56** (Suppl. 1) S106

——2003a Automatic feathering of split fields for step-and-shoot intensity modulated radiation therapy *Phys. Med. Biol.* **48** 1133–40

——2002a Improvement of dose distributions in abutment regions of intensity modulated radiation therapy and electron fields *Med. Phys.* **29** 38–44

Dogan N, Leybovich L B, Sethi A, Krasin M and Emami B 2000a A modified method of planning and delivery for dynamic multileaf collimator intensity-modulated radiation therapy *Int. J. Radiat. Oncol. Biol. Phys.* **47** 241–5

Donaldson S S, Boyer A L and Hendee W R 2000 New methods for precision radiation therapy exceed biological and clinical knowledge and institutional resources needed for implementation *Med. Phys.* **27** 2477–9

Dong D, Grant W, McGary J, Teh B and Butler E B 2000b Effect of beam energy in prostate serial tomotherapy *Proc. 42nd Annual ASTRO Meeting. Int. J. Rad. Oncol. Biol. Phys* **48** (Suppl.) 347

Dong L and Court L 2003 Intercomparison of three in-room imaging modalities for patient alignment: EPID, BAT, and CT-on-rails *Med. Phys.* **30** 1473

Dong L, Court L, Wang H, O'Daniel J and Mohan R 2004 Image-guided radiotherapy with in-room CT-on-rails *Proc. 14th Int. Conf. on the Use of Computers in Radiation Therapy (Seoul, May)* pp 18–19

Dong L, Liu H, Wang X, Zhang X, Tu S, Mohan R and Wu Q 2003 The effect of photon beam energy on the quality of IMRT plans *Med. Phys.* **30** 1485

Dong L, McGary J, Bellezza D, Berner B and Grant W 2000a Whole-body dose from Peacock-based IMRT treatment *Proc. 42nd Annual ASTRO Meeting. Int. J. Rad. Oncol. Biol. Phys* **48** (Suppl.) 342

Donovan E M, Bleackley N J, Evans P M, Reise S F and Yarnold J R 2002b Dose-position and dose-volume histogram analysis of standard wedged and intensity modulated treatments in breast radiotherapy *Brit. J. Radiol.* **75** 967–73

Donovan E M, Johnson U, Shentall G, Evans P M, Neal A J and Yarnold J R 2000 Evaluation of compensators in breast radiotherapy—a planning study using multiple static fields *Int. J. Radiat. Oncol. Biol. Phys.* **46** 671–9

Donovan E M, McNair H A, Eagle S, Shoulders B, Clements N, Evans P M, Symonds-Tayler J R N and Yarnold J R 2003 Minimizing cardiac irradiation in breast radiotherapy with an automatic breathing control device: Initial findings *Clinical Oncol.* **15** (Suppl. 2) S14

Donovan E M, McNair H, Evans P M, Symonds-Tayler R and Yarnold J R 2002a Initial experience with an automatic breathing control device in breast radiotherapy *Radiother. Oncol.* **64** (Suppl. 1) S45

Drzymala R, Low D and Klein E 2001 Geometric aspects of intensity modulated radiation therapy planning *Med. Phys.* **28** 1205

Duan J, Shen S, Fiveash J, Brezovich I, Popple R and Pareek P 2002 Impact of dose fractionation on dosimetric errors induced by respiratory motions during IMRT delivery: a 3-D volumetric dose measurement approach *Med. Phys.* **29** 1346

——2003b Dosimetric effect of respiration-gated beam on IMRT delivery *Med. Phys.* **30** 2241–52

Duan J, Shen S, Popple R, Ye S, Pareek P and Brezovich I 2003a Effect of the number of treatment fields on respiration-induced dose errors in single and multi-fraction IMRT delivery *Med. Phys.* **30** 1340

Dyke L, Emery R, Lam T and Berson A 2001 Combining respiratory gating and forward IMRT to treat patients with breast cancer of the left breast *Med. Phys.* **28** 1987

Earl M A, Naqvi S, Shepard D M and Yu C X 2003c A comparison between direct aperture optimization and two commercial IMRT systems *Med. Phys.* **30** 1335

Earl M, Shepard D, Li X A, Naqvi S and Xu C X 2002 Simplifying IMRT with direct aperture optimization *Med. Phys.* **29** 1305

Earl M A, Shepard D M, Naqvi S, Li X A and Yu C X 2003a Inverse planning for intensity-modulated arc therapy using direct aperture optimization *Phys. Med. Biol.* **48** 1075–89

Earl M A, Shepard D M, Naqvi S and Yu C X 2003b Jaws-only IMRT using direct aperture optimization *Med. Phys.* **30** 1487

Eberle K, Engler J, Hartmann G, Hofmann R and Hörandel J R 2003 First tests of a liquid ionisation chamber to monitor intensity modulated radiation beams *Phys. Med. Biol.* **48** 3555–64

Eisbruch A 2002 Intensity-modulated radiotherapy of head-and-neck cancer: encouraging early results *Int. J. Radiat. Oncol. Biol. Phys.* **53** 1–3

El Naqa I, Low D A, Nystrom M, Parikh P, Lu W, Deasy J O, Amini A, Hubenschmidt J and Wahab S 2004 An optical flow approach for automated breathing motion tracking in 4D computed tomography *Proc. 14th Int. Conf. on the Use of Computers in Radiation Therapy (Seoul, May)* pp 74–7

Engstrom P, Hansen H S, Haraldsson P, Landberg T, Engelholm S A and Nystrom H 2002 In vivo dose verification of IMRT treated head and neck patients *Radiother. Oncol.* **64** (Suppl. 1) S237

Erdi Y E, Yorke E D, Erdi A K, Braban L, Hu Y, Macapinlac H, Humm J L, Larson S M and Rosenzweig K 2000 Radiotherapy treatment planning for patients with non-small cell lung cancer using positron emission tomography (PET) *Proc. 42nd Annual ASTRO Meeting. Int. J. Radiat. Oncol. Biol. Phys.* **48** (Suppl.) 127

Erridge S C, Robar J L and Wu J S 2003 The use of electronic compensation with MLC in wide field irradiation for head and neck cancers *Clinical Oncol.* **15** (Suppl. 2) S13–14

Essers M, de Langen M, Dirkx M L P and Heijmen B J M 2001 Commissioning of a commercially available system for intensity-modulated radiotherapy dose delivery with dynamic multileaf collimation *Radiother. Oncol.* **60** 215–24

Esthappan J, Mutic S, Malyapa R S, Grigsby P W, Zoberi I, Dehdashti F, Miller T R, Bosch W R and Low D A 2004 Treatment planning guidelines regarding the use of CT/PET guided IMRT for cervical carcinoma with positive paraaortic lymph nodes *Int. J. Radiat. Oncol. Biol. Phys.* **58** 1289–97

Evans P M, Donovan E M, Partridge M, Childs P J, Convery D J, Eagle S, Hansen V N, Suter B L and Yarnold J R 2000 The delivery of intensity modulated radiotherapy to the breast using multiple fields *Radiother. Oncol.* **57** 79–89

Evans P M, Hansen V N and Swindell W 1997 The optimum intensities for multiple static multileaf collimator field compensation *Med. Phys.* **24** 1147–56

Evans P M, Partridge M and Symonds-Tayler J R N 2001 Sampling considerations for intensity-modulated radiotherapy verification using electronic portal imaging *Med. Phys.* **28** 543–52

Ezzell G A 2003 Clinical implementation of IMRT treatment planning *IMRT—the State of the Art: AAPM Medical Physics Monograph Number 29* ed J R Palta and T R Mackie (Madison: WI: Medical Physics Publishing) pp 475–93

Ezzell G and Chungbin S 2001 Low MU/segment issues with IMRT *Med. Phys.* **28** 1253–4

Ezzell G A, Galvin J M, Low D, Palta J R, Rosen I, Sharpe M B, Xia P, Xiao Y, Xing L and Yu C X 2003 Guidance document on delivery, treatment planning, and clinical implementation of IMRT: report of the IMRT subcommittee of the AAPM radiation therapy committee *Med. Phys.* **30** 2089–115

Ezzell G A, Schild S E and Wong W W 2000 Development of a treatment planning protocol for prostate treatments with IMRT *Proc. 42nd Annual ASTRO Meeting. Int. J. Radiat. Oncol. Biol. Phys.* **48** (Suppl.) 140

Falco T, Evans M D C, Shenouda G and Podgorsak E B 2001 Ultrasound imaging for external beam prostate treatment setup and dosimetric verification *Med. Phys.* **28** 1988

Faria S L, Ferrigno R and Souhami L 2002 Not treating all fields in every radiotherapy session. Questioning an old dogma *Radiother. Oncol.* **64** (Suppl. 1) S192

Fayos F, Saez Beltran M and Sanchez R 2001a Measurements for implementing IMRT with dMLC *Radiother. Oncol.* **61** (Suppl. 1) S107

——2001b Quality assurance in IMRT with dMLC *Radiother. Oncol.* **61** (Suppl. 1) S48

Fenton P and Lynch R 2002 A retrospective study on the impact of image registration use in radiation therapy treatment planning for primary brain tumour patients *Radiother. Oncol.* **64** (Suppl. 1) S210

Fielding A L, Clark C H and Evans P M 2003a Verification of patient position and delivery of IMRT by electronic portal imaging *Med. Phys.* **30** 1352

Fielding A L and Evans P M 2001 The use of electronic portal images to verify the delivery of intensity modulated radiotherapy beams produced with dynamic multi-leaf collimation *Radiother. Oncol.* **61** (Suppl. 1) S26

Fielding A L, Evans P M and Clark C H 2002a The use of electronic portal imaging to verify patient position during intensity modulated radiotherapy delivered by the dynamic MLC technique *Proc. IPEM Biennial Meeting (Southampton, 16–17 July)* pp 10–11

——2002b The use of electronic portal imaging to verify the patient set-up and delivery of intensity modulated radiotherapy *Radiother. Oncol.* **64** (Suppl. 1) S12

——2002c The use of electronic portal imaging to verify patient position during intensity modulated radiotherapy delivered by the dynamic MLC technique *Int. J. Radiat. Oncol. Biol. Phys.* **54** 1225–34

——2003b A new method for verification of patient position and delivery of IMRT using electronic portal imaging *Proc. World Congress in Medical Physics (Sydney)* CDROM and website only

Fitchard E, Olivera G, Kapatoes J, Ruchala K, Reckwerdt P and Mackie R 2000 Registration of tomotherapy animal study data *Proc. World Congress of Medical Physics and the AAPM Annual Congress, August 2000* paper WE-E313-07

Fiorino C, Cozzarini C, Fellin G, Fopplano F, Menegotti L, Piazzolla A, Sanguineti G, Vavassori V and Valdegni R 2002 Evidence of correlation between rectal DVHs and late bleeding in patients treated for prostate cancer *Radiother. Oncol.* **64** (Suppl. 1) S71

Fiorino C, Gianolini S and Nahum A E 2003 A cylindrical model of the rectum: comparing dose-volume, dose-surface and dose-wall histograms in the radiotherapy of prostate cancer *Phys. Med. Biol.* **48** 2603–16

Fix M K, Manser P, Born E J, Mini R and Rüegsegger P 2001a Monte Carlo simulation of dynamic MLC using a multiple source model *Med. Phys.* **28** 1197

——2001b Monte Carlo simulation of dynamic MLC using a multiple source model *Phys. Med. Biol.* **46** 3241–57

——2002 Influence of dynamic MLC on photon beam energy spectra *Med. Phys.* **29** 1270

Fogg R, Chetty I, DeMarco J, Agazaryan N and Solberg T 2000 Development and modification of a virtual source model for Monte Carlo based IMRT verification *Proc. World Congress of Medical Physics and the AAPM Annual Congress, August 2000* paper WE-B309-03

Fogliata A, Bolsi A and Cozzi L 2002 Critical appraisal of treatment techniques based on conventional photon beams, intensity modulated photon beams and proton beams for therapy of the intact breast *Radiother. Oncol.* **62** 137–45

——2003 Comparative analysis of intensity modulation inverse planning modules of three commercial treatment planning systems applied to head and neck tumour model *Radiother. Oncol.* **66** 29–40

Fong P M, Keil D C, Does M D and Gore J C 2001 Polymer gels for magnetic resonance imaging of irradiation dose distributions at normal room atmosphere *Phys. Med. Biol.* **46** 3105–14

Foppiano F, Fiorino C, Frezza G, Greco C, Valdagni R and the AIRO National Working Group on Prostate Radiotherapy 2003 The impact of contouring uncertainty on rectal 3D dose-volume data: Results of a dummy run in a multicenter trial (AIROPROS01-02) *Int. J. Radiat. Oncol. Biol. Phys.* **57** 573–9

Ford E, Chang J, Mueller K, Mageras G, Yorke E, Ling C and Amols H 2002b Low-dose megavoltage cone-beam CT with an amorphous silicon electronic portal imaging device *Med. Phys.* **29** 1241

Ford E C, Mageras G S, Yorke E, Rosenzweig K E, Wagman R and Ling C C 2002a Evaluation of respiratory movement during gated radiotherapy using film and electronic portal imaging *Int. J. Radiat. Oncol. Biol. Phys.* **52** 522–31

Forster K, Stevens C, Kitamura K, Starkschall G, Liu H, Liao Z, Chang J, Cox J, Jeter M, Guerrero T and Komaki R 2003 3D assessment of respiration induced NSCLC tumor motion *Med. Phys.* **30** 1364

Fraass B 2001 Optimization and IMRT: performing valid comparisons between techniques *Radiother. Oncol.* **61** (Suppl. 1) S51

Fraass B A, McShan D L, Vineberg K A, Balter J, Ten Haken R K and Eisbruch 2002 Generation of robust IMRT plans for head/neck cancer: replacing the PTV with direct accommodation of setup uncertainty and buildup region dosimetry *Radiother. Oncol.* **64** (Suppl. 1) S96

Francescon P, Cora S and Chiovati P 2001 Verification of an intensity modulated treatment plan with a Monte Carlo code *Radiother. Oncol.* **61** (Suppl. 1) S2

Fransson A, Backfrieder W, Lorang T, Dieckmann K, Bogner J, Nowotny R and Pötter R 2000a Treatment planning based on MR images *Radiother. Oncol.* **56** (Suppl. 1) S17

Fransson A, Bogner J, Dieckmann K and Pötter R 2000b Registration of MR and CT images using the BrainLab and Helax TMS treatment planning systems—a comparative study *Radiother. Oncol.* **56** (Suppl. 1) S208

Frazier R C, Vicini F A, Sharpe M B, Fayad J, Yan D, Martinez A A Wong J W 2000 The impact of respiration on whole breast radiotherapy: A dosimetric analysis using active breathing control *Proc. 42nd Annual ASTRO Meeting. Int. J. Radiat. Oncol. Biol. Phys.* **48** (Suppl.) 200

Frazier R C, Vicini F A, Sharpe M B, Yan D, Fayad J, Baglan K L, Kestin L L, Remouchamps V M, Martinez A A and Wong J W 2004 Impact of breathing motion on whole breast radiotherapy: A dosimetric analysis using active breathing control *Int. J. Radiat. Oncol. Biol. Phys.* **58** 1041–7

Freedman G M, Price Jnr R A, Mah D, Milestone B, Movsas B, Horwitz E, Mitra R and Hanks G E 2001 Routine use of MRI and CT simulation for treatment planning of intensity modulated radiation therapy (IMRT) in prostate cance *Int. J. Radiat. Oncol. Biol. Phys.* **51** (Suppl. 1) 301

Frencl L, Cernohuby P and Drnec I 2001 IMRT without MLC—When, how and why? *Radiother. Oncol.* **61** (Suppl. 1) S98

Fu W, Dai J, Hu Y, Han D and Song Y 2004 Delivery time comparison for intensity-modulated radiation therapy with/without flattening filter: a planning study *Phys. Med. Biol.* **49** 1535–47

Gaboriaud G, Papatheodorou S, Pontvert D and Rosenwald J 2000 Comparison of non-coplanar three-dimensional conformal radiation therapy and non-coplanar intensity-modulated radiation therapy in the case of brain tumours *Proc. World Congress of Medical Physics and the AAPM Annual Congress, August 2000* paper MO-B309-05

Gaede S, Wong E and Rasmussen H 2001 Evaluating alternative approaches to inverse planning optimization *Med. Phys.* **28** 1206

——2002 Evaluation of a systematic beam direction selection algorithm for IMRT *Med. Phys.* **29** 1335

——2004 An algorithm for systematic selection of beam directions for IMRT *Med. Phys.* **31** 376–88

Gagliardi G, Lax I, Ottolenghi A and Rutqvist L E 1996 Long term cardiac mortality after radiotherapy of beast cancer—applications of the relative seriality model *Brit. J. Radiol.* **69** 839–46

Gall K and Chang C 2003 Accuracy of the Accuray Cyberknife dose deposition algorithm *Med. Phys.* **30** 1352

Gallant G and Schreiner L 2000 TOMOPlan 1: Towards a 3D RTP optimization system for cobalt tomotherapy *Proc. World Congress of Medical Physics and the AAPM Annual Congress, August 2000* paper MO-CXH-12

Galvin J M 1999 The multileaf collimator—a complete guide *Proc. 1999 AAPM Annual Meeting*; see also *Med. Phys.* **26** 1092–3

Galvin J, Bednarz G, Lin L, Lally B and Komarnicky L 2000a Using segmented fields to treat the breast *Proc. World Congress of Medical Physics and the AAPM Annual Congress, August 2000* paper MO-GBR-05

Galvin J M, Chen X G and Smith R M 1993 Combining multileaf fields to modulate fluence distributions *Int. J. Radiat. Oncol. Biol. Phys.* **27** 697–705

Galvin J M, Ezzell G, Eisbruch A, Yu C, Butler B, Xiao Y, Rosen I, Rosenman J, Sharpe M, Xing L, Xia P, Lomax T, Low D A and Palta J 2004 Implementing IMRT in clinical practice: a joint document of the American society for therapeutic radiology and oncology and the American association of physicists in medicine *Int. J. Radiat. Oncol. Biol. Phys.* **58** 1616–34

Galvin J, Xiao Y, Bednarz G and Huq S 2000b A comparison of optimized forward planning and inverse planning for target with invaginations *Radiother. Oncol.* **56** (Suppl. 1) S115

Galvin J, Xiao Y, Michalski D, Censor Y, Houser C, Bednarz G and Huq M S 2001 Segmental inverse planning (SIP) that starts with a definition of allowable fields *Med. Phys.* **28** 1253

Garcia-Alvarez R, Liney G P and Beavis A W 2003a The use of functional MR for planning conformal avoidance radiotherapy delivered by IMRT *Clinical Oncol.* **15** (Suppl. 2) S13

——2003b Use of functional magnetic resonance imaging for planning intensity modulated radiotherapy *J. Radiother. Practice* **3** 66–72

George R, Keall P J, Kini V R, Vedam S S, Siebers J V, Wu Q, Lauterbach M H, Arthus D W and Mohan R 2003 Quantifying the effect of intrafraction motion during breast IMRT planning and dose delivery *Med. Phys.* **30** 552–62

Ghelmansarai F, Pouliot J, Zheng Z and Svatos M 2003 Low dose-high contrast megavoltage cone beam CT *Med. Phys.* **30** 1345

Gibbs I C, Sattah M, Guerrero T, Chang S D, Martin D, Kim D and Le Q T 2002 Cyberknife image-guided radiosurgery of the spine *Radiother. Oncol.* **64** (Suppl. 1) S300

Gibon D, Kulik C, Poupon L, Castelain B, Vasseur C and Rousseau J 2000 Optimization of micro-multileaf collimator irradiation parameters in conformal radiation therapy *Radiother. Oncol.* **56** (Suppl. 1) S211

Gierga D, Chen G, Kung J, Betke M, Lombardi J and Willett C 2003 Quantitative fluoroscopy of respiration-induced abdominal tumor motion on IMRT dose distributions *Med. Phys.* **30** 1340

——2004 Quantification of respiration-induced abdominal tumor motion and its impact on IMRT dose distributions *Int. J. Radiat. Oncol. Biol. Phys.* **58** 1584–95

Gierga D and Jiang S 2002 A Monte Carlo study on the interplay effect between MLC leaf motion and respiration-induced tumor motion in lung IMRT delivery *Med. Phys.* **29** 1266

Gillin M T 2003 Socio-economic issues of IMRT in *IMRT—the State of the Art: AAPM Medical Physics Monograph Number 29* cd J R Palta and T R Mackie (Madison, WI: Medical Physics Publishing) p 829–41

Giraud P, Helfre S, Servois V, Dubray B, Beigelman-Grenier C P, Zalcman-Liwartoski G A, Straus-Zelter C M, Neuenschwander S, Rosenwald J C and Cosset J M 2000 Evaluation of intrathoracic organs mobility using CT gated by a spirometer *Radiother. Oncol.* **56** (Suppl. 1) S39

Glatstein E 2003a The return of the snake oil salesmen *Int. J. Radiat. Oncol. Biol. Phys.* **55** 561–2

——2003b In response to Dr Amols *Int. J. Radiat. Oncol. Biol. Phys.* **56** 1507–8

Glendinning A G, Hunt S G and Bonnett D E 2001 Comparison of collimator position verification during IMRT using a commercial EPID and a novel strip ionisation chamber *Radiother. Oncol.* **61** (Suppl. 1) S17

Gluckman G R, Foraci J, Meek A G and Reinstein L E 2001b A system for comprehensive 3D validation of localization, planning and treatment delivery for IMRT *Int. J. Radiat. Oncol. Biol. Phys.* **51** 171–2

Gluckman G, Foraci J, Meek A, Reinstein L, Radford D and Followill D 2001a Adaptation of the RPC IMRT prostate phantom for full 3D dose verification using polymer gel and multiple-plane radiochromic film inserts *Med. Phys.* **28** 1269

Gluckman G and Reinstein L 2000 Evaluation of a new software package for validation of 3D-conformal treatment delivery *Proc. World Congress of Medical Physics and the AAPM Annual Congress, August 2000* paper MO-FXH-55

Goddu S, Vanek K N and Nelson S 2002 Moving target, dynamic treatment: is the respiratory gating a solution? *Med. Phys.* **29** 1270

Goitein M 2004 Organ and tumour motion: an overview *Seminars Radiat. Oncol.* **14** 2–9

Goldman S P, Chen J Z, Battista J J 2004 Fast inverse dose optimisation (FIDO) for IMRT via matrix inversion with no negative intensities *Proc. 14th Int. Conf. on the Use of Computers in Radiation Therapy (Seoul, May)* pp 112–15

Graham J D 2000 Geometrical uncertainties in prostate and bladder cancer *Radiother. Oncol.* **56** (Suppl. 1) S51

Greer P B, Beckham W A and Ansbacher W 2003 Improving the resolution of dynamic intensity modulated radiation therapy delivery by reducing the multileaf collimator sampling distance *Med. Phys* **30** 2793–803

Greer P B and van Doorn T 2000 A design for a dual assembly multileaf collimator *Med. Phys.* **27** 2242–55

Gregoire V 2002 Multimodality image coregistration for treatment planning in HNSCC *Radiother. Oncol.* **64** (Suppl. 1) S35

Gregoire V, Daisne J F, Duprez T, Geets X, Hamoir M, Lonneux M and Weynand B. 2003 Pathology-based verification of head and neck imaging: how far are we from the truth and what are the implications for 3D-CRT and IMRT? *Radiother. Oncol.* **68** (Suppl. 1) S16

Grigereit T, Nelms B, Dempsey J, Garcia-Ramirez J, Low D, Purdy J 2000 Compensating filters for IMRT 1: Material characterization and process verification *Proc. World Congress of Medical Physics and the AAPM Annual Congress, August 2000* paper TU-EBR-05

Grigorov G, Kron T, Wong E, Chen J, Sollazzo J and Rodrigues G 2003 Optimisation of helical tomotherapy treatment plans for prostate cancer *Phys. Med. Biol.* **48** 1933–43

Groll J, Bohsung J, Jahn U, Kurth C, Pfaender M, Grebe G, Budach V 2000 Dosimetric verification of the Varian IMRT system *Proc. World Congress of Medical Physics and the AAPM Annual Congress, August 2000* paper WE-E313-03

Grosser K-H 2000 Reduction of treatment time in IMRT step and shoot irradiations by means of a changed fraction scheme *Proc. 13th Int. Conf. on the Use of Computers in Radiation Therapy (Heidelberg, May)* ed W Schlegel and T Bortfeld (Heidelberg: Springer) pp 308–10

Grosser K H, Kober B and Thilmann C 2002 Minimizing field components for delivering of intensity-modulated radiotherapy step-and-shoot beam sequences *Radiother. Oncol.* **64** S213

Guan H 2002 Megavoltage portal CT using an amorphous silicon imaging device with comparison to a CCD based EPID system *Med. Phys.* **29** 1243

Guerrero T M, Crownover R L, Rodebaugh R F, Pawlicki T, Martin D P, Glosser G D, Whyte R I, Le Q T, Murphy M J, Shiomi H, Weinhaus M S and Ma C 2001 Breath holding versus real-time target tracking for respiratory motion compensation during radiosurgery for lung tumors *Int. J. Radiat. Oncol. Biol. Phys.* **51** (Suppl. 1) 26

Guerrero Urbano M T and Nutting C M 2004a Clinical use of intensity-modulated radiotherapy: part 1 *Brit. J. Radiol.* **77** 88–96

——2004b Clinical use of intensity-modulated radiotherapy: part 2 *Brit. J. Radiol.* **77** 177–82

Gull S F and Skilling J 1984 The maximum entropy method in image processing *IEE Proc. Special Issue on Communications, Radar and Signal Processing* **6** 646

Gulliford S L, Webb S, Rowbottom C G, Corne D W and Dearnaley D P 2004 Use of artificial networks to predict biological outcomes for patients receiving radical radiotherapy of the prostate *Radiother. Oncol.* **71** 3–12

Gum F, Bogner L, Scherer J, Bock M and Rhein B 2001 3D-dose verification of IMRT treatment plans by means of an inhomogeneous anthropomorphic Fricke-gel phantom *Med. Phys.* **28** 1207

Gunawardena A, Meyer R, D'Souza W, Shi L, Yang W and Naqvi S 2003 A new leaf-sequencing algorithm using difference matrices reduces aperture number in IMRT *Med. Phys.* **30** 1347

Gurgoze E and Rogers K 2002 Dosimetric replication of MLC by customized filters in the delivery of IMRT *Med. Phys.* **29** 1272

Gustafsson A 2000 A general algorithm for sequential multileaf modulation *Proc. World Congress of Medical Physics and the AAPM Annual Congress, August 2000* paper WE-E313-08

Gustavsson H, Back S A J, Lepage M, Rintoul L and Baldock C 2001a Development and optimisation of a 2-hydroxyethylacrylate MRI polymer gel dosimeter *Radiother. Oncol.* **61** S91

Gustavsson H, Back S A H and Olsson L E 2001b Verification of IMRT using gel dosimetry *Radiother. Oncol.* **61** (Suppl. 1) S5

Gustavsson H, Engstrom P, Haraldsson P, Magnusson P and Nystrom H 2000 3D dosimetric verification of IMRT using polymer gel and MRI *Radiother. Oncol.* **56** (Suppl. 1) S205

Gustavsson H, Haraldsson P, Karksson A, Engström P, Bäck S A J, Nyström H and Olsson L E 2002 IMRT verification using magic-type polymer gel *Radiother. Oncol.* **64** (Suppl. 1) S43

Ha S W 2004 Physician's perspective of use of computers in RT *Proc. 14th Int. Conf. on the Use of Computers in Radiation Therapy (Seoul, May)* pp 1–3

Hagekyriakou J, Song G L and Sephton R G 2000 The accuracy of an ultrasound guidance system for positioning the prostate at radiotherapy *Radiother. Oncol.* **56** (Suppl. 1) S38

Hajdok G, Kerr A and Schreiner L 2002 An investigation of megavoltage computed tomography using a Cobalt-60 gamma ray source for radiotherapy treatment verification *Med. Phys.* **29** 1341

Halperin E C 2000 Overpriced technology in radiation oncology *Int. J. Radiat. Oncol. Biol. Phys.* **48** 917–18

Hansen V N, Evans P M, Budgell G J, Mott J H L, Williams P C, Brugmans M J P, Wittkamper F W, Mijnheer B J and Brown K 1998 Quality assurance of the dose delivered by small radiation segments. *Phys. Med. Biol.* **43** 2665–75

Happersett L, Hunt M, Chong L, Chui C, Spirou S, Burman C and Amols H 2000 Intensity modulated radiation therapy for the treatment of thyroid cancer *Proc. 42nd Annual ASTRO Meeting. Int. J. Radiat. Oncol. Biol. Phys.* **48** 351

Happersett L, Mageras G S, Zelefsky M J, Burman C M, Leibel S A, Chui C, Fuks Z, Bull S, Ling C and Kutcher G J 2003 A study of the effects of internal organ motion on dose escalalation in conformal prostate treatments *Radiother. and Oncol.* **66** 263–70

Hardemark B, Rehbinder H and Löf J 2003 Rotating the MLC between segments improves performance in step-and-shoot IMRT delivery *Med. Phys.* **30** 1405

Harnisch G A, Willoughby T R and Kupelian P A 2000 Irradiation of dominant intraprostatic lesions with short-course intensity modulated radiotherapy: Dosimetry and planning *Proc. 42nd Annual ASTRO Meeting. Int. J. Radiat. Oncol. Biol. Phys.* **48** (Suppl.) 205

Hartkens T, Rueckert D, Schnabel J A, Hawkes D J and Hill D L G 2002 VTK CISG Registration Toolkit: an open source software for affine and non-rigid registration of single and multiple 3D images *BVM 2002* (Leipzig: Springer)

Hartmann M, Bogner L and Fippel M 2002a IMRT: Inaccuracies of a conventional inverse pencil beam TPS, obtained by a comparison with a Monte-Carlo based TPS *Radiother. Oncol.* **64** (Suppl. 1) S96

Hartmann M, Bogner L, Scherer J and Scherer S 2001 IMRT optimisation based on a new inverse Monte-Carlo code *Radiother. Oncol.* **61** (Suppl. 1) S47

Hartmann G H and Föhlisch F 2002 Dosimetric characterization of a new miniature multileaf collimator *Phys. Med. Biol.* **47** N171–7

Hartmann G, Pastyr O, Echner G, Hädinger U, Seeber S, Juschka J, Richter J and Schlegel W 2002b A new design for a mid-sized multileaf collimator (MLC) *Radiother. Oncol.* **64** (Suppl. 1) S214

Hawliczek R, Nespor W, Schmidt W and Pawlas K 2000 IMRT at an European community hospital—one year of experience *Radiother. Oncol.* **56** (Suppl. 1) S196

Hector C L 2001 The effect of patient motion on the delivery of intensity-modulated radiotherapy *PhD Thesis* London University

Heijmen B J M 2000 New applications for your electronic portal imaging device (EPID) *Radiother. Oncol.* **56** (Suppl. 1) S48

Henrys A, Taylor H, Bedford J and Tait D 2003 Optimizing the use of multileaf collimators in the treatment of colo-rectal cancer *Clinical Oncol.* **15** (Suppl. 2) S18

Héon J, Falco T, Shenouda G, Duclos M, Faria S and Souhami L 2002 Ultrasound imaging for external beam prostate treatment setup *Radiother. Oncol.* **64** (Suppl. 1) S45

Herman M G, Rosenzweig D P and Hendee W R 2003 Every patient receiving 3D or IMRT must have image-based target localization prior to turning on the beam *Med. Phys.* **30** 287–9

Higgins P D, Alaei P, Gerbi B J and Dusenbery K E 2003 *In vivo* diode dosimetry for routine quality assurance in IMRT *Med. Phys.* **30** 3118–23

Hills M, Audet C, Duzenli C and Jirasek A 2000a Polymer gel dosimetry using x-ray computed tomography: a feasibility study *Phys. Med. Biol.* **45** 2559–71

Hills M, Jirasek A, Audet C and Duzenli C 2000b X-ray CT polymer gel dosimetry: applications to stereotactic radiosurgery and proton therapy *Radiother. Oncol.* **56** (Suppl. 1) S80

Hoban P 2000 Conversion of leaf sequences to dose distributions in micro-IMRT *Proc. World Congress of Medical Physics and the AAPM Annual Congress, August 2000* paper MO-BBR-03

——2002 Use of a mini-multileaf collimator for simultaneous IMRT boost in prostate cancer treatment *Med. Phys.* **29** 1268

Hoban P, Short R, Biggs D, Rose A, Smee R and Schneider M 2001 Clinical use of a micro-IMRT system *Med. Phys.* **28** 1202–3

Höver K 2003 A quantitative comparison on characteristics of primary-, mini-and micro multileaf collimators *Proc. World Congress in Medical Physics (Sydney)* CDROM and website only

Höver K H, Haering P, Jahn U and Krohnholz H L 2001 Dosimetric characteristics of different MLCs an intercomparison *Radiother. Oncol.* **61** (Suppl. 1) S83

Höver K H, Haering P, Krohnholz H, Würm R and Jahn U 2000 A dosimetrical intercomparison of different micro multileaf collimators *Radiother. Oncol.* **56** (Suppl. 1) S97

Hofman P P, Nederveen A A J, van der Heide U U A, van Moorselaar J R J A, Dehnad H H and Lagendijk J J J 2002 Rectal dose in intensity modulated radiotherapy of the prostate: comparison with the previous conformal irradiation technique *Radiother. Oncol.* **64** (Suppl. 1) S269–70

Holloway L C and Hoban P W 2001 Combining a physically based optimization method with a biologically based tissue importance weight changing method *Int. J. Radiat. Oncol. Biol. Phys.* **51** (Suppl. 1) 72

——2002 Iterative optimization of beam positions for IMRT *Med. Phys.* **29** 1257

Holmberg O, Kasenter A, Verdugo V, Buckney S, Gribben A, Hollywood D and McClean B 2000 Temporal uncertainties of target volumes in step-and-shoot IMRT: a numerical simulation *Radiother. Oncol.* **56** (Suppl. 1) S108

Holmes T W 2001 A method to incorporate leakage and head scatter corrections into a tomotherapy inverse treatment planning algorithm *Phys. Med. Biol.* **46** 11–27

Holmes T W, Shepard D, Naqvi S, Li A, Ma L and Yu C X 2000 Improvements to pencil beam-based inverse treatment planning methodology *Proc. 42nd Annual ASTRO Meeting. Int. J. Radiat. Oncol. Biol. Phys.* **48** (Suppl.) 222

Hong L H, Alektiar K, Chui C, LoSasso T, Hunt M, Spirou S, Yang J, Amols H, Ling C, Fuks Z and Leibel S 2002 IMRT of large fields: Whole-abdomen irradition. *Int. J. Radiat. Oncol. Biol. Phys.* **54** 278–89

Hoogeman M S, Koper P C M, Heemsbergen W D, van Putten W, Boersma L J, Jansen P P, Levendag P C and Lebesque J V 2002 Correlation of 3D dose distribution and rectal complications after prostate treatment *Radiother. Oncol.* **64** (Suppl. 1) S71

Hoogeman M S, van Herk M, de Bois J, Boersma L J, Koper P C M and Lebesque J V 2003 Strategies to reduce the systematic error due to tumor motion in radiotherapy of prostate cancer *Phys. Medica.* **19** 54–5

Hoogeman M S, van Herk M, de Bois J, Muller-Timmermans P, Koper P C M and Lebesque J V 2004 Quantification of local rectal wall displacements by virtual rectum unfolding *Radiother. Oncol.* **70** 21–30

Horn B K P and Schunk B G 1981 Determining optical flow *Artificial Intelligence* **17** 185–204

Hossain M, Houser C and Galvin J 2001 The effect of switch rates on the dose output from an intensity modulating dynamic multileaf collimator *Med. Phys.* **28** 1283

Hou Q and Wang Y 2001 Molecular dynamics used in radiation therapy *Phys. Rev. Lett.* **87** 168101–14

Hou Q, Wang J, Chen Y and Galvin M 2003a An optimisation algorithm for intensity modulated radiotherapy—The simulated dynamics with dose-volume constraints *Med. Phys.* **30** 61–8

——2003b Beam orientation optimization for IMRT by a hybrid method of the genetic algorithm and the simulated dynamics *Med. Phys.* **30** 2360–7

Hou Q, Zhang C, Wu Z and Chen Y 2004 A method to improve spatial resolution and smoothness of intensity profiles in IMRT treatment planning *Med. Phys.* **31** 1339–47

Hough J and Jones D 2002 Robotic patient positioning for radiotherapy *Med. Phys.* **29** 1213

Hounsell A R, Mott J H L, Budgell G J and Wilkinson J M 2001 Dose model considerations for IMRT treatments using multileaf collimators *Phys. Medica.* **17** 161

Hsiung C-Y, Yorke E D, Chui C-S, Hunt M A, Ling C C, Huang E-Y, Wang C-J, Chen H-C, Yeh S-A, Hsu H-C and Amols H I. 2002 Intensity-modulated radiotherapy versus conventional three-dimensional conformal radiotherapy for boost or salvage treatment of nasopharyngeal carcinoma *Int. J. Radiat. Oncol. Biol. Phys.* **53** 638–47

Hu N, Downey D B, Fenster A and Ladak H M 2003 Prostate boundary segmentation from 3D ultrasound images *Med. Phys.* **30** 1648–59

Huang E, Dong L, Chandra A, Kuban D A, Rosen I I, Evans A and Pollack A 2002a Intrafraction prostate motion during IMRT for prostate cancer *Int. J. Radiat. Oncol. Biol. Phys.* **53** 261–8

Huang E, Dong L, Chandra A, Kuban D A, Rosen I I and Pollack A 2001b Intrafraction prostate motion during IMRT for prostate cancer *Int. J. Radiat. Oncol. Biol. Phys.* **51** 212–13

Huang J, Katz A, Haas J, Urrutia T, Lauritano J and MacMelville W 2002b IMRT treatment of prostate cancer using 6MV photon beam *Med. Phys.* **29** 1254

Huang J, Katz A, Haas J, Urrutia T, Lauritano J and Ortiz E 2001a Using beak slit with MIMiC in clinical IMRT *Med. Phys.* **28** 1202

Huang J, Urrutia T, Lauritano J and Katz A 2000 Application of gafchromic film in clinical QA of IMRT *Proc. World Congress of Medical Physics and the AAPM Annual Congress, August 2000* paper MO-CXH-04

Huang T-C, Guerrero T, Zhang G, Lin H-D, Kuban D and Lin K-P 2004 3D optical flow method: Speed, accuracy, and convergence in radiotherapy applications *Proc. 14th Int. Conf. on the Use of Computers in Radiation Therapy (Seoul, May)* pp 569–73

Hugo G D, Agazaryan N and Solberg T D 2002 An evaluation of gating window size, delivery method, and composite field dosimetry of respiratory-gated IMRT *Med. Phys.* **29** 2517–25

——2003a The effects of target motion on planning and delivery of respiratory-gated IMRT *Med. Phys.* **30** 1330

Hugo G, Solberg T, Demarco J, Paul T, Agazaryan N, Arellano A and Smathers J 2001 Preliminary dosimetric investigations into gated IMRT *Med. Phys.* **28** 1294

Hugo G, Tenn S and Agazaryan N 2003b An evaluation of intrafraction motion-induced error for fractionated IMRT delivery *Med. Phys.* **30** 1470

Huq M S, Das I J, Steinberg T and Galvin J M 2002 A dosimetric comparison of various multileaf collimators *Phys. Med. Biol.* **47** N159–70

Humm J, Erdi Y, Nehmeh S, Pugachev A, Zanzonico P and Ling C 2003 The potential role of PET in radiation treatment planning *Radiother. Oncol.* **68** (Suppl. 1) S28

Hunt M A, Hsiung C-Y, Spirou S V, Chui C-S and Amols H I 2000 Evaluation of concave dose distributions created using an inverse planning system *Proc. 42nd Annual ASTRO Meeting. Int. J. Radiat. Oncol. Biol. Phys.* **48** (Suppl.) 138

Hunt M A, Hsiung C-Y, Spirou S V, Chui C-S, Amols H I and Ling C C 2002 Evaluation of concave dose distributions created using an inverse planning system *Int. J. Radiat. Oncol. Biol. Phys.* **54** 953–62

Huntzinger C, and Hunt M 2000 A simple method to estimate IMRT monitor units *Proc. World Congress of Medical Physics and the AAPM Annual Congress, August 2000* paper MO-BBR-07

Hurkmans C W, Cho B C J, Damen E, Zijp L and Mijnheer B J 2002 Reduction of cardiac and lung complication probabilities after breast irradiation using conformal radiotherapy with and without intensity modulation *Radiother. Oncol.* **62** 163–71

Hutchison P and Halperin E C 2002 The hidden persuaders: subtle advertising in radiation oncology *Int. J. Radiat. Oncol. Biol. Phys.* **54** 989–91

Hwang A, Verhey L and Xia P 2001 Using a leaf sequencing algorithm to enlarge treatment field length in IMRT *Med. Phys.* **28** 1261

Intensity-modulated Radiation Therapy Collaborative Working Group 2001 Intensity-modulated radiotherapy: current status and issues of interest *Int. J. Radiat. Oncol. Biol. Phys.* **51** 880–914

Iori M, Paiusco M, Nahum A, Marini P, Iotti C, Polico R, Venturi A, Armaroli L and Borasi G 2003 IMRT: the Reggio Emilia experience *Radiother. Oncol.* **68** (Suppl. 1) S62

Ipe N, Roesier S, Jiang S and Ma C 2000 Neutron measurements for intensity modulated radiation therapy *Proc. World Congress of Medical Physics and the AAPM Annual Congress, August 2000* paper FR-EA328-01

Islam M, Ramaseshan R, Norrlinger B. Gutierrez, Alasti H and Heaton R 2001 Clinical implementation of a commercial IMRT system *Med. Phys.* **28** 1207

Iyadurai R, Subramanium B, Rao P, John S and Ayyangar K 2003 A comparison of manual MLC for telecobalt unit with linac-based MLC in conformal radiotherapy and intensity modulated radiotherapy *Proc. World Congress in Medical Physics (Sydney)* CDROM and website only

Jackson A 2002 Dose-volume relationships in rectal bleeding after conformal radiotherapy for prostate cancer *Radiother. Oncol.* **64** (Suppl. 1) S10

Jaffray D 2002 Flat-panel cone-beam CT: an emerging technology of image-guided radiation therapy *Radiother. Oncol.* **64** (Suppl. 1) S75

——2003 X-ray guided IMRT *IMRT—the State of the Art: AAPM Medical Physics Monograph Number 29* ed J R Palta and T R Mackie (Madison, WI: Medical Physics Publishing) pp 703–26

Jaffray D A, Siewerdsen J H, Wong J W and Martinez A A 2002 Flat-panel cone-beam computed tomography for image-guided radiation therapy. *Int. J. Radiat. Oncol. Biol. Phys.* **53** 1337–49

James H, Atherton S, Budgell G, Kirby M and Williams P 2000 Verification of dynamic multileaf collimation using an electronic portal imaging device *Phys. Med. Biol.* **45** 495–509

James H V, Crosbie J C, Poynter A J, Bulusu V R, LeVay J, Morgan J S, Scrase C D, and Hardy V 2002a IMRT in routine clinical use—practicalities within a small, non-academic hospital *Proc. IPEM Biennial meeting (Southampton, 16–17 July)* p 35

James H V, Poynter A J, Crosbie J C, MacKenzie L C, Boston S L and LeVay J 2002b Electronic compensation for CT planned breast treatments *Radiother. Oncol.* **64** (Suppl. 1) S133-S-4

James H V, Poynter A J, Crosbie J, Smith S, Le Vay J, Bulusu V R, Morgan J, Scrase C D and Hardy V 2003 Establishing IMRT in a small oncology centre *Clinical Oncol.* **15** (Suppl. 2) S6

James H V, Scrase C D and Poynter A J 2004 Practical experience with intensity-modulated radiotherapy *Brit. J. Radiol* **77** 3–14

Jayaraman S 2003 IMRT, 3D-CRT, inverse planning, etc., are new buzz words, but correspond to the same old historical concepts *Proc. World Congress in Medical Physics (Sydney)* CDROM and website only

Jeraj R, Balog J, Wenman D, Olivera G, Pearson D, Reckwerdt P and Mackie T R 2001 Monte Carlo simulation of the tomotherapy prototype *Radiother. Oncol.* **61** (Suppl. 1) S27

Jeraj R and Keall P 2000 The effect of statistical uncertainty on inverse treatment planning based on Monte Carlo dose calculation *Phys. Med. Biol.* **45** 3601–13

Jeraj R, Keall P J and Siebers J V 2002a The effect of dose calculation accuracy on inverse treatment planning *Phys. Med. Biol.* **47** 391–407

Jeraj R, Mackie T R, Balog J, Olivera G, Pearson D, Kapatoes J, Ruchala K and Reckwerdt P 2004 Radiation characteristics of helical tomotherapy *Med. Phys.* **31** 396–404

Jeraj R, Wu C and Mackie T R 2002b Clinical importance of optimisation convergence and local minima in inverse treatment planning *Radiother. Oncol.* **64** (Suppl. 1) S97

Jiang S 2004 Tracking tumor with dynamic MLC: Be SMART *Proc. 14th Int. Conf. on the Use of Computers in Radiation Therapy (Seoul, May)* p 47

Jiang S B, Bortfeld T, Trofimov A, Rietzel E, Sharp G, Choi N and Chen G T Y 2003a Synchronized moving aperture radiation therapy (SMART): Plan optimization and MLC leaf sequencing based on 4D CT data *Med. Phys.* **30** 1384

——2004a Synchronized moving aperture radiation therapy (SMART): Treatment planning using 4D CT data *Proc. 14th Int. Conf. on the Use of Computers in Radiation Therapy (Seoul, May)* pp 429–32

Jiang S B and Doppke K 2001b Dosimetric effect of respiratory motion on the treatment of breast cancer with tangential fields *Med. Phys.* **28** (6), 1228

Jiang S B, Kung J H, Zygmanski P and Chen G T Y 2001a A method to control the spatial distribution of target dose heterogeneities in IMRT inverse planning *Med. Phys.* **28** 1205

Jiang S B, Neicu T and Zygmanski P 2003b Synchronized moving aperture radiation therapy (SMART): Superimposing tumor motion on IMRT MLC leaf sequences under realistic delivery limitations *Med. Phys.* **30** 1399

Jiang S B, Pope C, Al Jarrah K M, Kung J H, Bortfeld T and Chen G T Y 2003c An experimental investigation on intra-fractional organ motion effects in lung IMRT treatments *Phys. Med. Biol.* **48** 1773–84

Jiang S, Rivinius C, Al Jarrah K, Kung J, Bortfeld T and Chen G 2002 An experimental investigation on interplay effect between DMLC leaf motion and respiration-induced tumor motion in lung IMRT treatment *Med. Phys.* **29** 1347

Jiang Z, Shepard D M, Earl M A and Yu C X 2004b Asymptotic limit of few field IMRT *Proc. 14th Int. Conf. on the Use of Computers in Radiation Therapy (Seoul, May)* pp 228–31

Johansen S, Eilertsen K and Furre T 2003 Comparison of treatment plans using conventional RT and IMRT for the treatment of 10 patients with different diagnosis *Radiother. Oncol.* **68** (Suppl. 1) S16

Jolly J, Ding M, Pawlicki T, Chu J and Ma C-M 2001 Monte Carlo simulation of the effect of thoracic motion on breast treatment *Med. Phys.* **28** 1228

Jones A O, Das I J and Jones F L 2003 A Monte Carlo study of IMRT beamlets in inhomogeneous media *Med. Phys.* **30** 296–300

Jones L and Hoban P 2000 The effect of movement on optimised IMRT dose distributions *Proc. World Congress of Medical Physics and the AAPM Annual Congress, August 2000* paper MO-CXH-03

——2002 A comparison of physically and radiobiologically based optimisation for IMRT *Med. Phys.* **29** 1447–55

Ju S G, Ahn Y C, Huh S J and Yeo I J 2002 Film dosimetry for intensity modulated radiation therapy: Dosimetric evaluation. *Med. Phys.* **29** 351–5

Jursinic P, Zhu R and Nelms B 2002 An implementation of solid filter (compensator)-based intensity modulated radiation therapy for clinical use *Med. Phys.* **29** 1266

Kagan A R, Schulz R J 2003 A commentary on dose escalation and bned in prostate cancer *Int. J. Radiat. Oncol. Biol. Phys.* **55** 1151

Kamath S, Sahni S, Li J, Palta J and Ranka S 2003 Leaf sequencing algorithms for segmented multileaf collimation *Phys. Med. Biol.* **48** 307–24

Kamath S, Sahni S, Palta J and Ranka S 2004a Algorithms for optimal sequencing of dynamic multileaf collimators *Phys. Med. Biol.* **49** 33–54

Kamath S, Sahni S, Palta J, Ranka S and Li J 2004b Optimal leaf sequencing with elimination of tongue-and-groove underdosage *Phys. Med. Biol.* **49** N7–19

Kapatoes J M, Olivera G H, Balog J P, Keller H, Reckwerdt P J and Mackie T R 2001a On the accuracy and effectiveness of dose reconstruction for tomotherapy *Phys. Med. Biol.* **46** 943–66

Kapatoes J M, Olivera G H, Lu W, Reckwerdt P J, Keller H, Balog J and Mackie T R 2000 Dose gradients as a tool in the optimization and verification of intensity modulated radiation therapy (IMRT) *Proc. 42nd Annual ASTRO Meeting. Int. J. Radiat. Oncol. Biol. Phys.* **48** 139

Kapatoes J, Olivera G, Reckwerdt P, Ruchala K, Jeraj R and Mackie R 2001b A strategy for rapidly adjusting dose distributions *Radiother. Oncol.* **61** (Suppl. 1) S16

Kapatoes J M, Olivera G H, Ruchala K J and Mackie T R 2001c On the verification of the incident energy fluence in tomotherapy IMRT *Phys. Med. Biol.* **46** 2953–65

Kapatoes J M, Olivera G H, Ruchala K J, Smilowitz J B, Reckwerdt P J and Mackie T R 2001d A feasible method for clinical delivery verification and dose reconstruction in tomotherapy *Med. Phys.* **28** 528–42

Kapatoes J, Ruchala K, Olivera G, Welsh J, Forrest L, Turek M, Mackie R, Jaradat H, Jeraj R and Reckwerdt P 2003 Internal 3D patient dose verification in image-guided radiotherapy *Proc. World Congress in Medical Physics (Sydney)* CDROM and website only

Kapulsky A, Mullokandov E, Gejerman G, Saini A, Rebo I and Cardel F 2001 An automated phantom-film QA procedure for IMRT treatments *Med. Phys.* **28** 1284

Kavanagh B D 2003 The emperor's new isodose curves *Med. Phys.* **30** 2559–60

Keall P 2004 4-dimensional computed tomography imaging and treatment planning *Seminars Radiat. Oncol.* **14** 81–90

Keall P, Kini V, Vedam S and Mohan R 2001b 4D IMRT: Dosimetric and practical considerations *Med. Phys.* **28** 1252

Keall P J, Siebers J V, Arnfield M, Lim J O and Mohan R 2001a Monte Carlo dose calculation for dynamic IMRT treatments *Phys. Med. Biol.* **46** 929–41

Keall P, Siebers J, Kim J and Mohan R 2002b The clinical implementation of Monte Carlo for IMRT *Radiother. Oncol.* **64** (Suppl. 1) S105

Keall P J, Starkshall G, Shukla H, Forster K M, Ortiz V, Stevens C W, Vedam S S, George R, Guerrero T and Mohan R 2004 Acquiring 4D thoracic CT scans using a multislice helical method *Phys. Med. Biol.* **49** 2053–67

Keall P, Vedam S, George R, Kinl V and Mohan R 2002a 4D IMRT: a phantom study *Radiother. Oncol.* **64** (Suppl. 1) S261

Keall P J and Williamson J F 2003 Clinical evidence that more precisely defined dose distributions will improve cancer survival and decrease morbidity *Med. Phys.* **30** 1281–2

Keall P, Wu Q, Wu Y and Kim J O 2003 Dynamic MLC IMRT *IMRT—the State of the Art: AAPM Medical Physics Monograph Number 29* ed J R Palta and T R Mackie (Madison, WI: Medical Physics Publishing) pp 319–71

Keane J T, Fontenla D P and Chui C S 2000 Applications of IMAT to total body radiation (TBI) *Proc. 42nd Annual ASTRO Meeting. Int. J. Radiat. Oncol. Biol. Phys.* **48** (Suppl.) 239

Keller H, Fix M K, Liistro L and Rüegsegger P 2000 Theoretical considerations to the verification of dynamic multileaf collimated fields with a SLIC-type portal image detector *Phys. Med. Biol.* **45** 2531–45

Keller-Reichenbecher M A, Bortfeld T, Levegrün S, Stein J, Preiser K and Schlegel W 1999 Intensity modulation with the 'step and shoot' technique using a commercial MLC: a planning study *Int. J. Radiat. Oncol. Biol. Phys.* **45** 1315–24

Kerambrun A, Chauvet I, Parent L, Rosenwald J, Torzsok O, Mazal A, Zefkili S, Drouard J, Gaboriaud G and Mijnheer B 2003 Dynamic multileaf collimator (DMLC) compensation for breast treatments: use of an anthropomorphic phantom to check the validity of dose computation and assess the improvement of dose distribution *Proc. World Congress in Medical Physics (Sydney)* CDROM and website only

Kermode R H and Lawrence G P 2001 Validation of the use of data from a treatment planning system in the design of simple lead compensators *Radiother. Oncol.* **61** (Suppl. 1) S100

Kermode R H, Lawrence G P and Lambert G D 2001a The use of compensators to improve dose distributions in the head and neck region *Radiother. Oncol.* **61** (Suppl. 1) S100

Kermode R H, Richmond N D and Lambert G D 2001b Dosimetric verification of multi-segmented treatment fields planned on Helax-TM *Phys. Medica.* **17** 164

Kerr A T, Salomons G J and Schreiner J 2001 Dose delivery accuracy of a scanned pencil beam for cobalt-60 tomotherapy studies *Med. Phys.* **28** 1986

Kessler M, Ten Haken R, Pickett B and Cao Y 2003 Use of MRI and PET data for CT-based treatment planning *Radiother. Oncol.* **68** (Suppl. 1) S28

Kestin L L, Sharpe M B, Frazier R C, Vicini F A, Yan D, Matter R C, Martinez A A and Wong J W 2000a Intensity modulation to improve dose uniformity with tangential breast radiotherapy: initial clinical experience *Int. J. Radiat. Oncol. Biol. Phys.* **48** 1559–68

——2000b Intensity modulation to improve dose uniformity with tangential breast radiotherapy: initial clinical experience *Int. J. Radiat. Oncol. Biol. Phys.* **48** (Suppl. 3) 295

Kim D J W, Murray B R, Halperin R and Roa W H Y 2001 Held-breath self-gating technique for radiotherapy of non-small-cell lung cancer: A feasibility study *Int. J. Radiat. Oncol. Biol. Phys.* **49** 43–9

Kim J, Siebers J and Keall P 2002 Dosimetric verification of IMRT fields using an amorphous silicon flat panel imager and Monte Carlo simulation *Med. Phys.* **29** 1197

Kim K-H, Oh Y-K, Cho M-J, Kim J-K, Shin K-C, Kim J-K and Jeong D-H 2004c The preliminary results of intensity-modulated radiation therapy using liquid shielding *Proc. 14th Int. Conf. on the Use of Computers in Radiation Therapy (Seoul, May)* pp 734–7

Kim S, Akpati H, Dempsey J and Palta J 2003 Time-integrated dose volume histogram (TDVH), a new paradigm for plan evaluation *Med. Phys.* **30** 1347

Kim S, Akpati H, Dempsey J, Palta J and Ye S-J 2004a Accounting for target localization uncertainty in DVHs: is patient-specific displacement data needed? *Proc. 14th Int. Conf. on the Use of Computers in Radiation Therapy (Seoul, May)* pp 710–13

Kim S, Akpati H C, Kielbasa J E, Li J G, Liu C, Amdur R J and Palta J R 2004b Evaluation of intrafraction patient movement for CNS and head and neck IMRT *Med. Phys.* **31** 500–6

Kim S and Palta J 2003 Multi-source intensity-modulated radiation beam delivery system and method *US Patent* 6,449,336

Kim S, Yi B, Choi E and Ha S 2003 Impact of CT scan time on reproducibility of respiratory movement at the time of radiotherapy for lung cancer *Proc. World Congress in Medical Physics (Sydney)* CDROM and website only

King G, Selvaraj R, Mogus R and Wu A 2002 Comparison of beam on time for sliding window and step and shoot IMRT treatment plans *Med. Phys.* **29** 1271

Kini V R, Keall P J, Vedam S S, Arthur D W, Kavanagh B D, Cardinale R M and Mohan R 2000 Preliminary results from a study of a respiratory motion tracking system: Underestimation of target volume with conventional CT simulation *Proc. 42nd Annual ASTRO Meeting. Int. J. Radiat. Oncol. Biol. Phys.* **48** (Suppl.) 164

Kini V R, Vedam S S, Keall P J and Mohan A R 2001 A dynamic non-invasive technique for predicting organ motion in respiratory-gated radiotherapy of the chest *Int. J. Radiat. Oncol. Biol. Phys.* **51** (Suppl. 1) 25

Kirby M C A and Shentall G S 2001 Why standard treatment margins may need to be modified for IMRT techniques *Radiother. Oncol.* **61** (Suppl. 1) S62

Kirby M C, Shentall G S, Bulmer N C, Glendinning A G and McKay D M 2002 Verifying step and shoot IMRT fluence maps using Elekta i View and i ViewGT electronic portal imaging devices (EPIDs) *Proc. IPEM Biennial Meeting (Southampton, 16–17 July)* pp 13–14

Kitamura K, Shirato H and Seppenwoolde Y 2001 3D intra-fractional movement of prostate measured during real-time tumor-tracking radiation therapy (RTRT) in supine and prone treatment positions *Int. J. Radiat. Oncol. Biol. Phys.* **51** (Suppl. 1) 213

Kitamura K, Shirato H, Seppenwoolde Y, Onimaru R, Oda M, Fujita K, Shimizu S, Shinohara N, Harabayashi T and Miyasaka K 2002 Three-dimensional intrafractional movement of prostate measured during real-time tumor-tracking radiotherapy in supine and prone treatment positions *Int. J Radiol Oncol. Biol. Phys.* **53** 1117–23

Klabbers B M, de Munck J C, Sloman B J, Langendijk J A, deBree R, Hoekstra O S, Boellaard R and Lammertsma A A 2002 Matching PET and CT scans of the head and neck area: development of method and validation *Radiother. Oncol.* **64** (Suppl. 1) S209

Klein E E and Low D A 2000 Interleaf leakage testing for 0.5 and 1.0 cm DMLC systems incorporating patient motion *Proc. 42nd Annual ASTRO Meeting. Int. J. Radiat. Oncol. Biol. Phys.* **48** 221

——2001 Interleaf leakage for 5 and 10 mm dynamic multileaf collimating systems incorporating patient motion *Med. Phys.* **28** 1703–10

Klein E E, Low D A, Sohn J W and Purdy J A 2000 Differential dosing of prostate and seminal vesicles using dynamic multileaf collimation *Int. J. Radiat. Oncol. Biol. Phys.* **48** 1447–56

Kneschaurek P, Geinitz H and Zimmermann F 2000 Daily position corrected dose histograms of the rectum *Proc. World Congress of Medical Physics and the AAPM Annual Congress, August 2000* paper MO-FXH-50

Knoos T, Karolak L and Nilsson P 2001 Quantification of dosimetry changes in a TPS after implementation of a new dose calculation algorithm: pencil beam vs collapsed cone *Radiother. Oncol.* **61** (Suppl. 1) S22

Koelbl O, Bratengeier K and Flentje M 2002 Intensity modulated radiotherapy (IMRT) versus 3-D conformal, conventional radiotherapy (3D-RT) in lung cancer *Radiother. Oncol.* **64** S259

Komanduri K 2002 Dosimetric considerations for breast cancer treatment using intensity modulated radiotherapy *Med. Phys.* **29** 1273

Korevaar E W, Huizenga H, Löf, Stroom J C, Leer J W H and Brahme A 2002 Investigation of the added value of high-energy electrons in intensity-modulated radiotherapy: Four clinical cases *Int. J. Radiat. Oncol. Biol. Phys.* **52** 236–53

Korreman S S, Pedersen A N, Specht L and Nyström H 2003 Breathing adapted radiotherapy (BART) for breast cancer *Radiother. Oncol.* **68** (Suppl. 1) S40

Kry S, Salehpour M, Followill D, Stovall M and Rosen I 2003 Risk assessment of secondary malignancies from IMRT treatments *Med. Phys.* **30** 1330

Kron T, Battista J, Bauman G and van-Dyk J 2004a On the role of helical tomotherapy in clinical practice: Results from planning studies *Proc. 14th Int. Conf. on the Use of Computers in Radiation Therapy (Seoul, May)* pp 5–9

Kron T, Chen J, Wong E, VanDyk J, Rodrigues G, Bauman G, Dar R, Yu E and Battista J 2002 Initial experience with tomotherapy treatment planning in London, Ontario *Radiother. Oncol.* **64** (Suppl. 1) S78

Kron T, Wong E, Yartsev S, Grigorov G, Rodrigues G, Yu E, Chen J and Van Dyk J 2004b Helical tomotherapy planning: Overlap between target and critical structures as a predictor of plan quality *Proc. 14th Int. Conf. on the Use of Computers in Radiation Therapy (Seoul, May)* pp 225–7

Kubo H D and Wang L 2000 Compatibility of Varian 2100C gated operations with enhanced dynamic wedge and IMRT dose delivery *Med. Phys.* **27** 1732–8

Kuefer K-H, Monz M, Scherrer A, Alonso F, Trinkaus H, Bortfeld T and Thieke C 2003 Real-time inverse planning using a precomputed multi-criteria plan database *Radiother. Oncol.* **68** (Suppl. 1) S76

Kulidzhanov F, Sabbas A, Trichter S, Wang J and Nori D 2003 Monitor unit linearity for the IMRT step and shoot technique *Med. Phys.* **30** 1494

Kung J H and Chen G T Y 2000a Intensity modulated radiotherapy dose delivery error from radiation field offset inaccuracy *Med. Phys.* **27** 1617–22

——2000b Monitor unit verification calculation in IMRT as a patient specific dosimetry QA. *Proc. 13th Int. Conf. on the Use of Computers in Radiation Therapy (Heidelberg, May)* ed W Schlegel and T Bortfeld (Heidelberg: Springer) pp 292–3

Kung J H, Chen G T Y and Kuchnir F K, 2000b A monitor unit verification calculation in intensity modulated radiotherapy as a dosimetry quality assurance *Med. Phys.* **27** 2226–30

Kung J, Reft C, Jackson W and Abdalla I 2000a Intensity modulated radiotherapy for a prostate patient with a prosthetic hip *Proc. World Congress of Medical Physics and the AAPM Annual Congress, August 2000* paper TU-GBR-04

Kung J H, Zygmanski P, Choi N and Chen G T Y 2003 A method of calculating a lung clinical target volume DVH for IMRT with intrafractional motion *Med Phys.* **30** 1103–9

Kung J H, Zygmanski P, Kooy H and Chen G 2000c Dose error in IMRT treatment of lung cancer without respiratory gating *Proc. 42nd Annual ASTRO Meeting. Int. J. Radiat. Oncol. Biol. Phys.* **48** (Suppl.) 344

Kupelian P A, Reddy C A, Klein E A and Willoughby T R 2001 Short-course intensity modulated radiotherapy (70 Gy at 2.5 Gy per fraction) for localized prostate cancer: Late toxicity and quality of life *Int. J. Radiat. Oncol. Biol. Phys.* **51** (Suppl. 1) 112

Kuriyama K, Onishi H, Sano N, Komiyama T, Aikawa Y, Tateda Y, Araki T and Uematsu M 2003 A new irradiation unit constructed of self-moving gantry-CT and linac *Int. J. Radiat. Oncol. Biol. Phys.* **55** 428–35

Kuterdem H G and Cho P S 2001 Leaf sequencing with secondary beam blocking under leaf positioning constraints for continuously modulated radiotherapy beams *Med. Phys.* **28** 894–902

Kwong D L, Pow E, McMillan A, Sham J and Au G 2003 Intensity-modulated radiotherapy for early stage nasopharyngeal carcinoma: preliminary results on parotid sparing *Int J. Radiat. Oncol. Biol. Phys.* **57** (Suppl.) S303

Lagendijk J J W and Bakker C J G 2000 MRI guided radiotherapy: a MRI based linear accelerator *Radiother. Oncol.* **56** (Suppl. 1) S60

Lagendijk J J W, Raaymakers B W, van der Heide U A, Topolnjak R, Dehnad H, Hofman P, Nederveen A J, Schulz I M, Welleweerd J and Bakker C J G 2002 MRI guided radiotherapy: MRI as position verification system for IMRT *Radiother. Oncol.* **64** (Suppl. 1) S75–6

Landau D, Adams E J, Webb S and Ross G 2001 Cardiac avoidance in breast radiotherapy: a comparison of simple shielding techniques with intensity modulated radiotherapy *Radiother. Oncol.* **60** 247–55

Langen K M and Jones D T L 2001 Organ motion and its management. *Int. J. Radiat. Oncol. Biol. Phys.* **50** 265–78

Langen K M, Pouliot J, Anezinos C, Aubin M, Gottschalk A R, Hsu I-C, Lowther D, Liu Y-M, Shinohara K, Verhey L J, Weinberg V and Roach M III 2004 Evaluation of ultrasound-based localization for image-guided radiotherapy *Int. J. Radiat. Oncol. Biol. Phys.* **57** 635–44

Langer M 2000 Application of coloring theory to reduce intensity modulated radiotherapy dose calculations *Med. Phys.* **27** 2077–83

——2003 What is different about IMRT? *IMRT—the State of the Art: AAPM Medical Physics Monograph Number 29* ed J R Palta and T R Mackie (Madison, WI: Medical Physics Publishing) pp 199-219

Langer M P, Thai V and Papiez L 2000 Achievable reductions in beam time and segment number with IMRT leaf sequences *Proc. 42nd Annual ASTRO Meeting. Int. J. Radiat. Oncol. Biol. Phys.* **48** (Suppl.) 342

——2001a Comparison of leaf sequencing algorithms against the efficient frontier of monitor unit and segment usage *Med. Phys.* **28** 1307

——2001b Improved leaf sequencing reduces segments or monitor units needed to deliver IMRT using multileaf collimators *Med. Phys.* **28** 2450–8

——2001c Tradeoffs between segments and monitor units are not required for static field IMRT delivery *Int. J. Radiat. Oncol. Biol. Phys.* **51** (Suppl. 1) 75

Large C, Brumley K and Ramsey C 2001 The dosimetric consequences of intra-fraction prostate motion in intensity modulation *Med. Phys.* **28** 1253

Lattanzi J 2000 Application of ultrasound imaging for conformal therapy in early prostate carcinoma *Radiother. Oncol.* **56** (Suppl. 1) S7

Lattanzi J, McNeeley S, Donnelly S, Palacio E, Hanlon A, Schultheiss T E and Hanks G E 2000 Ultrasound-based stereotactic guidance in prostate cancer—quantification of organ motion and set-up errors in external beam radiation therapy *Computer Aided Surgery* **5** 289–95

Lattanzi J, McNeeley S, Hanlon A, Schultheiss T E and Hanks G E 1999a Ultrasound-based stereotactic guidance of precision conformal external beam radiation therapy in clinically localized prostate cancer *Urology* **55** 73–8

Lattanzi J, McNeeley S, Pinover W, Horwitz E, Das I, Schultheiss T E and Hanks G E 1999b A comparison of daily CT localization to a daily ultrasound-based system in prostate cancer *Int. J. Radiat. Oncol. Biol. Phys.* **43** 719–25

Laub W, Alber M, Birkner M and Nüsslin F 2000a Monte Carlo dose computation for IMRT optimization *Phys. Med. Biol.* **45** 1741–54

Laub W and Bakai A 2001 Monte Carlo calculations and measurements for verification of an IMRT dose distribution in a thorax phantom *Med. Phys.* **28** 1283

Laub W U, Bakai Λ and Nüsslin F 2001 Intensity modulated irradiation of a thorax phantom: comparisons between measurement, Monte Carlo calculations and pencil beam calculations *Phys. Med. Biol.* **46** 1695–706

Laub W U, Yan D, Alber M, Nüsslin F, Martinez A and Wong J 2000b IMRT in the treatment of colo-rectal cancer: The influence of profile smoothing on the efficiency of delivery *Proc. 42nd Annual ASTRO Meeting. Int. J. Radiat. Oncol. Biol. Phys.* **48** (Suppl.) 340

Lauterbach M, Siebers J, Mohan R and Wu Q 2001 IMRT optimization based on deliverable intensities *Med. Phys.* **28** 1204

Leal A, Sanchez-Doblado F, Capote R, Carrasco E, Lagares J I, Arrans R, Rosello J V and Perucha M 2002 Monte Carlo based study of the influence of double-focusing on dose distribution in IMRT treatments *Radiother. Oncol.* **64** (Suppl. 1) S217

Lee C, Earl M, Shepard D and Yu C 2002b Acceleration of IMRT optimisation using objective function random sampling without replacement *Med. Phys.* **29** 1255

Lee K Y, Chau M C and Cheung K Y 2001b Design of an inexpensive phantom for IMRT certification *Radiother. Oncol.* **61** (Suppl. 1) S110

——2001c Virtual pre-treatment verification for IMRT treatments *Radiother. Oncol.* **61** (Suppl. 1) S25

Lee M C, Xia P, Lam S C P and Ma C-M 2001a Monte Carlo dose calculations for serial tomography *Med. Phys.* **28** 1267

Lee N, Akazawa C, Akazawa P, Quivey J M, Tang C, Verhey L J and Xia P 2004 A forward-planned treatment technique using multisegments in the treatment of head-and-neck cancer *Int. J. Radiat. Oncol. Biol. Phys.* **59** 584–94

Lee N and Xia P 2003 In response to Drs Clark *et al Int. J. Radiat. Oncol. Biol. Phys.* **55** 1152–3

Lee N, Xia P, Quivey J M, Sultanem K, Poon I, Akazawa C, Akazawa P, Weinberg V and Fu K K 2002a Intensity-modulated radiotherapy in the treatment of nasopharyngeal carcinoma: an update of the UCSF experience *Int. J. Radiat. Oncol. Biol. Phys.* **53** 12–22

Lee Y K, Bollet M, Charles-Edwards G, Flower M A, Leach M O, McNair H, Moore E, Rowbottom C and Webb S 2003 Radiotherapy treatment planning of prostate cancer using magnetic resonance imaging alone *Radiother. Oncol.* **66** 203–16

Lehmann J and Pawlicki T 2003 IMRT: Better plans with tuning structures *Med. Phys.* **30** 1485–6

Lehmann J and Xing L 2000 Role of the number of intensity levels in IMRT *Proc. World Congress of Medical Physics and the AAPM Annual Congress, August 2000* paper MO-CXH-20

Lepage M, Whittaker A K, Rintoul L, Bäck S A J and Baldock C 2001 The relationship between radiation-induced chemical processes and transverse relaxation times in polymer gel dosimeters *Phys. Med. Biol.* **46** 1061–74

Létourneau D, Gulam M, Yan D, Oldham M and Wong J W 2004 Evaluation of a 2D diode array for IMRT quality assurance *Radiother. Oncol.* **70** 199–206

Létourneau D, Jaffray D, Oldham M, Sharpe M and Wong J 2002 Clinical implementation of a kilovoltage imager for setup error verification in radiotherapy *Med. Phys.* **29** 1243

Levendag P, Wijers O, van Sornsen de Koste J, van der Est H and Nowak P 2000 Intensity-modulated radition therapy: preliminary multi-institutional experience in head and neck cancer *Radiother. Oncol.* **56** (Suppl. 1) S105

Levin S 2001 Accuray: tightly targeting tumours *Windhover's In Vivo The Business and Medicine Report* **19** 1–12

Levine R Y and Braunstein M 2002 Beam configurations for 3D tomographic intensity modulated radiation therapy *Phys. Med. Biol.* **47** 765–87

Levitt S H and Khan F M 2002 In regard to Purdy JA, Michalski JM. Does the evidence support the enthusiasm over 3D conformal radiation therapy and dose escalation in the treatment of prostate cancer? (*Int. J. Radiat. Oncol. Biol. Phys.* 2001 **51** 867–70) *Int. J. Radiat. Oncol. Biol. Phys.* **53** 1085

Lewellen T 2001 PET in radiation therapy planning *Med. Phys.* **28** (6) 1226

Lewis D G, Spezi E and Smith C W 2001 Portal verification and dosimetry of multileaf collimator fields by full Monte Carlo simulation *Radiother. Oncol.* **61** (Suppl. 1) S107

Levitt S H and Khan F M 2001 The rush to judgement: does the evidence support the enthusiasm over three-dimensional conformal radiation therapy and dose escalation in the treatment of prostate cancer? *Int. J. Radiat. Oncol. Biol. Phys.* **51** 871–9

Leybovich L, Dogan N, Sethi A, Krasin M and Emami B 2000a A method of correcting dose distributions produced by IMRT planning systems *Proc. World Congress of Medical Physics and the AAPM Annual Congress, August 2000* paper MO-EBR-08

Leybovich L B, Dogan N, Sethi A, Krasin M J and Emami B 2000b Improvement of tomographic intensity modulated radiotherapy dose distributions using periodic shifting of arc abutment regions *Med. Phys.* **27** 1610–16

Leybovich L, Sethi A, Dogan N and Glasgow G 2002 Is there any benefit in using small ion chambers for verification of target dose delivered by IMRT? *Radiother. Oncol.* **64** (Suppl. 1) S44

Li J S, Boyer A L and Ma C-M 2001a Verification of IMRT dose distributions using a water beam imaging system *Med. Phys.* **28** 2466–74

Li J, Boyer A and Xing L 2000b Intensity modulated arc therapy and its relation to fixed gantry IMRT *Proc. World Congress of Medical Physics and the AAPM Annual Congress, August 2000* paper MO-CXH-18

Li J G, Williams S S, Goffinet D R, Boyer A L and Xing L 2000a Breast-conserving radiation therapy using combined electron and intensity-modulated radiotherapy technique *Radiother. Oncol.* **55** 65–71

Li J and Xing L 2000b Inverse planning incorporating organ motion *Proc. World Congress of Medical Physics and the AAPM Annual Congress, August 2000* paper FR-EABR-10

——2000a Inverse planning incorporating organ motion *Med. Phys.* **27** 1573–8

Li J S, Freedman G M, Price R, Wang L, Anderson P, Chen L, Xiong W, Yang J, Pollack A and Ma C-M 2004c Clinical implementation of intensity-modulated tangential beam irradiation for breast cancer *Med. Phys.* **31** 1023–31

Li J S, Ozhasoglu C, Deng J, Boyer A L and Ma C-M 2000c Verification of IMRT dose distributions using a water beam imaging system *Proc. World Congress of Medical Physics and the AAPM Annual Congress, August 2000* paper TU-EBR-01

Li J S, Yang J, Price R, Chen L and Ma C-M 2003 Machine output linearity for breast IMRT *Med. Phys.* **30** 1493

Li J and Yu C 2001b Simultaneous optimization of beam intensity and beam orientation for IMRT *Int. J. Radiat. Oncol. Biol. Phys.* **51** (Suppl. 1) 73–4

Li K, Dai J and Ma L 2004a Simultaneous minimizing monitor units and number of segments without leaf end abutment for segmental intensity modulated radiation therapy *Med. Phys.* **31** 507–12

Li X A, Ma L, Naqvi S, Shih R and Yu C 2001c Monte Carlo dose verification for intensity-modulated arc therapy *Phys. Med. Biol.* **46** 2269–82

Li X A, Ma L, Yu C X, Naqvi S and Holmes T 2000d Monte Carlo dose verification for intensity modulated arc therapy *Proc. 42nd Annual ASTRO Meeting. Int. J. Radiat. Oncol. Biol. Phys.* **48** (Suppl.) 219

Li Y, Yao J and Yao D 2002 Direct optimization of segments for intensity-modulated radiotherapy *Med. Phys.* **29** (6) 1259–60

——2003a Aperture based optimization for static IMRT using genetic algorithm *Med. Phys.* **30** 1372

——2003b Genetic algorithm based automatic beam angles selection in IMRT planning *Med. Phys.* **30** 1373

——2003c Genetic algorithm based deliverable segments optimization for static intensity-modulated radiotherapy *Phys. Med. Biol.* **48** 3353–74

——2004b Automatic beam angle selection in IMRT planning using genetic algorithm *Phys. Med. Biol.* **49** 1915–32

Lian J, Cotrutz C and Xing L 2002 IMRT dose optimization with probabilistic dose prescription *Med. Phys.* **29** 1260

Lian J and Xing L 2004 Biological model based IMRT optimisation with inclusion of parameter uncertainty *Proc. 14th Int. Conf. on the Use of Computers in Radiation Therapy (Seoul, May)* pp 453–6

Lillicrap S C, Morgan H M and Shakeshaft J T 2000 X-ray leakage during radiotherapy *Brit. J. Radiol.* **73** 793–4

Ling C 2001a Potential of biological images for radiation therapy of cancer *Med. Phys.* **28** 1226

——2001b Potential of biological images for radiation therapy of cancer *Med. Phys.* **28** 1972

——2001c Current status of intensity-modulated radiotherapy:clinical aspects *Radiother. Oncol.* **61** (Suppl. 1) S30

—— 2003 UKRO Lecture: Intensity modulated radiotherapy and biological images *Clinical Oncol.* **15** (Suppl. 2) S4–5

Ling C C, Burman C, Chui C S, Kutcher G J, Leibel S A, LoSasso T, Mohan R, Bortfeld T, Reinstein L, Spirou S, Wang X H, Wu Q, Zelefsky M and Fuks Z 1996 Conformal radiation treatment of prostate cancer using inversely-planned intensity-modulated photon beams produced with dynamic multileaf collimation. *Int. J. Radiat. Oncol. Biol. Phys.* **35** 721–30

Linthout N, Verellen D, van Acker S, van de Vondel I, Coppens L and Storme G 2002 Assessment of the acceptability of the Elekta multileaf collimator (MLC) within the CORVUS planning system for static and dynamic delivery of intensity modulated beams (IMBs) *Radiother. Oncol.* **63** 121–4

Little D J, Dong L, Levy L B, Chandra A and Kuban D A 2003 Use of portal images and BAT ultrasonography to measure setup error and organ motion for prostate IMRT: Implications for treatment margins *Int. J. Radiat. Oncol. Biol. Phys.* **56** 1218–24

Litzenberg D W, Moran J M and Fraass B A 2002a Incorporation of realistic delivery limitations into dynamic MLC treatment delivery *Med.Phys.* **29** 810–20

——2002b Incorporation of realistic delivery limitations into dynamic MLC treatment delivery *Med. Phys.* **29** 1267

Liu C and Xing L 2000 Dosimetric effects of mechanical inaccuracy of MLC leaf positions on IMRT *Proc. World Congress of Medical Physics and the AAPM Annual Congress, August 2000* paper MO-GBR-01

Liu H H, Verhaegen F and Dong L 2001a Monte Carlo simulation of dynamic multi-leaf collimators *Radiother. Oncol.* **61** (Suppl. 1) S40

——2001b A method of simulating dynamic multileaf collimators using Monte Carlo techniques for intensity modulated radiation therapy *Phys. Med. Biol.* **46** 2283–98

Liu H H, Wang X, Dong L, Wu Q, Liao Z, Stevens C W, Guerrero T M, Komaki R, Cox J D and Mohan R 2004 Feasibility of sparing lung and other thoracic structures with intensity-modulated radiotherapy for non-small-cell lung cancer *Int. J. Radiat. Oncol. Biol. Phys.* **58** 1268–79

Liu T, Lizzi F L, Feleppa E J and Kutcher G J 2002 Ultrasonic tissue typing using spectrum analysis technique and the potential role in conformal radiation therapy *Radiother. Oncol.* **64** (Suppl. 1) S289

Liu W-C, Schulder M, Narra V, Kalnin A J, Cathcard C, Jacobs A, Lange G and Holodny A I 2000 Functional magnetic resonance imaging aided radiation treatment planning *Med. Phys.* **27** 1563–72

Livsey J, Mott J H, Wylie J P, Cowan R A, Khoo V S and Logue J P 2003 IMRT for carcinoma of the prostate *Clinical Oncol.* **15** (Suppl. 2) S6

422   *References*

Llacer J, Deasy J O, Bortfeld T R, Solberg T D and Promberger C 2003 Absence of multiple local minima effects in intensity modulated optimization with dose-volume constraints *Phys. Med. Biol.* **48** 183–210

Llacer J, Solberg T D and Promberger C 2001 Comparative behaviour of the dynamically penalized likelihood algorithm in inverse radiation therapy planning *Phys. Med. Biol.* **46** 2637–63

Locke J, Low D A, Grigireit T and Chao K S C 2002 Potential of tomotherapy for total scalp treatment *Int. J. Radiat. Oncol. Biol. Phys.* **52** 553–9

Lofroth P-O, Bergstrom P, Brannstrom K, Nystrom E, Zackrisson B and Widmark A 2000 Conformal radiotherapy of the prostate with high precision *Radiother. Oncol.* **56** (Suppl. 1) S108

Logue J 2000 Clinical implementation of IMRT *Radiother. Oncol.* **56** (Suppl. 1) S86

Loi G, Manfredda I, Secco C, Sacchetti G, Mones E, Natrella M, Bagnasacco P, Inglese E, Cotroneo A and Krengli M 2002 Fusion of CT-MR-SPECT images for the delineation of target volume in the treatment plan of high-grade gliomas. Preliminary results *Radiother. Oncol.* **64** (Suppl. 1) S300

Loof M, Engstrom P, Uttman K and Nystrom H 2001 Optimisation of IMRT: can geometrical uncertainties be incorporated into the optimisation process? *Radiother. Oncol.* **61** (Suppl. 1) S66

Lomax A J 2001 Methods for assessing the effects of inter- and intra-fraction positional errors in radiotherapy *Radiother. Oncol.* **61** (Suppl. 1) S22

Lopez Torrecilla J 2001 The introduction of IMRT in the clinic *Radiother. Oncol.* **61** (Suppl. 1) S31

LoSasso T L 2003 IMRT delivery system QA *IMRT—the State of the Art: AAPM Medical Physics Monograph Number 29* ed J R Palta and T R Mackie (Madison, WI: Medical Physics Publishing) pp 561–91

LoSasso T, Burman C, Chui C and Ling C 1998a QA and verification of intensity modulation radiotherapy. *Med. Phys.* **25** A206

LoSasso T, Chui C-S, and Ling C C 1998b Physical and dosimetric aspects of a multileaf collimation system used in the dynamic mode for implementing intensity modulated radiotherapy. *Med. Phys.* **25** 1919–27

——2001 Comprehensive quality assurance for the delivery of intensity modulated radiotherapy with a multileaf collimator used in the dynamic mode *Med. Phys.* **28** 2209–19

Lotz H, Remeijer P, van Herk M, Lebesque J, de Bois J and Zijp L 2003 A model to predict the bladder shape from changes in bladder and rectal volume *Radiother. Oncol.* **68** (Suppl. 1) S35

Love P A, Evans P M, Leach M O and Webb S 2003 Polymer gel measurement of dose homogeneity in the breast comparing MLC intensity modulation with standard wedged dosimetry *Phys. Med. Biol.* **48** 1065–74

Low D A 2001 Verification methods for IMRT *Med. Phys.* **28** 1271

——2003 Radiation shielding for IMRT *IMRT—the State of the Art: AAPM Medical Physics Monograph Number 29* ed J R Palta and T R Mackie (Madison, WI: Medical Physics Publishing) pp 401–14

Low D A, Bradley J D, Deasy J O, Laforest R, Parikh P J, Sohn J W, Hibbard L S, Dehdashti F and Mutic S 2001b Lung trajectory mapping *Int. J. Radiat. Oncol. Biol. Phys.* **51** (Suppl. 1) 208–9

Low D A, Dempsey J F, Markman J, Mutic S, Klein E E, Sohn J W and Purdy J A 2002a Toward automated quality assurance for intensity-modulated radiation therapy *Int. J. Radiat. Oncol. Biol. Phys.* **53** 443–52

Low D A, Dempsey J F, Markman J, Mutic S, Williamson J F and Purdy J A 2000b Applicator-guided intensity modulated radiation therapy *Proc. 42nd Annual ASTRO Meeting. Int. J. Radiat. Oncol. Biol. Phys.* **48** (Suppl.) 209

Low D A, Grigsby P W, Dempsey J F, Mutic S, Williamson J F, Markman J, Chao K S C, Klein E E and Purdy J A 2002b Applicator guided intensity modulated radiation therapy *Int. J. Radiat. Oncol. Biol. Phys.* **52** 1400–6

Low D A, Markman J, Dempsey J F and Mutic S 2000a Noise in polymer gel measurements using MRI *Med. Phys.* **27** 1814–17

Low D A, Parikh P J, Nystrom M, El Naqa I M, Nystrom M M, Lu W, Hubenschmidt J P, Wahab S H, Mutic S, Singh A K, Christensen G and Bradley J D, 2004 Quantitative 4-D CT using a multislice CT scanner *Proc. 14th Int. Conf. on the Use of Computers in Radiation Therapy (Seoul, May)* pp 57–61

Low D A, Sohn J W, Klein E E, Markman J, Mutic S and Dempsey J F 2001a Characterization of a commercial multileaf collimator used for intensity modulated radiation therapy *Med. Phys.* **28** 752–6

Lu T-X, Mai W-Y, Teh B S, Zhao C, Han F, Huang Y, Deng X-W, Lu L-X, Huang S-M, Zeng Z-F, Lin C-G, Lu H H, Chiu J K, Carpenter L S, Grant W H III, Woo S Y, Cui N-J and Butler E B 2004 Initial experience using intensity-modulated radiotherapy for recurrent nasopharyngeal carcinoma *Int. J. Radiat. Oncol. Biol. Phys.* **58** 682–7

Lu W and Mackie T R 2002 Tomographic motion detection and correction directly in sinogram space *Phys. Med. Biol.* **47** 1267–84

Lu X-Q, Burman C, Mychaiczak B, Hirsch A, Chui C-S, Leibel S and Ling C 2000 Feasibility study of using low-energy intensity modulated radiation for the treatment of localized prostate carcinoma to high doses *Proc. World Congress of Medical Physics and the AAPM Annual Congress, August 2000* paper MO-EBR-03

Lu Y, Spelbring R and Chen G T Y 1997 Functional dose-volume histograms for functionally heterogeneous normal organs *Phys. Med. Biol.* **42** 345–56

Luan S, Wang C, Chen D Z, Hu X S, Naqvi S A, Yu C X and Lee C L 2004 A new MLC leaf segmentation algorithm/software for step-and-shoot IMRT delivery *Med. Phys.* **31** 695–707

Luan S, Wang C, Naqvi S A, Chen D Z, Hu X S, Lee C L and Yu C X 2003 A new leaf sequencing algorithm/software for step and shoot IMRT delivery *Med. Phys.* **30** 1404

Lujan A E, Balter J M and Ten Haken R K 2003 A method for incorporating organ motion due to breathing into 3D dose calculations in the liver: Sensitivity to variations in motion *Med. Phys.* **30** 2643–9

Lujan A E, Roeske J C, Johnson L S, Reft C S, Laughton C, Kung J H and Pelizzari C A 2001a Verification of IMRT dose calculations using an independent monitor unit calculation *Med. Phys.* **28** 1282

Lujan A E, Roeske J C and Mundt A J 2001b Intensity-modulated radiation therapy as a means of reducing dose to bone marrow in gynaecologic patients receiving whole pelvic radiation therapy *Int. J. Radiat. Oncol. Biol. Phys.* **51** (Suppl. 1) 220

Luo C, Wu X, Chen Z, Shao H, Han H, Wolfson A and Markoe A 2000 Reducing dose to the contralateral breast in the conventional two-wedge tangential breast or chest wall radiotherapy *Proc. World Congress of Medical Physics and the AAPM Annual Congress, August 2000* paper MO-FXH-52

Luo C, Wu X, Shao H and Markoe A 2002 Linear algebraic approach for IMRT inverse planning *Med. Phys.* **29** 1260

Luxton G, Hancock S L and Boyer A L 2004 Dosimetry and radiobiologic model comparison of IMRT and 3D conformal radiotherapy in treatment of carcinoma of the prostate *Int. J. Radiat. Oncol. Biol. Phys.* **59** 267–84

Luxton G, Hancock S L, Chen Y, Xing L and Boyer A L 2000 Reduction of bowel dose in lymph node irradiation with IMRT treatment of prostate cancer *Proc. 42nd Annual ASTRO Meeting. Int. J. Radiat. Oncol. Biol. Phys.* **48** (Suppl.) 352

Lyatskaya Y, Chin L, Harris J and Lu H 2002 Accurate patient position control for breast cancer treatment using respiratory manoeuvres *Med. Phys.* **29** 1239

Ma C 2000 Monte Carlo for treatment planning and IMRT dose verification *Radiother. Oncol.* **56** (Suppl. 1) S79

——2003 Why Cyberknife? *Radiother. Oncol.* **68** (Suppl. 1) S39

Ma C-M, Ding M, Li J S, Lee M C, Pawlicki T and Deng J 2003c A comparative dosimetric study on tangential photon beams, intensity-modulated radiation therapy (IMRT) and modulated electron radiotherapy (MERT) for breast cancer treatment *Phys. Med. Biol.* **48** 900–24

Ma C M, Guerrero T, Pawlicki T, Luxton G and Adler J 2001b Treatment planning and dosimetry verification for the Cyberknife system *Radiother. Oncol.* **61** (Suppl. 1) S16

Ma C-M, Jiang S B, Pawlicki T, Chen Y, Li J S, Deng J and Boyer A L 2003a A quality assurance phantom for IMRT dose verification *Phys. Med. Biol.* **48** 561–72

Ma C, Jiang T, Pawlicki T, Chen Y, Li J, Deng J, Lee M, Mok E and Boyer A 2000a A quality assurance phantom for IMRT dose verification *Proc. World Congress of Medical Physics and the AAPM Annual Congress, August 2000* paper TU-EBR-06

Ma C, Li J, Chen L, Wang L, Price R, McNeeley S, Ding, M and Fourkal E 2002b Monte Carlo dose verification for advanced radiotherapy treatment techniques *Radiother. Oncol.* **64** (Suppl. 1) S105–6

Ma C M, Li J S, Pawlicki T, Jiang S B, Deng J, Lee M C, Koumrian T, Luxton M and Brain S 2002a A Monte Carlo dose calculation tool for radiotherapy treatment planning *Phys. Med. Biol.* **47** 1671–89

Ma C-M, Pawlicki T, Jiang S B, Li J S, Deng J, Mok E, Kapur A, Xing L, Ma L and Boyer A L 2000c Monte Carlo verification of IMRT dose distributions from a commercial treatment planning optimization system *Phys. Med. Biol.* **45** 2483–95

Ma C-M, Pawlicki T, Lee M C, Jiang S B, Li J S, Deng J, Yi B, Mok E and Boyer A L 2000e Energy and intensity-modulated electron beams for radiotherapy *Phys. Med. Biol.* **45** 2293–311

Ma C M, Pawlicki T, Li J S, Deng J, Shahine B, Lee M C and Boyer A L 2001a Quality assurance phantoms and Monte Carlo dose verification *Radiother. Oncol.* **61** (Suppl. 1) S4

Ma L 2002a Smoothing intensity-modulated radiotherapy delivery under hardware constraints *Med. Phys.* **29** 1304

——2002b Smoothing intensity-modulated treatment delivery under hardware constraints, *Med. Phys.* **29** 2937–45

Ma L, Phaisangittisakul N, Yu C X and Sarfaraz M 2003b A quality assurance method for analysing and verifying intensity modulated fields *Med. Phys.* **30** 2082–8

Ma L, Yu C and Sarfaraz M 2000b A dosimetric leaf-setting strategy for shaping radiation fields using a multileaf collimator *Med. Phys.* **27** 972–7

Ma L, Yu C, Sarfaraz M, Holmes T, Shepard D, Li A, DiBiase S, Amin P, Suntharalingam M and Mansfield C 2000d Simplified intensity modulated arc therapy for prostate cancer treatments *Proc. 42nd Annual ASTRO Meeting. Int. J. Radiat. Oncol. Biol. Phys.* **48** (Suppl.) 350

MacDougall N D, Pitchford W G and Smith M A 2002 A systematic review of the precision and accuracy of dose measurements in photon radiotherapy using polymer and Fricke MRI gel dosimetry *Phys. Med. Biol.* **47** R107–21

——2003 Letter to the Editor: Reply to 'Comment on A systematic review of the precision and accuracy of dose measurements in photon radiotherapy using polymer and Fricke MRI gel dosimetry' *Phys. Med. Biol.* **48** L19–22

MacKay R I and Amer A 2000 Radiobiological analysis of 3D motion *Radiother. Oncol.* **56** (Suppl. 1) S51

MacKay R I, Graham P A, Logue J P and Moore C J 2000 Patient positioning using detailed 3D surface data for patients undergoing conformal radiotherapy for carcinoma of the prostate: a feasibility study *Int. J. Radiat. Oncol. Biol. Phys.* **49** 225–30

MacKenzie M A and Robinson D M 2002 Intensity modulated arc deliveries approximated by a large number of fixed gantry position sliding window dynamic multileaf collimator fields *Med. Phys.* **29** 2359–65

Mackie T 2001 Tomotherapy *Med. Phys.* **28** 1299

Mackie T R, Balog J, Becker S, Tomé W, Reckwerdt P J, Kapatoes J, Ruchala K, Smilowitz J and Olivera G 2002 Can helical tomotherapy do simple radiotherapy procedures? *Med. Phys.* **29** 1211

Mackie T R, Kapatoes J, Ruchala K, Lu W, Wu C, Olivera G, Forrest L, Tomé W, Welsh J, Jeraj R, Herari P, Reckwerdt P, Paliwal B, Ritter M, Keller H, Fowler J and Mehta M 2003a Image guidance for precise conformal radiotherapy *Int. J. Radiat. Oncol. Biol. Phys.* **56** 89–105

Mackie T R, Olivera G H, Kapatoes J M, Ruchala K J, Balog J R, Tomé W A, Hui S, Kissick M, Wu C, Jeraj R, Reckwerdt P J, Harari P, Ritter M, Forrest L, Welsh J and Mehta M P 2003b Helical tomotherapy *IMRT—the State of the Art: AAPM Medical Physics Monograph Number 29* ed J R Palta and T R Mackie (Madison, WI: Medical Physics Publishing) pp 247–84

Mackie T R, Ruchala K, Olivera G, Kapatoes J, Lu W, Fang G, Murray D, Reckwerdt P, Hughes J, Kissick M and Jeraj R 2004 Overview of helical tomotherapy *Proc. 14th Int. Conf. on the Use of Computers in Radiation Therapy (Seoul, May)* pp 200–3

Macklis R, Weinhous M and Harnisch G 2000 Intensity-modulated radiotherapy: rethinking basic treatment planning paradigms *Int. J. Radiat. Oncol. Biol. Phys.* **48** 317–18

Macpherson M, Crljenko T, Vo D, El-Hakim S and Gerig L 2002 A machine-vision based radiotherapy gating device *Med. Phys.* **29** 1213

Maes A, Weltens C, Huyskens D and van den Bogaert W 2002 Conformal and intensity-modulated radiotherapy for preservation of salivary function *Radiother. Oncol.* **64** (Suppl. 1) S110

Maes A, Weltens C, van Esch A, Kutcher G, Huyskens D and van den Bogaert W 2000 The potential of IMRT for parotid sparing in head and neck cancer *Radiother. Oncol.* **56** (Suppl. 1) S109

Mageras G S 2001 Respiratory gating in conformal radiotherapy *Radiother. Oncol.* **61** (Suppl. 1) S35

——2002 Respiration correlated CT techniques for gated treatment of lung cancer *Radiother. Oncol.* **64** (Suppl. 1) S75

——2004 When is a deep inspiration breath-hold technique during external beam radiotherapy beneficial? *Proc. 14th Int. Conf. on the Use of Computers in Radiation Therapy (Seoul, May)* p 52

Mageras G, Joshi S, Davis B, Pevsner A, Hertanto A, Yorke E, Rosenzweig K, Erdi Y, Nehmeh S, Humm J, Larson S M and Ling C C 2004b Evaluation of an automated deformable matching method for quantifying lung tumor motion in respiration correlated CT images *Proc. 14th Int. Conf. on the Use of Computers in Radiation Therapy (Seoul, May)* pp 364–7

Mageras G S, Sillanpaa J, Seppi E, Chang J, Ling C C and Amols H 2004a Preclinical studies of megavoltage cone beam CT systems for tumor localization in gated radiotherapy of lung cancer *Proc. 14th Int. Conf. on the Use of Computers in Radiation Therapy (Seoul, May)* pp 20–1

Mageras G S and Yorke E 2004 Deep inspiration breath hold and respiratory gating strategies for reducing organ motion in radiation treatment *Seminars Radiat. Oncol.* **14** 65–75

Mah D, Freedman G, Milestone B, Movsas B, Mitra R, Horwitz E and Hanks G E 2001 Dancing prostates: Measurement of intra-fractional prostate motion using MRI *Int. J. Radiat. Oncol. Biol. Phys.* **51** (Suppl. 1) 212

Mah D, Hanley J, Rosenzweig K E, Yorke E, Braban L, Ling C C, Leibel S A and Mageras G 2000 Technical aspects of the deep inspiration breath-hold technique in the treatment of thoracic cancer *Int. J. Radiat. Oncol. Biol. Phys.* **48** 1175–85

Mah K, Danjoux C E and Caldwell C B 2000 Is spiral CT too fast for radiation therapy planning of thoracic neoplasms? *Radiother. Oncol.* **56** (Suppl. 1) S38

Malden C H, Meehan C A, Mubata C D and Bidmead A M 2001 Dose verification of dynamic IMRT for treatment of the prostate and pelvic nodes *Radiother. Oncol.* **61** (Suppl. 1) S108

Malsch U, Frühling C and Bendl R 2004 Methods for elastic adaptation of segmented volumes of interest for adaptive radiotherapy planning *Proc. 14th Int. Conf. on the Use of Computers in Radiation Therapy (Seoul, May)* pp 69–73

Manning M A, Wu Q, Cardinale R M, Mohan R, Lauve A D, Kavanagh B D, Morris M M and Schmidt-Ullrich R K 2001 The effect of setup uncertainty on normal tissue sparing with IMRT for head-and-neck cancer *Int. J. Radiat. Oncol. Biol. Phys.* **51** 1400–9

Manning M A, Wu Q, Mohan R, Cardinale R M, Kavanagh B D, Morris M M and Schmidt-Ullrich R K 2000 The effect of set-up uncertainty on normal tissue sparing with IMRT for head and neck *Proc. 42nd Annual ASTRO Meeting. Int. J. Radiat. Oncol. Biol. Phys.* **48** (Suppl.) 193

Mapelli M, Fumagalli M, Pellegrini R, Salehi N, Milanesi I and Fariselli L 2001 Physical and dosimetric characterization of a micro-multileaf collimator for dynamic conformal radiotherapy *Radiother. Oncol.* **61** (Suppl. 1) S97

Marcié S, Martin E, Dalmasso C, Bensadoun R J, Serrano B, Hérault J, Costa A, Bensadoun R J, Hachem S and Gérard J P 2002 IMRT implementation at the NICE-CAL: from theoretical to clinical cases *Radiother. Oncol.* **64** (Suppl. 1) S215

Markman J, Low D A, Beavis A W and Deasy J O 2002 Beyond bixels: Generalizing the optimisation parameters for intensity modulated radiation therapy. *Med. Phys.* **29** 2298–304

Martens C, Budgell G J and Claus F 2001a Improved delivery efficiency for step and shoot intensity-modulated radiotherapy using a fast-tuning magnetron *Radiother. Oncol.* **61** (Suppl. 1) S83

Martens C, Claeys I, de Wagter C and de Neve 2002 The value of radiographic film for the characterization of intensity-modulated beams *Phys. Med. Biol.* **47** 2221–34

Martens C, de Gersem W, de Neve W and de Wagter C 2001b Combining the advantages of step-and-shoot and dynamic delivery of intensity-modulated radiotherapy by interrupted dynamic sequences *Int. J. Radiat. Oncol. Biol. Phys.* **50** 541–50

Martens C, Reynaert N, de Wagter C, Nilsson P, Palmans H, Thierens H and de Neve W 2001c Underdosage of the mucosa for small fields as used in IMRT: a comparison between radiochromic film measurements, Monte Carlo simulations, and collapsed cone convolution calculations *Radiother. Oncol.* **61** (Suppl. 1) S25

Martin E and Hachem A 2003 Validation of intensity modulation on a commercial treatment planning system *Med. Phys.* **30** 925–36

Martineau A and Manens J P 2001 Dosimetry of small photon fields for conformal radiotherapy *Phys. Medica.* **17** 113

Martinez A A, Yan D, Lockman D, Brabbins D, Kota K, Sharpe M, Jaffray D A, Vicini F and Wong J 2001 Improvement in dose escalation using the process of adaptive radiotherapy combined with three-dimensional conformal or intensity-modulated beams for prostate cancer *Int. J. Radiat. Oncol. Biol. Phys.* **50** 1226–34

Mather M L, Whittaker A K and Baldock C 2002 Ultrasound evaluation of polymer gel dosimeters *Phys. Med. Biol.* **47** 1449–58

Mavroidis P, Lind B K, van Dijk J, Koedooder K, de Neve W, de Wagter C, Planskoy B, Rosenwald J-C, Proimos B, Kappas C, Danciu C, Benassi M, Chierego G and Brahme A 2000 Comparison of conformal radiation therapy techniques within the dynamic radiotherapy project 'Dynarad' *Phys. Med. Biol.* **45** 2459–81

Mayles W P M, Wolff T, Cassapi L, Clements R W, Fenwick J D, Gately A, Martin L, and Mayles H 2002 Introduction of IMRT as a routine clinical treatment *Proc. IPEM Biennial Meeting (Southampton, 16–17 July)* pp 11–12

Mazurier J, Castelain B and Lartigau E 2001a Dosimetric evaluation of IMRT at Centre Oscar Lambret *Radiother. Oncol.* **61** (Suppl. 1) S64

——2001b Dosimetric evaluation of intensity modulated treatments in radiotherapy at Oscar Lambret Centre *Phys. Medica.* **17** 114

Mazurier J, Kouto H, Poupon L, Castelain B and Lartigau E 2002 Implementation and quality control of IMRT for head and neck cancer *Radiother. Oncol.* **64** (Suppl. 1) S215

McBain C A, Sykes J S, Buckley D L, Amer A, Moore C J, Cowan R A and Khoo V S 2003 Optimizing bladder radiotherapy: MR assessment of time-dependent organ motion *Clinical Oncol.* **15** (Suppl. 2) S13

McClean B and Rock L 2001 Characterisation of small MLC-defined fields for use in IMRT treatment techniques *Phys. Medica.* **17** 161

McCurdy B, Malkoske K and Furutani K 2002 An automated compensator exchanging (ACE) device to facilitate the cost-effective delivery of IMRT *Med. Phys.* **29** 1306

McKenzie A L 2000 How should breathing motion be combined with other errors when drawing margins around clinical target volumes? *Brit. J. Radiol.* **73** 973–7

——2003a Optimum radiotherapy planning margins *RAD Magazine* **29** (334) 31–2

——2003b Protocols for target volumes and organs at risk *Clinical Oncol.* **15** (Suppl. 2) S12

McKenzie A L, van Herk M and Mijnheer B 2000 The width of margins in radiotherapy plans *Phys. Med. Biol.* **45** 3331–42

——2002 Margins for geometric uncertainty around organs at risk in radiotherapy *Radiother. Oncol.* **63** 299–307

McNair H 2002 Clinical IMRT: Impact on a radiotherapy department *Radiother. Oncol.* **64** (Suppl. 1) S79

McNair H A, Adams E J, Clark C H, Miles E A and Nutting C M 2003b Implementation of IMRT in the radiotherapy department *Brit. J. Radiol.* **76** 850–6

McNair H, Christian J, Knowles C, Symonds-Tayler J and Brada M 2003a Can the active breathing co-ordinator be used for patients receiving radical lung radiotherapy? *Radiother. Oncol.* **68** (Suppl. 1) S47

McNair H A, Francis G and Balyckyi J 2004 Clinical implementation of dynamic intensity-modulated radiotherapy: radiographers' perspectives *Brit. J. Radiol.* **77** 493–8

McNeeley S, Buyyounouski M, Price R, Horwitz E and Ma C 2003 The use of ultrasound in target localization *Radiother. Oncol.* **68** (Suppl. 1) S16

McShan D, Lynn K, Vineberg K and Fraass B 2002 Radiotherapy plan optimization accounting for set-up and motion uncertainty using a multiple instance geometry approximation *Med. Phys.* **29** 1257

McShan D, Ten Haken R, Kessler M, Balter J, Lam K and Fraass B 2001 Radiotherapy plan optimisation accounting for set-up and motion uncertainty using a multiple instance geometry approximation *Radiother. Oncol.* **61** (Suppl. 1) S13

Mechalakos J G, Mageras G S, Zelefsky M J, Lyass O, van Herk M, Kooy H M, Leibel S A and Ling C C 2002 Time trends in organ position and volume in patients receiving prostate three-dimensional conformal radiotherapy *Radiother. Oncol.* **62** 261–5

Mechalakos J G, Yorke E, Mageras G S, Hertano A, Jackson A, Obcemea C, Rozenzweig K and Ling C C 2004 Dosimetric effect of respiratory motion in external beam radiotherapy of the lung *Radiother. Oncol.* **71** 191–200

Meedt G, Alber and Nuesslin F 2001 A method of 3D beam direction optimisation in IMRT *Radiother. Oncol.* **61** (Suppl. 1) S64

——2002 Beam direction optimisation in head and neck IMRT *Radiother. Oncol.* **64** (Suppl. 1) S10–11

——2003 Non-coplanar beam direction optimisation for intensity-modulated radiotherapy *Phys. Med. Biol.* **48** 2999–3019

Meeks S L, Tomé W A, Bouchet L G, Haas A C, Buatti J M and Bova F J 2003 Patient positioning using optical and ultrasound techniques *IMRT- the state of the art: AAPM Medical Physics Monograph Number 29* ed J R Palta and T R Mackie (Madison, WI: Medical Physics Publishing) pp 727–48

Meyer J 2003 Beam direction optimization for anatomy-based IMRT using beam's-eye-view metrics *Clinical Oncol.* **15** (Suppl. 2) S30

Meyer J, Mills J A, Haas O C L, Burnham K J and Parvin E M 2001 Accommodation of couch constraints for coplanar intensity modulated radiation therapy *Radiother. Oncol.* **61** 23–32

Meyer J, Phillips M H, Cho P S, Kalet I and Doctor J N 2004 Application of influence diagrams to prostate intensity-modulated radiation therapy plan selection *Phys. Med. Biol.* **49** 1637–53

Miften M, Das S, Su M and Marks L 2004a Functional equivalent uniform dose (fEUD) for reporting and comparing functional image-guided dose distributions *Proc. 14th Int. Conf. on the Use of Computers in Radiation Therapy (Seoul, May)* pp 159–62

——2004b Incorporation of functional imaging data in the evaluation of dose distributions using the generalized concept of equivalent uniform dose *Phys. Med. Biol.* **49** 1711–21

Mihailidis D and Gibbons J 2002 Optimum number of treatment fields for inversely planned prostate irradiation: a planning comparison *Med. Phys.* **29** 1335

Mijnheer B 2001a Possibilities and limitations of the use of radiographic film for the verification of IMRT fields *Radiother. Oncol.* **61** (Suppl. 1) S3

Miles E A, Clark C H, Guerrero Urbano M T, Dearnaley D P and Nutting C M 2003 How routine can IMRT become in daily clinical practice? *Clinical Oncol.* **15** (Suppl. 2) S30

Miller T R and Grigsby P W 2003 In response to Dr. Beitler *Int. J. Radiat. Oncol. Biol. Phys.* **55** 282

Minken A W H and Mijnheer B J 2001 Possibilities and limitations in the use of radiographic film for the evaluation of dose delivery in small fields *Med. Phys.* **28** 1282

Mitra R, Bayouth J and Das I 2001a Startup characteristics and dosimetry of small monitor unit (MU) segments in step and shoot IMRT delivery for two types of digital linear accelerators *Med. Phys.* **28** 1202

Mitra R, Mah D, Das I, Horwitz E and Hanks G 2001b Evaluation of dosimetric characteristics of a virtual micro-multileaf collimator—Siemens HD270 MLC *Med. Phys.* **28** 1277

Mittal B, Kepka A, Mahadevan A, Kies M, Pelzer H, List M, Rademaker A and Logemann J 2001 Use of IMRT to reduce toxicity from concomitant radiation and chemotherapy for advanced head and neck cancers *Int. J. Radiat. Oncol. Biol. Phys.* **51** (Suppl. 1) 82–3

Mock U, Kodym R, Weitmann H, Wolff U, Jetzek S, Pittner I and Potter R 2000 Influence of set-up displacements in conformal radiotherapy of prostate cancer on changes in dose-volume histograms (DVH), tumour control probabilities (TCP) and normal tissue complication probabilities (NTCP) of the bladder and rectum *Radiother. Oncol.* **56** (Suppl. 1) S109

Moeckli R, Jeanneret Sozzi W, Valley J F and Mirimanoff R O 2002 The reasons for discrepancies in target volume delineation: 1. Physical part *Radiother. Oncol.* **64** (Suppl. 1) S208–9

Mohan R 2002 IMRT present and future prospects *Radiother. Oncol.* **64** (Suppl. 1) S64

Mohan R, Antolak J and Hendee W R 2001a Monte Carlo techniques should replace analytical methods for estimating dose distributions in radiotherapy treatment planning *Med. Phys.* **28** 123–6

Mohan R, Arnfield M, Tong S, Wu Q and Siebers J 2000 The impact of fluctuations in intensity patterns on the number of monitor units and the quality and accuracy of intensity modulated radiotherapy *Med. Phys.* **27** 1226–37

Mohan R, Wang X, Jackson A, Bortfeld T, Boyer A L, Kutcher G J, Leibel S A, Fuks Z and Ling C C 1994 The potential and limitations of the inverse radiotherapy technique *Radiotherapy and Oncology* **32** 232–48

Mohan R, Wu Q, Morris M, Lauve A, Tong S, Benedict S and Schmidt-Ullrich 2001b Simultaneous integrated boost (SIB) IMRT of advanced head and neck squamous cell carcinomas—dosimetric analysis *Int. J. Radiat. Oncol. Biol. Phys.* **51** (Suppl. 1) 180–1

Moore C J and Graham P A 2000 3D dynamic surface sensing and CT-body matching: a tool for patient set-up and monitoring in radiotherapy *Computer Aided Surgery* **5** 234–45

Moran J M, Fraass B A, Lam K L and Litzenberg D L 2001a An integrated system for DMLC (IMRT) fluence verification *Med. Phys.* **28** 1284

Moran J M, Litzenberg D L, Nurushev T S and Fraass B A 2001b Dosimetric comparisons of static, SMLC, and DMLC IMRT delivery techniques *Radiother. Oncol.* **61** (Suppl. 1) S17

Morgan A 2001 Commissioning and implementation of a simple IMRT technique Abstracts of the Elekta IMRT User Meeting 2001 in NKI Amsterdam, p 14

Morita K 2000 Keynote address—conformal RT and conformation RT *Int. J. Radiat. Oncol. Biol. Phys.* **48** 431–4

Mornex F, Chapet O, Chapuy S, Dimcovski Z, Morignat E, Romestaing P, Sentenac I and Gerard J P 2000 IRIS, A dynamic portal imaging system to visualize and control treatment in radiotherapy *Radiother. Oncol.* **56** (Suppl. 1) S61

Morr J, DiPetrillo T, Tsai J-S, Engler M and Wazer D E 2002 Initial experience with ultrasound localization for positioning prostate cancer patients for external beam radiotherapy. *Int. J. Radiat. Oncol. Biol. Phys.* **53** 1124–9

Morr J, DiPetrillo T, Tsai J and Wazer D 2000 Clinical use of daily transabdominal ultrasound localization with intensity modulated radiation therapy (IMRT) for prostate cancer *Proc. 42nd Annual ASTRO Meeting. Int. J. Radiat. Oncol. Biol. Phys.* **48** 308

Morrill S, Bayouth J and Mehta S 2001 A comparison of inverse IMRT treatment plans *Med. Phys.* **28** (Suppl.) 1206

Mosleh-Shirazi M A 2002 Measurement of the field shaping properties of the Elekta high-resolution MLC (beam modulator) *Radiother. Oncol.* **64** (Suppl. 1) S212–13

Mosleh-Shirazi M A, Evans P M, Swindell W, Webb S and Partridge M 1998 A cone-beam megavoltage CT scanner for treatment verification in conformal radiotherapy. *Radiother. Oncol.* **48** 319–28

Mott J H L, Budgell G F and Williams P C 2001 Treatment planning for IMRT using pre-determined intensity modulations for particular target geometries—a feasibility study *Radiother. Oncol.* **61** (Suppl. 1) S32

Mott J H, Livsey J E and Logue J P 2004 Development of a simultaneous boost IMRT class solution for hypofractionated prostate cancer protocol *Brit. J. Radiol.* **77** 377–86

Moya Garcia F 2001 The need for IMRT: a PET point of view *Radiother. Oncol.* **61** (Suppl. 1) S1

Muench R K, Seiler P G and Verwey J 2001 Application of magnetic tracking for the continuous precise measurement of tumour positions in conformal radiotherapy *Phys. Medica.* **17** 164

Mundt A J, Lujan A E, Rotmensch J, Waggoner S E, Yamada S D, Fleming G and Roeske J C 2002 Intensity-modulated whole pelvic radiotherapy in women with gynecologic malignancies *Int. J. Radiat. Oncol. Biol. Phys.* **52** 1330–7

Mundt A J, Roeske J C and Lujan A E 2001 Clinical experience with intensity-modulated whole pelvic radiation therapy (IM-WPRT) in patients with gynaecologic malignancies *Int. J. Radiat. Oncol. Biol. Phys.* **51** (Suppl. 1) 220

Munro P, Seppi E, Johnsen S, Shapiro E, Tognina C, Jones D, Pavkovich J, Webb C, Mollov I, Partain L and Colbeth R E 2002 Megavoltage cone-beam computed tomography using a high quantum efficiency image receptor *Med. Phys.* **29** 1340

Münter M W, Thilmann C, Hof H, Didinger B, Rhein B, Nill S, Schlegel W, Wannenmacher M and Debus J 2003 Stereotactic intensity modulated radiation therapy and inverse treatment planning for tumours of the head and neck region: clinical implementation of the step and shoot approach and first clinical results *Radiother. Oncol.* **66** 313–21

Murphy M 2002 Image-guided radiosurgery using the Cyberknife *Radiother. Oncol.* **64** (Suppl. 1) S22

——2004 Tracking moving organs in real time *Seminars Radiat. Oncol.* **14** 91–100

Murphy M J, Chang S D, Gibbs I C, Le Q-T, Hai J, Kim D, Martin D P and Adler J R 2003 Patterns of patient movement during frameless image-guided radiosurgery *Int. J. Radiat. Oncol. Biol. Phys.* **55** 1400–8

Murphy M J, Isaksson M, Jalden J and Ozhasoglu C 2001 Realtime compensation of breathing motion during radiotherapy *Int. J. Radiat. Oncol. Biol. Phys.* **51** (Suppl. 1) 388

Murshed H, Liu H H, Liao Z, Barker J L, Wang X, Tucker S L, Chandra A, Guerrero T, Stevens C, Change J Y, Jeter M, Cox J D, Komaki R and Mohan R 2004 Dose and volume reduction for normal lung using intensity-modulated radiotherapy for advanced-stage non-small-cell lung cancer *Int. J. Radiat. Oncol. Biol. Phys.* **58** 1258–67

Mutic S, Low D A, Klein E E, Dempsey J F and Purdy J A 2001 Room shielding for intensity-modulated radiation therapy treatment facilities *Int. J. Radiat. Oncol. Biol. Phys.* **50** 239–46

Mutic S, Malyapa R S, Grigsby P W, Dehdashti F, Miller T R, Zoberi I, Bosch W R, Esthappan J and Low D A 2003 PET-guided IMRT for cervical carcinoma with positive para-aortic lymph nodes—a dose-escalation treatment planning study *Int. J. Radiat. Oncol. Biol. Phys.* **55** 28–35

Myrianthopoulos L, Turian J, Hamilton R and Baird C 2001 Dose rate effects on the calculated fluence maps of IMRT fields *Med. Phys.* **28** 1203

Nabid A, Bujold R, Bénard F, Verreault J and Turgeon P P 2002 Evaluation of the size of tumours reported by total body PET scans and pathology reports for resected tumors *Radiother. Oncol.* **64** (Suppl. 1) S209–10

Nahum A E. 2003 Tumour local-control probability prediction (TLCP) and other radiobiological issues in the use of IMRT *Radiother. Oncol.* **68** (Suppl. 1) S5

Nakagawa K, Aoki Y, Akanuma A, Onogi Y, Terahara A, Sakata K, Muta N, Sasaki Y, Kawakami H and Hanakawa K 1994 Real-time beam monitoring in dynamic conformation therapy *Int. J. Radiat. Oncol. Biol. Phys.* **30** 1233–8

Nakagawa K, Aoki Y, Tago M, Terahara A and Ohtomo K 2000a Megavoltage CT-assisted stereotactic radiosurgery for thoracic tumours: original research in the treatment of thoracic neoplasms *Int. J. Radiat. Oncol. Biol. Phys.* **48** 449–57

Nakagawa K, Aoki Y and Ohtomo K 2000b Dynamic conical conformal therapy using a C-arm mounted accelerator *Proc. 42nd Annual ASTRO Meeting. Int. J. Radiat. Oncol. Biol. Phys.* **48** (Suppl.) 220

Naqvi S A, Holmes T W, Sarfaraz M, Li X A and Yu C X 2000 Improving the accuracy of monitor unit calculation for complex leaf sequences in IMRT delivery *Proc. 42nd Annual ASTRO Meeting. Int. J. Radiat. Oncol. Biol. Phys.* **48** (Suppl.) 221

Naqvi S, Holmes T and Yu C 2001a Verifying static-to-dynamic conversion of IMRT leaf sequences by raytracing *Med. Phys.* **28** 1268

Naqvi S A, Sarfaraz M, Holmes T, Yu C X and Li X A 2001b Analysing collimator structure effects in head-scatter calculations for IMRT class fields using scatter raytracing *Phys. Med. Biol.* **46** 2009–28

Nederveen A J, van der Heide U A, Dehnad H, Jeroen R, van Moorselaar A, Hofman P and Lagendijk J J W 2002 Measurements and clinical consequences of prostate motion during a radiotherapy fraction *Int. J. Radiat. Oncol. Biol. Phys.* **53** 206–14

Neicu T, Papiez L and Jiang S B 2003 Synchronized moving aperture radiation therapy (SMART): Dynamic MLC leaf sequencing with tumor motion *Med. Phys.* **30** 1405

Neicu T, Shirato H, Seppenwold Y and Jiang S B 2002 Gated motion adaptive therapy (GMAT): average tumour trajectory for lung patients *Med. Phys.* **29** 1309–10

Nguyen N, Higgins P and Alaei P 2002 Evaluation of Kodak EDR2 film and its application in IMRT dose verification *Med. Phys.* **29** 1269

Nill S, Tücking T, Münter M and Oelfke U 2002 IMRT with MLCS of different leaf widths: a comparison of achievable dose distributions *Radiother. Oncol.* **64** (Suppl. 1) S107–8

Nioutsikou E, Bedford J L, Webb S and Christian J 2004 Segmentation of IMRT plans for radical lung radiotherapy delivery with the step-and-shoot technique *Radiother. Oncol.* **31** 892–901

Nishioka T, Shiga T, Shirato H, Tsukamoto E, Tsuchiya K, Kato T, Ohmori K, Yamazaki A, Aoyama H, Hashimoto S, Chang T-C and Mikyasaka K 2002 Image fusion between [18]FDG-PET and MRI/CT for radiotherapy planning of oropharyngeal and nasopharyngeal carcinomas *Int. J. Radiat. Oncol. Biol. Phys.* **53** 1051–7

Nishioka T, Shirato H, Karo T, Watanabe Y, Yamazaki A, Ohmori K, Aoyama H, Shiga T, Tsukamoto E, Hashimoto S, Tsuchiya K and Miyasaka K 2000 Impact of [18]FDG-PET and CT/MRI image fusion in radiotherapy planning of head-and-neck tumours *Proc. 42nd Annual ASTRO Meeting. Int. J. Radiat. Oncol. Biol. Phys.* **48** (Suppl.) 260

Novotny J Jnr, Dvorak P, Spevacek V, Tintera J, Novotny J, Cechak T and Liscak R 2002 Polymer-gel dosimetry as a tool for quality control of the stereotatic radiosurgery procedure *Radiother. Oncol.* **64** (Suppl. 1) S44

Nüsslin F 2001 IMRT and beyond—perspective in radiation therapy *Phys. Medica.* **17** 163–4

Nüsslin F, Alber M, Baum C, Birkner M and Buck D 2003 Early steps toward IGRT: Development of imaging tools for IMRT treatment planning radiotherapy *Phys. Medica.* **19** 45

Nüsslin F, Alber M, Birkner M and Fippel M 2003 Hyperion—a novel Monte-Carlo IMRT treatment planning program based on isoeffect dose optimisation *Proc. World Congress in Medical Physics (Sydney)* CDROM and website only

Nutting C 2003 Intensity-modulated radiotherapy (IMRT): the most important advance in radiotherapy since the linear accelerator? *Brit. J. Radiol.* **76** 673

Nutting C and Dearnaley D 2001 Intensity-modulated radiation therapy (IMRT) for tumours of the head and neck, pelvis and thorax: pre-clinical evaluation and implementation *Compendium of the 3rd IMRT Winter School, Heidelberg (London, Jan 26/27)* pp C2.1–C2.3

——2002 Prostate cancer—the Royal Marsden conformal experience *Normal Tissue Reactions in Radiotherapy and Oncology* ed W Dörr, R Engenhart-Cabillic and J S Zimmermann (Basel: Karger) pp 196–9

Nutting C and Harrington K 2004 (in regards to Dr Glatstein 2003 *Int. J. Radiat. Oncol. Biol. Phys.* **55** 561–2 and Dr Amols 203 *Int. J. Radiat. Oncol. Biol. Phys.* **56** 1507) *Int. J. Radiat. Oncol. Biol. Phys.* **58** 1316–17

Nutting C M, Rowbottom C, Cosgrove V P, Convery D J, Webb S and Dearnaley D P 2000 Evaluation of intensity modulated radiotherapy for tumours of the pelvis, head and neck, and thorax: Optimisation and clinical implementation *Proc. 42nd Annual ASTRO Meeting. Int. J. Radiat. Oncol. Biol. Phys.* **48** (Suppl.) 139

O'Brien P, Woo M and Nico A 2002 Beam energy selection for intensity modulated radiation therapy *Radiother. Oncol.* **64** (Suppl. 1) S212

O'Daniel J C 2001 Comparison of 3D conformal and intensity-modulated radiation therapy for treatment of prostate cancer *Med. Phys.* **28** 1292

O'Daniel J, Dong L, Kuban D A, Liu H, Schechter N, Tucker S L and Rosen I 2004 The delivery of IMRT with a single physical modulator for multiple fields: a feasibility study for paranasal sinus cancer *Int. J. Radiat. Oncol. Biol. Phys.* **58** 876–87

O'Daniel J, Dong L, Rosen I and Kuban D 2002 Comparison of planning quality and delivery efficiency of two commercial IMRT planning systems *Med. Phys.* **29** 1283

Oelfke U 2001 Delivery techniques of IMRT: MLCs *Radiother. Oncol.* **61** (Suppl. 1) S1

Oelfke U, Rhein B and Häring P 2001 Dosimetric procedures and problems of IMRT *Proc. Topical Meeting on 'Medical Radiation Physics and Engineering'*, Lisbon, Portugal, Nov 20–22, 2000 *Phys. Medica.* **17** (Suppl. 4) 30–2

Oelfke U, Siemer N and Nill S 2002 Optimization of IMRT treatment plans with EUD-based objective functions *Radiother. Oncol.* **64** (Suppl. 1) S51

Oh Y, Ji Y, Kim K, Cho M, Kim J, Jeong D, Shin K and Kim J 2003 The realization of intensity-modulated radiation therapy using the liquid shielding material *Proc. World Congress in Medical Physics (Sydney)* CDROM and website only

Olch A 2001 Clinical implementation of IMRT in a small US clinic *Radiother. Oncol.* **61** (Suppl. 1) S74

Oldham M and Kim L 2004 Optical-CT gel-dosimetry 2: optical artifacts and geometrical distortion *Med. Phys.* **31** 1093–104

Oldham M, McJury M, Baustert I B, Webb S and Leach M O 1998 Improving calibration accuracy in gel-dosimetry *Phys. Med. Biol.* **43** 2709–20

Oldham M, Siewerdsen J H, Kumar S, Wong J and Jaffray D A 2003 Optical-CT gel-dosimetry 1: basic investigations *Med. Phys.* **30** 623–34

Oldham M, Siewerdsen J H, Shetty A and Jaffray D A 2001 High resolution gel-dosimetry by optical and MR scanning *Med. Phys.* **28** 1436–45

Oliver L, Phillips R, Vial P, Hunt P and Mallik A 2002 Conformal versus IMRT treatment plans *Radiother. Oncol.* **64** (Suppl. 1) S78

Olivera G, Fitchard E, Reckwerdt P, Ruchala K and Mackie R 2000b Modifying a plan delivery without re-optimization to account for patient offset in tomotherapy *Proc. World Congress of Medical Physics and the AAPM Annual Congress, August 2000* paper MO-EBR-01

Olivera G, Kapatoes J, Mackie T, Lu W, Reckwerdt P, Jeraj R, Balog J and Ruchala K 2002 Use of helical tomotherapy optimization from very simple to very complex treatment planning *Radiother. Oncol.* **64** (Suppl. 1) S77

Olivera G H, Mackie T R, Ruchala K J, Kapatoes J M, Reckwerdt P J, Jeraj R, Balog J P, Lu W, Smilowitz and Keller H 2001 Tomotherapy: a tool for image guided adaptive radiotherapy *Radiother. Oncol.* **61** (Suppl. 1) S24

Olivera G, Reckwerdt P, Shepard D M and Mackie R 2000a Large-scale helical tomotherapy optimization: Four clinical case studies *Proc. World Congress of Medical Physics and the AAPM Annual Congress, August 2000* paper FR-B309-04

Onishi H, Kuriyama K, Komiyama T, Ueki J and Araki T 2001 A new irradiation system for lung cancer: patient's self-breath-hold and self-switching radiation—beam on and off without any respiratory monitoring devices *Int. J. Radiat. Oncol. Biol. Phys.* **51** (Suppl. 1) 26

Orton N and Tomé 2003 Dosimetric impct of daily shifts on prostate IMRT dose distributions *Med. Phys.* **30** 1339

Otto K, Alkins R and Clark B G 2003 Limitations of IMRT planning with low resolution pencil beam kernels *Radiother. Oncol.* **68** (Suppl. 1) S43

Otto K and Clark B G 2002 Enhancement of IMRT delivery through MLC rotation *Phys. Med. Biol.* **47** 3997–4017

Ozhasoglu C, Christie N, Luketich J, Burton S, Gerszten P, Wu A and Kalnicki S 2003 Target immobilization with natural breath-holding: Possible applications to IMRT and body radiosurgery *Med. Phys.* **30** 1338

Ozhasoglu C, Murphy M J and Adler J R 2001 Real time compensation of breathing motion in radiotherapy *Med. Phys.* **28** 1295

Paliwal B, Balog J, Smilowitz J, Das R and Mackie T 2002a Commissioning dosimetry for a UW clinical tomotherapy unit *Med. Phys.* **29** 1211–12

Paliwal B, Balog J, Tomé W, Das R, Richardson S, Mackie R and Mehta S 2002b Tomotherapy: initial University of Wisconsin experience *Radiother. Oncol.* **64** (Suppl. 1) S22

Paliwal B, Balog J, Tomé W, Richardson S and Mackie T 2001 Dosimetry verification for IMRT/Tomotherapy *Radiother. Oncol.* **61** (Suppl. 1) S16

Paliwal B R, Brezovich I A and Hendee W R 2004 IMRT may be used to excess because of its higher reimbursement from medicare *Med. Phys.* **31** 1–3

Paliwal B Tomé W, Richardson S and Mackie T R 2000 A spiral phantom for IMRT and tomotherapy treatment delivery verification *Med. Phys.* **27** 2503–7

Palta J R and Mackie T R (eds) 2003 *IMRT—the State of the Art: AAPM Medical Physics Monograph Number 29* (Madison, WI: Medical Physics Publishing)

Papatheodorou S, Castellanos M, Zefkili S, Gaboriaud G and Rosenwald J 2000b Dynamic collimation for missing tissue compensation *Proc. World Congress of Medical Physics and the AAPM Annual Congress, August 2000* paper MO-CXH-02

Papatheodorou S, Rosenwald J-C, Zefkili S, Murillo M-C, Drouard J and Gaboriaud G 2000a Dose calculation and verification of intensity modulation generated by dynamic multileaf collimators *Med. Phys.* **27** 960–71

Papanikolaou N 2001 Dose calculation algorithms in the IMRT era *Radiother. Oncol.* **61** (Suppl. 1) S12

Papanikolaou N, Gutierrez L, Yan Y, Wu C, Penagaricano J and Ratanatharathorn V 2003 The effect of tissue inhomogeneity in dose based and biologically based IMRT *Radiother. Oncol.* **68** (Suppl. 1) S42

Papanikolaou N, Stathakis S, Tsougo Y and Kappas C 2001 The effect of tissue inhomogeneity on IMRT *Med. Phys.* **28** 1207

Papanikolaou N, Yan Y, Penagaricano J and Ratanatharathorn V 2002 Application of IMRT for the treatment of medulloblastoma *Med. Phys.* **29** 1215

Papiez L, Thai V and Langer M 2001 Optimal step-and-shoot segmentation algorithm for unidirectional motion of leaves *Med. Phys.* **28** 1204

Parker B, Shiu A and Comiskey R 2000 Comparison of dose distributions from geometric conformation and intensity modulation for intracranial lesions *Proc. World Congress of Medical Physics and the AAPM Annual Congress, August 2000* paper MO-BBR-05

Parker C C, Damyanovich A, Haycocks T, Haider M, Bayley A and Catton C N 2003 Magnetic resonance imaging in the radiation treatment planning of localized prostate cancer using intra-prostatic fiducial markers for computed tomography co-registration *Radiother. Oncol.* **66** 217–24

Parker W and Brodeur M 2002 Measurement of 3D dose distributions and DVH of IMRT treatment plans using MR based gel dosimetry *Med. Phys.* **29** 1263–4

Parker W, Hristov D, Moftah B, Vuong T, Tsien C and Podgorsak E 2000 Intensity modulated radiation therapy for thyroid malignancies *Proc. World Congress of Medical Physics and the AAPM Annual Congress, August 2000* paper TU-GBR-02

Parsai H, Cho P S, Phillips M H, Giansiracusa R S and Axen D 2003 Random and systematic beam modulator errors in dynamic intensity modulated radiotherapy *Phys. Med. Biol.* **48** 1109–21

Parsai H, Phillips M H and Cho P S 2001a Dosimetry verifications of dynamic IMRT: An *in-vivo* study in canine subjects *Med. Phys.* **28** 1208

——2001b Verification of dynamic intensity-modulated beam deliveries in canine subjects *Med. Phys.* **28** 2198–208

Partridge M 2000 Reconstruction of megavoltage photon spectra from electronic portal imager derived transmission measurements *Phys. Med. Biol.* **45** N115–31

Partridge M, Aldridge S, Donovan E and Evans P M 2001a An intercomparison of IMRT delivery techniques: a case study for breast treatment *Phys. Med. Biol.* **46** N175–85

Partridge M, Christian J A, Flux G, McNair H A, Cronin B, Cook G and Courbon F 2003 Accurate co-registration of CT and SPECT perfusion images for lung radiotherapy planning *Radiother. Oncol.* **68** (Suppl. 1) S67

Partridge M, Ebert M and Hesse B-M 2002 IMRT verification by three-dimensional dose reconstruction from portal beam measurements *Med. Phys.* **29** 1847–58

Partridge M, Evans P M, van Herk M, Ploeger L S, Budgell G J and James H V 2000a Leaf position verification during dynamic beam delivery: a comparison of three applications using electronic portal imaging *Med. Phys.* **27** 1601–9

Partridge M and Hesse B M 2002 IMRT dose verification *in-vivo* in 3D *Radiother. Oncol.* **64** (Suppl. 1) S123

Partridge M, Hesse B M, Groh B, Ebert M and Bortfeld T 2001b Verification of 3D dose delivery at treatment time *Radiother. Oncol.* **61** (Suppl. 1) S4

Partridge M, Symonds-Tayler J and Evans P 2000b Dynamic MLC verification with portal imaging *Radiother. Oncol.* **56** (Suppl. 1) S89

Partridge M, Symonds-Tayler J R N and Evans P M 2000c IMRT verification with a camera-based electronic portal imaging system *Phys. Med. Biol.* **45** N183–96

Paskalev K, Ma C, Jacob R and Price R 2003 Daily target localization based on 3D image correlation *Med. Phys.* **30** 1337

Paskalev K, Ma C-M, Jacob R, Price R, McNeeley S, Wang L, Movsas B and Pollack A 2004 Daily target localization for prostate patients based on 3D image correlation *Phys. Med. Biol.* **49** 931–9

Pastyr O, Echner G, Hartmann G, Richter J and Schlegel W 2001 Dynamic edge focussing: A new MLC-design to deliver IMRT with a double focussing high precision multi-leaf-collimator *Radiother. Oncol.* **61** (Suppl. 1) S24

Pasma K L, Vieira S C, de Langen M, Dirkx M L P, Storchi P and Heijmen B J M 2001 Dosimetric verification of dynamic multileaf collimation using a fluoroscopic electronic portal imaging device (EPID) *Radiother. Oncol.* **61** (Suppl. 1) S16

Patel R R, Tomé W A, Tannehill S P, Harari P M, Paliwal B R and Mehta M P 2002 Optically guided intensity modulated radiotherapy for the head and neck *Radiother. Oncol.* **64** (Suppl. 1) S248

Paulino A C and Johnstone P A S 2004 FDG-PET in radiotherapy treatment planning: Pandora's box? *Int. J. Radiat. Oncol. Biol. Phys.* **59** 4–5

Paulsen F, Alber M, Eschmann S M, Plasswilm L, Budach W, Machulla H J, Nüsslin F, Bares R and Bamberg M 2002 Using PET imaging and IMRT to generate a smart hypoxia boost for head and neck tumours *Radiother. Oncol.* **64** (Suppl. 1) S92

Pavel L, Rowbottom C G, Hector C, Partridge M and Bortfeld T 2001 Estimation of the effects of movements on dose distribution for prostate patients *Radiother. Oncol.* **61** (Suppl. 1) S36

Pavel-Mititean L M, Rowbottom C G, Hector C L, Partridge M, Bortfeld T and Schlegel W 2004 A geometrical model for evaluating the effects of inter-fraction rectal motion during prostate radiotherapy *Phys. Med. Biol.* **49** 2613–29

Pawlicki T, Jiang S B, Li J, Deng J and Ma C M 2000a Including setup uncertainty in Monte Carlo dose calculation for IMRT *Proc. World Congress of Medical Physics and the AAPM Annual Congress, August 2000* paper MO-B309-02

——2000b The role of set-up uncertainty in intensity modulated treatment planning *Proc. 42nd Annual ASTRO Meeting. Int. J. Radiat. Oncol. Biol. Phys.* **48** (Suppl.) 192

Pawlicki T, Luxton G, Le Q-T, Findley D and Ma C-M 2004 Lens dose in MLC-based IMRT treatments of the head and neck *Int. J. Radiat. Oncol. Biol. Phys.* **59** 293–9

Pelizzari C A 2003 Medical imaging in IMRT planning *IMRT—the State of the Art: AAPM Medical Physics Monograph Number 29* cd J R Palta and T R Mackie (Madison, WI: Medical Physics Publishing) pp 139–81

Peltola S 2003 Intensity modulation radiotherapy using compensators—dose calculations using CADPLAN HELIOS dose planning *Proc. World Congress in Medical Physics (Sydney)* CDROM and website only

Pemler P, Besserer J, Lombriser N, Pescia R and Schneider U 2001 Influence of respiration-induced organ motion on dose distributions in treatments using enhanced dynamic wedges *Med. Phys.* **28** 2234–40

Perrin B, Budgell G J and Mackay R I 2003 A tool for quantitative analysis of IMRT verification measurements *Clinical Oncol.* **15** (Suppl. 2) S32

Petersen J B, Spejlborg H, Hansen A T, Thomsen M S, Tanderup K, Vestergaard A, Heyer M, Lindegaard J C, Ulso N and Grau C 2002 Implementation of intensity modulated radiotherapy in Aarhus, Denmark *Radiother. Oncol.* **64** (Suppl. 1) S216

Phillips M 2002 IMRT of lung cancer *Med. Phys.* **29** 1258–9

Phillips M H, Cho P S and Parsaei H 2001a Validation of an IMRT Program from A(lgorithm) to Z(ooRadiotherapy) *Radiother. Oncol.* **61** (Suppl. 1) S1

Phillips M H, Cho P S, Parsaei H, Kippenes H and Gavin P 2000 Verification of dynamic IMRT with *in vivo* and phantom dosimetry *Radiother. Oncol.* **56** (Suppl. 1) S89

Phillips M H, Meyer J, Cho P S, Kalet I J and Doctor J N 2004 Bayesian decision making applied to IMRT *Proc. 14th Int. Conf. on the Use of Computers in Radiation Therapy (Seoul, May)* pp 108-111

Phillips M, Parsaei H and Cho P 2001b Modelling dynamic IMRT dose distributions *Med. Phys.* **28** 1293

Pickett B, Pirzkall A, Kurhanawicz J, Verhey L and Roach M 2000b Radiosurgical intensity modulated radiotherapy for prostate cancer *Proc. 42nd Annual ASTRO Meeting. Int. J. Radiat. Oncol. Biol. Phys.* **48** (Suppl.) 138

Pickett B, Xia P, Woods P, Kurhanawicz J and Roach M 2000a Most dramatic magnetic resonance spectroscopy imaging changes correlate with high dose regions following intensity modulated radiotherapy for prostate cancer *Radiother. Oncol.* **56** (Suppl. 1) S106

Pignoli E, Serretiello S, Somigliana A, Zonca G, Pellegrini R, Mongioj V and Marchesini R 2000 Dosimetric verification of a commercial 3D treatment planning system for conformal radiotherapy with a dynamic multileaf collimator *Phys. Med. Biol.* **45** N77–84

Pirzkall A, Carol M, Lohr F, Hoess A, Wannenmacher M and Debus J 2000 Comparison of intensity-modulated radiotherapy with conventional conformal radiotherapy for complex-shaped tumours *Int. J. Radiat. Oncol. Biol. Phys.* **48** 1371–80

Pirzkall A, Carol M, Pickett B, Roach III M, Verhey L 2000a The effect of beam energy and number of fields on photon-based IMRT for deep seated targets *Proc. World Congress of Medical Physics and the AAPM Annual Congress, August 2000* paper MO-EBR-10

Pirzkall A. Carol M P, Pickett B, Xia P, Roach M and Verhey L J 2002 The effect of beam energy and number of fields on photon-based IMRT for deep-seated targets *Int. J. Radiat. Oncol. Biol. Phys.* **53** 434–42

Pirzkall A, Debus J, Haering P, Rhein B, Grosser K, Höss A and Wannenmacher M 2003 Intensity modulated radiotherapy (IMRT) for recurrent, residual, or untreated skull-base meningiomas: preliminary clinical experience *Int. J. Radiat. Oncol. Biol. Phys.* **55** 362–72

Plasswilm L, Alber M, Paulsen F, Bakai A, Birkner M, Glocker S, Christ G, Buck C, Nüsslin F and Bamberg M 2002 First clinical experience with image-guided intensity modulated radiation treatment of prostate cancer patients *Radiother. Oncol.* **64** (Suppl. 1) S271

Ploeger L S, Smitsmans M H P, Gilhuijs K G A and van Herk M 2002 A method for geometrical verification of dynamic intensity modulated radiotherapy using a scanning electronic portal imaging device *Med. Phys.* **29** 1071–9

Pollack A and Price R A 2003 IMRT for prostate cancer *IMRT—the State of the Art: AAPM Medical Physics Monograph Number 29* ed J R Palta and T R Mackie (Madison, WI: Medical Physics Publishing) pp 617–30

Pollack A, Zagars G K, Antolak J A, Kuban D A and Rosen I I 2002b In response to Drs. Kagan and Schulz *Int. J. Radiat. Oncol. Biol. Phys.* **55** 1151–2

Pollack A, Zagars G K, Smith L G, Lee J, von Eschenbach A C, Antotak J A, Starkschall G and Rosen I 2000 Preliminary results of a randomized radiotherapy dose escalation study comparing 70 Gy with 78 Gy for prostate cancer *J. Clin. Oncol.* **18** 3904–11

Pollack A, Zagars G K, Starkshall G, Antotak J A, Lee J, Huang E, von Eschenbach A C, Kuban D A and Rosen I 2002a Prostate cancer radiation dose response: results of the M D Anderson Phase-3 trial *Int. J. Radiat. Oncol. Biol. Phys.* **53** 1097–105

Popple R, Prellop P, Spencer S, Fiveash J, Duan J and Brezovich I 2003 Simultaneous optimisation of sequential IMRT plans *Radiother. Oncol.* **68** (Suppl. 1) S76

Portelance L, Chao K S C, Grigsby P W, Bennet H and Low D 2001 Intensity-modulated radiation therapy (IMRT) reduces small bowel, rectum and bladder doses in patients with cervical cancer receiving pelvic and para-aortic irradiation *Int. J. Radiat. Oncol. Biol. Phys.* **51** 261–6

Potter L, Chang S and Cullip T 2003a Intensity modulation via a 4-bank micro-MLC system *Med. Phys.* **30** 1348

Potter L D, Chang S X, Cullip T J and Siochi A C 2002 A quality and efficiency analysis of the IMFAST segmentation algorithm in head and neck 'step and shoot' IMRT treatments *Med. Phys.* **29** 275–83

Potters L, Steinberg M, Wallner P and Hevezi J 2003b How one defines intensity-modulated radiation therapy *Int. J. Radiat. Oncol. Biol. Phys.* **56** 609–10

Pouliot J, Aubin M, Verhey L, Bani-Hashemi A, Mitschke M, Hernandez P and Hughes J 2002 Low dose megavoltage CT cone beam reconstruction for patient alignment *Med. Phys.* **29** 1241

Pouliot J, Xia P, Aubin M, Verhey L, Langen K, Bani-Hashemi A, Svatos M, Ghelmansarai F and Mitchke M 2003 Dose-guided radiation therapy using low-dose megavoltage cone-beam CT *Med. Phys.* **30** 1337–8

Poynter A 2003 Intensity modulated radiotherapy—a clinical experience in a small oncology centre *RAD Magazine* **29** (334) 35–6

Price P 2002 Functional imaging to define radiotherapy target volumes—the future *Radiother. Oncol.* **64** (Suppl. 1) S35

Price R, Ma C, McNeeley S, Chen L, Fourkal E, Li J, Horwitz E and Pollack A 2003 IMRT at Fox Chase Cancer Center *Radiother. Oncol.* **68** (Suppl. 1) S63

Price R, Murphy S, McNeeley S and Ma C 2002 A method for increased dose conformity and segment reduction for SMLC delivered IMRT treatment of the prostate *Med. Phys.* **29** 1257

Price R A, Murphy S, McNeeley S W, Ma C-M, Horwitz E, Movsas B, Raben A and Pollack A 2003 A method for increased dose conformity and segment reduction for sMLC delivered IMRT treatment of the prostate *Int. J. Radiat. Oncol. Biol. Phys.* **57** 843–52

Pugachev A B, Boyer A L and Xing L 2000a Beam orientation optimization in intensity-modulated radiation treatment planning *Med. Phys.* **27** 1238–45

Pugachev A B, Hancock S, Le Q T, Boyer A L and Xing L 2000b Pseudo beam's-eye-view (BEV) applied to beam orientation selection for IMRT *Proc. 42nd Annual ASTRO Meeting. Int. J. Radiat. Oncol. Biol. Phys.* **48** (Suppl.) 353

Pugachev A, Li J, Boyer A and Xing L 2000c Role of non-coplanar beams in IMRT *Proc. World Congress of Medical Physics and the AAPM Annual Congress, August 2000* paper MO-EBR-05

Pugachev A, Li M, Boyer A L, Hancock S L, Le Q-T, Donaldson S S, and Xing L 2001 Role of beam orientation optimization in intensity-modulated radiation therapy *Int. J. Radiat. Oncol. Biol. Phys.* **50** 551–60

Pugachev A and Xing L 2001a Two stage beam orientation selection method for IMRT *Med. Phys.* **28** 1204

——2001b Computer assisted selection of coplanar beam orientations in intensity-modulated radiation therapy *Phys. Med. Biol.* **46** 2467–76

——2001c Pseudo beam's-eye-view as applied to beam orientation selection in intensity-modulated radiation therapy *Int. J. Radiat. Oncol. Biol. Phys.* **51** 1361–70

——2001d Computer assisted selection of beam energy and orientation in IMRT *Int. J. Radiat. Oncol. Biol. Phys.* **51** 75–6

——2002a Examination of the degree of validity of class-solution approach for IMRT *Med. Phys.* **29** 1284

——2002b Incorporating prior knowledge into beam orientation optimisation in IMRT. *Int. J. Radiol. Oncol. Biol. Phys.* **54** 1565–74

Purdy J A and Michalski J M 2001 Does the evidence support the enthusiasm over 3D conformal radiation therapy and dose escalation in the treatment of prostate cancer? *Int. J. Radiat. Oncol. Biol. Phys.* **51** 867–70

——2002 In response to Drs Levitt and Khan *Int. J. Rad. Onc. Biol. Phys.* **53** 1085

Que W 1999 Comparison of algorithms for multileaf collimator field segmentation. *Med. Phys.* **26** 2390–6

Que W, Kung J and Dai J 2004 'Tongue-and-groove' effect in intensity modulated radiotherapy with static multileaf collimator fields *Phys. Med. Biol.* **49** 399–405

Raaymakers B W, Lagendijk J J W, van der Heide U A, Overweg J, Brown K, Topolnjak R, Dehnad H, Jürgenliemk-Schulz I M, Welleweerd J and Bakker C J G 2004 Integrating a MRI scanner with a radiotherapy accelerator: a new concept of precise on line radiotherapy guidance and treatment monitoring *Proc. 14th Int. Conf. on the Use of Computers in Radiation Therapy (Seoul, May)* pp 89–92

Raaymakers B W, Lagendijk J J W, van der Heide U A and Topolnak R 2003 The development of a Linac-MRI combination for on-line position verification *Phys. Medica.* **19** 45–6

Raben A, Yang J, Murphy S, Stapleton T, Chen H, Grebler A, Geltzeller J, Simmons D and Gollamudi S 2002 Feasibility and safety of high dose intensity modulated radiotherapy (IMRT) for localized prostate cancer in a community practice: preliminary results on acute toxicity *Radiother. Oncol.* **64** (Suppl. 1) S277

Radford D L, Followill D and Hanson W 2000 Design of an anthropomorphic intensity modulated radiation therapy quality assurance phantom *Proc. World Congress of Medical Physics and the AAPM Annual Congress, August 2000* paper MO-CXH-01

——2001 A standard method of quality assurance for intensity modulated radiation therapy of the prostate *Med. Phys.* **28** 1211

Ramaseshan R, Cho Y, Purdie T, Alasti H, Wong R, Kohli K, Chan B, Heaton R and Islam M 2003 Breathing motion in lung: is external surface surrogate for internal structure? *Radiother. Oncol.* **68** (Suppl. 1) S54

Ramsey C 2002 The effects of respiration motion on dynamic IMRT *Med. Phys.* **29** 1260–1

Ramsey C, Dube S and Hendee W R 2003 It is necessary to validate each individual IMRT treatment plan before delivery *Med. Phys.* **30** 2271–3

Ramsey C R, Large C, Scaperoth D, Arwood D and McDonald W 2001b The dosimetric consequences of intra-fraction prostate motion in intensity modulation *Int. J. Radiat. Oncol. Biol. Phys.* **51** 214

Ramsey C R, Scaperoth D and Arwood D 2000 Clinical experience with a commercial respiratory gating system *Proc. 42nd Annual ASTRO Meeting. Int. J. Radiat. Oncol. Biol. Phys.* **48** (Suppl.) 164

Ramsey C R, Spencer K M, Alhakeem R and Oliver A L 2001a Leaf position error during conformal dynamic arc and intensity modulated arc treatments *Med. Phys.* **28** 67–72

Remouchamps V, Claus F, Vermael S, de Gersem W, Vakaet L and de Neve W 2000 Intensity modulated radiation therapy for large breast: physical wedges versus segmented tissue compensation *Radiother. Oncol.* **56** (Suppl. 1) S107

Remouchamps V M, Vicini F A, Martinez A A, Sharpe M B, Kestin L L and Wong J W 2002 Reductions in heart and lung doses using deep inspiration breath hold

440   *References*

with active breathing control and IMRT for patients treated with locoregional breast irradiation *Radiother. Oncol.* **64** (Suppl. 1) S139

Remouchamps V M, Vicini F A, Sharpe M B, Kestin L L, Martinez A A and Wong J W 2003 Significant reductions in heart and lung doses using deep inspiration breath hold with active breathing control and intensity-modulated radiation therapy for patients treated with locoregional breast irradiation *Int. J. Radiat. Oncol. Biol. Phys.* **55** 392–406

Renner W D, Sarfaraz M, Earl M A and Yu C X 2003 A dose delivery verification method for conventional and intensity-modulated radiation therapy using measured field fluence distributions *Med. Phys.* **30** 2996–3005

Rhein B and Häring P 2001 The IMRT phantom verification procedure in Heidelberg *Radiother. Oncol.* **61** (Suppl. 1) S2

Richardson S L , Tomé W A, Orton N P, McNutt T R and Paliwal B R 2003 IMRT delivery verification using a spiral phantom *Med. Phys.* **30** 2553–8

Richter J, Hadinger U, Kolbl O, Neumann M and Flentje M 2000 IMRT versus quasi dynamic rotation using MLC *Radiother. Oncol.* **56** (Suppl. 1) S195

Rietzel E, Chen G T Y, Adams J A, Hernandez A H and Willett C G 2004 Treatment planning including respiratory motion based on 4D computed tomography data *Proc. 14th Int. Conf. on the Use of Computers in Radiation Therapy (Seoul, May)* pp 27-31

Rietzel E, Doppke K, Pan T, Choi N, Willett C and Chen G 2003 4D computer tomography for radiation therapy *Med. Phys.* **30** 1365–6

Rivard M, Tsai J-S and Engler M 2002 Implementation of inversely-planned, intensity modulated radiation therapy using doubly-dynamic MLC *Med. Phys.* **29** 1268

Rock L, McClean B, McCavana P and Buckney S 2001 Comparison of calculated and measured data for small MLC generated fields defined on Varian and Elekta accelerators *Radiother. Oncol.* **61** (Suppl. 1) S24

Rodebaugh R, Crownover R, Meziane M, Rice T, Graham R, Mehta A, Bodduluri M, Glosser G, Wang J, Chen Q S, Macklis R and Weinhous M 2001a Chasing lung lesions with the Cyberknife *Med. Phys.* **28** 1219

Rodebaugh R F, Crownover R L, Weinhous M S, Meziane M, Graham R I, Glosser G, Wang J, Bodduluri M, Whyte R I and Rice T W 2001b The accuracy of tracking lung tumors with the Cyberknife. *Int. J. Radiat. Oncol. Biol. Phys.* **51** (Suppl. 1) P24–5

Roeske J, Lujan A, Rotmensch J and Mundt A 2000a A feasibility study of IMRT for the treatment of cervical cancer patients unable to receive intracavitary brachytherapy *Proc. World Congress of Medical Physics and the AAPM Annual Congress, August 2000* paper MO-EBR-07

Roeske J C, Lujan A, Rotmensch J, Waggoner S E, Yamada D and Mundt A J 2000b Intensity-modulated whole pelvic radiation therapy in patients with gynaecologic malignancies *Int. J. Radiat. Oncol. Biol. Phys.* **48** 1613–21

Rohlfing T, Maurer C R, O'Dell W G and Zhong J 2004 Modeling liver motion and deformation during the respiratory cycle using intensity-based nonrigid registration of gated MR images *Med. Phys.* **31** 427–32

Romeijn H E, Ahuja R K, Dempsey J F, Kumar A and Li J G 2003 A novel linear programming approach to fluence map optimization for intensity modulated radiation therapy treatment planning *Phys. Med. Biol.* **48** 3521–42

Rose A, Short R, Hoban P and Biggs D 2001 A stereotactically located phantom for verification of IMRT patient treatments *Radiother. Oncol.* **61** (Suppl. 1) S109

Rosello J, Ferrando I, Gonzalez Lopez A, Perez Calatayud J, Lopez Torrecilla J, Dualde Beltran D, Gonzalez Sanchis A and Hernandez Mechancoses A 2000 Implementation of intensity modulated radiotherapy (IMRT) with HELAX-TMS inverse dose planning system and PRIMUS-SIEMENS linear accelerator: problems and solutions *Radiother. Oncol.* **56** (Suppl. 1) S196

Rosello J, Gonzalez A, Sanchez-Doblado F, Leal A, Arrans R, Nunez L and Brualla L 2001 Weak points of an inverse treatment planning system *Radiother. Oncol.* **61** (Suppl. 1) S110

Rosu M, Dawson L A, Balter J M, McShan D L, Lawrence T S and Ten Haken R K 2003 Alterations in normal liver doses due to organ motion *Int. J. Radiat. Oncol. Biol. Phys.* **57** 1472–9

Rousseau V, Grandjean P, Diaz J C, Kafrouni H, Berre F and Lefkopoulos D 2002 Comparison between two inverse methods for IMRT *Radiother. Oncol.* **64** (Suppl. 1) S217

Rowbottom C G 1998 Optimisation of beam orientations in conformal radiotherapy treatment planning *PhD Thesis* London University

Rowbottom C G, Nutting C M and Webb S 2001 Beam optimization of intensity-modulated radiotherapy: clinical application to parotid gland tumours *Radiother. Oncol.* **59** 169–77

Ruchala K, Olivera G, Forrest L, Mehta M, Kapatoes J, Welsh J and Mackie T 2002c Megavoltage CT for image-guided radiotherapy *Radiother. Oncol.* **64** (Suppl. 1) S12

Ruchala K, Olivera G, Kapatoes J, Balog J, Welsh J, Reckwerdt P, Jeraj R, Lu W and Mackie T 2003 On-board CT image quality for image-guided radiotherapy *Proc. World Congress in Medical Physics (Sydney)* CDROM and website only

Ruchala K, Olivera G, Kapatoes J, Jeraj R, Reckwerdt P and Lu W 2002b Multi-margin optimization with daily selection (MMODS) for image-guided radiotherapy *Radiother. Oncol.* **64** (Suppl. 1) S213

Ruchala K, Olivera G, Kapatoes J, Lu W, Reckwerdt P, Jeraj R and Mackie T 2002a A solution for the 'Cassandra Complex' of image-guided radiotherapy *Med. Phys.* **29** 1240

Ruchala K J, Olivera G H, Kapatoes J M, Reckwerdt P J and Mackie T R 2002d Methods of improving limited field-of-view radiotherapy reconstructions using imperfect *a priori* images *Med. Phys.* **29** 2590–605

Ruchala K J, Olivera G H, Kapatoes J M, Schloesser E A, Reckwerdt P J and Mackie T R 2000a Megavoltage CT image reconstruction during tomotherapy treatments *Phys. Med. Biol.* **45** 3545–62

——2000b Megavoltage CT imaging as a by-product of multileaf collimator leakage *Phys. Med. Biol.* **45** N61–70

Ryu J, Winter K, Michalski J, Purdy J A, Markoe A M, Earle J D, Perez C A, Roach M III, Sandler H M, Pollack A and Cox J D 2001 Interim report of toxicity from 3D conformal radiation therapy for prostate cancer on 3DOG/RTOG 9406, Level III (79.2 Gy) *ASCO Proc. abstract 725, J. Clin. Oncol.*

Salter B 2001 NOMOS Peacock IMRT utilizing the Beak post collimation device *Medical Dosim.* **26** 37–45

Salter B J, Fuss M, Vollmer D G, Sadeghi A, Bogaev C A, Cheek D A, Herman T S and Hevezi J M 2001 The talon removable head frame system for stereotactic radiosurgery/radiotherapy: measurement of the repositioning accuracy *Int. J. Radiat. Oncol. Biol. Phys.* **51** 555–62

Salter B, Voeltz L and Fuss M 2003 Characterizing the dosimetric impact of BAT ultrasound alignment *Med. Phys.* **30** 1339

Samuelsson A and Johansson K 2003 Intensity modulated radiotherapy treatment planning for dynamic multileaf collimator delivery: influence of different parameters on dose distributions *Radiother. Oncol.* **66** 19–28

Samuelsson A, McShan D L, Ten Haken R K, Eisbruch A, Johansson K A and Fraass B A 2001 Set-up uncertainties in optimised IMRT plans for the head and neck: plan degradation and correction *Radiother. Oncol.* **61** (Suppl. 1) S47

Samuelsson A, Mercke C and Johansson K-A 2003 Systematic set-up errors for IMRT in the head and neck region: effect on dose distribution *Radiother. Oncol.* **66** 303–11

Sánchez Doblado F, Capote R, Leal A, Lagares J, Arráns R and Hartmann G 2003 Absolute dosimetry of narrow irregular centred IMRT beamlet *Proc. World Congress in Medical Physics (Sydney)* CDROM and website only

Sanchez-Doblado F, Leal A, Arrans R, Rosello J, Carrasco E and Perucha M 2001 Automated assessment of IMRT treatments *Radiother. Oncol.* **61** (Suppl. 1) S101

Sanchez-Nieto B 2000 Practical use of TCP in treatment plan optimization *Radiother. Oncol.* **56** (Suppl. 1) S10

Sanford R, Al-Ghazi M, Chung H and Yakoob R 2000 Dosimetric validation of clinical serial tomotherapy delivery *Proc. World Congress of Medical Physics and the AAPM Annual Congress, August 2000* paper WE-E313-01

Saw C, Ayyangar K, Zhen W, Shen B and Enke C 2001 Assessment of phantom size for IMRT *Med. Phys.* **28** 1209

Saw C B, Li S, Ayyangar K M, Yoe-Sein M, Pillai S, Enke C A and Celi J C 2003 Dose linearity and uniformity of a linear accelerator designed for implementation of multileaf collimation system-based intensity modulated radiation therapy *Med. Phys.* **30** 2253–6

Sawada A, Yoda K, Kokubo M, Nagata Y and Hiraoka M 2002 Non-invasive respiratory gated radiation treatment system based on a 3D ultrasound device and a 3D digital localizer *Med. Phys.* **29** 1310

——2004 A technique for noninvasive respiratory gated radiation treatment system based on a real time 3D ultrasound image correlation: a phantom study *Med. Phys.* **31** 245–50

Schaefer M, Münter M W, Thilmann C, Sterzing F, Haering P, Combs S E and Debus J 2004 Influence of intra-fractional breathing movement in step-and-shoot IMRT *Phys. Med. Biol.* **49** N175–9

Schaly B, Kempe J A, Bauman G S, Battista J J and van Dyke J 2004 Tracking the dose distribution in radiation therapy by accounting for variable anatomy *Phys. Med. Biol.* **49** 791–805

Schefter T, Kavanagh BH, Newell S and McCourt S 2002 Solid-phase compensator - versus multi-leaf collimator-based intensity-modulated radiotherapy for cervix cancer: dosimetric comparison, resource utilization, and financial considerations *Radiother. Oncol.* **64** (Suppl. 1) S217

Schlegel W 2003 Implementation of IMRT *Clinical Oncol.* **15** (Suppl. 2) S5

Schlegel W, Oelfke U, Thieke C, Bortfeld T and Küfer K H 2003 Inverse planning for IMRT with Konrad *Phys. Medica.* **19** 53

Schmuecking M, Plichta K, Lopatta E C, Przetak C, Leonhardi J, Gottschild D, Wendt T G and Baum R P 2000 Image fusion of F-18 FDG PET and CT—is there a role in

3D-radiation treatment planning of non-small cell lung cancer? *Proc. 42nd Annual ASTRO Meeting. Int. J. Radiat. Oncol. Biol. Phys.* **48** 130

Schmidt W F O, Krauss H, Poljanc K, Buechler F and Wirth P 2001 IMRT in routine: questions to be answered *Radiother. Oncol.* **61** (Suppl. 1) S58

Schmidt W, Nespor W, Wirth P, Beuchier F and Hawliczek R 2000 IMRT in routine - Experiences from a feasibility study *Proc. World Congress of Medical Physics and the AAPM Annual Congress, August 2000* paper MO-CXH-21

Schreibmann E, Lahanas M, Xing L and Baltas D 2004c Multiobjective evolutionary optimization of the number of beams, their orientations and weights for intensity-modulated radiation therapy *Phys. Med. Biol.* **49** 747–70

Schreibmann E, Theodorou K, Kappas C and Xing L 2004a A software package for dose visualization in IMRT *Proc. 14th Int. Conf. on the Use of Computers in Radiation Therapy (Seoul, May)* pp 700–3

Schreibmann E and Xing L 2004b Score-based beam selection in IMRT with inclusion of EUD constraints *Proc. 14th Int. Conf. on the Use of Computers in Radiation Therapy (Seoul, May)* pp 704–6

Schreiner L J 2001 The potential for intensity modulated radiation therapy with cobalt-60 *Insights* **3** 3

Schulz R J, Deye J A and Hendee W R 2001 Through the preoccupation with new technical developments, physicists have lost sight of the realities of cancer care and statistics *Med. Phys.* **28** 2185–7

Schulz R J and Kagan A R 2002a (In regard to 'research in medical physics' *Int. J. Radiat. Oncol. Biol. Phys.* **49** 891–95) *Int. J. Radiat. Oncol. Biol. Phys.* **52** 274

——2002b On the role of intensity modulated radiation therapy in radiation oncology *Med. Phys.* **29** 1473–80

——2003a Commentary on IMRT and cancer of the prostate *Int. J. Radiat. Oncol. Biol. Phys.* **55** 851–2

——2003b More precisely defined dose distributions are unlikely to affect cancer mortality *Med. Phys.* **30** 276

Schwarz M, Bos L, Lebesque J V, Mijnheer B J and Damen E M F 2002 The importance of accurate dose calculations outside the beam edges in IMRT treatment planning *Radiother. Oncol.* **64** (Suppl. 1) S97

Schwarz M, Lebesque B J, Mijnheer B J and Damen E M F 2003 Influence of the volume dependence parameter on treatment planning optimisation using the equivalent uniform dose (EUD) *Radiother. Oncol.* **68** (Suppl. 1) S75

Schweikard A and Adler J R 2000 Motion tracking for respiratory motion in radiosurgery *Proc. 42nd Annual ASTRO Meeting. Int. J. Radiat. Oncol. Biol. Phys.* **48** (Suppl.) 166

Schweikard A, Glosser G, Bodduluri M, Murphy M and Adler J R 2000 Robotic motion compensation for respiratory motion in radio-surgery *Computed Aided Surgery* **5** 263–77

Scielzo G, Stasi M, Chauvie S, Baiotto B, Malinverni G, Macias V, Munoz F and Gabriele P 2002 IMRT sliding window with a 120-leaf MLC: dosimetry and first clinical applications *Radiother. Oncol.* **64** (Suppl. 1) S236

Seco J, Evans P M and Webb S 2001 Analysis of the effects of the delivery technique on an IMRT plan: comparison for multiple static field, dynamic and NOMOS MIMiC collimation *Phys. Med. Biol.* **46** 3073–87

Seco J, Verhaegen F and Bidmead M 2003 Development and commissioning of a Monte Carlo dose calculation engine for clinical IMRT beams *Proc. World Congress in Medical Physics (Sydney)* CDROM and website only

Seddon B, Bidmead M, Wilson J, Khoo V and Dearnaley D 2000 Target volume definition in conformal radiotherapy for prostate cancer: quality assurance in the MRC RT-01 trial *Radiother. Oncol.* **56** 73–83

Seiler P G, Blattmann H, Kirsch S, Muench R K and Schilling Ch 2000 A novel tracking technique for the continuous precise measurement of tumour positions in conformal radiotherapy *Phys. Med. Biol.* **45** N103–10

Sen A, Kelly D and Flynn J 2000 Verification of leaf patterns of a MIMiC collimator in IMRT treatment *Proc. World Congress of Medical Physics and the AAPM Annual Congress, August 2000* paper MO-CXH-13

Seppenwoolde Y, Shirato H, Kitamura K, Shimizu S, van Herk M, Lebesque J V and Miyasaka K 2001 3D Tumor motion in lung due to breathing and heartbeat, measured during real-time tumor tracking radiation therapy *Int. J. Radiat. Oncol. Biol. Phys.* **51** (Suppl. 1) 24

——2002 Precise and real-time measurement of 3D tumor motion in lung due to breathing and heartbeat, measured during radiotherapy *Int. J. Radiat. Oncol. Biol. Phys.* **53** 822–34

Seppi E J, Munro P, Johnsen S W, Shapiro E G, Tognina C, Jones D, Pavkovich J M, Webb C, Mollov I, Partain L D and Colbeth R E 2003 Megavoltage cone-beam computed tomography using a high-efficiency image receptor *Int. J. Radiat. Oncol. Biol. Phys.* **55** 793–803

Serago C F, Chungbin S J, Buskirk S J, Ezzell G A, Collie A C and Vora S A 2002 Initial experience with ultrasound localization for positioning prostate cancer patients for external beam radiotherapy *Int. J. Radiat. Oncol. Biol. Phys.* **53** 1130–8

Sethi A, Leybovich L, Dogan N and Emami B 2001a Matching tomographic IMRT fields with static photon fields *Med. Phys.* **28** 2459–65

Sethi A, Leybovich L B, Dogan N and Jeswani S 2001b The effect of heterogeneity corrections on dose distributions when IMRT is used for lung cancer *Radiother. Oncol.* **61** (Suppl. 1) S108

Sethi A, Leybovich L, Dogan N, Krasin M and Emami B 2000 Improvement of target and critical structure doses in Corvus IMRT treatment plans *Proc. World Congress of Medical Physics and the AAPM Annual Congress, August 2000* paper TU-GBR-05

Sethi A, Mohideen N, Leybovich L, Dogan N and Mulhall J 2002 Role of IMRT in reducing corporal bodies doses in dose escalation of prostate cancer *Radiother. Oncol.* **64** (Suppl. 1) S277

Sethi A, Mohideen N, Leybovich L and Mulhall J 2003 Role of IMRT in reducing penile doses in dose escalation for prostate cancer *Int. J. Radiat. Oncol. Biol. Phys.* **55** 970–8

Sham E, Hristov D and Fallone B 2001a Inverse planning by simulated annealing minimization of dose-volume based objective function *Med. Phys.* **28** 1206

Sham E, Hristov D, Stavrev P and Fallone G B 2001b Inverse treatment planning by simulated annealing minimization of a dose-volume cost function *Med. Phys.* **28** 1987

Sharp G, Jiang S B, Ruan D, Castanon and Shirato H 2003 Evaluation of prediction methods for real-time tumour tracking during treatment *Med. Phys.* **30** 1346

Sharpe M B 2003 Commissioning and quality assurance for IMRT treatment planning *IMRT- the state of the art: AAPM Medical Physics Monograph Number 29* ed J R Palta and T R Mackie (Madison, WI: Medical Physics Publishing) pp 449–73

Sharpe M B, Miller B M and Wong J W 2000a Compensation of x-ray beam penumbra in conformal radiotherapy *Med. Phys.* **27** 1739–45

Sharpe M B, Miller B M, Yan D and Wong J W 2000b Monitor unit settings for intensity modulated beams delivered using a step-and-shoot approach *Med. Phys.* **27** 2719–25

Shentall G S, Kirby M C, McKay D M, Harris F W and Read G 2001 Using biological models as a way of determining the efficacy of IMRT for the prostate *Radiother. Oncol.* **61** (Suppl. 1) S53

Shepard D, Earl M, Li X, Holmes T and Yu C 2001 Direct aperture optimization for step and shoot IMRT *Med. Phys.* **28** 1308

Shepard D M, Earl M A, Li X A, Naqvi S and Yu C 2002b Direct aperture optimisation: A turnkey solution for step-and-shoot IMRT *Med. Phys.* **29** 1007–18

Shepard D, Earl M, Naqvi S and Yu C 2002a An inverse planning tool for intensity modulated arc therapy *Med. Phys.* **29** 1336

Shepard D M, Earl M A, Yu C X and Xiao Y 2003a Aperture-based inverse planning *IMRT—the State of the Art: AAPM Medical Physics Monograph Number 29* ed J R Palta and T R Mackie (Madison, WI: Medical Physics Publishing) pp 115–37

Shepard D M, Jiang Z, Earl M A, Ferris M C, Lim J and Naqvi S 2003b A toolbox for intensity modulated radiation therapy optimisation *Med. Phys.* **30** 2320–2

Sherouse G W 2002 In regard to intensity-modulated radiotherapy collaborative working group (*Int. J. Radiat. Oncol. Biol. Phys.* 2001 **51** 880–914) *Int. J. Radiat. Oncol. Biol. Phys.* **53** 1088–9

Shih R, Li X A and Hsu W-L 2001 Dosimetric characteristics of dynamic wedged fields: a Monte Carlo study *Phys. Med. Biol.* **46** N281–92

Shimizu S, Shirato H, Kitamura K, Ogura S, Akita-Dosaka H, Tateishi U, Watanabe Y, Fujita K Shimizu T and Miyasaka K 2000 Fluoroscopic real-time tumour-tracking radiation treatment (RTRT) can reduce internal margin (IM) and set-up margin (SM) of planning target volume (PTV) for lung tumours *Proc. 42nd Annual ASTRO Meeting. Int. J. Radiat. Oncol. Biol. Phys.* **48** (Suppl.) 166

Shimizu S, Shirato H, Ogura S, Akita-Dosaka H, Kitamura K, Nishioka T, Kagei K, Nishimura M and Miyasaka K 2001 Detection of lung tumour movement in real-time tumour-tracking radiotherapy *Int. J. Radiat. Oncol. Biol. Phys.* **51** 304–10

Shimoga K B, Kalend A M, Clark K and Greenberger J S 2001 SmartGate: A fuzzy-logic based intelligent system for respiratory gating *Int. J. Radiat. Oncol. Biol. Phys.* **51** (Suppl. 1) 29

Shiomi H, Inoue T, Nakamura S and Inoui T 2001a Quality Assurance for an image-guided frameless radiosurgery system using radiochromic film *Radiation Medicine* **18** 107–13

Shiomi H, Inoue T, Nakamura S, Shimamoto S, Inoui T, Adler Jr. J R and Bodduluri M 2001b CyberKnife *Japan J. Med. Phys.* **21** 10–16 (in Japanese)

Shirato H, Seppenwoolde Y, Kitamura K, Onimura R and Shimizu S 2004 Intrafractional tumor motion: lung and liver *Seminars Radiat. Oncol.* **14** 10–18

Shirato H, Shimizu S, Kitamura K, Nishioka T, Kagei K, Hashimoto S, Aoyama H, Kuneida T, Shinohara N, Dosaka-Akita H and Miyasaka K 2000b Four-dimensional treatment planning and fluoroscopic real-time tumor tracking radiotherapy for moving tumour *Int. J. Radiat. Oncol. Biol. Phys.* **48** 435–42

Shirato H, Shimizu S, Kunieda T, Kitamura K, van Herk M, Kagei K, Nishioka T, Hashimoto S, Fukota K, Aoyama H, Tsuchita K, Kudo K and Miyasaka K 2000a Physical aspects of a real-time tumour-tracking system for gated radiotherapy *Int. J. Radiat. Oncol. Biol. Phys.* **48** 1187–95

Short R, Hoban P, Biggs D, Rose A and Smee B 2001 Pre-treatment dosimetric verification of IMRT *Radiother. Oncol.* **61** (Suppl. 1) S108

Shou Z and Xing L 2004 Improve IMRT distribution by using spatially non-uniform important factors *Proc. 14th Int. Conf. on the Use of Computers in Radiation Therapy (Seoul, May)* pp 120–3

Siebers J, Keall P, Kim J O, Arnfield M and Mohan R 2000 Dynamic IMRT Monte Carlo dose calculation *Proc. World Congress of Medical Physics and the AAPM Annual Congress, August 2000* paper FR-EA309-05

Siebers J V, Lauterbach M, Keall P J and Mohan R 2002b Incorporating multi-leaf collimator leaf sequencing into iterative IMRT optimisation *Med. Phys.* **29** 952–9

Siebers J, Lauterbach M, Tong S, Wu Q and Mohan R 2001a Reducing dose calculation time for accurate iterative IMRT planning *Med. Phys.* **28** 1308

——2002a Reducing dose calculation time for accurate iterative IMRT planning *Med. Phys.* **29** 231–7

Siebers J V and Mohan R 2003 Monte Carlo and IMRT *IMRT- the state of the art: AAPM Medical Physics Monograph Number 29* ed J R Palta and T R Mackie (Madison, WI: Medical Physics Publishing) pp 531–60

Siebers J V, Tong S, Lauterbach M, Wu Q and Mohan R 2001b Acceleration of dose calculations for intensity-modulated radiotherapy *Med. Phys.* **28** 903–10

Singh I, Ravindran B, Subramaniam B and John S 2002 Manual multi-leaf collimator for ^{60}Co beam shaping—a feasibility study *Med. Phys.* **29** 1213

Siochi R A C 1999 Minimising static intensity modulation delivery time using an intensity solid paradigm. *Int. J. Radiat. Oncol. Biol. Phys.* **43** 671–80

——2000a Simultaneous minimization of segments and junctions in virtual micro IMRT *Proc. World Congress of Medical Physics and the AAPM Annual Congress, August 2000* paper MO-EBR-04

——2000b Getting started with IMRT *RAD Magazine* **26** (304) 47–8

——2000c Virtual micro-intensity modulated radiation therapy *Med. Phys.* **27** 2480–93

Sixel K E, Aznar M C and Ung Y C 2001 Deep inspiration breath hold to reduce irradiated heart volume in breast cancer patients *Int. J. Radiat. Oncol. Biol. Phys.* **49** 199–204

Sohn J, Dempsey J, Drzymala R, Klein E, Grigereit T and Purdy J 2000b Analysis of small beamlets for IMRT using radiochromic films *Proc. World Congress of Medical Physics and the AAPM Annual Congress, August 2000* paper MO-GBR-02

Sohn J, Dempsey J, Low D, Klein E, Dryzmala R, Kong F and Mansur D 2001a IMRT to breast tissue and associated lymph nodes with consideration of breathing motion *Med. Phys.* **28** 1253

Sohn J W, Dempsey J F, Suh T S and Low D A 2003 Analysis of various beamlet sizes for IMRT with 6 MV photons *Med. Phys.* **30** 2432–9

Sohn J W, Low D A and Deasy J O 2001b Motion-induced distribution artifacts in lung cancer IMRT *Int. J. Radiat. Oncol. Biol. Phys.* **51** 347–8

Sohn J W, Low D A and Klein E E 2000a Dynamic multileaf collimator performance for intensity modulated radiation therapy *Proc. 42nd Annual ASTRO Meeting. Int. J. Radiat. Oncol. Biol. Phys.* **48** (Suppl.) 341

Solberg T, Agazaryan N, Tenn S, Wink N, DeMarco J, Medin P, Lee S and Parker R 2003 Initial experience with intensity modulated radiotherapy for prostate cancer delivered using a micro-multileaf collimator. *Proc. World Congress in Medical Physics (Sydney)* CDROM and website only

Solberg T, Paul T, Boone R, Agazaryan N, Urmanita T, Arellano A, Llacer J, Fogg R, DeMarco J and Smathers J 2000 Feasibility of gated IMRT *Proc. World Congress of Medical Physics and the AAPM Annual Congress, August 2000* paper TH-EBR-02

Sonke J-J, Ploeger L S, Brand B, Smitsmans M H P and van Herk M 2003 Geometrical verification of dynamic intensity modulated radiotherapy using an amorphous silicon flat panel imager *Med. Phys.* **30** 1352

——2004 Leaf trajectory verification during dynamic intensity modulated radiotherapy using an amorphous silicon flat panel imager *Med. Phys.* **31** 389–95

Sorvari P, Hamalainen T, Sievinen J, Pyykkonen J and Tenhunen M 2000 Commissioning of the dynamic multileaf collimator system for clinical use in a 6 MV linear accelerator *Radiother. Oncol.* **56** (Suppl. 1) S194

Soubra M, Stablein C, Dalope M G, Elbadawi N and El-Ghandour O 2001a Prostate immobization and an alignment system for IMRT patients *Radiother. Oncol.* **61** (Suppl. 1) S109

——2001b IMRT prostate patient immobilization and alignment system *Med. Phys.* **28** 1209

Spirou S V and Chui C S 1994 Generation of arbitrary intensity profiles by dynamic jaws or multileaf collimators *Med. Phys.* **21** 1031–41

——1996 Generation of arbitrary intensity profiles by combining the scanning beam with dynamic multileaf collimation *Med. Phys.* **23** 1–8

——1998 A gradient inverse-planning algorithm with dose-volume constraints *Med. Phys.* **25** 321–33

Spirou S V, Fournier-Bidoz N, Yang J, Chui C-S and Ling C C 2001 Smoothing intensity-modulated beam profiles to improve the efficiency of delivery. *Med. Phys.* **28** 2105–12

Spirou S, Yang J, Chui C and Ling C, 2000 Smoothing intensity modulated beam profiles to facilitate delivery *Proc. World Congress of Medical Physics and the AAPM Annual Congress, August 2000* paper MO-EBR-02

Spitters-Post I, van der Vight L P, van Lin E N J T and Visser A G 2002 The use of fiducial (gold) markers for monitoring prostate movements during radiation therapy *Radiother. Oncol.* **64** (Suppl. 1) S44

Staffurth J, Dearnaley D, Convery D, Adams E, Clark C, Huddart R and Webb S 2003a A planning study to compare the plan quality and delivery efficiency of three IMRT planning and delivery system *Clinical Oncol.* **15** (Suppl. 2) S33–4

Staffurth J, Dearnaley D, McNair H, Convery D, Adams E, Clark C, Bidmead M, Warrington A P, Huddart R, Webb S and Horwich A 2003b Early results of a Phase 2 trial of pelvic nodal irradiation in prostate cancer IMRT *Clinical Oncol.* **15** (Suppl. 2) S17

Staffurth J, Dearnaley D, McNair H, Convery D, Adams E, Clark C, Vaarkamp J, Huddart R, Webb S and Horwich A 2002 Early results of a phase 1 trial of pelvic nodal irradiation in prostate cancer with intensity modulated radiotherapy (IMRT) *Radiother. Oncol.* **64** (Suppl. 1) S284

Starkshall G, Sherouse G W and Hendee W R 2001 The future will not need clinical therapy physicists *Med. Phys.* **28** 865–7

Stasi M, Baiotto B, Malinverni G, Gabriele P and Scielzo G 2000 Multiple static field technique in prostate cancer: a pilot study using a traditional 3D treatment planning system *Radiother. Oncol.* **56** (Suppl. 1) S110

Stavrev P V, Hristov D H and Sham E 2001 IMRT inverse treatment planning optimisation based on physical constraints and biological objectives *Med. Phys.* **28** 1987

Steenbakkers R J H M, Deurloo K E I, Nowak P J C M, van Herk M and Rasch C R N 2002 Comparison of CT and MRI delineation for 3-D conformal treatment planning of the prostate *Radiother. Oncol.* **64** (Suppl. 1) S274

Stein J, Bortfeld T, Dörschel B and Schlegel W 1994 Dynamic x-ray compensation for conformal radiotherapy by means of multileaf collimation *Radiother. Oncol.* **32** 163–73

Stein J, Mohan R, Wang X-H, Bortfeld T, Wu Q, Preiser K, Ling C C and Schlegel W 1997 Number and orientation of beams in intensity-modulated radiation treatments *Med. Phys.* **24** 149–60

Steinberg T, Allmendinger K, Dudas A and Calderon E 2002 Dose stability for low monitor unit treatments *Med. Phys.* **29** 1212

Sternick E S, Bleier A R, Carol M P, Curran B H, Holmes T W, Kania A A, Lalonde R and Larson L S 1997 Intensity modulated radiation therapy: what photon energy is best? *Proc 12th International Conference on the Use of Computers in Radiation Therapy, Salt Lake City, Utah, May 1997* ed D D Leavitt and G Starkschall (Madison, WI: Medical Physics Publishing) pp 418–19

Stevens C W 2003 Lung cancer radiotherapy *IMRT—the State of the Art: AAPM Medical Physics Monograph Number 29* ed J R Palta and T R Mackie (Madison, WI: Medical Physics Publishing) pp 645–62

Strom E A 2002 Breast IMRT: New tools leading to new vision *Int. J. Radiat. Oncol. Biol. Phys.* **54** 1297–8

Stromberg J S, Sharpe M B, Kim L H, Kini V J, Jaffray D A, Martinez A A and Wong J W 2000 Active breathing control (ABC) for Hodgkin's disease: reduction in normal tissue irradiation with deep inspiration and implications for treatment *Int. J. Radiat. Oncol. Biol. Phys.* **48** 797–806

Subramanian T R, Gibbons J P and Hendee W R 2002 Linear accelerators used for IMRT should be designed as small field high intensity intermediate energy units *Med. Phys.* **29** 2526–8

Suh Y, Yi B, Ahn S, Kim J, Lee S, Shin S and Choi E 2004 Aperture manoeuver with compelled breath (AMC) for moving tumors: a feasibility study using moving phantom *Med. Phys.* **31** 760–6

Suh Y, Yi B, Ahn S, Lee S, Kim J, Shin S and Choi E 2003 Adaptive field shaping to moving tumor with the compelled breath control: a feasibility study using moving phantom *Proc. World Congress in Medical Physics (Sydney)* CDROM and website only

Suh Y, Yi B, Kim J, Ahn S, Lee S, Shin S and Choi E 2002 Prediction of the tumour location in the lung from the skin motion *Med. Phys.* **29** 1240

Suit H 2002 The Gray Lecture 2001: coming technical advances in radiation oncology *Int. J. Radiat. Oncol. Biol. Phys.* **53** 798–809

Sun X and Xia P 2004 A new smoothing procedure to reduce delivery segments for static MLC-based IMRT planning *Med. Phys.* **31** 1158–65

Svatos M, Rosenman J, Cullip T, Verhey L and Hughes J 2000 Mixing electrons with intensity modulated photon beams to reduce integral dose *Proc. World Congress of Medical Physics and the AAPM Annual Congress, August 2000* paper WE-E313-10

Svensson R, Källman P and Brahme A 1994 Analytical solution for the dynamic control of multileaf collimators *Phys. Med. Biol.* **39** 37–61

Swerdloff S, Mackie T R and Holmes T 1994a Method and apparatus for radiation therapy *US Patent* 5,317,616

——1994b Multi-leaf radiation attenuator for radiation therapy *US Patent* 5,351,280

Swindell W, Simpson R G and Oleson J R 1983 Computed tomography with a linear accelerator with radiotherapy applications *Med. Phys.* **10** 416–20

Swinnen A, van Esch A and Huyskens D P 2002 Comparison of static and dynamic delivery of intensity-modulated radiation therapy *Radiother. Oncol.* **64** (Suppl. 1) S107

Szanto J, Malone S, Gerig L, Punkari and Hube P 2002 A system for image-guided prostate radiotherapy *Radiother. Oncol.* **64** (Suppl. 1) S278

Tanner S F, Finnigan D J, Khoo V S, Mayles P, Dearnaley D P, and Leach M O 2000 Radiotherapy planning of the pelvis using distortion corrected MR images: the removal of system distortion *Phys. Med. Biol.* **45** 2117–32

Tao L, Rakfal S and Wu A 2002 Comparison of integral doses in conventional 2D, conformal 3D and IMRT plans *Med. Phys.* **29** 1211

Teh B S, Bastasch M D, Wheeler T M, Mai W-Y, Frolov A, Uhl M, Lu H H, Carpenter L S, Chiu J K, McGary J, Woo S Y, Grant W H III and Butler E B 2003 IMRT for prostate cancer: defining target volume based on correlated pathologic volume of disease *Int. J. Radiat. Oncol. Biol. Phys.* **56** 184–91

Teh B S, Mai W Y, Huang E, Carpenter L S, Lu H H, Chiu J K, Woo S Y, Grant W H and Butler E B 2001b Late gastrointestinal (GI) and genitourinary (GU) toxicity following intensity modulated radiation therapy (IMRT) for prostate cancer *Int. J. Radiat. Oncol. Biol. Phys.* **51** 310

Teh B S, Mai W-Y, Uhl B M, Augspurger M E, Grant W H III, Lu H H, Woo S Y, Carpenter S L, Chiu J K and Butler E B 2001a Intensity-modulated radiation therapy (IMRT) for prostate cancer with the use of a rectal balloon for prostate immobilization: acute toxicity and dose-volume analysis *Int. J. Radiat. Oncol. Biol. Phys.* **49** 705–12

Tepper J 2000 Form and function: the integration of physics and biology *Int. J. Radiat. Oncol. Biol. Phys.* **47** 547–8

Teslow T N 2002 A software program that evaluates treatment plans for intensity modulated radiotherapy *Radiother. Oncol.* **64** (Suppl. 1) S51

Thieke C, Bortfeld T and Kuefer K-H 2003a New optimization concepts in inverse treatment planning *Radiother. Oncol.* **68** (Suppl. 1) S2

Thieke C, Bortfeld T, Niemierko A, Kuefer K-H and Nill S 2003b Multicriteria optimisation in inverse radiotherapy planning *Radiother. Oncol.* **68** (Suppl. 1) S44

Thieke C, Bortfeld T, Niemierko A and Nill S 2003c From physical dose constraints to equivalent uniform dose constraints in inverse radiotherapy planning *Med. Phys.* **9** 2332–9

Thieke C, Nill S, Oelfke U and Bortfeld T 2001 Acceleration of IMRT dose calculations by importance sampling of calculation matrices *Radiother. Oncol.* **61** (Suppl. 1) S46

——2002 Acceleration of intensity-modulated radiotherapy dose calculation by importance sampling of the calculation matrices *Med. Phys.* **29** 676–81

Thilmann C, Krempien R, Hoess A, Sroka-Perez G, Schramm O, Wannenmacher M and Debus J 2002 Inversely planned intensity modulated radiotherapy (IMRT) of the breast and the locoregional lymph nodes *Radiother. Oncol.* **64** (Suppl. 1) S56

Thomas M D R and Symonds-Tayler J R N 2003 Characterization of a transmission ionisation beam-imager for radiotherapy verification *Phys. Med. Biol.* **48** 2633–44

Ting J Y, Butker E, Ghavidel S and Davis L W 2000a A 'stop and shoot' leaf-sequencing algorithm for IMRT *Proc. World Congress of Medical Physics and the AAPM Annual Congress, August 2000* paper MO-CXH-15

Ting J, Iwinski A, Papanikolaou N, Ghavidel S and Davis L 2000b A comparison between 'step-and-shoot' and 'sliding windows' leaf sequencing algorithms in IMRT *Proc. 42nd Annual ASTRO Meeting. Int. J. Radiat. Oncol. Biol. Phys.* **48** (Suppl.) 220

Tomé W A, Meeks S L, McNutt T R, Buatti J M, Bova F J, Friedman W A and Mehta M 2001 Optically guided intensity modulated radiotherapy *Radiother. Oncol.* **61** 33–44

Tomé W A, Mehta M P, Meeks S L, Buatti J M, Bova F J and Friedman W A 2000 Image guided fractionated stereotactic radiotherapy *Radiother. Oncol.* **56** (Suppl. 1) S60

Toner S, Shih J, Boccuzzi D, Chiu-Tsao S, Gliedman P and Harrison L 2000 The use of external MR- and CT-compatible fiducial markers for image fusion in stereotactic treatment planning *Radiother. Oncol.* **56** (Suppl. 1) S17

Topolnjak R, van der Heide U A and Lagendijk J J W 2004a IMRT sequencing for six-bank multi-leaf collimator system *Proc. 14th Int. Conf. on the Use of Computers in Radiation Therapy (Seoul, May)* pp 322–4

Topolnjak R, van der Heide U A, Raaymakers B W, Kotte A N T J and Lagendijk J J W 2003b IMRT sequencing for a six-bank multi-leaf system *Radiother. Oncol.* **68** (Suppl. 1) S114–15

Topolnjak R, van der Heide U A, Raaymakers B, Kotte A, Welleweerd J and Lagendijk J J W 2002 Six banks multi-leaf system, high resolution and large field size MLC *Radiother. Oncol.* **64** (Suppl. 1) S107

——2003a Six-bank multi-leaf system for IMRT *Phys. Medica.* **19** 59

——2004b A six-bank multi-leaf system, high precision shaping of large fields *Phys. Med. Biol.* **49** 2645–56

Trapp J V 2003 Imaging and radiation interactions of polymer gel dosimeters *PhD Thesis* University of Queensland, Australia

Trapp J V, Michael G, Evans P M, Baldock C, Leach M O and Webb S 2004b Dose resolution in gel dosimetry: effect of uncertainty in the calibration function *Phys. Med. Biol.* **49** N139–46

Trapp J, Partridge M, Hansen V, Childs P, Bedford J, Warrington J, Leach M O and Webb S 2004a The use of gel dosimetry for verification of electron and photon treatment plans in carcinoma of the scalp *Phys. Med. Biol.* **49** 1625–35

Trichter F and Ennis R D 2001 Patient positioning in prostate IMRT: Advantages of BAT over EPID *Med. Phys.* **28** 1292

Tsai J-S, Rivard M J and Engler M J 2000 Dependence of linac output on the switch rate of an intensity-modulated tomotherapy collimator *Med. Phys.* **27** 2215–35

Tsunashima Y, Sakae T, Shioyama Y, Kagei K, Terunuma T and Akine Y 2003 Correlation between respiratory wave form using a respiratory sensor and the 3D tumor motion in respiratory gated radiotherapy *Proc. World Congress in Medical Physics (Sydney)* CDROM and website only

Tücking T, Nill S, Thilmann C and Oelfke U 2003 Application of a new external MMLC for high precision IMRT—a feasibility study *Radiother. Oncol.* **68** (Suppl. 1) S75

Turian J and Smith B 2002 Evaluation of dose differences between multiple static segment (MSS) and sliding window (SW) IMRT techniques *Med. Phys.* **29** 1261

Twyman N and Thomas S 2002 HD-270—turning 1 cm multileaf collimators into 2 mm multileaf collimators *RAD Magazine* **28** (330) 14–15

Unkelbach J and Oelfke U 2003 Inclusion of stochastic organ movements in IMRT treatment planning *Radiother. Oncol.* **68** (Suppl. 1) S101–2

——2004 Organ movements in IMRT treatment planning: inverse planning based on probability distributions *Proc. 14th Int. Conf. on the Use of Computers in Radiation Therapy (Seoul, May)* pp 104–7

Ung Y C, Caldwell C B, Mah K, Danjoux C E, Balogh J B, Ganguli S N, Tirona R and Ehrlich L E 2000 Fusing ^{18}flourodeoxyglucose (FDG)-hybrid PET to CT images significantly alters treatment planning in the radical treatment of non-small cell carcinoma of the lung *Proc. 42nd Annual ASTRO Meeting. Int. J. Radiat. Oncol. Biol. Phys.* **48** (Suppl.) 327

Uy N W, Woo S Y, Teh B S, Mai W-Y, Carpenter L S, Chiu J K, Lu H H, Gildenberg P, Trask T, Grant W H and Butler E B 2002 Intensity-modulated radiation therapy (IMRT) for meningioma *Int. J. Radiat. Oncol. Biol. Phys.* **53** 1265–70

Vaarkamp J, Warrington A P, and Dearnaley D P 2003 A forward planned conformal class solution to generate concave dose distribution to treat prostate and pelvic nodes *Clinical Oncol.* **15** (Suppl. 2) S34

Valinta D, Dejean C, Erenaud S and Floiras J L 2001a Quality control of the IMRT *Radiother. Oncol.* **61** (Suppl. 1) S110

Valinta D, Dejean C and Floiras JL 2001b Inverse planning and IMRT delivery quality control *Phys. Medica.* **17** 114

Valinta D, Migdal S, de Carlan L and Floiras J L 2000 First clinical application of IMRT using an inverse plan in St. Cloud *Radiother. Oncol.* **56** (Suppl. 1) S196

Van Asselen B, Dehnad H, Raaijmakers C P J, Terhaard C H J and Lagendijk J J W 2001a Intensity modulated radiotherapy for oropharyngeal cancer: a class solution *Radiother. Oncol.* **61** (Suppl. 1) S32

Van Asselen B, Dehnad H, Terhaard C H J, Lagendijk J J W and Raaijmakers C P J, 2004 Segmental IMRT for oropharyngeal cancer in a clinical setting *Radiother. Oncol.* **69** 259–66

Van Asselen B, Raaijmakers C P J, Hofman P and Lagendijk J J W 2001b An improved breast irradiation technique using three-dimensional geometrical information and intensity modulation *Radiother. Oncol.* **58** 341–7

Van Dalen S, Keijzer M, Huizenga H and Storchi P R M 2000 Optimization of multileaf collimator settings for radiotherapy treatment planning *Phys. Med. Biol.* **45** 3615–25

Van den Heuvel F, Powell T, Seppi E, Littrupp P, Khan M, Wang Y and Forman JD 2003 Independent verification of ultrasound based image-guided radiation treatment, using electronic portal imaging and implanted gold markers *Med. Phys.* **30** 2878–87

Van der Heide U A, Hoekstra A, Kotte A, van Rijk P P, Lagendijk J J W, Terhaard C H J 2003 The use of PET imaging in radiotherapy dose painting *Phys. Medica.* **19** 44

Van der Heide U, Nederveen A, Hofman P, Welleweerd J and Lagendijk J 2001 The microboost approach to IMRT of the prostate *Med. Phys.* **28** 1261

Van Dieren E B, Nowak P J C M, Wijers O B, van Sörnsen de Koste J R, van der Est H, Binnekamp D P, Heijmen B J M and Levendag P C 2000 Beam intensity modulation using tissue compensators or dynamic multileaf collimation in three-dimensional conformal radiotherapy of primary cancers of the oropharynx and larynx, including the elective neck *Int. J. Radiat. Oncol. Biol. Phys.* **47** 1299–309

Van Dyk J, Kron T, Grigorov G, Chen J, Wong E, Yu E, Lock M, Rodrigues G and Bauman G 2003 Early experience with helical tomotherapy *Radiother. Oncol.* **68** (Suppl. 1) S39

Van Esch A, Sommer P D T, Verstraete J and Huyskens D 2001b QA of IMRT and DMLC by means of the Varian PortalVision system: the clinical implementation. *Radiother. Oncol.* **61** (Suppl. 1) S4

Van Esch A, Vanstraelen B, Verstraete J, Kutcher G and Huyskens D 2001a Pre-treatment dosimetric verification by means of a liquid-filled electronic portal imaging device during dynamic delivery of intensity modulated treatment fields *Radiother. Oncol.* **60** 181–90

Van Herk M 2000 The role of multimodality imaging in radiotherapy *Radiother. Oncol.* **56** (Suppl. 1) S16

——2003 Applications of image registration in radiotherapy *Phys. Medica.* **19** 44–5

Van Herk M, Artignan X, de Bola J, Smitsmans M H P, Lebesque J V and Jaffray D A 2002b Automatic prostate localization for image-guided radiotherapy of the prostate *Radiother. Oncol.* **64** (Suppl. 1) S287

Van Herk M, Betgen A, Brand B, Jaffray D, Remeijer P, Smitsmans M, Sonke J-J and Zijp L 2004 An integrated system for on-line cone-beam CT image-guided radiotherapy software aspects *Proc. 14th Int. Conf. on the Use of Computers in Radiation Therapy (Seoul, May)* pp 504–6

Van Herk M, de Bois J, de Munck J, de Jaeger K, Lebesque J and Rasch C 2003 Use of multi-modality imaging to improve target volume delineation accuracy *Clinical Oncol.* **15** (Suppl. 2) S12

Van Herk M, Remeijer P, Rasch C and Lebesque J V 2000 The probability of correct target dosage: dose-population histograms for deriving treatment margins in radiotherapy *Int. J. Radiat. Oncol. Biol. Phys.* **47** 1121–35

Van Herk M, Scheider C, Sonke J J, Damen E and DeJaeger K 2002a Respiration-correlated CT of lung cancer patients *Radiother. Oncol.* **64** (Suppl. 1) S80–1

Van Nimwegen A, Lagerwaard F, de Pan C, Nowak P and Levedag P 2001 IMRT for the nasopharynx, a planning study *Radiother. Oncol.* **61** (Suppl. 1) S47

Vanregemorter J, de Ost B and van den Weyngaert D 2000 Forward IMRT planning: a phantom study *Radiother. Oncol.* **56** (Suppl. 1) S97

Vanregemorter J, Van Gestel D and van den Weyngaert D 2002 Introduction of multi-modality imaging for conformal prostate radiation treatment *Radiother. Oncol.* **64** (Suppl. 1) S268

Van Santvoort J P C and Heijmen B J M 1996 Dynamic multileaf collimation without 'tongue-and-groove' underdosage effects *Phys. Med. Biol.* **41** 2091–105

Van Sörnsen de Koste J R, Lagerwaard F J, de Boer H C J, Nijssen-Visser M R J and Senan S 2003a Are multiple CT scans required for planning curative radiotherapy in lung tumours of the lower lobe? *Int. J. Radiat. Oncol. Biol. Phys.* **55** 1394–9

Van Sörnsen de Koste, Lagerwaard F J, Nijssen-Visser M R J, Graveland W J and Senan S 2003b Tumor location cannot predict the mobility of lung tumours: a 3D analysis of data generated from multiple CT scans *Int. J. Radiat. Oncol. Biol. Phys.* **56** 348–54

Van Vliet-Vroegindeweij C, Louwe R, Dwarswaard M, van Ingen K, Schneider C, Minken A, van Herk M and Mijnheer B 2001 Dosimetric verification of IMRT using an electronic portal imaging device *Radiother. Oncol.* **61** (Suppl. 1) S107

Vedam S S, Keall P J, Kini V R and Mohan R 2001 Determining parameters for respiration-gated radiotherapy *Med. Phys.* **28** 2139–46

Vedam S, Keall P, Kini V, Mostafavi H and Mohan R 2003b Predicting breathing motion for 4D radiotherapy (WIP) *Med. Phys.* **30** 1472

Vedam S S, Kini V R, Keall P J, Ramakrishnan V, Mostafavi H and Mohan R 2003a Quantifying the predictability of diaphragm motion during respiration with a non-invasive external marker *Med. Phys.* **30** 505–13

Verellen D 2001 Delivery techniques of IMRT: tomotherapy and miniMLCs *Radiother. Oncol.* **61** (Suppl. 1) S1

Verellen D, Linthout N, Soete G, van Acker S, de Roover P and Storme G 2002 Considerations on treatment efficiency of different conformal radiation therapy techniques for prostate cancer *Radiother. Oncol.* **63** 27–36

Verellen D, Linthout N and Storme G 2000 Comparison between different approaches for conformal treatment of the prostate: A theoretical and dosimetric evaluation *Proc. World Congress of Medical Physics and the AAPM Annual Congress, August 2000* paper FR-EABR-05

Verellen D, Soete G, Linthout N, Van Acker S and Storme G 2001 Improved target localization and set-up for CFRT/IMRT of the prostate by combining real-time infrared tracking and stereoscopic X-ray imaging *Med. Phys.* **28** 1295

Vergote K, de Deene Y, Duthoy W, de Gersem W, de Neve W, Achten E and de Wagter C 2004 Validation and application of polymer gel dosimetry for the dose verification of an intensity modulated arc therapy (IMAT) treatment *Phys. Med. Biol.* **49** 287–305

Verhaegen F and Liu H H 2001 Incorporating dynamic collimator motion in Monte Carlo simulations: an application in modelling a dynamic wedge *Phys. Med. Biol.* **46** 287–96

Verhey L 2000 IMRT: applications and benefits in comparison with 3D conformal radiotherapy *Radiother. Oncol.* **56** (Suppl. 1) S86

——2001 IMRT with multileaf collimators *Med. Phys.* **28** 1241

Vervoort C, Verstraete J, Vanstraelen B, Vandevelde G, Van Esch A, Scherpenberg J and Huyskens D 2001 Implementation of a patient set-up verification and correction protocol with an on-line portal imaging in a IMRT procedure for localised prostate cancer *Radiother. Oncol.* **61** (Suppl. 1) S47

Vial P, Bromley R and Oliver L 2003 Commissioning of a dynamic multileaf collimator delivery system *Proc. World Congress in Medical Physics (Sydney)* CDROM and website only

Vicini F A, Sharpe M, Kestin L, Martinez A, Mitchell C K, Wallace M F, Matter R and Wong J 2002 Optimising breast cancer treatment efficacy with intensity modulated radiotherapy *Int. J. Radiat. Oncol. Biol. Phys.* **54** 1336–44

Vicini F A, Sharpe M, Kestin L, Wong J, Remouchamps V and Martinez A 2004 The use of intensity modulated radiation therapy in the treatment of breast cancer: evolving definition, misdirected criticism, and untoward effects *Int. J. Radiat. Oncol. Biol. Phys.* **58** 1642–4

Vieira S C, Dirkx M L P, Pasma K L and Heijmen B J M 2001 Fast and accurate verification of leaf calibration for dynamic multileaf collimation with an EPID *Radiother. Oncol.* **61** (Suppl. 1) S17

Vineberg K A, Eisbruch A, Coselmon M M, McShan D L, Kessler M L and Fraass B A 2002 Is uniform target dose possible in IMRT plans in the head and neck? *Int. J. Radiat. Oncol. Biol. Phys.* **52** 1159–72

Vineberg K A, Eisbruch A, Kessler M L, McShan D L, Martel M K and Fraass B F 2000 Parotid sparing without sacrificing target dose uniformity using optimized beamlet IMRT *Proc. 42nd Annual ASTRO Meeting. Int. J. Radiat. Oncol. Biol. Phys.* **48** (Suppl.) 174

Vineberg K A, McShan D L, Kessler M L and Fraass B A 2001 Comparison of dose, dose-volume and biologically-based cost functions for IMRT plan optimization *Int. J. Radiat. Oncol. Biol. Phys.* **51** (Suppl. 1) 71–2

Visser A G, Stroom J C and Huizenga H 2000 Target volume margins based on measured uncertainties *Radiother. Oncol.* **56** (Suppl. 1) S54

Visser A G, van Lin E N J T, Welmers A, Futterer J, Barentsz J O, van der Vight L P and van der Meazen R W M 2002 Fiducial markers for MRI-based target volume definition and treatment verification *Radiother. Oncol.* **64** (Suppl. 1) S41

Vuong T, Belliveau P, Falco T, Meterissian S and Wexler M 2002 Clinical application of IMRT in the treatment of patients with unresectable T4 or recurrent rectal or rectosigmoid cancer: feasibility and toxicity data *Radiother. Oncol.* **64** (Suppl. 1) S56

Wagner T H, Meeks S L, Bova F J, Friedman W A, Buatti J M and Bouchet L G 2001 Isotropic beam bouquets for shaped linear accelerator radiosurgery *Phys. Med. Biol.* **46** 2571–86

Wagman R, Yorke E, Giraud P, Ford E, Sidhu K, Mageras G, Minsky B and Rosenzweig K 2001 Reproducibility of organ position with respiratory gating for liver tumours: Use in dose-escalation *Int. J. Radiat. Oncol. Biol. Phys.* **51** (Suppl.) 28

Wahab S, Low D, El Naqa I, Parikh P, Nystrom M, Bradley J, Mutic S and Dempsey J 2003 Use of four-dimensional computed tomography in conformal therapy planning for lung cancer *Med. Phys.* **30** 1364

Wahab S H, Malyapa R S, Mutic S, Grigsby P W, Deasy J O, Miller T R, Zoberi I and Low D A 2004a Intensity modulated radiation therapy for cervical cancer: a treatment planning comparison against brachytherapy *Proc. 14th Int. Conf. on the Use of Computers in Radiation Therapy (Seoul, May)* p 692–5

Wahab S H, Parikh P, Nystrom M, El Naqa I, Hubenschmidt J, Pierburg B, Lu W, Bradley J D and Low D A 2004b Treatment planning using four dimensional computed tomography *Proc. 14th Int. Conf. on the Use of Computers in Radiation Therapy (Seoul, May)* pp 62–6

Walker V 2001 Conformal radiotherapy treatment delivery *RAD Magazine* **27** (233) 25–6

Wang C, Dai J and Hu Y 2003a Optimization of beam orientations and beam weights for conformal radiotherapy using mixed integer programming *Phys. Med. Biol.* **48** 4065–76

Wang D, Hill R W and Lam S 2004 A new algorithm for determining collimator angles that favour efficiency in MLC based IMRT delivery *Med. Phys.* **31** 1249–53

Wang J, Sabbas A, Kulidzhanov F, Presser J, Vallejo A and Nori D 2002b Comparison study of IMRT treatment with 3D-RTP in treating ethmoid cancer *Med. Phys.* **29** 1260

Wang L, Jacob R, Movsas B, Chen L, Fourkal E, Ma C and Pollack A 2003b Dosimetric comparison of different multileaf collimator widths in the treatment of prostate cancer with intensity modulated radiotherapy *Med. Phys.* **30** 1405

Wang L, Jacob R, Movsas B, Fourkal E, Chen L, Konoski A, Feigenberg S P A and Ma C 2003c Dosimetric advantage of using small-width multi-leaf collimator in the IMRT of prostate cancer *Radiother. Oncol.* **68** (Suppl. 1) S99

Wang L, Movsas B, Jacob R and Ma C 2003d Dosimetric advantage and clinical implication of using a micro multileaf collimator in the treatment of prostate cancer with intensity modulated radiotherapy *Proc. World Congress in Medical Physics (Sydney)* CDROM and website only

Wang L, Yorke E and Chui C-S 2002a Monte Carlo evaluation of 6 MV intensity modulated radiotherapy plans for head and neck and lung treatments *Med. Phys.* **29** 2705–17

Warkentin B, Steciw S, Rathee S and Fallone B G 2003 Dosimetric IMRT verification with a flat-panel EPID *Med. Phys.* **30** 3143–55

Warrington A P and Adams E J 2001 Conformal and intensity modulated radiotherapy using cobalt-60 and 6 MV X-ray beams: a treatment planning comparison of different sites *Radiother. Oncol.* **61** (Suppl. 1) S73

——2002 Cobalt-60 teletherapy for cancer: a revived treatment modality for the 21st century *Proc. Seminar on Appropriate Medical Technology for Developing Countries; IEE Feb 6th* pp 21.1–21.19

Warrington A P, Robinson M, Smith N, Adams E J, Hansen V N, and Partridge M 2002 Conformal and intensity modulated radiotherapy on a cobalt unit *Proc. IPEM Biennial Meeting (Southampton, 16–17 July)* p 37

Wavelength 2000a IMRT update from ASTRO99 *Wavelength* **4** (1) 1–4

——2000b Phased field technique enables virtual 160 leaf MLC for finer field shaping *Wavelength* **4** (1) 12–17

——2000c Intensity modulated arc therapy looks promising in world's first clinical trial *Wavelength* **4** (2) 1–5

——2000d IMRT international symposium provides state-of-the technique update *Wavelength* **4** (2) 8–11

——2001a New technology promises to redefine precision in radiotherapy *Wavelength* **5** (2) 7–14

——2001b Ghent university hospital's class solution approach streamlines IMRT for ethmoid sinus cancer *Wavelength* **5** (3) 1–4

——2001c Collaboration results in new IMRT solution *Wavelength* **5** (3) 9–10

——2001d University medical centre Utrecht begins IMRT protocol for prostate cancer *Wavelength* **5** (3) 5–6

——2001e Elekta users to demonstrate IMRT employing commercial solutions *Wavelength* **5** (3) 11–13

——2002a Active breathing coordinator brings greater precision to radiotherapy *Wavelength* **6** (1) 11–12

——2002b Hospitals integrating Active Breathing Coordinator into clinical protocols *Wavelength* **6** (2) 1–7

——2002c Christie hospital among first to evaluate new Elekta technology *Wavelength* **6** (3) 1–4

——2002d Active breathing coordinator freezes targets in extracranial stereotactic radioablation *Wavelength* **6** (3) 10–13

——2003a Beaumont work results in breakthrough Elekta Synergy system *Wavelength* **7** (1) 1–5

——2003b Clatterbridge Centre for Oncology pursues goal of better prostate cancer treatment with IMRT *Wavelength* **7** (2) 1–8

——2003c PrecisePLAN—making the complex simple. Aperture-based inverse planning reduces MUs, segments *Wavelength* **7** (2) 1–4

——2003d Clinical feasibility studies continue to show value of active breathing coordinator *Wavelength* **7** (2) 14–18

——2003e *Wavelength* **7** (3) 1–16

——2004 *Wavelength* **8** (1) 1–19

Webb S 1989 Optimisation of conformal radiotherapy dose distributions by simulated annealing *Phys. Med. Biol.* **34** 1349–69

——1991a Optimisation by simulated annealing of three-dimensional conformal treatment planning for radiation fields defined by a multileaf collimator *Phys. Med. Biol.* **36** 1201–26

——1991b Optimisation of conformal radiotherapy dose distributions by simulated annealing 2: inclusion of scatter in the 2D technique *Phys. Med. Biol.* **36** 1227–37

——1992 Optimisation by simulated annealing of three-dimensional conformal treatment planning for radiation fields defined by a multileaf collimator: 2. Inclusion of two-dimensional modulation of the x-ray intensity *Phys. Med. Biol.* **37** 1689–704

——1993 *The Physics of Three-Dimensional Radiation Therapy: Conformal Radiotherapy, Radiosurgery and Treatment Planning* (Bristol: IOP Publishing)

——1995 The problem of isotropically orienting N converging vectors in space with applications to radiotherapy planning *Phys. Med. Biol.* **40** 945–54

——1997 *The Physics of Conformal Radiotherapy: Advances in Technology* (Bristol: IOP Publishing)

——1998a Configuration options for intensity-modulated radiation therapy using multiple-static fields shaped by a multileaf collimator *Phys. Med. Biol.* **43** 241–60

——1998b Configuration options for intensity-modulated radiation therapy using multiple-static fields shaped by a multileaf collimator. 2—constraints and limitations on 2D modulation *Phys. Med. Biol.* **43** 1481–95

——1999 Conformal intensity-modulated radiotherapy (IMRT) delivered by robotic linac—testing IMRT to the limit? *Phys. Med. Biol.* **44** 1639–54

——2000a Optimisation of IMRT: can the ultimate be established and what are the consequences of approximation? *Radiother. Oncol.* **56** (Suppl. 1) S96

——2000b Regarding x-ray leakage during IMRT *Brit. J. Radiol.* **73** 1339

——2000c A new concept of multileaf collimator (the shuttling MLC)—an interpreter for high-efficiency IMRT *Phys. Med. Biol.* **45** 3343–58

——2000d *Intensity Modulated Radiation Therapy* (Bristol: IOP Publishing)

——2000e Conformal intensity-modulated radiotherapy (IMRT) delivered by robotic linac—conformality versus efficiency of dose delivery *Phys. Med. Biol.* **45** 1715–30

——2001a Concepts for shuttling multileaf collimators for intensity-modulated radiation therapy *Phys. Med. Biol.* **46** 637–51

——2001b A simple method to control aspects of fluence modulation in IMRT planning *Phys. Med. Biol.* **46** N187–95

——2001c The philosophy of IMRT *Radiother. Oncol.* **61** (Suppl. 1) S12

——2002a Intensity-modulated radiation therapy using only jaws and a mask *Phys. Med. Biol.* **47** 257–75

——2002b Intensity-modulated radiation therapy using only jaws and a mask 2: a simplified concept of single bixel attenuators *Phys. Med. Biol.* **47** 1869–79

——2002c Fewer segments for IMRT generated by modulation splitting *Phys. Med. Biol.* **47** N217–22

——2002d Some snapshots from the history of radiotherapy physics *SCOPE* **11** 8–12

——2002e The future of external beam radiotherapy: the dream and the reality *Phys. Medica.* **17** 207–15

——2003a The history of IMRT *IMRT—the State of the Art: AAPM Medical Physics Monograph Number 29* ed J R Palta and T R Mackie (Madison, WI: Medical Physics Publishing) pp 1–23

——2003b The physical basis of IMRT and inverse planning *Brit. J. Radiol.* **76** 678–89

——2004a A gravity-oriented-device for IMRT can never rival other IMRT delivery methods *Phys. Med. Biol.* **49** L5–9

——2004b Direct aperture optimisation for a variable aperture collimator for intensity modulated radiation therapy *Phys. Med. Biol.* **49** N47–55

Webb S, Bortfeld T, Stein J and Convery D 1997 The effect of stair-step leaf transmission on the 'tongue-and-groove' problem' in dynamic radiotherapy with a multileaf collimator *Phys. Med. Biol.* **42** 595–602

Webb S, Convery D J and Evans P M 1998 Inverse planning with constraints to generate smoothed intensity modulated beams *Phys. Med. Biol.* **43** 2785–94

Webb S, Cosgrove V and Evans P M 2001 Royal Marsden Hospital in Phase 1 IMRT clinical trial for prostate cancer *Wavelength* **5** (1) 1–5

Webb S, Hartmann G, Eschner G and Schlegel W 2003 Intensity modulated radiation therapy using a variable aperture collimator *Phys. Med. Biol.* **48** 1223–38

Webb S and Nahum A E 1993 The biological effect of inhomogeneous tumour irradiation with inhomogeneous clonogenic cell density *Phys. Med. Biol.* **38** 653–66

Webb S and Oldham M 1996 A method to study the characteristics of 3D dose distributions created by superposition of many intensity-modulated beams via a slit aperture with multiple absorbing vanes *Phys. Med. Biol.* **41** 2135–53

West M and Jones L 2000 Using MLCs to deliver compensated fields *Proc. World Congress of Medical Physics and the AAPM Annual Congress, August 2000* paper MO-CXH-14

Welch M E and Harlow S 2001 The effect of having one multileaf collimator in a department where it is used for conformal fields *Radiother. Oncol.* **61** (Suppl. 1) S80

Welsh J, Mehta M, Olivera G, Forrest L, Forouzannia A, Kapatoes J, Ruchala K, Lu W, Reckwerdt P and Mackie T 2002b Helical tomotherapy: an innovative technology for delivery of intensity modulated radiotherapy *Radiother. Oncol.* **64** (Suppl. 1) S211

Welsh J, Olivera G, Hui S, Henderson D, Forouzannia A, Kapatoes J, Ruchala K, Lu W, Reckwerdt P and Mackie R 2002a Helical tomotherapy with conformal avoidance appears superior to conventional radiotherapy, 3-D CRT, and IMRT for treatment of complex tumor volumes *Radiother. Oncol.* **64** (Suppl. 1) S124

Wierzbicki J and Blackmore L 2003 Comparison of IMRT plans for different delivery systems *Proc. World Congress in Medical Physics (Sydney)* CDROM and website only

Wiesmeyer M D and Beavis A W 2003 A method for using biological imaging data to modulate equivalent uniform dose based inverse planning *Med. Phys.* **30** 1335

Wilks R J and Bliss P 2002 The use of a compensator library to reduce dose inhomogeneity in tangential radiotherapy of the breast *Radiother. Oncol.* **62** 147–57

Williams P 2002a The practical implementation of IMRT *Proc. IPEM Biennial Meeting (Southampton, 16–17 July)* 2002 pp 8–9

——2002b Image guided radiotherapy—The next step in conformal radiotherapy *Radiother. Oncol.* **64** (Suppl. 1) S26

——2003a Integration of imaging and radiation therapy on the linear accelerator *Phys. Medica.* **19** 47

——2003b IMRT: delivery techniques and quality assurance *Brit. J. Radiol* **76** 766–76

Williams P C and Cooper P 2000 High-resolution field shaping using a masked multileaf collimator *Phys. Med. Biol.* **45** 2313–29

Williams P C and Hounsell A R 2001 X-ray leakage considerations for IMRT *Brit. J. Radiol.* **74** 98–100

Willoughby T R, Weinhous M S and Kupelian P A 2000 Daily transabdominal ultrasound localization of the prostate gland with the BAT system: 2705 alignments from 100 patients *Proc. 42nd Annual ASTRO Meeting. Int. J. Radiat. Oncol. Biol. Phys.* **48** (Suppl.) 308

Wilson E, Williams J, Lyn E and Aird E 2001 Active breathing control (ABC) in the treatment of non-small cell lung cancer (NSCLC) with chart week end less (CHARTWEL) *Radiother. Oncol.* **61** (Suppl. 1) S63

Wilson E M, Williams F J, Lyn B E, Wong J J and Aird E G A 2004 Validation of active breathing control in patients with non-small-cell lung cancer to be treated with CHARTWEL *Int. J. Radiat. Oncol. Biol. Phys.* **57** 864–74

Wong E, Chen J Z and Greenland J 2002a Intensity-modulated arc therapy simplified *Int. J. Radiat. Oncol. Biol. Phys.* **53** 225–35

Wong J 2000 Prospects of PET imaging in radiation oncology *Radiother. Oncol.* **56** (Suppl. 1) S7

——2001 Quality assurance and verification of IMRT. *Phys. Medica.* **17** 160

——2002 Methods to reduce the effects of respirator motion in radiation treatment *Radiother. Oncol.* **64** (Suppl. 1) S41–2

——2003 Methods to manage respiratory motion in radiation treatment *IMRT- the state of the art: AAPM Medical Physics Monograph Number 29* ed J R Palta and T R Mackie (Madison, WI: Medical Physics Publishing) pp 663–702

Wong J W, Armour E, Oldham M, Jaffray D A, Gu R and Jiang L 2002b An image guided small animal radiation research platform *Radiother. Oncol.* **64** (Suppl. 1) S61

Wong J, Jaffray D, Siewerdsen J and Sharpe M 2001 Respiratory gated radiation therapy with active breathing control *Med. Phys.* **28** 1220

Woo M 2000 Verification of MLC field shapes in IMRT treatment using portal imaging *Proc. World Congress of Medical Physics and the AAPM Annual Congress, August 2000* paper WE-FXH-40

Woo M K, Lightstone A W, Shan G, Kumaraswamy L and Li Y 2003 Automatic verification of step-and-shoot IMRT field segments using portal imaging *Med. Phys.* **30** 348–51

Woo S Y, Augspurger M, Teh B S, Lee A G, Uhl B, Grant W and Butler E B 2002 Conformal intensity modulated radiation therapy for the treatment of primary optic nerve sheath meningloma *Radiother. Oncol.* **64** (Suppl. 1) S301

Woo S Y, Grant W III, McGary J E, Teh B S and Butler E B 2003 The evolution of quality assurance for intensity-modulated radiation therapy (IMRT): sequential tomotherapy *Int. J. Radiat. Oncol. Biol. Phys.* **56** 274–86

Woudstra E, Heijmen B J M and Storchi P R M 2002 Automated sequential selection of beam orientations *Radiother. Oncol.* **64** (Suppl. 1) S10

Woudstra E and Storchi P R M 2000 Constrained treatment planning using sequential beam selection *Phys. Med. Biol.* **45** 2133–49

Wu C, Jeraj R, Lu W and Mackie T R 2004 Fast treatment plan modification with an over-relaxed Cimmino algorithm *Med. Phys.* **31** 191–200

Wu C, Jeraj R and Mackie T R 2003a The method of intercepts in parameter space for the analysis of local minima caused by dose-volume constraints *Phys. Med. Biol.* **48** N149–57

Wu C, Jeraj R, Olivera G and Mackie T R 2001a Comparison between a deterministic and a stochastic optimizer in radiotherapy optimization *Med. Phys.* **28** 1204

——2002a Re-optimization in adaptive radiotherapy *Phys. Med. Biol.* **47** 3181–95

Wu Q, Arnfield, Tong S, Wu Y and Mohan R 2000c Dynamic splitting of large intensity-modulated fields *Phys. Med. Biol.* **45** 1731–40

Wu Q, Djajaputra D, Wu Y, Zhou J, Liu H H and Mohan R 2003b Intensity-modulated radiotherapy optimisation with gEUD-guided dose-volume objectives *Phys. Med. Biol.* **48** 279–91

Wu Q and Mohan R 2000 Algorithms and functionality of an intensity modulated radiotherapy optimization system *Med. Phys.* **27** 701–11

——2001 Multiple local minima traps in IMRT optimization *Med. Phys.* **28** 1260

——2002 Multiple local minima in IMRT optimisation based on dose-volume criteria *Med. Phys.* **29** 1514–27

Wu Q, Mohan R, Morris M, Tong S and Schmidt-Ulrich R 2001e Simultaneous integrated boost IMRT of advanced head and neck squamous cell carcinomas using dynamic multi-leaf collimators *Int. J. Radiat. Oncol. Biol. Phys.* **51** (Suppl. 1) 180

Wu Q, Mohan R and Niemierko A 2000a IMRT optimization based on the generalized equivalent uniform dose (EUD) *Proc. World Congress of Medical Physics and the AAPM Annual Congress, August 2000* paper MO-G313-05

Wu Q, Mohan R, Niemierko A and Schmidt-Ullrich R 2002b Optimization of intensity-modulated radiotherapy plans based on the equivalent uniform dose *Int. J. Radiat. Oncol. Biol. Phys.* **52** 224–35

Wu X, Chen D, Hu X, Luan S, Zhang L and Yu C 2001c A new leaf-sequencing algorithm for intensity-modulated arc therapy *Med. Phys.* **28** 1252

Wu X, Dibiase S J, Gullapalli R and Yu C X 2004 Deformable image registration for use of magnetic resonance spectroscopy in prostate treatment planning *Int. J. Radiat. Oncol. Biol. Phys.* **58** 1577–83

Wu X and Zhu Y 2001a An optimization method for importance factors and beam weights based on genetic algorithms for radiotherapy treatment planning *Phys. Med. Biol.* **46** 1085–99

——2001b A maximum-entropy method for the planning of conformal radiotherapy *Med. Phys.* **28** 2241–6

Wu X, Zhu Y, Dai J and Wang Z 2000b Selection and determination of beamweights based on genetic algorithms for conformal radiotherapy treatment planning *Phys. Med. Biol.* **45** 2547–58

Wu Y, Wu Q and Mohan R 2002c A method of adaptive filtering and retro-mapping of intensity-modulated beams to improve delivery efficiency *Med. Phys.* **29** 1304

Wu Y, Yan D, Sharpe M, Miller B and Wong J 2001f Implementing multiple static field delivery for intensity modulated beams *Med. Phys.* **28** 2188–97

Xia P, Chuang C F and Verhey L J 2002a Communication and sampling rate limitations in IMRT delivery with a dynamic multileaf collimator system *Med. Phys.* **29** 412–23

Xia P, Fu K K, Wong G W, Akazawa C and Verhey L J 2000a Comparison of treatment plans involving intensity-modulated radiotherapy for nasopharyngeal carcinoma *Int. J. Radiat. Oncol. Biol. Phys.* **48** 329–37

Xia P, Hwang A B and Verhey L J 2002c A leaf sequencing algorithm to enlarge treatment field length in IMRT *Med. Phys.* **29** 991–8

Xia P, Pickett B, Vigneault E, Verhey L J and Roach M III 2001 Forward or inversely planned segmental multileaf collimator IMRT and sequential tomotherapy to treat multiple dominant intraprostatic lesions of prostate cancer to 90 Gy *Int. J. Radiat. Oncol. Biol. Phys.* **51** 244–54

Xia P and Verhey L 1998 Multileaf collimator leaf-sequencing algorithm for intensity modulated beams with multiple static segments *Med. Phys.* **25** 1424–34

Xia P, Wong G, Somers J and Verhey L J 2000b Dosimetric considerations in step and shoot IMRT delivery *Proc. World Congress of Medical Physics and the AAPM Annual Congress, August 2000* paper MO-GBR-04

Xia P, Yu N, Xing L, Sun X and Verhey L J 2004 Investigation of a new objective function for inverse planning optimisation *Proc. 14th Int.Conf. on the Use of Computers in Radiation Therapy (Seoul, May)* pp 116–19

Xia P, Yu N, Xing L and Verhey L 2002b Investigation of a variable power objective function for inverse planning optimization in IMRT *Med. Phys.* **29** 1257

Xiao Y, Bednarz G and Galvin J 2000b A 'sequential' optimization technique for forward treatment planning for IMRT *Proc. World Congress of Medical Physics and the AAPM Annual Congress, August 2000* paper WE-CXH-67

Xiao Y, Censor Y, Michalski D and Galvin J 2002b Intrinsic smooth intensity pattern for beamlet-based IMRT inverse planning from a simultaneous projection algorithm *Med. Phys.* **29** 1254

——2001 An IMRT optimization approach that minimizes the total MLC segments *Med. Phys.* **28** 1262

——2002a The least-intensity feasible solution for aperture-based inverse planning in radiation therapy *Private Communication*

Xiao Y, Galvin J, Hossain M and Valicenti R 2000a An optimized forward-planning technique for intensity modulated radiation therapy *Med. Phys.* **27** 2093–9

Xiao Y, Michalski D, Censor Y and Galvin J 2003 Smoothness evaluation of intensity pattern resulting from simultaneous subgradient projection method for dose-volume objective satisfaction *Med. Phys.* **30** 1488

Xing L 2001 Computer verification of MLC leaf sequences for IMRT *Med. Phys.* **28** 1268–9

Xing L, Cotrutz C, Hunjan S, Boyer A L, Adalsteinsson E and Spielman D 2002a Inverse planning for functional image-guided intensity-modulated radiation therapy *Phys. Med. Biol.* **47** 3567–78

Xing L, Crooks S, Li J G, Ozhasoglu C, Chen Y, Mehta V, Goffinet D, Beavis A and Boyer A 2000b Incorporating respiratory motion into the design of intensity maps in IMRT treatment of breast cancer *Proc. 42nd Annual ASTRO Meeting. Int. J. Radiat. Oncol. Biol. Phys.* **48** (Suppl.) 199

Xing L, Hamilton R J, Spelbring D, Pelizzari C A, Chen G T Y and Boyer A L 1998 Fast iterative algorithms for three-dimensional treatment planning *Med. Phys.* **25** 1845–9

Xing L, Hunjan S, Cotrutz C, Yang Y, Boyer A, Adalsteinsson E and Spielman D 2002b Inverse planning for functional imaging-guided IMRT *Med. Phys.* **29** 1285

Xing L, Hunjan S, Lian J, Yang Y, Shou Z, Cotrutz C, Schreibmann E, Adalsteinsson E, Spielman D and Boyer A 2004 Toward biologically conformal radiotherapy: Functional and molecular image-guided intensity modulated radiation therapy *Proc. 14th Int. Conf. on the Use of Computers in Radiation Therapy (Seoul, May)* pp 36–9

Xing L and Li J G 2000 Computer verification of fluence map for intensity modulated radiation therapy *Med. Phys.* **27** 2084–92

Xing L, Li J, Pugachev A, Le Q and Boyer A 2000a Computer-assisted selection of importance factors in inverse planning *Proc. World Congress of Medical Physics and the AAPM Annual Congress, August 2000* paper FR-B309-01

Xing L, Yang Y, Li J, Song Y, Pawlicki T, Boyer A and Le Q 2003 Dose matching of an IMRT plan with an electron or 3D conformal treatment plan *Proc. World Congress in Medical Physics (Sydney)* CDROM and website only

Xu B, Yan D, Kota K, Brabbins D S and Martinez A A 2000 The influence of inter and intra-patient rectum variation on treatment planning for prostate cancer radiotherapy *Proc. 42nd Annual ASTRO Meeting. Int. J. Radiat. Oncol. Biol. Phys.* **48** (Suppl.) 191

Xu T 2001 An *in vivo* verification method for fixed gantry IMRT *Med. Phys.* **28** 1293–4

——2002 A new physical compensation technique for intensity modulated radiation therapy *Med. Phys.* **29** 1305

Xu T, Al-Ghazi M, Chung H and Molloi S 2003 Re-shapeable physical modulator for intensity modulated radiation therapy: A treatment planning evaluation *Med. Phys.* **30** 1406

Xu T, Shikhaliev P M, Al-Ghazi M and Molloi S 2002 Reshapable physical modulator for intensity modulated radiation therapy *Med. Phys.* **29** 2222–8

Yan H 2003 3D correlation analysis between body marker and target motion for breath gating control *Med. Phys.* **30** 1338

Yan H, Yin F-F, Guan H-Q and Kim J H 2003 AI-guided parameter optimisation in inverse treatment planning *Phys. Med. Biol.* **48** 3565–80

Yan H, Yin F-F and Kim J H 2004 Correlation analysis between skin-marker motions and internal target motions *Proc. 14th Int. Conf. on the Use of Computers in Radiation Therapy (Seoul, May)* pp 413–16

Yang C, Chen H, Gollamudi S and Raben A 2002 Physics characteristics and dose comparison for micro multi-leaf collimator and standard multi-leaf collimator in whole body conformal radiosurgery *Med. Phys.* **29** 1212

Yang Y and Xing L 2003 Incorporating leaf transmission and head scatter corrections into step and shoot leaf sequences for IMRT *Int. J. Radiat. Oncol. Biol. Phys.* **55** 1121–34

——2002 An algorithm to incorporate leaf transmission and head scatter corrections into step-and-shoot leaf sequences for IMRT *Med. Phys.* **29** 1336

Yarnold J R, Donovan E, Bleackley N, Reise S, Regan J, Denholm E, Patel S, Ross G, Tait D and Evans P 2002 Randomised trial of standard 2D radiotherapy (RT) versus 3D intensity modulated radiotherapy (IMRT) in patients prescribed breast radiotherapy *Radiother. Oncol.* **64** (Suppl. 1) S15

Yenice K, Hunt M and Amols H 2001 Advantages of intensity modulated stereotatic radiosurgery using a mini-multileaf compared with static conformal linac radiosurgery *Med. Phys.* **28** 1256–7

Yi B Y 2004 To follow AMC: Aperture manoeuvre with controlled breath *Proc. 14th Int. Conf. on the Use of Computers in Radiation Therapy (Seoul, May)* pp 48–9

Yi B Y, Chol E K, Lee S W, Ahn S D, Kim J H, Ji Y H and Vahc Y W 2002 Rtp research tool box (RTB) for IMRT optimization *Radiother. Oncol.* **64** (Suppl. 1) S210

Yin F, Guan H, Ratkewicz A and Kim J 2000 A new technique for 3-D dose reconstruction using IMRT beams *Proc. World Congress of Medical Physics and the AAPM Annual Congress, August 2000* paper TU-GBR-06

Yin F-F, Yan H, Guan H-Q and Kim J-H 2004 Intelligent parameter optimisation technique in inverse treatment planning *Proc. 14th Int. Conf. on the Use of Computers in Radiation Therapy (Seoul, May)* pp 128–31

Yoda K and Aoki Y 2003 A multiportal compensator system for IMRT delivery *Med. Phys.* **30** 880–6

Yorke E 2001 Advantages of IMRT for dose escalation in radiation therapy of lung cancer *Med. Phys.* **28** 1291–2

Young C D, Speight J L, Akazawa P F, Pickett B, Verhey L J and Roach M 2000 Improved conformal coverage of the prostate with an IMRT potency-sparing technique *Proc. 42nd Annual ASTRO Meeting. Int. J. Radiat. Oncol. Biol. Phys.* **48** (Suppl.) 351

Yu C X 1995 Intensity modulated arc therapy with dynamic multileaf collimation: an alternative to tomotherapy *Phys. Med. Biol.* **40** 1435–49

Yu C, Earl M and Shepard D 2004 A scheme for optimising hybrid IMRT *Proc. 14th Int. Conf. on the Use of Computers in Radiation Therapy (Seoul, May)* pp 135–8

Yu C and Li J 2001 Delivery optimization with combined rotational and fixed-gantry IMRT *Med. Phys.* **28** 1306

Yu C X, Li X A, Ma L, Chen D, Naqvi S, Shepard D, Sarfaraz M, Holmes T W, Suntharalingam M and Mansfield C M 2002 Clinical implementation of intensity-modulated arc therapy. *Int. J. Radiat. Oncol. Biol. Phys.* **53** 453–63

Yu C X, Li X A, Ma L, Shepard D, Sarfaraz M, Holmes T, Suntharalingam M and Mansfield C M 2000 Clinical implementation of intensity-modulated arc therapy *Proc. 42nd Annual ASTRO Meeting. Int. J. Radiat. Oncol. Biol. Phys.* **48** (Suppl.) 219

Zackrisson B, Karlsson M and Lofroth P 2000 Biological effect of prolonged fraction delivery time in intensity modulated radiotherapy (IMRT)—in vitro simulation and model evaluation *Radiother. Oncol.* **56** (Suppl. 1) S195

Zelefsky M J 2003 Tumour control and tolerance of high dose IMRT in the treatment of prostate cancer *Radiother. Oncol.* **68** (Suppl. 1) S62

Zelefsky M J, Fuks Z, Happersctt L, Lee H J, Ling C C, Burman C M, Hunt M, Wolfe T, Venkatraman E S, Jackson A, Skwarchuk M and Leibel S A 2000 Clinical experience with intensity modulated radiation therapy (IMRT) in prostate cancer *Radiother. Oncol.* **55** 241–9

Zelefsky M J, Fuks Z, Hunt M, Yamada Y, Marion C, Ling C C, Amols H, Venkatraman E S and Leibel S A 2002 High-dose intensity modulated radiation therapy for prostate cancer: early toxicity and biochemical outcome in 772 patients *Int. J Rad. Oncol. Biol. Phys.* **53** 1111–16

Zelefsky M J and Leibel S A 2003 In response to Drs Schulz and Kagan *Int. J. Radiat. Oncol. Biol. Phys.* **55** 852–3

Zhang G, Guerrero T, Huang T-C, Fitzpatrick M J, Ibbott G, Lin K-P and Starkschall G 2004a 3D optical flow method implementation for mapping of 3D anatomical structure contours across 4D CT data *Proc. 14th Int. Conf. on the Use of Computers in Radiation Therapy (Seoul, May)* pp 356–60

Zhang T, Jeraj R, Keller H, Lu W, Olivera G H, McNutt T R, Mackie T R and Paliwal B 2004c Treatment plan optimization incorporating respiratory motion *Med. Phys.* **31** 1576–86

Zhang T, Keller H, Jeraj R, Mannon R, Welsh J, Patel R, Mehta M, Mackie R and Paliwal B 2003a Radiation treatment of the targets with respiratory motion using Breathing Synchronized Delivery (BSD)—A feasibility study *Radiother. Oncol.* **68** (Suppl. 1) S74–5

Zhang T, Keller H, O'Brien M J, Mackie T R and Paliwal B 2003b Application of the spirometer in respiratory gated radiotherapy *Med. Phys.* **30** 3165–71

Zhang X, Liu H, Wang X, Dong L, Wu Q and Mohan R 2004b Speed and convergence properties of gradient algorithms for optimisation of IMRT *Med. Phys.* **31** 1141–52

Zheng Z, Cardarelli G, Shearer D, Wazer D, DiPetrillo T and Chougule P 2002a Peregrine as a verification tool of Corvus treatment planning system *Med. Phys.* **29** 1231–2

Zheng Z, Cardarelli G, Shearer D, Wazer D, DiPetrillo T, Chougule P, Puthawala Y 2002b Dosimetry reproducibility of intensity-modulated radiation therapy using a commercially available system *Med. Phys.* **29** 1263

Zietman A L 2002 Dose escalation in localized prostate cancer: make no assumptions *Int. J. Radiat. Oncol. Biol. Phys.* **54** 635–6

Zijp L, Sonke J-J and van Herk M 2004 Extraction of the respiratory signal from sequential thorax cone-beam x-ray images *Proc. 14th Int. Conf. on the Use of Computers in Radiation Therapy (Seoul, May)* pp 507–9

Zinkin H D, Rivard M J, Mignano J E and Wazer D E 2004 Analysis of dose conformity and normal-tissue sparing using two different IMRT prescription methodologies for irregularly shaped CNS lesions irradiated with the Beak and 1-cm MIMiC collimators *Int. J. Radiat. Oncol. Biol. Phys.* **59** 285–92

Zurlo A 2002 Regarding 'A treatment planning comparison of 3D conformal therapy, intensity modulated photon therapy and proton therapy for treatment of advanced head and neck tumours' *Radiother. Oncol.* **63** 231

Zygmanski P and Kung J H 2001 Method of identifying dynamic multileaf collimator irradiation that is highly sensitive to a systematic MLC calibration error *Med. Phys.* **28** 2220–6

Zygmanski P, Kung J H and Jiang S B 2001b Dependence of IMRT fluence pattern on DMLC positional tolerance and root-mean-square values (RMS) *Med. Phys.* **28** 1208

Zygmanski P, Kung J, Jiang S and Chen G 2001a Measured dose errors in IMRT delivered with DMLC in the presence of organ motion: A phantom study *Med. Phys.* **28** 1283–4

——2001c A novel method for calculating 3D dose error in IMRT treatment of lung cancers with and without respiratory gating *Med. Phys.* **28** 1203

Zygmanski P, Kung J, Jiang S and Chin L 2003b Dependence of fluence errors in dynamic IMRT on leaf-positional errors varying with time and leaf number *Med. Phys.* **30** 2736–49

Zygmanski P, Kung J H, Jiang S B, Choi N and Chen G 2001d Is IMRT of lung possible without respiratory gating? Clinical evaluation of a true dose received by a moving target volume *Int. J. Radiat. Oncol. Biol. Phys.* **51** (Suppl. 1) 27

Zygmanski P, Kung J, Kooy H, Urribarri J, Bues M and Chen G 2000 A figure of merit indicating potential dose error in IMRT from inaccurate MLC calibration *Proc. World Congress of Medical Physics and the AAPM Annual Congress, August 2000* paper WE-E313-04

Zygmanski P, Que W, Chen G and Kung J 2003a DMLC sequence that are least sensitive to intrafractional organ motion *Med. Phys.* **30** 1493

Index

Numbers in **bold** refer to **figures**

Printed and bound by CPI Group (UK) Ltd, Croydon, CR0 4YY

23/10/2024

01778238-0015